OT 42
Operator Theory: Advances and Applications
Vol. 42

Birkhäuser Verlag
Basel · Boston · Berlin

Toeplitz Operators and Spectral Function Theory

Essays from the Leningrad Seminar on Operator Theory

Edited by

N. K. Nikolskii

1989

Birkhäuser Verlag
Basel · Boston · Berlin

Volume Editorial Office:

Steklov Mathematical Institute
Leningrad Branch
Fontanka 27
191 011 Leningrad
USSR

CIP-Titelaufnahme der Deutschen Bibliothek

Toeplitz operators and spectral function theory: essays from the Leningrad Seminar on Operator Theory / ed. by. N. K. Nikolskii. [Vol. ed. office: Steklov Math. Inst., Leningrad Branch]. –
Basel ; Boston ; Berlin : Birkhäuser, 1989
(Operator theory ; Vol. 42)
ISBN 3-7643-2344-2
NE: Nikol'skij, Nikolaj N. [Hrsg.]; Matematičeskij Institut Imeni V. A. Steklova <Moskva> / Leningradskoe Otdelenie; GT

Printed in Germany on acid-free paper
ISBN 3-7643-2344-2
ISBN 0-8176-2344-2

Contents

Preface 7

N.K. Nikolskii
Multicyclicity phenomenon. I. An introduction and maxiformulas 9

V.V. Peller
When is a function of a Toeplitz operator close to a Toeplitz operator? 59

B.M. Solomyak and A.L. Volberg
Multiplicity of analytic Toeplitz operators 87

B.M. Solomyak and A.L. Volberg
Operator of multiplication by an analytic matrix-valued function 193

S.R. Treil'
Geometric methods in spectral theory of vector-valued functions: Some recent results 209

V.I. Vasyunin
Formula for multiplicity of contractions with finite defect indices 281

D.V. Yakubovich
Riemann surface models of Toeplitz operators 305

Author Index 417

Subject Index 421

PREFACE

The volume contains selected papers of the Spectral Function Theory seminar, Leningrad Branch of Steklov Mathematical Institute. The papers are mostly devoted to the theory of Toeplitz and model operators. These subjects are considered here from various points of view.

Several papers concern the relationships of Toeplitz operators to weighted polynomial approximation. Namely, two papers by B. Solomyak and A. Volberg intensively treat the problem of spectral multiplicity for analytic Toeplitz operators (which are, in fact, multiplication operators) and my paper can serve as an introduction to the problem. This theme of multiplicities is continued in a paper by V. Vasyunin where the multiplicity of the spectrum is computed for Hilbert space contractions with finite defect indices.

V. Peller's paper deals with a perturbation theory problem for Toeplitz operators.

In a paper by D. Yakubovich a new similarity model for a class of Toeplitz operators is constructed.

S. Treil' presents a survey of a part of spectral function theory for vector valued function (Szegö-Kolmogorov extreme problems for operator weights, bases of vector rational functions, estimations of Hilbert transform with respect to operator weights, the operator corona problem).

As a concluding remark I dare only note that the whole collection convinces us once more without a doubt of the fruitfullness of the natural union of operator theory and complex analysis (if at all the union of these fields is at all different from their intersection).

Nikolai K. Nikolskii

Operator Theory:
Advances and Applications, Vol. 42

MULTICYCLICITY PHENOMENON. I.

AN INTRODUCTION AND MAXI-FORMULAS

Nikolai K.Nikolskii

CONTENTS

Chapter 0. Introduction 11
0.1. General subject description. 11
0.2. Main definitions, notation and conventions. 12
0.3. Main problems and applications. 14
Chapter 1. General properties and maxi-formulas. 18
1.0. Preliminaries 18
1.29. Deformations and quasi-similarities. 24
1.32. Estimations through decompositions. 25
1.42. Splittings lemma. 31
1.43. Algebras generated by a single operator. 32
1.45. Splittings and maxi-formulas in spectral terms. . 35
1.61. More general maxi-formulas. 38
1.78. Corollary on weighted polynomial approximation. . 49
1.80. Joining of eigenvectors. 50
1.82. Comment on common cyclic vectors. 52
1.88. Herrero's inequality. 54
1.89. Concluding remarks. 54

Chapter 2. Multicyclicity of multiplication operators *)
Normal and Dunford scalar operators
Weighted shifts
Toeplitz operators
Subnormal operators

Chapter 3. Multicyclicity through function model
Multiplicity of class C_0.
Influence of shifts on the multiplicity
Contractions with finite deficiency indices

Chapter 4. Overlappings of spectral multiplicities.

Chapter 5. Noncyclic functions of translation groups (mean periodic functions)

Chapter 6. Miscellanea and unsolved problems

---

*) Chapters 2 through 6 will be published elsewhere

CHAPTER 0. INTRODUCTION

0.1. GENERAL SUBJECT DESCRIPTION. The notion of the spectral multiplicity plays an important role in the theory of self-adjoint and normal operators. It serves a complete unitary invariant for operators of these classes, and in fact forms a background of the whole theory. As the main object it is proved to be the multiplicity function $\mu_N(\cdot)$ of a normal operator N on a separable Hilbert space H, i.e. the dimension function

$$\mu_N(\zeta) = \dim H(\zeta)$$

of the spectral decomposition of N

$$H \cong \int_{\mathbb{C}} \oplus H(\zeta)\, d\lambda(\zeta),$$

$$Nf(\zeta) = \zeta f(\zeta), \qquad \lambda\text{ - a.e. } \zeta$$

where $\cong$ stands for a unitary equivalence and λ for a Borel (scalar) spectral measure of N. In principle, all properties of N can be derived from properties of the multiplicity function $\mu_N(\cdot)$. In the case of a pure point spectrum one has simply

$$\mu_N(\zeta) = \dim \operatorname{Ker}(T - \zeta I), \qquad \lambda\text{- a.e. } \zeta .$$

The same formula can serve as a definition of a "local multiplicity" for one more important case, namely for operators on finite dimensional spaces, $T : X \to X$, $\dim X < \infty$:

$$\mu_T(\zeta) = \dim \operatorname{Ker}(T - \zeta I), \quad \zeta \in \mathbb{C} .$$

In fact, $\mu_T(\zeta)$ is equal to the number of Jordan blocks over a spectral point ζ, and together with an indication on the sizes of these blocks forms a complete system of similarity invariants for finite dimensional operators. Unfortunately, no analogue of such a "local multiplicity function" exists for operators of more or less general nature. However, it does exist, for instance, for Dunford scalar operators, the very close relatives of normal ones (see Chapter 2 of this paper).

In general case, instead of a multiplicity functions we are forced to consider a variety of more rough numerical invariants like global (multi)cyclicity number $\mu(T)$,

$$\mu(T) = \min\{\mathrm{card}\, A : \mathrm{span}\,(T^n A : n \geqslant 0) = X\}$$

(span $(\cdot)$ stands for the closed linear hull of $(\cdot)$), rational multicyclicity $\mu_R(T)$, commutant and bicommutant multicyclicities $\mu'(T)$, $\mu''(T)$, strict multicyclicity and so on. The collection of known facts about all these multiplicities can not be labeled as a completely formed theory (compare with a B.Beauzamy remark, [1] page vii) but it contains many interesting observations and computations as well as has important applications. We try to outline some of them in Sec.0.3 below together with a few general goals of the theory.

0.2. MAIN DEFINITIONS, NOTATION AND CONVENTIONS. The word space always means a separable Banach space (over the field of complex numbers) if the context does not imply something else. A subspace is a closed linear subset.

Operator means a linear bounded operator on a Banach space (or from one space to another); the algebra of all operators on a space X will be denote by $L(X)$.

If $\mathfrak{A} \subset L(X)$ we denote by $\mathrm{Lat}\,\mathfrak{A}$ the collection (lattice) of all subspaces $E \subset X$ invariant with respect to each operator from $\mathfrak{A}$: $TE \subset E$, $T \in \mathfrak{A}$.

If $C \subset X$ and $\mathfrak{A} \subset L(X)$ the smallest $\mathfrak{A}$-invariant subspace containing C will be denoted by $E_{\mathfrak{A}}(C)$:

$E_{\mathfrak{A}}(C) = \cap E$, $\quad E$ runs over all $E \in \mathrm{Lat}\,\mathfrak{A}$, $C \subset E$.

A subspace (or a set) $C \subset X$ is said to be cyclic for $\mathfrak{A}$, $\mathfrak{A} \subset L(X)$ if $E_{\mathfrak{A}}(C) = X$;

$$\mathrm{Cyc}\,\mathfrak{A} \overset{\mathrm{def}}{=\!=} \{C : C \text{ is linear}, \dim C < \infty,\ E_{\mathfrak{A}}(C) = X\}.$$

<u>The multicyclicity</u> (<u>multiplicity</u>) of a set $\mathfrak{A}$, $\mathfrak{A} \subset L(X)$ is defined by the equality

$$\mu(\mathfrak{A}) = \min\{\dim C : C \in \mathrm{Cyc}\,\mathfrak{A}\}$$

(the minimum of an empty set is equal to $+\infty$).

Many algebras related to a given operator $T \in L(X)$ will be used:

$\mathfrak{A}(T)$ - the weak closed algebra generated by T (and the identity operator I which is always included in an operator algebra);

$\mathfrak{A}_R(T)$ - the weak closed algebra generated by the resolvent $R_\lambda \overset{\mathrm{def}}{=\!=} (\lambda I - T)^{-1}$, $\lambda \in \mathbb{C} \setminus \sigma(T)$, where $\sigma(T)$ stands for the spectrum of T;

$\{T\}'$ - the commutant of $T\ (= \{A \in L(X) : AT = TA\})$;

$\{T\}''$ - the double commutant of $T\ (= \{A \in L(X) : AB = BA$ for every $B \in \{T\}'\})$;

$\mathfrak{A}^*(T)$ - the star weak closed algebra generated by T (for a Hilbert space operator T).

The corresponding multiplicities will be denoted as follows.

$$\mu(T) = \mu(\mathfrak{A}(T))$$

$$\mu_R(T) = \mu(\mathfrak{A}_R(T))$$

$$\mu'(T) = \mu(\{T\}')$$

$$\mu''(T) = \mu(\{T\}'')$$

$$\mu^*(T) = \mu(\mathfrak{A}^*(T)).$$

We also will shorten in a similar way some other notation, e.g. $\operatorname{Cyc} T = \operatorname{Cyc} \mathfrak{A}(T)$, $E(C) = E_{\mathfrak{A}(T)}(C)$, $E_R(C) = E_{\mathfrak{A}_R(T)}(C)$ and so on.

Sometimes we are considering some variations of the main multiplicities, e.g. the upper and lower multiplicities,

$$\bar{\mu}(T) = \sup\{\mu(T \mid E) : E \in \operatorname{Lat} T\},$$

$$\underline{\mu}(T) = \inf\{\mu(U) : U \in L(\mathcal{X}),\ \mathcal{X} \supset X \text{ and } X \in \operatorname{Lat} U,\ T = U|X\},$$

and so-called disc-characteristic of an operator,

$$\operatorname{disc} T = \sup_{C \in \operatorname{Cyc} T} \min\{\dim C' : C' \subset C,\ C' \in \operatorname{Cyc} T\}.$$

It is important to note that (despite of the notation) the multicyclicities $\mu(T)$, $\mu_R(T)$ etc. depend not on T itself but only on $\operatorname{Lat} T$, $\operatorname{Lat}_R T$, etc.

To finish the section let us recall that the symbol span$(\cdot)$ will stand for the closed linear hull of $(\cdot)$; $\mathcal{L}in(\cdot)$ means the linear hull of $(\cdot)$.

0.3. MAIN PROBLEMS AND APPLICATIONS. Main general goals of the field seem to be the following:

- to produce some recipes for calculating the mentioned multiplicities for general or specificly important classes of operators;
- to relate the multiplicities with other characteristics of operators (e.g. eigenvectors; spectral subspaces; direct, triangular and other decompositions, and so on);
- to describe all cyclic subsets (subspaces) of a given

operator.

To disclose a little the second mentioned desire let us recall that among the most general and important operator theory tools we can see a trick of a decomposition of an operator into a direct sum $T = T_1 + \dots + T_n + \dots$ of simpler summands (eigen- and rootvectors decompositions, various types of spectral decomposition, etc.). It is the more useful the more operator characteristics we can restore by their traces on the summands. The same is true for one more general method consisting in considering intertwining relations $\mathcal{D}T = T'\mathcal{D}$ for a given operator T and in an investigation of its properties through properties of T'. These well-known truisms play an important role in all known attempts to compute or estimate the multicyclicities. We widely use them in this paper too.

APPLICATIONS. Now I list some (possible) applications of multiplicity notions, both inside and outside of the operator theory.

0.4. INVARIANTS FOR VARIOUS CLASSIFICATIONS. There are several known equivalence relations between operators, say T, T' : unitary equivalence ($TU = UT'$ for a unitary operator U ; has a meaning for Hilbert space operators T, T'); similarity ($TV = VT'$ for a linear homeomorphism V); quasi-similarity ($TX = XT'$, $YT = T'Y$ for two deformations*) X, Y , i.e. injective operators with dense ranges Range X , Range Y). It is easy to see that all kind of multiplicities introduced above are invariant with respect to the similarity (and hence, the unitary equivalence). For the quasisimilarity case see 1.29-1.31 below.

0.5. EXISTENCE OF NON-TRIVIAL INVARIANT SUBSPACES. The problem is still open for Hilbert space operators and for operators on most of Banach spaces. It is obvious that $\operatorname{Lat} T$ is trivial (i.e. consists of $\{\mathbb{0}\}$, X only) iff $C \in \operatorname{Cyc} T$ for every subspace $C \neq \{\mathbb{0}\}$.

*) Also called quasiaffinitet.

0.6. HYPONORMAL AND, IN PARTICULAR, SUBNORMAL HILBERT SPACE OPERATORS. The finite multiplicity property $\mu_R(T) < \infty$ plays an important role for operators with positive self-commutators $[T^*, T] = T^*T - TT^* \geq \mathbb{O}$ (hyponormal). Namely, by the well-known Berger-Shaw theorem $\mathrm{trace}\,[T^*T] \leq \leq \pi^{-1}\mu_R(T)\cdot \mathrm{Area}\,(\sigma(T))$, see [2]. For subnormal operators (i.e. for restrictions $T = N\,|\,E$ of normals onto invariant subspaces, $E \in \mathrm{Lat}\,N$) the simplicity of the spectrum ($\mu(T)=1$) leads to a unitary equivalence $T \cong z$ to the multiplication operator $f \mapsto zf$ on $H^2(\nu) = \mathrm{clos}_{L^2(\nu)} P_A$ where ν stands for a Borel measure on $\mathbb{C}$ with compact support, P_A is the set of all complex polynomials.

0.7. OPERATOR ALGEBRAS use multiplicities in several ways, e.g. in the form of transitivity property of an algebra $\mathfrak{A}$: $\mathrm{clos}\,\mathfrak{A}x = X$ for every $x \neq \mathbb{O}$. See [3].

0.8. WEIGHTED POLYNOMIAL AND RATIONAL APPROXIMATION. If $T_z f(z) = = zf(z)$ on a space X of holomorphic functions over a domain $\Omega \subset \mathbb{C}$ one can consider a cyclic function $f \in \mathrm{Cyc}\,T$ as a weight enjoyed polynomial approximation: $f \in \mathrm{Cyc}\,T_z$ iff P_A is dense in the space $f^{-1}X$ endowed with the weighted norm $\|g\|_f = \|gf\|_X$. <u>Example</u>: $X = L^2_A(\Omega, \mu) = \{f : f$ is holomorphic in Ω, $\int_\Omega |f(z)|^2\, d\mu(z) < \infty\}$, $f^{-1}X = = L^2_A(\Omega, |f|^2 d\mu)$ (the non-vanishing property $f(z) \neq 0$, $z \in \Omega$), is assumed since it is necessary for $f \in \mathrm{Cyc}\,T_z$). Similar relations take place between $\mathrm{Cyc}\,\mathfrak{A}_R(T)$ and weighted rational approximation. For direct products $T_z \times T_z$ of the multiplication operators on different spaces X, Y the problem of cyclic vectors becomes a problem of simultaneous weighted approximation by polynomials (or rational functions); see [4], [5]. Multiplications by other functions T_f (not by the "independent variable") lead to approximation problems too.

0.9. CONTROL THEORY. A linear dynamical system can be described by an equation

(0.10) $$\frac{dx(t)}{dt} = Ax(t) + Bu(t), \quad t \geq 0$$

with some operators $A \in L(X)$, $B = B|U$ where $B \in L(\mathcal{U}, X)$ and X and $\mathcal{U}$ are Banach spaces (the state space and control space, respectively) and $x(t) \in X$, $u(t) \in U$. The first problem of control consists in searching for spaces U with the following controllability property: for every state $x \in X$, every $\varepsilon > 0$ and every time moment $t > 0$ there exists a (smooth) control function $u(\cdot)$ allowing the system to reach the ε-neighbourhood of x, $\|x - x(t)\| < \varepsilon$, starting from a fixed initial state $x(0)$ (say, $x(0) = \mathbb{O}$). The well-known Kalman theorem says that the system $A, B|U$ is controllable if and only if $BU \in \mathrm{Cyc}\,A$. Hence, the minimal dimension of control spaces for (0.10) is equal to $\mu(A)$. More sofisticated problems of control theory are related to other characteristics of the transfer operator A; see [6], [7]. One of them is $\mathrm{disc}\,A$ introduced above; see [8], [10], [11].

0.11. MANY SPECIAL PROBLEMS AND THEORIES, in fact, need multiplicity numbers like μ, μ_R, μ' etc. For example, the notion of the vacuum (vacuum vector) in the field theory includes the requirement of the existence of a cyclic vector for the field operators; see [12].

CHAPTER 1. GENERAL PROPERTIES AND MAXI-FORMULAS

1.0. PRELIMINARIES. As it is mentioned in Chapter 0 the multiplicity ($\equiv$ multicyclicity) $\mu(\mathfrak{A})$ of any set $\mathfrak{A}\subset \mathcal{L}(X)$ of operators on a (Banach) space X depends not on $\mathfrak{A}$ itself but on the lattice $\operatorname{Lat}\mathfrak{A}$ of all subspaces E, $E\subset X$ invariant with respect to $\mathfrak{A}$: $AE\subset E$ for every $A\in\mathfrak{A}$. In fact, for any lattice $\mathcal{L}$ of subspaces of X (i.e. a collection of subspaces containing $\{\mathbb{O}\}$ and X and closed with respect to intersections and spans of any subcollections of $\mathcal{L}$) one can define the $\mathcal{L}$-hull of a set $C\subset X$ by the equality

$$E_{\mathcal{L}}(C) = \cap E \qquad \text{over all } E\in\mathcal{L},\ C\subset E,$$

and then the set of all $\mathcal{L}$-cyclic subspaces

$$\operatorname{Cyc}\mathcal{L} = \{C : C\subset X,\ \dim C<\infty,\ E_{\mathcal{L}}(C)=X\}$$

and at last the multiplicity ($\equiv$ multicyclicity) of $\mathcal{L}$

$$\mu(\mathcal{L}) = \min\{\dim C : C\in\operatorname{Cyc}\mathcal{L}\}.$$

It is obvious that for a set of operators $\mathfrak{A}\subset\mathcal{L}(X)$ and for $\mathcal{L}=\operatorname{Lat}\mathfrak{A}$ one has

$$\mu(\mathcal{L}) = \mu(\mathfrak{A}) \geqslant \mu(\operatorname{alg}\mathcal{L})$$

where $\operatorname{alg}\mathcal{L}$ stands for the algebra of all operators T with $\mathcal{L}\subset\operatorname{Lat}T$.

In what follows we usually deal with multiplicities of operator families but mention sometimes their lattice analogues too.

Let us start with some elementary observations on the multiplicity behaviour with respect to restrictions onto invariant subspaces and quotients. Let X be a Banach space, $\mathfrak{A}$ be a set (usually, an algebra) of operators on X. If $T\in L(X)$ and $E\in \operatorname{Lat} T$, denote by $T|E$ the restriction of T onto E and by T/E, or $\dot{T}_E$, or $\dot{T}$ the quotient operator acting on X/E according to the natural rule $\dot{T}\dot{x}=(Tx)^{\cdot}$ where $\dot{x}$ stands for a coset $\dot{x}=x+E$, $x\in X$. Symbols $\mathfrak{A}|E$, $\mathfrak{A}/E$ have a similar meaning (for $E\in \operatorname{Lat}\mathfrak{A}$).

Letter $\mathcal{L}$ (maybe with indices) will be used for a lattice of subspaces of X.

Several first propositions are obvious.

1.1. $\mu(\mathcal{L}_1)\leqslant\mu(\mathcal{L}_2)$ if $\mathcal{L}_1\subset\mathcal{L}_2$.

1.2. $\mu(T)\geqslant\mu_R(T)\geqslant\mu''(T)\geqslant\mu'(T)$

because of

$$\mathfrak{A}(T)\subset\mathfrak{A}_R(T)\subset\{T\}''\subset\{T\}'$$

and hence

$$\operatorname{Lat} T\supset \operatorname{Lat}\mathfrak{A}_R(T)\supset \operatorname{Lat}\{T\}''\supset \operatorname{Lat}\{T\}'.$$

For a Hilbert space operator onecan add the following inequality

1.3. $\mu''(T)\geqslant\mu^*(T)$

because of

$$\operatorname{Lat}\mathfrak{A}^*(T)\subset \operatorname{Lat}\{T\}''$$

(in fact, if $P=P_E$ is the orthogonal projection onto a subspace E, $E\in\operatorname{Lat}\mathfrak{A}^*(T)$, one gets $PT=TP$ and hence $PA=AP$ for every $A\in\{T\}''$; this implies $E\in\operatorname{Lat}\{T\}''$).

1.4. LEMMA. (a) <u>Let</u> $\mathfrak{A}\subset L(X)$ <u>and</u> $E\subset\operatorname{Lat}\mathfrak{A}$. <u>Then</u>

(1.5) $\mu(\mathfrak{A}/_E) \le \mu(\mathfrak{A})$

and moreover, $C \in Cyc\,\mathfrak{A}$ implies $\dot{C} \in Cyc\,\dot{\mathfrak{A}}$ where $\dot{C} = \{\dot{x} : x \in C\}$ stands for the canonical projection of C into $X/_E$.

(b) If $T \in L(X)$ and $E \in Lat\,\mathfrak{A}_R$,

(1.6) $\mu_R(T/_E) \le \mu_R(T)$,

with a supplement on cyclic sets similar to (a).

(c) If $T \in L(X)$ and $E \in Lat\{T\}'$ we have

(1.7) $\mu'(T/_E) \le \mu'(T)$

with a similar supplement on cyclic sets.

PROOF. (a) Obvious.

(b) Since $\sigma(T/_E) \subset \sigma(T)$ for $E \in Lat\,\mathfrak{A}_R(T)$ we have $\mathfrak{A}_R(T)/_E \subset \mathfrak{A}_R(T/_E)$, and then apply (1.5): $\mu_R(T/_E) = \mu(\mathfrak{A}_R(T/_E)) \le \mu(\mathfrak{A}_R(T)/_E) \le \mu(\mathfrak{A}_R(T)) = \mu_R(T)$.

(c) Similar reasoning. ●

1.8. COROLLARY. $\mu_R(T) \ge \sup_{\lambda\in\mathbb{C}} \dim Ker(T^*-\lambda I)$.

Indeed, if $E = clos(T-\lambda I)X$ the quotient operator $T/_E$ is scalar λI . Moreover $E \in Lat\,\mathfrak{A}_R(T)$ (in fact, $E \in Lat\{T\}'$) and so $\dim Ker(T^*-\lambda I) = \dim X/_E = \mu_R(T/_E) \le \mu_R(T)$. ●

1.9. REMARK. More general inequality $\mu(\mathfrak{A}) \ge \dim \bigcap_{A\in\mathfrak{A}} Ker(A^*-\lambda_A I)$ can be similarly prove for any choice of $\lambda_A \in \mathbb{C}$, $A \in \mathfrak{A}$. In particular, $\mu^*(T) \ge \sup_{\lambda\in\mathbb{C}} \dim(Ker(T-\lambda I) \cap Ker(T^*-\bar{\lambda} I))$ for a Hilbert space operator T .

1.10. SUPPLEMENT. Proposition 1.8 (and 1.9) has the following lattice generalization: if E is a "co-atom" of a lattice $\mathcal{L}$ (i.e. $E' \supset E \Longrightarrow E' \in \mathcal{L}$) one has (by an obvious lattice analogue of (1.5)): $\mu(\mathcal{L}/_E) \le \mu(\mathcal{L})$ and $\mu(\mathcal{L}/_E) = \dim X/_E$

because $\mathcal{L}/E$ is the lattice of all subspaces of X/E. Hence

$\mu(\mathcal{L}) \geq \sup \operatorname{codim} E$, E runs over all "co-atoms" of $\mathcal{L}$. ●

1.11. COMMENT. For simply invariant subspaces (i.e. for $E \in \operatorname{Lat} T \setminus \operatorname{Lat} \mathfrak{A}_R(T)$) the both sides of inequality (1.6) have a meaning but the inequality fails to be true. A reason for this effect is in the absence of the relations between the algebras $\mathfrak{A}_R(T)/E$ and $\mathfrak{A}_R(T/E)$ as it is illustrated by Example 1.12 below. However, for every $E \in \operatorname{Lat} T$ we get $\mu_R(T/E) \leq \mu(T/E) \leq \mu(T) \leq \mu_R(T)+1$ (for the last inequality see 1.88 below).

1.12. EXAMPLE. $T = \mathcal{S} \oplus S_A$ where $\mathcal{S}$ stands for the bilateral shift operator $\mathcal{S}f = zf$ on the space $L^2(\mathbb{T})$, $\mathbb{T} = \{z : |z| = 1\}$ and S_A for a shift operator over an annulus $A = \{z : r_1 < |z| < r_2\}$, $r_2 < 1$, $S_A f = zf$, $f \in H^2(A)$. Then $\mu_R(T) = 1$. If $E = H^2 \oplus \mathbb{O}$ the quotient operator T/E is unitary equivalent to $S^* \oplus S_A$ where $S \overset{def}{=} \mathcal{S}|H^2$. This gives $\mu_R(T/E) = \mu(T/E) \geq \mu(S_A) = 2$. Hence,

$$\mu_R(T/E) > \mu_R(T). \quad ●$$

1.13. LEMMA. $\mu(\mathfrak{A}) \leq \mu(\mathfrak{A}/E) + \mu(\mathfrak{A}|E)$ <u>for</u> $E \in \operatorname{Lat} \mathfrak{A}$.

Indeed, if $C_1 \in \operatorname{Cyc}(\mathfrak{A}/E)$, $C_2 \in \operatorname{Cyc}(\mathfrak{A}|E)$ one can first approximate an arbitrary element of X by $\mathfrak{A}C_1$ "up to the subspace E" and **then approximate the rest** by $\mathfrak{A}C_2$. Hence, $C_1 + C_2 \in \operatorname{Cyc} \mathfrak{A}$ and $\mu(\mathfrak{A}) \leq \dim C_1 + \dim C_2$. ●

1.14. COROLLARY. <u>If</u> $X_i \in \operatorname{Lat} \mathfrak{A}$, $i = 1,2$ <u>and</u> $X = X_1 \dot{+} X_2$ <u>(direct sum) one has</u>

$$(1.15) \qquad \max_{i=1,2} \mu(\mathfrak{A}|X_i) \leq \mu(\mathfrak{A}) \leq \mu(\mathfrak{A}|X_1) + \mu(\mathfrak{A}|X_2).$$

In fact, $\mathfrak{A}|X_1$ is isomorphic (similar) to $\mathfrak{A}/X_2$, and (1.5) implies the left hand inequality. The right hand one is obvious. ●

1.16. REMARK. It is obvious that the right hand inequality of (1.15) is valid under more general assumptions; namely, if $X_k \in \operatorname{Lat} \mathfrak{A}$, $1 \le k \le n$, and $X = \operatorname{span}(X_1, \dots, X_n)$ one has

$$\mu(\mathfrak{A}) \le \sum_{k=1}^{n} \mu(\mathfrak{A} | X_k) .$$

1.17. DEFINITION. If $\mathfrak{X} = \{X_k : k \in K\}$ is a family of subspaces, $X_k \subset X$ and if $\sigma \subset K$, we set $X_\sigma = \operatorname{span}(X_k : k \in \sigma)$. $\mathfrak{X}$ is said to be: complete if $X_K = X$; minimal (or topologically free) if $X_k \cap X_{K \setminus \{k\}} = \{\mathbb{O}\}$ for every $k \in K$; an unconditional basis if $X_\sigma \dotplus X_{K \setminus \sigma} = X$ for every $\sigma \subset K$.

1.18. COROLLARY. If $X_k \in \operatorname{Lat} \mathfrak{A}$, $k \ge 1$ and if $\{X_k : k \ge 1\}$ is a minimal and complete family one has

(1.19) $\sup \mu(\mathfrak{A} | X_\sigma) \le \mu(\mathfrak{A}) \le \min(\mu(\mathfrak{A} | X_\sigma) + \mu(\mathfrak{A} | X_{\mathbb{N} \setminus \sigma}))$

where the both sup and min are taken over all finite subsets σ of $\mathbb{N}$.

For the proof it is enough only to mention that $X_\sigma \dotplus X_{\mathbb{N}\setminus\sigma} = X$ for every finite σ, and then apply (1.15). ●

1.20. REMARK. If $\{X_k : k \ge 1\}$ forms an unconditional basis in X the same reasoning shows that (1.19) is valid with sup and min taken over all subsets $\sigma \subset \mathbb{N}$.

1.21. COROLLARY. Let $T \in L(X)$ and $\{X_k : k \ge 1\}$ be a complete minimal family of T-invariant subspaces. Then

(1.22) $\sup \mu_R(T | X_\sigma) \le \mu_R(T)$

(1.23) $\sup \mu'(T | X_\sigma) \le \mu'(T)$

(1.24) $\sup \mu''(T | X_\sigma) \le \mu''(T)$

where - as in (1.19) - sup's are taken over all finite subsets

$\sigma \subset \mathbb{N}$.

In fact, if $\mathcal{E}_\sigma$ stands for the projection onto X_σ generated by a direct decomposition $X = X_\sigma \dot{+} X_{\mathbb{N}\setminus\sigma}$, one can see $\mathcal{E}_\sigma T = T\mathcal{E}_\sigma$, and hence

$$X_\sigma \in \operatorname{Lat}\{T\}'' \subset \operatorname{Lat} \mathfrak{A}_R(T) .$$

Thus, we result in (1.22) and (1.23) by using of (1.6) and (1.7) respectively and taking into account that $T|X_\sigma$ is isomorphic to $T/X_{\mathbb{N}\setminus\sigma}$. It is also easy to see that $\{T\}''|X_\sigma \subset \{T|X_\sigma\}''$ and hence $\mu''(T|X_\sigma) \leq \mu(\{T\}''|X_\sigma) \leq \mu''(T)$ by (1.15) for $\mathfrak{A} = \{T\}''$. ●

1.25. COMMENT. A natural analogues of the right hand inequalities of (1.15) and (1.19) are valid for μ' instead of μ , but fail to be true for μ_R (see the next section). About the left hand inequality of (1.15) see 1.27 and 1.28.

1.26. EXAMPLE. Let $T = S \oplus S_A$ where S stands for the shift operator on the Hardy space $H^2 (= H^2(\mathbb{D}))$, $Sf = zf$, $f \in H^2$ and S_A for the same multiplication operator but on $H^2(A)$ of an annulus $A = \{z : r_1 < |z| < r_2\}$, $0 < r_1 < r_2 \leq 1$. Then $\mu_R(S) = 1$, $\mu_R(S_A) = 1$ but $\mu_R(T) = \mu(T) = \mu S + \mu S_A = 1 + 2 = 3$; about the second equality see [13], Lemma 2 or Chapter 3 of this paper. Hence, $\mu_R(S \oplus S_A) > \mu_R(S) + \mu_R(S_A)$. ●

1.27. EXAMPLE. Instead of the left hand inequality of (1.15) one could conjecture more general one: $\mu(\mathfrak{A}|E) \leq \mu(\mathfrak{A})$ <u>for every</u> $E = \operatorname{Lat}\mathfrak{A}$. However, it does not true without some extra hypotheses. Indeed, let S_n^* be the backward shift "of multiplicity n " (they say so!), i.e. $S_n^* \{x_k\}_{k\geq 0} = \{x_{k+1}\}_{k\geq 0}$, $x = \{x_k\}_{k\geq 0} \in \ell^2(\mathbb{C}^n)$, Then $\mu(S_n^*) = 1$ as it is well-known (and will be shown later on, see 1.81, 1.82), although $S_n^*|E = \mathbb{O}$ for $E = \{x \in \ell^2(\mathbb{C}^n) : x_k = \mathbb{O},\ k \geq 1\}$ and hence $\mu(S_n^*|E) = n$. Moreover, there exist subspaces $E_k \in \operatorname{Lat} S_n^*$ with $\mu(S_n^*|E_k) = k$, $1 \leq k \leq n$. ●

1.28. EXAMPLE. A slightly complicated construction is need to get two "quasi-complementary" invariant subspaces X_K, $K=1,2$ (i.e. $X_K \in \mathrm{Lat}\,T$; $X_1 \cap X_2 = \{\mathbb{O}\}$, $\mathrm{clos}(X_1 + X_2) = X$) with

(1.28a) $\mu(T|X_K) > \mu(T)$, $K=1,2$.

To this aim, let's consider the backward shift "of infinite multiplicity" S_∞^* on the space $\ell^2(E)$ where E is a (separable) Hilbert space, $\dim E = \infty$. Let A be an operator on E obeying the following properties:

(1.28b) $\mathrm{clos}\, AE = E$,

but there exist two quasi-complementary subspace $E_K \subset E$, $E_1 \cap E_2 = \{\mathbb{O}\}$, $\mathrm{clos}(E_1 + E_2) = E$ such that

(1.28c) $AE_K \subset E_K$, $\mathrm{clos}\, AE_K \neq E_K$ $K = 1,2$.

If so, the weighted backward shift $T = AS_\infty^*$, $T\{x_i\}_{i\geq 0} = \{Ax_{i+1}\}_{i\geq 0}$ for $x \in \ell^2(E)$ satisfies desired property (1.28a) with $X_K = \ell^2(E_K) \overset{def}{=\!=} \{x = \{x_i\}_{i\geq 0} \in \ell^2(E) : x_i \in E_K , i \geq 0\}$, $K=1,2$. In fact, $\dim \mathrm{Ker}(T|X_K)^* = \infty$ and hence $\mu(T|X_K) = \infty$, $K = 1,2$. On the other hand, as it is not hard to check (see 1.89) $\mu(T) = 1$.

As for properties (1.28b), (1.28c) we can put first $E = L^2(\mathbb{T})$, $\mathbb{T}$ stands for the unit circle $\mathbb{T} = \{\zeta \in \mathbb{C} : |\zeta| = 1\}$, then $Af = zf$ (the unitary multiplication operator by the identity function $z(\zeta) \equiv \zeta$) and at last $E_1 = H^2$ (the Hardy subspace of L^2) $E_2 = \varepsilon H^2$ where ε is a measurable function on $\mathbb{T}$, $|\varepsilon| = 1$ a.e. on $\mathbb{T}$ (with respect to Lebesgue measure) which is not of the form f_1/f_2 , $f_i \in H^2$ (for example, $\varepsilon = \pm 1$ on $\mathbb{T} \cap \{\zeta : \pm \mathrm{Im}\,\zeta > 0\}$). With this choice we get the quasi-complementary property $E_1 \cap E_2 = \{\mathbb{O}\}$, $\mathrm{clos}(E_1 + E_2) = L^2$ (see, [14] Ch.1) and, obviously , properties (1.28b), (1.28c). ●

1.29. DEFORMATIONS AND QUASI-SIMILARITIES. The multiplicity of the spectrum is an invariant with respect to quasi-similarity. Let $A \in L(X)$, $B \in L(Y)$ and $\mathcal{D} \in L(X,Y)$; the operator $\mathcal{D}$ is said to be a <u>deformation</u> if $\mathrm{Ker}\,\mathcal{D} = \{\mathbb{O}\}$, $\mathrm{clos}\,\mathcal{D}X = Y$;

we say B __is a deformation of__ A if there exists a deformation $\mathcal{D}$ such that $\mathcal{D}A = B\mathcal{D}$; in sign, $B \prec A$ (note that, for instance, in [15] the above situation is described in different terms: $\mathcal{D}$ is a quasi-affine transformation and A is quasi-affine transformation of B , $A \prec B$). If $B \prec A$ and $A \prec B$, the operators are said to be __quasi-similar,__ $A \sim B$. Obviously, if A and B are similar (i.e. $\mathcal{D}A = B\mathcal{D}$ with an invertible operator $\mathcal{D}$) they are quasi-similar too. In fact, for multiplicity applications a weaker relation is also useful. Namely, we say B is a d-__deformation__ of A , $B \overset{d}{\prec} A$, if $\mathcal{D}A = B\mathcal{D}$ with an operator $\mathcal{D} \in L(X,Y)$ of a dense range: $clos\,\mathcal{D}X = Y$.
It is obvious, that $B \prec A$ implies $B \overset{d}{\prec} A$.

1.30. LEMMA. (a) __If__ $\mathcal{D}$ __is a__ d-__deformation,__ $\mathcal{D}A = B\mathcal{D}$ __and__ $C \in Cyc\,A$ __we have__ $\mathcal{D}C \in Cyc\,B$, __and hence__ $\mu(B) \le \mu(A)$.

(b) __Let__ A', B' __be__ WOT -__limits of a sequence of polynomials__ $A' = \lim_n p_n(A)$, $B' = \lim_n p_n(B)$ __and let__ $B \overset{d}{\prec} A$ __then__ $B' \overset{d}{\prec} A'$ __and__ $\mu(B') \le \mu(A')$.

(c) $\mu(A) = \mu(B)$ __provided__ $A \sim B$.

THE PROOF is obvious since $\mathcal{D}A^n = B^n\mathcal{D}$, $n \ge 0$ and $\mathcal{D}p(A) = p(B)\mathcal{D}$ for every polynomial p . ●

COROLLARY. __Let__ $T \in L(X)$, X_k __be__ T-__invariant subspaces__ $k = 1,2,\ldots$, $X = span(X_k : k \ge 1)$. __Consider__ ℓ^1-__type direct sum of__ X_k, $\ell^1(X_k) \overset{def}{=\!=} \{\{x_k\}_{k\ge1} : x_k \in X_k , \sum_{k\ge1} \|x_k\| < \infty\}$ __and an operator__ $\widetilde{T}$ __on__ $\ell^1(X_k)$ __defined by__ $\widetilde{T}\{x_k\}_{k\ge1} = \{Tx_k\}_{k\ge1}$. __Then,__ $T \overset{d}{\prec} \widetilde{T}$ __and hence__ $\mu(T) \le \mu(\widetilde{T})$.

In fact, let $\mathcal{D}\{x_k\}_{k\ge1} = \sum_{k\ge1} x_k$. Then $\mathcal{D}$ is a d-deformation, $\mathcal{D}\widetilde{T} = T\mathcal{D}$ and the corollary follows from 1.30. ●

1.32. ESTIMATIONS THROUGH DECOMPOSITIONS. Now, we start to develop one of a few general methods for estimations of spectral multiplicities. It consists first in looking for a decomposition of the whole space X into a sum (finite or not, direct or not) of $\mathfrak{A}$- invariant subspaces X_k , e.g. a "decomposition" in the weakest form

$$X = \mathrm{span}\,(X_k : k \geq 1),$$

and then for additional hypotheses on such a decomposition guaranteeing the transformation of one of the inequalities of (1.19) into an equality. In fact, in the whole this Chapter we aim at so called <u>maxi-formulas</u> $\mu(\mathfrak{A}) = \sup_k \mu(\mathfrak{A}|X_k)$ which mean a kind of independence (not subordination to each other) of summands $\mathfrak{A}|X_k$, $k \geq 1$. Sometimes, we will choose X_k's to be spectral subspaces of $\mathfrak{A}$ (in a sense) and express the independence mentioned in a form of spectral independence (e.g. as the disjointness of the spectra $\sigma(T|X_k)$, $\sigma(T|X_i)$, $k \neq i$, $T \in \mathfrak{A}$, or something so). The estimations of $\mu(\mathfrak{A})$ will be the better the finer decompositions we deal with. Of course, we mean the algebras $\mathfrak{A} = \mathfrak{A}(T)$, $\mathfrak{A}_R(T)$ etc. generated by a single operator T as the main partial cases. If, in the contrary, a part of such an operator, say $T|X_1$, is dominating over other ones we have to get an equality of the form $\mu(T) = \mu(T|X_1) + \ldots$ (here is a rough example of a domination which one can have in mind: the mapping $p \mapsto p(T)y$, $p \in \mathcal{P}_A$ is continuous in the norm $\|p\|_x = \|p(T)x\|$ for every $x \in X_1 \setminus \{\mathbb{O}\}$ and $y \in \mathrm{span}\,(X_k : k \geq 2)$; example: $T = T_1 \oplus T_2$, $T_1 = S$ on H^2, $\sigma(T_2) \subset \mathbb{D}$). An analysis of the influence of such kind of dependences between $T|X_k$ to the value of multicyclicity $\mu(T)$ will be given in Chapters 2 through 4 of the paper.

Returning to 1.18 let us consider the decomposition $X = X_\sigma \dot{+} X_{\mathbb{N}\setminus\sigma}$ (σ stands for a finite subset of $\mathbb{N}$) and the corresponding projection

$$\mathcal{E}_\sigma(x_\sigma + x_{\mathbb{N}\setminus\sigma}) = x_\sigma .$$

It follows from 1.4 (a) that $\mathcal{E}_\sigma C \in \mathrm{Cyc}(\mathfrak{A}|X_\sigma)$ for every finite σ and every $C \in \mathrm{Cyc}\,\mathfrak{A}$. If the converse is true one says that $\mathrm{Cyc}\,\mathfrak{A}$ splits.

1.33. DEFINITION. Let $\mathfrak{A}$ be an algebra of operators on X, and let $\mathfrak{X} = \{X_k : k \geq 1\}$ be a complete topologically free family of $\mathfrak{A}$-invariant subspaces. We say $\mathrm{Cyc}\,\mathfrak{A}$ <u>splits</u> (with respect to $\mathfrak{X}$) and write

$$\operatorname{Cyc}\mathfrak{A} = \oplus \operatorname{Cyc}(\mathfrak{A}|X_k)$$

if $C \in \operatorname{Cyc}\mathfrak{A} \Longleftrightarrow \mathcal{E}_k C \in \operatorname{Cyc}(\mathfrak{A}|X_k), k \geq 1$; $\mathcal{E}_k$ stands for the natural projection onto X_k (parallel to all other X_i , $i \neq k$). We say $\operatorname{Cyc}\mathfrak{A}$ <u>weakly splits</u> if the inclusions $\mathcal{E}_k C \in \operatorname{Cyc}(\mathfrak{A}|X_k)$, $k \geq 1$ imply $\mathcal{E}_\sigma C \in \operatorname{Cyc}(\mathfrak{A}|X_\sigma)$ for every finite set σ , $\sigma \subset \mathbb{N}$. And at last, let us say $\operatorname{Cyc}\mathfrak{A}$ <u>restrictedly</u> (<u>weakly restrictedly</u>) <u>splits</u> if the above properties hold when being restricted to sets C with $\dim \mathcal{E}_k C = \mu(\mathfrak{A}|X_k)$, $k \geq 1$, only.

Obviously, for a finite family $\mathfrak{X}$ the splitting property of $\operatorname{Cyc}\mathfrak{A}$ coincides with the weak splitting property. Differences between these splittings as well as between them and their restricted versions will be seen in what follows. An advantage of these notions is supported, in particular, by the following two simple propositions.

1.34. LEMMA. <u>Let</u> $\mathfrak{A} \subset L(X)$ <u>and</u> $\mathfrak{X} = \{X_k : k \geq 1\}$ <u>be a complete minimal family of</u> $\mathfrak{A}$<u>-invariant subspaces</u>. <u>If</u> $\operatorname{Cyc}\mathfrak{A}$ (<u>restrictedly</u>) <u>splits, or even only weakly restrictedly splits, one has</u>

$$(1.35) \qquad \mu(\mathfrak{A}) = \sup_k \mu(\mathfrak{A}|X_k).$$

PROOF. The inequality $\geq$ follows from (1.19). Proving the converse, one can suppose $\mu \overset{def}{=\!=} \sup_k \mu(\mathfrak{A}|X_k) < \infty$. Let $\mu = \mu(\mathfrak{A}|X_n)$ and let $C_k \in \operatorname{Cyc}(\mathfrak{A}|X_k)$, $\dim C_k = \mu(\mathfrak{A}|X_k)$, $k = 1, 2, \ldots$. Then for every $k \geq 1$ there exists a linear operator B_k such that $B_k C_n = C_k$ and $\sum_{k\geq 1} \|B_k\| < \infty$. Let $C = \{\sum_{k \geq 1} B_k x : x \in C_n\}$.

Supposing the (restricted) splitting property of $\operatorname{Cyc}\mathfrak{A}$ we get immediately $C \in \operatorname{Cyc}\mathfrak{A}$, $\dim C = \mu$. So, the inequality $\mu(\mathfrak{A}) \leq \mu$ follows.

If $\operatorname{Cyc}\mathfrak{A}$ (restrictedly) weakly splits only, we need to use a more rapidly decreasing sequence $\{B_k\}_{k \geq 1}$ which will be constructed in the next lemma generalizing this one. ●

1.36. GENERALIZATION. In fact, one can use splitting like arguments without minimality assumptions. Now, we give a version of 1.34 both eliminating these assumptions and using a weaker form of splitting properties.

1.37. LEMMA. Let $\mathfrak{A}$ be an algebra of operators on a Banach space X and let $X_k \in \operatorname{Lat}\mathfrak{A}$, $k \geq 1$, and $X = \operatorname{span}(X_k : k \geq 1)$. Let further C_k be $\mathfrak{A}$-cyclic subspaces of X_k, $\dim C_k = \mu(\mathfrak{A}|X_k) < \infty$, and $\dim C_n$ be a maximum of $\dim C_k$, $k \geq 1$. Expressing C_k in a form $C_k = B_k C_n$ with a linear operator $B_k : C_n \to C_k$, let us assume the set

$$\text{(1.37a)} \quad \left\{ \sum_{k=1}^{N} \lambda_k B_k x : x \in C_n \right\}$$

be $\mathfrak{A}$-cyclic in $\operatorname{span}(X_k : 1 \leq k \leq N)$ for every choice of $\lambda_k \neq 0$, $1 \leq k \leq N$, and for every $N \geq 1$. Then

$$\mu(\mathfrak{A}) \leq \max_k \mu(\mathfrak{A}|X_k).$$

PROOF. If the considered sequence $\{X_k\}_{k \geq 1}$ is finite there is nothing to prove. If not, let us define a sequence of non-zero scalars λ_k by the following inductive rules. Let e_j, $1 \leq j \leq \mu = \dim C_n$ be a basis for C_n and let $\lambda_1 = 1$.

If $\lambda_1, \ldots, \lambda_N$ are already choiced, let $A_{N,m,j,i}$ be operators from $\mathfrak{A}$ satisfying

$$\left\| \sum_{i=1}^{\mu} A_{N,m,j,i} \sum_{k=1}^{N} \lambda_k B_k e_i - B_m e_j \right\| < 1/N$$

for all j, m with $1 \leq m \leq N$, $1 \leq j \leq \mu$. Now, consider a complex number λ_{N+1} with

$$0 < |\lambda_{N+1}| \sum_{1 \leq m, s \leq N} \sum_{1 \leq i, j \leq \mu} \|B_{N+1} e_i\| \, \|A_{s,m,j,i}\| < 1/N^2 .$$

Prove now the set

$$C = \{ \sum_{k\geq 1} \lambda_k B_k x : x \in C_n \}$$

is cyclic for $\mathfrak{A}$ (on X). Fix m, j ($m \geq 1, 1 \leq j \leq \mu$) and consider $N \geq m$. One has

$$\| \sum_{i=1}^{\mu} A_{N,m,j,i} \sum_{k\geq 1} \lambda_k B_k e_i - B_m e_j \| \leq$$

$$\leq \| \sum_{i=1}^{\mu} A_{N,m,j,i} \sum_{k=1}^{N} \lambda_k B_k e_i - B_m e_j \| +$$

$$+ \sum_{k>N} |\lambda_k| \sum_{i=1}^{\mu} \|B_k e_i\| \|A_{N,m,j,i}\| < 1/N + \sum_{k>N} 1/k^2 .$$

Passing to the limit for $N \longrightarrow \infty$, we get $B_m e_j \in E_{\mathfrak{A}}(C)$ for all m, j. Hence, $C_k \subset E_{\mathfrak{A}}(C)$ and $X_k \subset E_{\mathfrak{A}}(C)$ for $k \geq 1$ which implies $E_{\mathfrak{A}}(C) = X$. ●

1.38. COMMENT. In fact, we proved that there exists a sequence $\{\varepsilon_k\}_{k\geq 1}$ such that the inequalities $0 < |\lambda_k| < \varepsilon_k$, $k \geq 1$, imply the set $\{\sum_{k\geq 1} \lambda_k B_k x : x \in C_n\}$ is $\mathfrak{A}$-cyclic. We use this remark later on (see 1.82 where the case $\dim X_k = 1$, $k \geq 1$, is discussed).

One more remark is about condition (1.37a). Namely, if every finite part $\{X_k : 1 \leq k \leq N\}$ of the sequence $\{X_k\}_{k\geq 1}$ is minimal (i.e. $X_j \cap \operatorname{span}(X_k : 1 \leq k \leq N, k \neq j) = \{\mathbb{O}\}$ for every j) it is enough to require that (1.37a) to be cyclic even though for one choice of $\lambda_k \neq 0$. (Indeed, in this case any diagonal map $\mathcal{D}: \sum_{i=1}^{N} x_i \mapsto \sum_{i=1}^{N} a_i x_i$, $x_i \in X_i$ with $a_i \neq 0$ is an isomorphism of $\operatorname{span}(X_k : 1 \leq k \leq N)$ into itself, and $\mathcal{D}T = T\mathcal{D}$ ●).

In order to make easier the use of maxi-formulas (estimations) of 1.34 and 1.37 one can try to express splitting properties of $\operatorname{Cyc} \mathfrak{A}$ in terms of some properties of operators from

$\mathfrak{A}$ or/and of the lattice $\mathrm{Lat}\,\mathfrak{A}$. One of the possibilities consists in using of the following notions.

1.39. DEFINITION. Let $\mathfrak{A}$ be an algebra of operators on a Banach space X and let $\{X_k : k \geqslant 1\}$ be a complete minimal family of $\mathfrak{A}$-invariant subspaces. We say <u>the lattice</u> $\mathrm{Lat}\,\mathfrak{A}$

- <u>weakly splits</u> if $\mathcal{E}_k E \subset E$ for every $k \geqslant 1$ and every $E \in \mathrm{Lat}\,\mathfrak{A}$ (i.e. if $\mathcal{E}_k \in \mathrm{alg}\,\mathrm{Lat}\,\mathfrak{A} \overset{def}{=} \{A \in L(X) : \mathrm{Lat}\,A \supset \mathrm{Lat}\,\mathfrak{A}\}$);
- <u>splits</u> if

$$E = \mathrm{span}(\mathcal{E}_k E : k \geqslant 1) \qquad \text{for every } E \in \mathrm{Lat}\,\mathfrak{A};$$

- <u>finitely splits</u> if $\mathrm{Lat}(\mathfrak{A}\,|\,X_\sigma)$ splits($\equiv$ weakly splits) for every finite set σ, $\sigma \subset \mathbb{N}$.

(Here $\mathcal{E}_k$ stands for the projection onto X_k parallel to all X_i , $i \neq k$ and $X_\sigma = \mathrm{span}(X_i : i \in \sigma)$).

<u>An algebra</u> $\mathfrak{A}$ (by definition)

- <u>weakly splits</u> if $\mathcal{E}_k \in \mathfrak{A}$ for every $k \geqslant 1$;
- <u>splits</u> if for any choice of $A_k \in \mathfrak{A}$ satisfying $\sup_k \|A_k | X_k\| < \infty$ there exists $A \in \mathfrak{A}$ such that

$$A \,|\, X_k = A_k \,|\, X_k, \quad k \geqslant 1;$$

- <u>finitely splits</u> if $\mathrm{clos}_{WOT}(\mathfrak{A}\,|\,X_\sigma)$ splits ($\equiv$ weakly splits) for every finite set σ , $\sigma \subset \mathbb{N}$.

1.40. COMMENT. Obviously the all three splitting properties - both for $\mathrm{Lat}\,\mathfrak{A}$ and $\mathfrak{A}$ - coincide in the case of a finite (minimal) family $\{X_k, k \geqslant 1\}$. Let us also note that if $\mathrm{Lat}\,\mathfrak{A}$ weakly splits one has $\mathcal{E}_k E = E \cap X_k \in \mathrm{Lat}(\mathfrak{A}\,|\,X_k)$, $k \geqslant 1$ for every $E \in \mathrm{Lat}\,\mathfrak{A}$.

In few below lemmas we try to clarify the dependence of introduced splitting properties on $\mathfrak{A}$ as well as their dependences on each other. For a symmetry one can (formally) introduce one more splitting notion for Cyc: we say $\mathrm{Cyc}\,\mathfrak{A}$ <u>finitely splits</u> if $\mathrm{Cyc}(\mathfrak{A}\,|\,X_\sigma)$ splits for every finite set σ, $\sigma \subset \mathbb{N}$. In fact, this last property coincides with the weak splitting property of $\mathrm{Cyc}\,\mathfrak{A}$ introduced in 1.33. On the other hand, all other relationships between splitting properties - conceivable but not marked in 1.41 and 1.42 below - fail, already for diagonal Hilbert space operators.

1.41. LEMMA. Let X be a Banach space, $\mathfrak{A}_1$ and $\mathfrak{A}_2$ be algebras of operators on X, and

$$\mathfrak{A}_1 \subset \mathfrak{A}_2 .$$

Let $\mathfrak{X} = \{X_k : k \geq 1\}$ be a complete minimal family of $\mathfrak{A}_2$-invariant subpsaces. Then

(a) Lat $\mathfrak{A}_2$ (weakly finitely) splits if Lat $\mathfrak{A}_1$ (weakly finitely) splits.

(b) $\mathfrak{A}_2$ weakly (finitely) splits if $\mathfrak{A}_1$ weakly (finitely) splits.

(c) If X is a Hilbert space and if $\mathfrak{A}_2$ is closed in the weak operator topology (WOT), the splitting property for $\mathfrak{A}_1$ implies the same for $\mathfrak{A}_2$.

PROOF. (a) Obvious since $\operatorname{Lat}\mathfrak{A}_1 \supset \operatorname{Lat}\mathfrak{A}_2$.

(b) Obvious.

(c) Considering scalar operators $A_k = \lambda_k I \mid X_k$, $\sup|\lambda_k| < \infty$ and using the splitting property of $\mathfrak{A}_1$ one can see that $\mathfrak{X}$ forms an unconditional basis in X. Hence (for a Hilbert space X) series $\sum_{k \geq 1} x_k$, $x_k \in X_k$ converges iff $\sum_{k \geq 1} \|x_k\|^2 < \infty$. Now, if $A_k \in \mathfrak{A}_2$ and $\sup_k \|A_k \mid X_k\| < \infty$ series $Ax \overset{def}{=} \sum_{k \geq 1} A_k \mathcal{E}_k x$ converges for every $x \in X$. Since $\mathcal{E}_k \in \mathfrak{A}_1 \subset \mathfrak{A}_2$ we get $A_k \mathcal{E}_k \in \mathfrak{A}_2$ and then $A \in \mathfrak{A}_2$. The assertion follows. ●

1.42. LEMMA. In the notation of 1.33 and 1.39 the following implications hold.

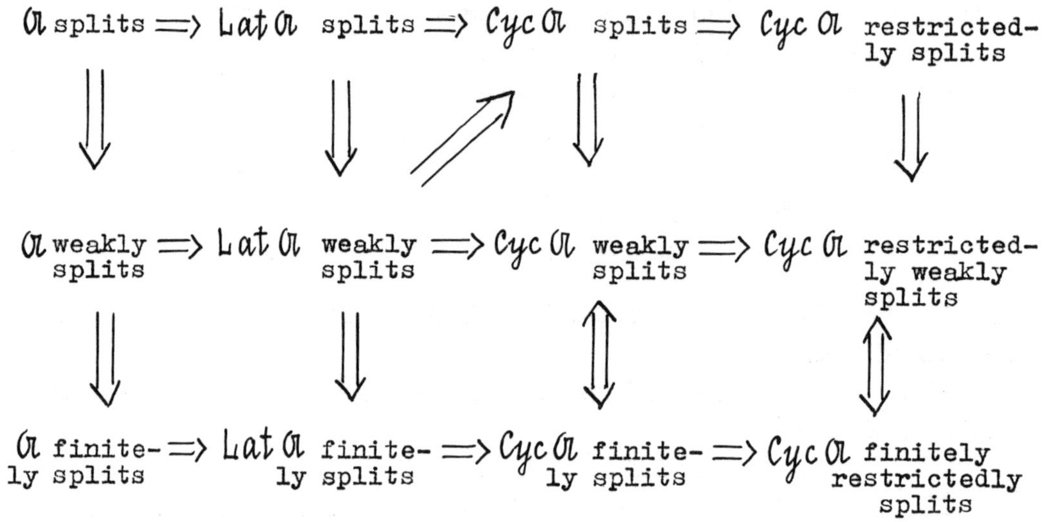

PROOF. The vertical implications are easy (and was mentioned above).

The upper row. If $\mathfrak{A}$ splits and if $E \in \operatorname{Lat} \mathfrak{A}$ we have successively $\mathcal{E}_k \in \mathfrak{A}$, $\mathcal{E}_k E \subset E$ for every $k \geq 1$, $E = \operatorname{span}(\mathcal{E}_k E : k \geq 1)$ (use that $\mathfrak{X}$ is a basis, see proof of 1.41 (c)). Thus, $\operatorname{Lat} \mathfrak{A}$ splits.

The second upper row implication follows from the inclined one. Let us check the latter. Let $C \subset X$ and $\mathcal{E}_k C \in \operatorname{Cyc}(\mathfrak{A} | X_k)$, $k \geq 1$. One has to check that $C \in \operatorname{Cyc} \mathfrak{A}$. Let $E = E_{\mathfrak{A}}(C)$. Since $\operatorname{Lat} \mathfrak{A}$ weakly splits, $\mathcal{E}_k E \subset E$ and then $\mathcal{E}_k C \subset E$ and $X_k \subset E$ for every $k \geq 1$. This means $E = X$, i.e. $C \in \operatorname{Cyc} \mathfrak{A}$. So, $\operatorname{Cyc} \mathfrak{A}$ splits.

The middle row. We only need to prove the first implication but it is immediate from the definitions.

The lower row. All follows from the definitions and the upper row inclusions. ●

1.43. ALGEBRAS GENERATED BY A SINGLE OPERATOR are of particular interest for the theory. We mean $\mathfrak{A}(T)$, $\mathfrak{A}_R(T)$, $\{T\}''$, $\{T\}'$ for an operator $T \in L(X)$ on a Banach space X and plus the star algebra $\mathfrak{A}^*(T)$ for a Hilbert space operator. The main relations between splitting properties of these algebras follow from lemma 1.41 and the inclusions $\mathfrak{A}(T) \subset \mathfrak{A}_R(T) \subset \{T\}'' \subset \{T\}'$ and $\operatorname{Lat} \mathfrak{A}^*(T) \subset \operatorname{Lat} \{T\}''$ (for the Hilbert space case). Now, we list some comments on properties mentioned above (with respect to a fixed minimal family $\mathfrak{X} = \{X_k\}_{k \geq 1}$ of $\mathfrak{A}$-invariant subspaces).

<u>Algebra</u> $\mathfrak{A}(T)$. The weak splitting property ($\mathcal{E}_k \in \mathfrak{A}(T)$, $k \geq 1$) is a kind of polynomial approximation: $\mathcal{E}_k = \lim_\alpha p_\alpha(T)$ (WOT - limit). The possibility of such approximation depends on special spectral properties of T (nonobvious even for Hilbert space normal operators). The finite splitting requirement is much weaker than the weak one because this case we need only in approximations $\mathcal{E}_k | X_\sigma = \lim_\alpha p_\alpha(T) | X_\sigma$, $k \in \sigma$ for finite σ's. For instance, this is the case if the spectra $\sigma(T | X_k)$, $k \geq 1$ are finite and pairwise disjoint. Even for normal Hilbert space operators this last property is compatible with the failure of the weak splitting property for $\mathfrak{A}(T)$, see e.g. [16].

In the Hilbert space case: the (strict) splitting property = (weak splitting) + ($\mathfrak{X}$ forms an unconditional basis).

The weak splitting of $\operatorname{Lat} \mathfrak{A}(T)$ is still a kind of approximation (but of a non easy nature); in order to split such a lattice one has to add some extra assumptions similar to a very weak form of the property to be a basis (it is enough to require of so called hereditary completeness of $\mathfrak{X}$: $x \sim \sum_{k \geq 1} \mathcal{E}_k x$ (formal Fourier series) implies $x \in \operatorname{span}(\mathcal{E}_k x : k \geq 0)$, see [16], [17]).

<u>Algebra</u> $\mathfrak{A}_R(T)$. Here the same comments can be given (with a change of polynomial approximations to rational ones). However, it should be noted that the

splitting properties of $\mathfrak{A}_R(T)$ influence on the multiplicities $\mu(\mathfrak{A}_R(T)|X_i)$ only, not on $\mu(\mathfrak{A}_R(T|X_i)) = \mu_R(T|X_i)$. In fact, even for a family $\{X_i\}_{i\geq 1}$ which forms an orthogonal basis of a Hilbert space one can have $\sigma(T) \neq \mathrm{clos} \bigcup_{i\geq 1} \sigma(T|X_i)$, and hence the algebras $\mathfrak{A}_R(T|X_i)$ and $\mathrm{clos}_{WOT}(\mathfrak{A}_R(T)|X_i)$ can be very different, even under assumptions $\sigma(T|X_i) \cap \sigma(T|X_j) = \emptyset$, $i \neq j$.

We consider $\mathfrak{A}_R$ and μ_R in 1.46, 1.53, 1.57, 1.59, 1.60, 1.88 below.

<u>Algebra</u> $\{T\}''$. Note, first, it follows from $TX_k \subset X_k$, $k \geq 1$, that $T\mathcal{E}_k = \mathcal{E}_k T$, $k \geq 1$ (and vice versa) and hence $X_k \in \mathrm{Lat}\{T\}''$ for every $k \geq 1$. The weak splitting property ($\mathcal{E}_k \in \{T\}''$, $k \geq 1$) is equivalent to the inclusions $X_k \in \mathrm{Lat}\{T\}'$ (i.e. X_k has to be hyperinvariant for all $k \geq 1$). The same is true for the finite splitting. It is important to note that the inclusions $X_k \in \mathrm{Lat}\{T\}'$, $k \geq 1$, imply $\{T\}''|X_k = \{T|X_k\}''$ for all $k \geq 1$.

<u>Algebra</u> $\{T\}'$. The commutant $\{T\}'$ weakly splits for any operator T (inclusions $\mathcal{E}_k \in \{T\}'$ are obvious for <u>complete</u> family of T-invariant subspaces). Hence $\mathrm{Lat}\{T\}'$ and $\mathrm{Cyc}\{T\}'$ split too. We also have $\{T\}'|X_k = \{T|X_k\}'$, $k \geq 1$.

<u>Algebra</u> $\mathfrak{A}^*(T)$. Since $\mathrm{Lat}\,\mathfrak{A}^*(T) \subset \mathrm{Lat}\{T\}''$ (see 1.3 above) the weak (finite) splitting property for $\mathrm{Lat}\{T\}''$ implies the same one for $\mathfrak{A}^*(T)$. Other properties of $\mathfrak{A}^*(T)$ and $\mu^*(T)$ we discuss in Chapter 2 for Hilbert space normal operators.

We close this section with two remarks. First, the inclusions $X_k \in \mathrm{Lat}\,T$, $k \geq 1$, not imply (generally speaking) that $X_k \in \mathrm{Lat}\{T\}'$, $k \geq 1$ (even for an orthogonal basis $\mathfrak{X}$ of a Hilbert space X). Second, $\mathrm{Lat}\,\mathfrak{A}(T) \not\subset \mathrm{Lat}\,\mathfrak{A}_R(T)$, $\mathrm{Lat}\{T\}'' \not\subset \mathrm{Lat}\{T\}'$ despite of $(TX_k \subset X_k, \forall_k) \Rightarrow (\{T\}''X_k \subset X_k, \forall_k)$

for any <u>complete</u> minimal family $\{X_k : k \geq 1\}$.

And at last a corollary of the preceding discussion.

1.44. COROLLARY. <u>If $\{X_k : k \geq 1\}$ is a complete minimal family of T-hyperinvariant subspaces (i.e. if $X_k \in \mathrm{Lat}\{T\}'$, $k \geq 1$) we have</u>

$$\mu'(T) = \sup_k \mu'(T \mid X_k)$$

$$\mu''(T) = \sup_k \mu''(T \mid X_k) .$$

In fact, the corollary follows from 1.34, 1.42 and remarks of sec.1.43. ●

1.45. **SPLITTINGS AND MAXI-FORMULAS IN SPECTRAL TERMS.** A kind of spectral independence of summands $T \mid X_k$, $k \geq 1$, is sufficient in order for splitting properties to be valid for the algebras $\mathfrak{A}(T)$ and $\mathfrak{A}_R(T)$ generated by an operator T . Recall, that splittings for $\mathfrak{A}(T)$ imply a maxi-formula for the multiplicity of the spectrum $\mu(T)$ whereas relationships between $\mu_R(T)$ and splittings for $\mathfrak{A}_R(T)$ need in an extra investigation. Spectral independences mentioned above will be expressed in terms of the spectra or polynomially convex hulls of the spectra. Recall, the <u>polynomially convex hull</u> of a compact set σ, $\sigma \subset \mathbb{C}$, is the set

$$\mathrm{pch}\,\sigma = \{ \lambda \in \mathbb{C} : |p(\lambda)| \leq \max_\sigma |p| \quad \text{for all polynomials } p \}.$$

We start with a simple observation (lemma 1.46 below), then discuss the behaviour of $\mu_R(T)$ in its dependence on $\mu_R(T \mid X_k)$ and next prove a general maxi-formula for the simple spectra.

1.46. LEMMA. <u>Let $\mathfrak{X} = \{X_k : k \geq 1\}$ be a complete minimal family of T-invariant subspaces</u>, $T \in L(X)$. <u>Then</u>

(a) <u>The algebra $\mathfrak{A}(T)$ weakly splits if</u>

$$(1.47) \quad \mathrm{pch}\,\sigma(T \mid X_k) \cap \mathrm{pch}\,\sigma(T \mid X_{\mathbb{N} \setminus \{k\}}) = \emptyset , \quad k \geq 1 ,$$

<u>and finitely splits if</u>

$$(1.48)\quad \operatorname{pch}\sigma(T|X_k)\cap \operatorname{pch}\sigma(T|X_j)=\emptyset,\quad k\neq j.$$

<u>In the both cases</u>

$$(1.49)\quad \mu(T)=\sup_k \mu(T|X_k).$$

(b) <u>The algebra</u> $\mathfrak{A}_R(T)$ <u>weakly splits if</u>

$$(1.50)\quad \sigma(T|X_k)\cap\sigma(T|X_{\mathbb{N}\setminus\{k\}})=\emptyset,\quad k\geq 1,$$

<u>and finitely splits if</u>

$$(1.51)\quad \sigma(T|X_k)\cap\sigma(T|X_j)=\emptyset,\quad k\neq j.$$

<u>In the case of finite family</u> $\mathcal{X}$ <u>we have</u>

$$(1.52)\quad \mu_R(T)=\max_k \mu_R(T|X_k).$$

PROOF. (a) Fixing any $k\geq 1$ and assuming (1.47) be valid one can find a sequence of polynomials $\{p_n\}_{n\geq 1}$ uniformly tending to zero on a neighbourhood of $\sigma(T|X_{\mathbb{N}\setminus\{k\}})$ and to one on a neighbourhood of $\sigma(T|X_k)$. This implies $\mathcal{E}_k = =\lim_n p_n(T)\in\mathfrak{A}(T)$, and the weak splitting follows.

If (1.48) is fulfilled we conclude $\operatorname{pch}\sigma(T|X_k)\cap \cap \operatorname{pch}\sigma(T|X_{\sigma\setminus\{k\}})=\emptyset$ for every $k\geq 1$ and every finite set σ, $\sigma\subset\mathbb{N}$. For such a set σ this results in polynomials like previous ones with a change of $X_{\mathbb{N}\setminus\{k\}}$ for $X_{\sigma\setminus\{k\}}$. Hence $\mathcal{E}_k|X_\sigma=\lim_n p_n(T)|X_\sigma\in \operatorname{clos}_{WOT}\mathfrak{A}(T)|X_\sigma$ for every $k\in\sigma$. This means $\mathfrak{A}(T)$ finitely splits.

(b) The same proof for splittings but with suitable rational functions instead of polynomials. For a finite direct decomposition $X=X_1\dot{+}X_2\dot{+}\cdots\dot{+}X_n$ we have $\sigma(T)=\bigcup_{i=1}^n\sigma(T|X_i)$, and moreover, none of connected component of $\mathbb{C}\setminus\sigma(T|X_i)$ is covered by $\sigma(T|X_j)$, $j\neq i$. Hence, $\mathfrak{A}_R(T|X_i)= =\operatorname{clos}_{WOT}(\mathfrak{A}_R(T)|X_i)$ and $\mu_R(T)=\mu(\mathfrak{A}_R(T))=\max_i(\mathfrak{A}_R(T)|X_i)=$

$= \max_i \mu_R(T|X_i)$. ●

1.53. COMMENT. In fact, it will be proved later on that on the simple spectra (i.e. for $\mu(T|X_k) = 1$, $k \geq 1$) condition (1.51) already implies maxi-formula (1.49). We turn to this matter in a few moments, but first we discuss maxi-formula (1.52) for μ_R.

In order for (1.52) be valid for an infinite (complete and minimal) family $\mathfrak{X}$ we need only equalities $\mathfrak{A}_R(T|X_k) = \operatorname{clos}_{WOT} \mathfrak{A}_R(T)|X_k$, $k \geq 1$. The latter depends on relations between the spectrum $\sigma(T)$ and the spectra $\sigma(T|X_k)$, $k \geq 1$. The inclusion $\sigma(T) \supset \operatorname{clos} \bigcup_{k \geq 1} \sigma(T|X_k)$ is obvious. If

(1.54) $\sigma(T) = \operatorname{clos} \bigcup_{k \geq 1} \sigma(T|X_k)$

one says T admits the <u>localization of the spectrum</u> (with respect to $\mathfrak{X}$). This last property depends on metric properties of $T|X_k$, $k \geq 1$, not on the geometry of the sequence $\mathfrak{X} = \{X_k : k \geq 1\}$. Indeed, in [4] the following example is constructed: a sequence of operators $Q_n \in L(\mathbb{C}^n)$ with disjoint and unimodular spectra, $\sigma(Q_n) \subset \mathbb{T}$ and such that

(1.55) $\sigma(Q) = \operatorname{clos} \mathbb{D}$, $Q = \sum_{n \geq 1} \oplus Q_n$.

The following example shows how to use the nonlocalization property to avoid maxi-formula (1.52) for μ_R. On the other hand, assuming the localization of the spectrum we prove maxi-formulas for μ_R (see 1.57 below).

1.56. EXAMPLE. Consider an annulus $A = \{z : r_1 < |z| < r_2\}$ with $0 < r_1 < r_2 < 1$ and the shift operator S_A on the Hardy space $H^2(A)$: $S_A f = zf$. Let further Q_n be operators from (1.55), and $T = S_A \oplus Q_1 \oplus Q_2 \oplus \dots$. Then $\sigma(T) = \operatorname{clos} \mathbb{D}$ and

$\mu_R(T) = \mu(T) \geq \mu(S_A) = 2 > 1 = \max(\mu_R(S_A), \mu_R(Q_n))$.

One can note that in this case condition (1.48) is fulfilled. ●

1.57. LEMMA. Let $\mathfrak{X}=\{X_k : k\geq 1\}$ be a complete minimal family of T-invariant subspaces, $T\in L(X)$ and let disjointness property (1.51) be fulfilled. Then

(a) $\mu_R(T)\leq 1+\sup_{k\geq 1}\mu_R(T\,|\,X_k)$.

(b) If $\mu_R(T\,|\,X_k)=\mu(T\,|\,X_k)$ for every $k\geq 1$, we get

(1.58) $\mu_R(T)=\sup_{k\geq 1}\mu_R(T\,|\,X_k)$.

(c) If none of connected component of $\mathbb{C}\setminus\sigma(T\,|\,X_k)$, $k\geq 1$ is contained in $\sigma(T)$, maxi-formula (1.58) is valid.

(d) If localization property (1.54) takes place and if none of connected component of $\mathbb{C}\setminus\sigma(T\,|\,X_k)$, $k\geq 1$, is densely covered by $\sigma(T\,|\,X_j)$, $j\neq k$, we have the same maxi-formula.

PROOF. In fact, (a) depends on a result in D.A. Herrero [18] mentioned above (always $\mu\leq\mu_R+1$; see 1.88 for the proof). Namely, by 1.46 (b), 1.42 and 1.34 we have $\mu_R(T)=$ $=\sup_{k\geq 1}\mu(\mathfrak{A}_R(T)\,|\,X_k)$, and further $\mu(\mathfrak{A}_R(T)\,|\,X_k)\leq\mu(T\,|\,X_k)\leq$ $\leq\mu_R(T\,|\,X_k)+1$.

Part (b) follows from the same remarks and inequality (1.22).

As to assertions (c) and (d), it is easy to see that under their assumptions we get $\mathfrak{A}_R(T\,|\,X_k)=\operatorname{clos}_{WOT}(\mathfrak{A}_R(T)|X_k)$ for every $k\geq 1$. All other follows from 1.46 (b) and 1.42. ●

1.59. COROLLARY. If condition (1.50) is fulfilled the maxi-formula is valid for μ_R. ●

1.60. EXAMPLE. An additional assumption of 1.57 (d) is essential. For instance, let $T=T_1\oplus T_2\oplus T_3\oplus\ldots$ where $T_1=S_A$ and $T_i=\lambda_i I$, $i\geq 2$, acting on $\mathbb{C}$, and $\{\lambda_i : i\geq 2\}$ be a dense subset of the hole of the annulus A. Then, $\mu_R(T)=\mu(T)\leq$ $\geq\mu(S_A)=2>1=\max_{i\geq 1}\mu_R(T_i)$. ●

1.61. MORE GENERAL MAXI-FORMULAS. Now, we proceed to discuss maxi-formulas under the weakest possible hypothesis of spectral independence type (see (1.51)): $\sigma(T|X_k) \cap \sigma(T|X_j) = \emptyset$, $k \neq j$. In fact, we will see later on that sometimes this condition implies maxi-formula (1.49) and some splitting properties for $\mathrm{Cyc}\,T$. On the other hand, even if the intersection $\sigma(T_1) \cap \sigma(T_2)$ is a singleton the maxi-formula can fail to be true.

Our main results about maxi-formulas (1.69 , 1.71, 1.72, 1.75) will be deduced from the following proposition on simultaneous polynomial approximation (in fact, a partial case of 1.69).

1.62. LEMMA. Let K be a compact subset of the plain $\mathbb{C}$ and Y be a normed space containing the set $\mathcal{P}_A$ of all polynomials in z as a dense subset. If $f \mapsto zf$, $f \in \mathcal{P}_A$, is a continuous operator on Y and if

(1.63) $\sigma(z) \cap \mathrm{pch}\, K = \emptyset$,

polynomials are dense in the product space $Y \times \mathcal{P}(K)$, $\mathcal{P}(K) \overset{def}{=} \mathrm{clos}_{C(K)} \mathcal{P}_A$ stands for the closure of $\mathcal{P}_A$ in the space $C(K)$ of all continuous functions endowed with the uniform norm $\|f\|_C = \max_K |f|$ (formally speaking, the diagonal set $\{(p,p) : p \in \mathcal{P}_A\}$ is dense in $Y \times \mathcal{P}(K)$).

PROOF. Since the Hahn-Banach theorem it is enough to prove that for every functionals $\varphi \in Y^*$ and $\nu \in C(K)^*$ the identity

(1.64) $$\varphi((z-\lambda)^{-1}) = \int_K \frac{d\nu(z)}{z-\lambda}, \quad |\lambda| > R$$

implies $\varphi = \mathbb{O}$.

Define the Cauchy transform $C\psi$ of a functional $\psi \in Y^*$ by

$$C\psi(\lambda) = \psi((z-\lambda)^{-1}), \qquad \lambda \in \mathbb{C} \setminus \sigma(z) \overset{def}{=} \Omega .$$

It is easy to see that C is a continuous injective mapping from Y^* to $H = \mathrm{Hol}_0(\Omega)$, the space of all functions holomorphic on Ω and vanishing at infinity, endowed with the topology of the uniform convergence on compact sets. The adjoint operator

z^* is transformed by C into an operator A,

(1.65) $A = Cz^*C^{-1}$; $Af = zf - (zf)(\infty)$, $f \in CY^*$.

Moreover, the operator A is defined (by the same expression) on the whole space H and, in fact, is a continuous endomorphism of H into itself. It is not hard to check that for every $\mu \in \Omega$ the operator $A - \mu I$ is invertible and

$$(A - \mu I)^{-1} f = \frac{f - f(\mu)}{z - \mu}, \quad f \in H.$$

Now, consider the closed A-invariant subspace $E = E_f$ generated by $f = C\varphi$:

$$E_f = span_H (A^n f : n \geqslant 0).$$

In view of (1.65) we have $(A - \mu I)^{-1} | CY^* = C(z^* - \mu I)^{-1} C^{-1}$ for every $\mu \in \Omega = \mathbb{C} \setminus \sigma(z)$. By the lemma assumption there exist polynomials p_n such that $(z-\mu)^{-1} = \lim_n p_n$ (in the space Y). Hence, applying the continuous operator $(z - \mu I)^{-1}$ and then the functional φ, we get $(A - \mu I)^{-1} f \in E_f$, $\mu \in \Omega$. Thus, the subspace E is not only invariant but Ω-rationally invariant:

$$(A - \mu I)^{-1} E \subset E, \quad \mu \in \Omega.$$

Let us show that the last property contradicts identity (1.64). To this aim, consider the adjoint space $Hol_0(\Omega)^* = Hol(\sigma)$ consisted of all (germs of) analytic functions on the compact set $\sigma = \sigma(z) = \mathbb{C} \setminus \Omega$. The duality between H and $Hol(\sigma)$ is given by the following bilinear form

$$\langle f, g \rangle = \frac{1}{2\pi i} \int_{\partial\Gamma} f(z) g(z)\, dz,$$

$f \in Hol_0(\Omega)$, $g \in Hol(\sigma)$ where Γ stands for a compact with the piecewise smooth oriented boundary $\partial\Gamma$ and such that

$\sigma \subset int\,\Gamma$. It is easy to see that with respect to this pairing the adjoint operator A^* acts by the formula $A^* g = zg$, $g \in Hol(\sigma)$.

Consider the annulator $E^\perp$ of E ,

$$E^\perp = \{ g \in Hol(\sigma) : \langle f, g \rangle = 0,\ f \in E \}.$$

It is obvious that $A^* E^\perp \subset E^\perp$ and even $(A^* - \mu I)^{-1} E^\perp \subset E^\perp$ for every $\mu \in \mathbb{C} \setminus \sigma$. Hence, by Runge theorem, $E^\perp$ is an ideal of the algebra $Hol(\sigma)$. And it is well-known (and easy to see) that being non-zero an ideal is of the form $q \cdot Hol(\sigma)$ with a polynomial $q \not\equiv 0$. In this case, one has $codim\, E^\perp < \infty$ and $dim\, E < \infty$. The last property implies $dim\, Y < \infty$, and hence the spectrum $\sigma(z)$ is a finite set. Now, (1.64) and (1.63) give $C\varphi \equiv 0$, then $\varphi = \mathbb{O}$.

It remains to prove that the equality $E^\perp = \{\mathbb{O}\}$, i.e. $E = Hol_0(\Omega)$, is impossible for a subspace $E = E_f$ generated by a function $f = C\varphi$ satisfying (1.64). From (1.64) we only use that the restriction $f \mid \Omega_\infty$ permits an analytic continuation into $(\mathbb{C} \setminus K)_\infty$ where $(\cdot)_\infty$ stands for the unbounded component of an open set with compact complement. The continuation, say f_∞ , being defined on $(\mathbb{C} \setminus K)_\infty = \mathbb{C} \setminus pch\, K$ is analytic in a neighbourhood of $\sigma = \sigma(z)$ (due to (1.63)).

Further denote by $\Omega_1, \Omega_2, \dots$ all bounded connected components of Ω , $\Omega = \Omega_\infty \cup \Omega_1 \cup \Omega_2 \cup \dots$ Since $pch\, K \subset \Omega$, and hence $min\{ dist(z, \partial\Omega) : z \in pch\, K \} > 0$ there exists an integer N such that $pch\, K \subset \Omega_\infty \cup (\bigcup_{i=1}^{N} \Omega_i)$. Consider the intersections $\Omega_j \cap pch\, K$, $j = \infty, 1, 2, \dots, N$, and choose in Ω_j smooth closed Jordan curves γ_j encircling these intersections*) in such a way that $\Omega_j \cap pch\, K \subset Int\, \gamma_j \subset \Omega_j$ (here $Int\, \gamma$ stands for the bounded component of the complement $\mathbb{C} \setminus \gamma$

*) We can first consider a number of curves which encircle the compact and are included in an ε-neighbourhood of the compact set $\Omega_j \cap pch\, K$, and then join them by some slits (or very thin passages) still remaining in Ω_j .

of a simple closed Jordan curve). Thus, in view of (1.63), the complement $\overline{\mathbb{D}}_R \setminus (\bigcup \mathrm{Int}\,\gamma_j) \overset{def}{=\!=} \Gamma$, $\overline{\mathbb{D}}_R = \{z : |z| \leq R\}$, can serve as a compact with an oriented boundary enjoying the duality between $Hol_0(\Omega)$ and $Hol(\sigma)$.

Now, we are in a position to show a non-zero function from $Hol(\sigma)$ orthogonal to all $z^n f$, $n \geq 0$. To this aim, let us denote $f_j = f|\Omega_j$, $j = 1, \dots, N$. Then, for every $g \in Hol(\mathrm{clos}\,\Gamma)$ we have

$$\int_{\partial\Gamma} z^n f g\, dz = \int_{|z|=R} z^n f_\infty g\, dz + \sum_{j=1}^{n} \int_{\gamma_j} z^n f_j\, g\, dz =$$

$$= \sum_{j=1}^{n} \int_{\gamma_j} z^n (f_j - f_\infty)\, g\, dz .$$

By lemma 1.66 below we can write the difference $f_j - f_\infty$ as a product $f_j - f_\infty = F_j F_j^\infty$ of functions $F_j \in Hol(\mathrm{clos\,Int}\,\gamma_j)$ and $F_j^\infty \in Hol_0(\mathbb{C} \setminus \mathrm{Int}\,\gamma_j)$. Moreover, $F_j^\infty(\zeta) \neq 0$, $\zeta \in \mathbb{C} \setminus \mathrm{Int}\,\gamma_j$. In particular, $1/F_j^\infty$ is holomorphic on $\mathrm{clos\,Int}\,\gamma_k$ for $k \neq j$. Thus, putting $g = \prod_{k=1}^{N} (1/F_k^\infty)$ we get

$$\int_{\partial\Gamma} z^n f g\, dz = \sum_{j=1}^{n} \int_{\gamma_j} z^n F_j \prod_{k \neq j} (1/F_k^\infty)\, dz = 0$$

for $n = 0, 1, \dots$. ●

1.66. LEMMA. <u>Let</u> γ <u>be a smooth simple closed Jordan curve and let</u> $F \in Hol(\gamma)$ (<u>i.e.</u> F <u>is holomorphic on a neighbourhood of</u> γ). <u>Then there exist functions</u> $F_1 \in Hol(\mathrm{clos\,Int}\,\gamma)$ <u>and</u> $F_\infty \in Hol_0(\mathbb{C} \setminus \mathrm{Int}\,\gamma)$ <u>such that</u> $F = F_1 F_\infty$ <u>and</u> $F_\infty(\zeta) \neq 0$, $\zeta \in \mathbb{C} \setminus \mathrm{Int}\,\gamma$.

PROOF. Let $z_1, \dots, z_n$ be all zeros of F on γ (multiplicities induced) and let $F = p \cdot G$ where $p = \prod_{k=1}^{n} (z - z_k)$, $G(\zeta) \neq 0$ on γ. Let Γ be a compact with an oriented (smooth) boundary and let $\gamma \subset \operatorname{Int}\gamma$. Without loss of generality we assume that $\gamma_+ \overset{def}{=\!=} \partial\Gamma \cap \operatorname{Int}\gamma$ and $\gamma_- \overset{def}{=\!=} \partial\Gamma \cap (\mathbb{C} \setminus \operatorname{clos}\operatorname{Int}\gamma)$ be Jordan curves homotopic inside of Γ to each other (up to orientations) and to γ . Further, let $N = \operatorname{wind}_\gamma G$ be the increment of $\arg G$ along with γ , and let λ be any point of $(\operatorname{Int}\gamma) \setminus \Gamma$. Then $G = (z-\lambda)^N G^\circ$, $\operatorname{wind}_\gamma G^\circ = 0$. Hence, $G^\circ = \exp h$ with a function h from $\operatorname{Hol}(\Gamma)$. Putting

$$h_\pm = \frac{1}{2\pi i}\int_{\gamma_\pm} \frac{h(\zeta)}{\zeta - z}\,d\zeta, \quad G_1 = (z-\lambda)^{q+1}\exp h_-, \quad G_\infty = (z-\lambda)^{N-q-1}\exp h_+$$

where $q = \max(N, 0)$, and

$$F_1 = pG_1 \ , \quad F_\infty = G_\infty$$

we finish the proof. ●

1.67. COMMENT. One can apply lemma 1.6.2 in the following way. Let T_1 , T_2 be operators acting on X_1 , X_2 respectively. In order to prove a maxi-formula for $\mu(T_1 \oplus T_2)$ it is enough to find a sequence of polynomials $\{p_n\}_{n\geq 1}$ in such a way that for dense subsets $x_i \in X_i$, $p_n(T_i)x_i$, $n = 1, 2, \ldots$ converge to x_i or $\mathbb{O}$ according to whether $i=1$ or $i=2$. In fact, one can restrict himself to $x_i \in C_i$, $C_i \in \operatorname{Cyc} T_i$, and so, to produce the polynomials mentioned above, can try to use Lemma 1.6 2. It is a little surprising but the result depends rather on $\dim C_i$ than on the multiplicities $\mu(T_i)$. To be specific, it depends on a possibility to choose (say, for $i=1$) a cyclic subset $C_1' \subset C_1$ of the minimal dimension $\dim C_1' = \mu(T_1)$. The corresponding property (or its absence) can be described by a particular quantity introduced in [8] - [11] and called "disc" (dimension of input subspace of control). By definition,

$$(1.68) \quad \operatorname{disc} T = \sup_{C \in \operatorname{Cyc} T} \min\{\dim C' : C' \subset C,\ C' \in \operatorname{Cyc} T\}.$$

An arbitrary couple of naturals m , n with $m \geq n$ can serve as the disc and the multiplicity of an operator: $m = \operatorname{disc} T$, $n = \mu(T)$. In [8] - [11] several classes of operators are considered for the purpose of analysis of the relations between disc(·) and $\mu(\cdot)$.

We will see below how $\operatorname{disc} T$ influences the splitting properties for $T_1 \oplus T_2$ (see 1.71, 1.73, 1.75). Let us start with a corollary of lemma 1.62.

1.69. COROLLARY. Let $T_i \in L(X_i)$, X_i be Banach spaces, $i=1,2$, and let C be a subspace of $X_1 \oplus X_2$ such that $\mathcal{E}_i C \in Cyc\, T_i$ $(i=1,2)$, $\dim \mathcal{E}_1 C = 1$ (where as usual $\mathcal{E}_i$ stands for the projection onto X_i, $i=1,2$). Claim: $C \in Cyc(T_1 \oplus T_2)$ provided

(1.70) $\sigma(T_1) \cap pch\, \sigma(T_2) = \emptyset$.

In particular, $Cyc(T_1 \oplus T_2)$ restrictedly splits if $\mu(T_1) = 1$ and (1.70) is satisfied.

In fact, consider a (semi)norm $\|p\|_Y = \|p(T_1)x_1\|$ on the set $\mathcal{P}_A$ of all polynomials, and let Y be the closure (completion) of $\mathcal{P}_A$ with respect to this norm; here x_1 stands for a non-zero vector from $\mathcal{E}_1 C$. Applying 1.62, we get a sequence of polynomials such that $\lim_n p_n(T_1)x_1 = x_1$, $p_n \rightrightarrows 0$ on a neighbourhood of $pch\, \sigma(T_2)$ and hence $\lim_n \|p_n(T_2)\| = 0$. It follows, $\mathcal{E}_1 C \oplus \{\mathbb{O}\} \subset E_C = span((T_1 \oplus T_2)^n C : n \geq 0)$, hence $\{\mathbb{O}\} \oplus \mathcal{E}_2 C \subset E_C$ and then $X_1 \oplus \{\mathbb{O}\}$, $\{\mathbb{O}\} \oplus X_2 \subset E_C$, i.e. $E_C = X_1 \oplus X_2$. ●

1.71. COROLLARY. Let T_i be Banach space operators, $i=1,2$, $disc\, T_1 = 1$ and (1.70) is satisfied. Then $Cyc(T_1 \oplus T_2)$ splits and $\mu(T_1 \oplus T_2) = max(\mu(T_1), \mu(T_2)) = \mu(T_2)$.

Indeed, one can easily reduce 1.71 to 1.69: starting with a subset C , $C \subset X_1 \oplus X_2$, with cyclic projections $\mathcal{E}_i C$ let us choose a vector $x \in C$ with $\mathcal{E}_1 x \in Cyc\, T_1$, and put $X_2' = E_{\mathcal{E}_2 x} = span(T_2^n \mathcal{E}_2 x : n \geq 0)$. Since $\sigma(T_2 | X_2') \subset pch\, \sigma(T_2)$ one can apply 1.69 to $T_1 \oplus (T_2 | X_2')$ and conclude that $X_1 \oplus \{\mathbb{O}\} \subset E_x \subset E_C$. The corollary follows. ●

1.72. COROLLARY. If $\mu(T_1) = 1$ and (1.70) is satisfied, we have $\mu(T_1 \oplus T_2) = max(\mu(T_1), \mu(T_2)) = \mu(T_2)$.

Obvious from 1.69. ●

1.73. EXAMPLES. Let us show that a natural analogue of 1.71 with a change of the request of $disc\, T_1 = 1$ to $\mu(T_1) = 1$ is false (and so, 1.69 becomes false if it happens $dim\, \mathcal{E}_1 C > 1\ (= \mu(T_1))$, even if $\mu(T_2) = disc\, T_2 = 1$) . There exists a similar example showing that corollary 1.69 can fail to be true if $dim\, \mathcal{E}_1 C = \mu(T_1)$ but $\mu(T_1) > 1$.

(a) <u>There exist Hilbert space operators</u> T_i <u>with</u> $\mu(T_i) = 1$, $i = 1, 2$, <u>with disjointness property</u> (1.70), <u>and a non-cyclic subspace</u> C, $dim\, C = 2$ <u>such that</u> $\mathcal{E}_i C \in Cyc\, T_i$, $i = 1, 2$. <u>This means</u> $Cyc(T_1 \oplus T_2)$ <u>does not split whereas</u> $\mu(T_1 \oplus T_2) = max(\mu(T_1), \mu(T_2)) = 1$.

Let $T_1 = \mathbb{S}$ be the (unitary) shift operator $f \mapsto zf$ on $L^2(\mathbb{T})$ and $T_2 = r S^* \overset{def}{=\!=} r(\mathbb{S}|H^2)^*$, $0 < r < 1$. Let C be the linear hull of $1 \oplus \mathbb{O}$ and $f \oplus g \in L^2 \oplus H^2$ where

$g(z) = -rz^{-1} f(rz^{-1}) = -exp((z-2)^{-1})$, $f(z) = z^{-1} exp(z(r-2z)^{-1})$.

Then $\sigma(T_1) = \mathbb{T}$; $\sigma(T_2) = pch\, \sigma(T_2) = r\overline{\mathbb{D}}$, $\overline{\mathbb{D}} = clos\, \mathbb{D}$; $\mu(T_1) = \mu(T_2) = disc\, T_2 = 1$. Moreover, $\mathcal{E}_1 C$ is $\mathbb{S}$-cyclic subspace: if $E \overset{def}{=\!=} E_{\mathcal{E}_1 C}$ is rationally invariant, E obviously is equal to $L^2(\mathbb{T})$; or else one can apply known Beurling-Helson theorem (see [14], chapter 1) and conclude $E = \varepsilon \cdot H^2$ with a measurable unimodular function ε. Thus, 1, $f \in \varepsilon H^2$ and these inclusions imply $f = f_1/f_2$, with $f_i \in H^2$, $i = 1, 2$. It is well-known (see [14], Chapter 2) that such a function f admitting an analytic continuation into $\{z : |z| > r/2\}$ has to be rational one. But for f this is not the case. Finally, $E = L^2(\mathbb{T})$, and hence $\mathcal{E}_1 C \in Cyc\, \mathbb{S}$. The inclusion $\mathcal{E}_2 C \in Cyc\, T_2$ follows in a similar way.

On the other hand, we can easily show an element of $L^2 \oplus H^2$ orthogonal to E_C. For instance, $\bar{z} \oplus 1 \perp T^n C$, $n \geq 0$:

$$(z^n f, \bar{z}) + (r^n S^{*n} g, 1) = \hat{f}(-n-1) + r^n \hat{g}(n) = 0$$

for $n \geqslant 0$, and (obviously) $\bar{z} \oplus 1 \perp T^n(1 \oplus \mathbb{O})$, $n \geqslant 0$. ●

(b) <u>There exist Hilbert space operators</u> T_i <u>with</u> $\mu(T_1) = 2$, $\mu(T_2) = 1$ <u>and such that (1.70) is satisfied but</u> $Cyc(T_1 \oplus T_2)$ <u>does not restrictedly split whereas</u> $\mu(T_1 \oplus T_2) = max(\mu(T_1), \mu(T_2)) = 2$.

Let us modify the previous example setting $T_1 = S_A$, the shift operator $S_A f = zf$ on the Hardy space $H^2(A)$ of an annulus $A = \{z : r_1 < |z| < 1\}$, and $T_2 = r_2 S^*$, $0 < r_2 < r_1$. Then $\sigma(T_1) = \operatorname{clos} A$, $\sigma(T_2) = \operatorname{pch} \sigma(T_2) = r_2 \overline{\mathbb{D}}$, $\mu(T_1) = 2$ and $\mu(T_2) = \operatorname{disc} T_2 = 1$. Moreover, it is easy to see that S_A is unitarily equivalent to the operator $\tilde{T}_1 = r_1 S | H^2_- + S | H^2$ on $L^2(\mathbb{T})$. Let $x = 1 \oplus \mathbb{O}$ and let E_x be $(\tilde{T}_1 \oplus T_2)$-invariant subspace generating by x. Since (obviously) $E_x = H^2 \oplus \{\mathbb{O}\}$ the resting quotient operator $\tilde{\tilde{T}}_1 \oplus T_2 / E_x$ is unitarily equivalent to $\tilde{T} = r_1 S^* \oplus r_2 S^*$. Now, we can choose S^*-cyclic functions $f, g \in H^2$ in such a way that $r_1^n \hat{f}(n) = -r_2^n \hat{g}(n)$, $n \geqslant 0$ (for instance, $g = \exp((z-2)^{-1})$, $f(z) = -g(r_2 r_1^{-1} z)$). Then $\tilde{T}^n (f \oplus g) \perp 1 \oplus 1$ for all $n \geqslant 0$, and we constructed (up to a unitary equivalence) a subspace C (= the linear hull of $1 \oplus \mathbb{O}$ and $f \oplus g$) of $H^2(A) \oplus H^2$ such that $C \notin \operatorname{Cyc}(T_1 \oplus T_2)$; $\mathcal{E}_i C \in \operatorname{Cyc} T_i$, $\dim \mathcal{E}_i C = \mu(T_i)$ for $i = 1, 2$.

The equality $\mu(T_1 \oplus T_2) = 2$ follows from forthcoming results (see 1.86), or can be checked simply by showing any cyclic vector for $\tilde{T}$ (examples are well-known [14], for instance, $f \oplus g$ with f and g having "very lacunary" and rapidly convergent Fourier series). ●

1.74. COMMENT. Operators T_1 and T_2 of the above examples are "independent" over some pairs of vectors $x \oplus y$ and "dependent" over some others. Our idea consisted in considering a pair of the first kind, say $x_1 \oplus \mathbb{O}$, and then looking for cyclic vectors $\dot{f} \in X_1 / E_{x_1}$, $g \in X_2$ such that a functional $p \mapsto \langle p(T_2) g, \varphi \rangle$, $p \in \mathcal{P}_A$, happens to be continuous with respect to the quotient norm $\| p(\dot{T}_1) \dot{f} \|$ for a non-zero $\varphi \in X_2^*$. For this, the assumption $\sigma(T_2) \subset \operatorname{pch} \sigma(T_1)$ is essential (in our examples, even $\sigma(T_2) \subset \sigma(\dot{T}_1)$). We could be free of particular choice of g and φ if a sharper estimation was held: $\| p(T_2) \| \leqslant \operatorname{const} \cdot \| p(\dot{T}_1) \dot{f} \|$, $p \in \mathcal{P}_A$. In our main construction from 1.73 we have as T_1 the shift operator S on $L^2(\mathbb{T})$ with $x_1 = 1$ and as T_2 an operator with $\sigma(T_2) \subset r \overline{\mathbb{D}}$, $0 < r < 1$. Thus, the desired sharp estimation involves Hankel operators, for instance in the form $\max_{|\lambda|\le r} |p(\lambda)| \le$ $\le const \|P_- f p\|_2$, $p \in \mathcal{P}_A$. (By the way, the existence of such r , $0<r<1$ is equivalent to the inequality $\Lambda_r^* \Lambda_r \le$ $\le const\, H_f^* H_f$ where $H_f p = P_- f p$ stands for a Hankel operator, and Λ_r for a diagonal one: $\Lambda_r z^n = r^n z^n$, $n \ge 0$.) Unfortunately, the last kind of inequalities can't to be true, because it implies successively $|p(0)| \le const \|P_- f p\|_2$, $p \in \mathcal{P}_A$; $1 \in H_f^*(H_-^2)$; $f = \bar{\theta} h$ with an inner function θ and $h \in H^\infty$; and finally $|p(\lambda)| \le const \|P_\theta p\|_2$, $p \in \mathcal{P}_A$, where $|\lambda| \le r$ and P_θ denotes the orthogonal projection onto $K_\theta = H^2 \ominus \theta H^2$. But the last estimation means $(1-\bar{\lambda} z)^{-1} \in$ $\in K_\theta$, i.e. $\theta(\lambda)=0$ for $|\lambda| \le r$. Contradiction. ●

Now, we proceed to the next step of the program and replace condition (1.70) of corollary 1.69 by a weaker one, more natural in this question.

1.75. THEOREM. Let T_i be Banach space operators, $T_i \in L(X_i)$, $1 \le i \le n$, let $\mu(T_i)=1$ for every i and

$$\sigma(T_i) \cap \sigma(T_j) = \emptyset , \quad i \ne j .$$

Then $Cyc\, T$, $T = T_1 \oplus T_2 \dots \oplus T_n$ restrictedly splits, i.e. $x = \sum_{i=1}^n \oplus x_i \in Cyc\, T$ provided $x_i \in Cyc\, T_i$, $1 \le i \le n$. In particular, $\mu(T) = \max \mu(T_i) = 1$.

PROOF. In fact, we need only some considerations of plane geometry type in order to reduce the theorem to corollary 1.69 (in fact, to lemma 1.62). Let us devide these considerations into several steps.

Consider pairwise disjoint compact sets σ_i which are neighbourhoods of the corresponding spectra $\sigma(T_i)$ (i.e. $\sigma(T_i) \subset$ $\subset \operatorname{int} \sigma_i$, $i=1,\dots,n$) and have boundaries $\partial\sigma_i$ consisted of finite union of smooth pairwise disjoint closed Jordan curves. Let σ_{ij} be all connected components of σ_i (finite collection for every i). It is obvious that all of $\{\sigma_{ij}\}$ are pairwise disjoint.

If $(i,j) \ne (k,\ell)$ the corresponding set σ_{ij} is entirely contained in one connected component of $\mathbb{C}\setminus\sigma_{k,\ell}$, bounded or not. Thus, the hulls $\mathrm{pch}\,\sigma_{ij}$ and $\mathrm{pch}\,\sigma_{k\ell}$ either are disjoint or included in each other, and from here

(1.76) $$\mathrm{pch}\bigcup_I \sigma_{ij} = \bigcup_I \mathrm{pch}\,\sigma_{ij}$$

where (i,j) runs over an arbitrary subset I of the whole indices set.

Now, let $x_i \in \mathrm{Cyc}\,T_i$, $i=1,\dots,n$. Consider a decomposition of X_i into the direct sum of spectral subspaces, corresponding to the closed-open subsets $\sigma(T_i)\cap\sigma_{ij}$ of the spectrum: $X_{ij}=X_i(\sigma_{ij})$, $X_i = X_{i1}\dot{+}\dots$ (finite sum). Let x_{ij} be the corresponding projections of x_i onto the components. Then, by 1.4 and 1.14, x_{ij} is cyclic for $T_{ij}=T_i\,|\,X_{ij}$. To prove the theorem it is enough to check that $\mathrm{Cyc}\,T$ restrictedly splits with respect to the decomposition $X=\Sigma_{ij}X_{ij}$. On the other hand, the (restricted) splitting property is transitive, i.e. if $\mathrm{Cyc}(A\dot{+}B)$, $\mathrm{Cyc}(C\dot{+}\mathcal{D})$ (restrictedly) split and if $B=C\dot{+}\mathcal{D}$, $\mathrm{Cyc}(A\dot{+}C\dot{+}\mathcal{D})$ (restrictedly) splits too. So, one can use inductive arguments, and simply can prove that $\mathrm{Cyc}(T_{ij}\dot{+}T'_{ij})$ restrictedly splits for a choice of (i,j), where T'_{ij} stands for the (direct) sum of all other $T_{k,\ell}$: $T'_{ij}=\Sigma T_{k\ell}$, $(k,\ell)\neq(i,j)$.

Let us consider the boundary $\Gamma=\partial\,\mathrm{pch}\,(\bigcup_{i=1}^{n}\sigma_i)=$ $=\partial\,\mathrm{pch}\bigcup_{i,j}\sigma_{ij}=\partial\bigcup\mathrm{pch}\,\sigma_{ij}\subset\bigcup\partial\,\mathrm{pch}\,\sigma_{ij}\subset\bigcup\partial\sigma_{ij}$ and any two indices i,j for which $\Gamma\cap\partial\sigma_{ij}\neq\emptyset$. Then σ_{ij} cannot be included in any bounded component of $\mathbb{C}\setminus\sigma_{k\ell}$, and hence, by (1.76), $\sigma_{ij}\cap\mathrm{pch}(\bigcup\sigma_{k\ell})=\emptyset$ where (k,ℓ) runs over all pairs, $(k,\ell)\neq(i,j)$.

Making use of corollary 1.69, one can conclude that $\mathrm{Cyc}(T_{ij}\dot{+}T'_{ij})$ restrictedly splits. All is proved. ●

1.77. COROLLARY. <u>Let</u> $T\in L(X)$, X <u>be a Banach space and let</u> $\mathfrak{X}=\{X_k : k\geqslant 1\}$ <u>be a complete minimal family of</u> T-<u>invariant subspaces</u>. <u>If</u>

$$\sigma(T\,|\,X_i)\cap\sigma(T\,|\,X_j)=\emptyset,\quad i\neq j$$

<u>and if</u> $\mu(T\,|\,X_i)=1$ <u>for every</u> $i\geq 1$, $\mathrm{Cyc}\,T$ <u>weakly restrictedly splits. In particular,</u> $\mu(T)=1$.

It follows from theorem 1.75, definition 1.33 and lemma 1.34. ●

1.78. PARTIAL CASE. Let μ be a finite Borel measure on $\mathbb{C}$ with the compact support $\operatorname{supp}\mu$, and let $\operatorname{supp}\mu=\bigcup_{i=1}^{n}\sigma_i$, are closed and pairwise disjoint $\sigma_i\cap\sigma_j=\emptyset$, $i\neq j$. Then complex polynomials $\mathcal{P}_A$ are dense in $L^p(\mu)$ if (and only if) they are dense in $L^p(\chi_{\sigma_i}\mu)$, $1\leq p\leq\infty$, for every i, $i=1,\dots,n$.

Of course, the case $p=\infty$ is trivial (even if we replace $L^\infty(\chi_{\sigma_i}\mu)$ by $C(\sigma_i)$ and $L^\infty(\mu)$ by $C(\bigcup_1^n\sigma_i$). In any case, we set $Tf=zf$, $f\in L^p(\mu)$, choose cyclic vectors $1\cdot\chi_{\sigma_i}\in L^p(\chi_{\sigma_i}\mu)$ and apply 1.75. ●

1.79. REMARKS. Certainly, the most interesting case for the theory of polynomial approximations is the one when $\{\sigma_i\}$ is a Borel (<u>not closed</u>) partition of $\operatorname{supp}\mu$. There is no general condition for the (restricted) splitting in this case, but many special cyclicity conditions have been considered in the literature; see [19] for strongest results and for an account.

The spaces $L^{p_i}(\chi_{\sigma_i}\mu)$ with different p_i, $1\leq p_i\leq\infty$, can be considered in the same way, as well as any other function spaces X_i contained polynomials as a dense subset.

One more remark is about the (strong) splitting property in 1.77 instead of the weak one. In general, this is not longer the case even in the restricted form. The simplest EXAMPLES concern again with polynomial approximations. For instance, let m be (normalized) Lebesque measure on $\mathbb{T}=\{\zeta\in\mathbb{C}:|\zeta|=1\}$ and σ_i be closed disjoint arcs such that $m(\mathbb{T}\setminus\bigcup_{i\geq 1}\sigma_i)=0$. Then, polynomials are dense in $L^p(\chi_{\sigma_i}m)$ for every $i\geq 1$, but are not in $L^p(m)$. ● It is of interest to note that in this case corollary 1.77 being applied produces a cyclic vector, say f , for the shift operator $Tf=zf$, $f\in L^p(m)$, $1\leq p<\infty$.

Recall, that the well-known necessary and sufficient conditions to this consist in $f(\zeta)\neq 0$ m - a.e. $\zeta\in\mathbb{T}$, $\int_{\mathbb{T}}\log|f(\zeta)|\,dm(\zeta)=-\infty$.

1.80. JOINING OF EIGENVECTORS. Several subsequent sections are devoted to the influence of eigenvectors and rootvectors to the multicyclicity. Let $\sigma\subset\mathbb{C}$. We shall say that an operator $T\in L(X)$ is σ-complete if there exists a family $\{x_\lambda\}$ of eigenvectors generating the whole space:

(1.80a) $T x_\lambda=\lambda x_\lambda,\ \lambda\in\sigma,\ X=\operatorname{span}(x_\lambda:\lambda\in\sigma)$.

An operator T is said to be generalized σ-complete if there exists a family $\{x_{k,\lambda}: 1\leqslant k<n_\lambda,\ \lambda\in\sigma\}$ of rootvectors generating X:

(1.80b) $(T-\lambda I)x_{k,\lambda}=x_{k-1,\lambda},\ 1\leqslant k<n_\lambda\leqslant+\infty,\ \lambda\in\sigma,$
$X=\operatorname{span}\{x_{k,\lambda}\}$;

here $x_{0,\lambda}=\mathbb{O}$, $\lambda\in\sigma$. In view of the separability assumption one can think (in these definitions and in what follows) on at most countable sets σ only.

A number of next propositions shows that under a kind of spectral independence condition eigen- and rootvectors give no impact to the multicyclicity. We start with an elementary lemma generalizing a proposition from [10]; in this lemma we denote by k_f the multiplicity function (divisor) of a holomorphic function f, $f\not\equiv 0$:

$$k_f(\lambda)=\max\{s\geqslant 0: f^{(k)}(\lambda)=0,\ 0\leqslant k<s\}.$$

1.81. LEMMA. Let T be a generalized σ-complete operator for a set σ, $\sigma\subset\mathbb{C}$. Then $\mu(T)=1$. Moreover,
(a) If T is σ-complete there exists a family $\{\varepsilon_\lambda:\lambda\in\sigma\}$ of positive reals such that every vector $x=\sum_{\lambda\in\sigma}a_\lambda x_\lambda$ with $0<|a_\lambda|<\varepsilon_\lambda$, $\lambda\in\sigma$, where x_λ's are from (1.80a), is T-cyclic.

(b) If T is generalized σ-complete and $\{Q_n\}$ is any sequence of polynomials with $\lim_n k_{Q_n}(\lambda) \geq n_\lambda$, $\lambda \in \sigma$ (we borrow the notation from (1.80b)) there exists a cyclic vector x with

$$\lim_n \| Q_n(T) x \| = 0 .$$

PROOF. (a) This is a partial case of 1.37.

(b) Let us renumber the whole (countable) family $\{x_{k,\lambda} : 1 \leq k < n_\lambda, \lambda \in \sigma\}$ as $\{y_1, y_2, \dots\}$ and try to find a cyclic vector x in the form $x = \sum_{j \geq 1} a_j y_j$, $a_j \neq 0$. Let $a_1 = 1$, and let N_k be naturals such that $N_k < N_{k+1}$ and $Q_n(T) y_j = \mathbb{O}$ for $1 \leq j \leq k$, $n \geq N_k$. Note that if $a_1, \dots, a_k$ are chosen.. there exist polynomials p_{ik} such that

$$\Big\| p_{ik}(T) \sum_{j=1}^{k} a_j y_j - y_i \Big\| < \frac{1}{k} \qquad \text{for } 1 \leq i \leq k .$$

Using these polynomials let us subject the subsequent a_j's to following requirements:

$$\| Q_n(T) \| \sum_{j \geq k+1} |a_j| \cdot \| y_j \| < 1/k \qquad \text{for } n < N_{k+1} ,$$

$$\| p_{ik}(T) \| \sum_{j \geq k+1} |a_j| \cdot \| y_j \| < 1/k \qquad \text{for } 1 \leq i \leq k .$$

Then for $N_k \leq n < N_{k+1}$ we have

$$\| Q_n(T) x \| = \Big\| Q_n(T) \sum_{j \geq k+1} a_j y_j \Big\| < 1/k ,$$

and hence $\lim_n \| Q_n(T) x \| = 0$. Moreover, for every $k \geq 1$ and $1 \leq i \leq k$ we have

$$\| p_{ik}(T) x - y_i \| \leq$$

$$\leq \Big\| p_{ik}(T) \sum_{j=1}^{k} a_j y_j - y_i \Big\| + \Big\| p_{ik}(T) \sum_{j \geq k+1} a_j y_j \Big\| < 1/k + 1/k ,$$

and hence $y_i \in E(x)$ for all $i \geq 1$. This means $x \in \mathrm{Cyc}\,T$. ●

1.82. COMMENT. In fact, Proposition 1.81 (a) was noted even in [20] in a form of so-called individual spectral synthesis. On the other hand, making use of a deeper technique of unicity theorems for series of rational fractions $\sum_{k\geq 1} a_k(z-\lambda_k)^{-1}$ (see [21], [22], [23]) one can derive the following. If $\{\lambda_k\}_{k\geq 1}$ is a bounded sequence of complex numbers, $\lambda_k \neq \lambda_j$ $(k\neq j)$, if $\sum_{k\geq 1} \lambda_k^n a_k = 0$ for $n = 0,1,\ldots$ and if $\sup_k |a_k| e^{\varepsilon k} < \infty$ for a positive ε, one can claim $a_k = 0$ for all $k=1,2,\ldots$.

COROLLARY: The vector $x = \sum_{k\geq 1} (k!\, \|e_k\|)^{-1} e_k$ is cyclic for $T\,|\,\operatorname{span}(e_k : k\geq 1)$, provided $Te_k = \lambda_k e_k$, $\lambda_k \neq \lambda_j$ $(k\neq j)$.

PROOF: If $\varphi \in X^*$ and $0 = \langle T^n x, \varphi\rangle = \sum_{k\geq 1} \lambda_k^n (k!\,\|e_k\|)^{-1} \langle e_k, \varphi\rangle$ for $n \geq 0$, we get $\langle e_k, \varphi\rangle = 0$, $k \geq 0$. ●

Further, it is not hard to see that a similar reasoning leads to the following: if $\{e_k\}_{k\geq 1}$ is "overcomplete", i.e. $X = \operatorname{span}(e_k : k \geq n)$ for every $n \geq 1$, and if eigenvalues λ_k converge fast enough to a limit (say, $\lim_k |\lambda_{k+1}\lambda_k^{-1}| = 0$), we have $x \in \operatorname{Cyc} f(T)$ for every $x = \sum_{k\geq 1} a_k \|e_k\|^{-1} e_k$ with $a_k \neq 0$, $\sum_{k\geq 1} |a_k| < \infty$, and for every non-zero analytic function $f(T)$ of T. For some concrete applications of the last kind assertions see [24], [25].

In the next corollaries we show that under some independence conditions eigen- and rootvectors of an operator do not influence its multiplicity.

1.83. COROLLARY. Let $T \in L(X)$ and $\sigma \subset \mathbb{C}$. Let further X_1, $X_2 \in \operatorname{Lat} T$ and

$$\operatorname{clos}(T-\lambda I)X_1 = X_1, \quad \lambda \in \sigma$$

(1.84)

$T\,|\,X_2$ is generalized σ-complete.

Then $\mu(T\,|\,E) \leq \mu(T\,|\,X_1)$ where $E = \operatorname{span}(X_1, X_2)$.

In fact, let $X_1 = E(C_1)$, $d \overset{def}{=} \dim C_1 < \infty$ (otherwise it is nothing to prove). The assumption implies that the set $p(T)C_1$ is cyclic for $T|X_1$, provided a polynomial p, $p \not\equiv 0$, has its zeros in σ. Hence, fixing any basis $e_1, \dots, e_d$ of C_1 one can choose polynomials $p_{j,m,n}$ in such a way that

$$\Big\| \sum_{j=1}^{d} p_{j,m,n}(T)\, e_j - e_m \Big\| < 1/n$$

for $m = 1, \dots, d$ and the zeros divisors $k_{p_{j,m,n}}(\lambda)$ tend to n_λ for every j, m and every $\lambda \in \sigma$ (n_λ is from (1.80b) applied to $T \mid X_2$).

Let $\{Q_n\}$ be any enumeration of the polynomials $\sum_{j=1}^{d} p_{j,m,s}$ ($1 \le m \le d$, $s = 1, 2, \dots$) and let $x \in Cyc(T \mid X_2)$ be a cyclic vector from 1.81(b). Consider $f_j = e_j + x$, $1 \le j \le d$. We have

$$\Big\| \sum_{j=1}^{d} p_{j,m,s}(T) f_j - e_m \Big\| \le \Big\| \sum_{j=1}^{d} p_{j,m,s}(T) e_j - e_m \Big\| + \| Q_n(T) x \| ,$$

and thus for every m, $1 \le m \le d$, and s tending to infinity we get $e_m \in E(C)$, where $C = \mathcal{L}in(f_j : 1 \le j \le d)$. This implies $E(C) \supset E(C_1) = X_1$, and then $x \in E(C)$ and at last $E(x) = X_2 \subset E(C)$. The corollary follows since $C \in Cyc(T \mid E)$ and $\mu(T \mid E) \le \dim C = \dim C_1 = \mu(T \mid X_1)$. ●

1.85. COROLLARY. <u>Let</u> $T_i \in L(X_i)$, $i = 1,2$ <u>and</u> $X = X_1 \oplus X_2$, $T = T_1 \oplus T_2$ <u>and let</u> (1.84) <u>be fulfilled</u>. <u>Then</u>

$$\mu(T_1 \oplus T_2) = max(\mu(T_1), \mu(T_2)) = \mu(T_2). \; ●$$

1.86. COMMENT. Two last corollaries show that generalized σ-complete operators are very "**weak**" ones and give no impact to the multicyclicity under a weak spectral independence condition. By the way, this implies the known D.Herrero inequality [18] (for an independent proof see [26]; our proof in 1.88 below is different from the both mentioned).

Now we list a few examples to 1.85; on its possible generalizations see 1.89 below.

1.87. EXAMPLES. We use a partial case of 1.85 with $\sigma = \{0\}$

(i.e. when T_1 has a dense range and T is generalized $\{0\}$ - complete, $X_2 = span(e_k : k \geq 1)$, $Te_k = e_{k-1}$, $T_2 e_1 = \mathbb{O}$).

(a) If T_1 has dense range we have $\mu(T_1 \oplus S^*) = \mu T_1$, where S^* is the backward shift operator on H^2.

(b) In particular the maxi-formula holds for $S_A \oplus S^*$, S_A being the shift operator $S_A f = zf$, on $H^2(A)$ on an annulus $A = \{z : 0 < r_1 < |z| < r_2\}$. In a sense, the same is true even in the limit case $r_1 = 0$ if we change $H^2(A)$ to a space of functions holomorphic on $0 < |z| < 1$ admitting a growth at $z = 0$. E.g., $\mu(T_z \oplus S^*) = \mu(T_z)$ for the shift operator $T_z f = zf$ on the space $\{f : f \in Hol(z : 0 < |z| < 1)$, $\iint \rho(|z|)|f(z)|^2 dx\,dy < \infty\}$, where $\lim_{r \to 0} r^{-n}\rho(r) = 0$ for every $n \geq 1$.

(c) In contrast with (b) we have $\mu(S \oplus S^*) > \max(\mu(S), \mu(S^*)) = 1$ for the shift S on H^2.

1.88. THEOREM. Let $T \in L(X)$ and C be a subspace of X. There exists a subspace $C' \supset C$, $\dim C' \leq \dim C + 1$ such that

$$E(C') = E_R(C) \left(\overset{def}{=\!=} span(R_\lambda C : \lambda \notin \sigma(T))\right).$$

In particular,

$$\mu(T) \leq \mu_R(T) + 1 .$$

PROOF. Let us note that the cosets $\dot{x}_\lambda = R_\lambda x + E(C)$, $x \in C$ are eigenvectors (if non-zeros) of the quotient operator (say, $\dot{T}$) of $T | E_R(C)$ with respect to $E(C)$: $(\dot{T} - \lambda I)\dot{x}_\lambda = \dot{\mathbb{O}}$, $\lambda \in \mathbb{C} \setminus \sigma(T)$. By the definition of $E_R(C)$ the operator $\dot{T}$ is $(\mathbb{C} \setminus \sigma(T))$ -complete. It remains to apply 1.13 and 1.81 to $T | E_R(C)$:

$$\mu(T | E_R(C)) \leq \mu(T | E(C)) + \mu(\dot{T}) \leq \dim C + 1. \quad \bullet$$

1.89. GENERALIZATIONS AND CONCLUDING REMARKS. A result related to 1.81 is contained in [10]: if $T \in L(X)$ and $X = span(Ker\, T^n : n \geq 0)$ we have $\mu(T) = \max(1, \dim Ker\, T^*)$. This implies, for instance, that for matrix valued antianalytic Toeplitz operators T_{G^*}, $T_{G^*}f = P_+G^*f$ $(f\in H^2(\mathbb{C}^n)$, $G\in H^\infty(\mathbb{C}^n\to\mathbb{C}^n))$ we have $\mu(T_{G^*})=1$, provided $\det(G-\lambda I)\not\equiv 0$ for every $\lambda\in\mathbb{C}$, and $\mu(T_{G^*})=\infty$ otherwise (see [27] for the proof). The same techniques show that for every compact operator $T\in L(X)$ with $X=\operatorname{span}(\operatorname{Ker}(T-\lambda I)^n:$ $n\geq 0$, $\lambda\in\sigma(T)\setminus\{0\})$ the maxi-formula holds

$$\mu(T)=\sup_{\lambda\in\sigma(T)\setminus\{0\}}\dim\operatorname{Ker}(T-\lambda I),$$

see [10] for the proof. In a sense more general spectral independence condition than (1.84) (which implies maxi-formulas too) is considered in [28]. A result close to 1.85 is contained in [4] (Sec.3.3, Cor.4): if $\sup_{n\geq 0}\|T^n\|<\infty$, $T=T_1\oplus T_2$, where T_2 is $\mathbb{T}$-complete and $\sigma_p(T_1)\cap\sigma_p(T_2)=\emptyset$ then $\operatorname{Lat}(T_1\oplus T_2)$ splits and so $\mu(T_1\oplus T_2)=\max(\mu(T_1),\mu(T_2))$.

We will return to a discussion of maxi-formulas in Chapters 2-4 where the general multicyclicity overlappings will be described.

REFERENCES

1. Beauzamy, B.: Introduction to operator theory and invariant subspaces, Publications de l'Université Paris VII, 1988, pp.339.
2. Conway, J.B.: Subnormal operators, Pitman, Boston, 1981.
3. Naimark, M.A., Loginov, A.I., Shulman, V.S.: Non-selfadjoint algebras of operators in Hilbert space, Itogi Nauki: Mat.Anal., 12, VINITI, Moscow, 1974, 413-465. (Russian). English transl. in Soviet Math.Surveys 5 (1976).
4. Nikol'skii, N.K.: Selected problems of weighted approximation and spectral analysis, (Proceedings Steklov Inst. of Math., 120 (1974)), AMS, 1976.
5. Khrushchev, S.V.: Problem of simultaneous approximations and removing of singularities of Cauchy integrals, Pro-

ceedings Steklov Inst. of Math., 130 (1978), 124-195; AMS, 1979, Issue 4.

6. Wonham, M.W.: Linear Multivariable Control, Springer-Verlag, Berlin, 1974.
7. Fuhrmann, P.A.: Linear Systems and Operators in Hilbert Space, McGraw-Hill, 1981.
8. Nikolskii, N.K. and Vasyunin, V.I.: Control subspaces of minimal dimension and spectral multiplicities, Invariant Subspaces, Other Topics, Birkhäuser Verlag, 1981, 163-179.
9. Nikolskii, N.K. and Vasyunin, V.I.: Control subspaces of minimal dimension. Elementary introduction. Discotheca, Zap.Nauchn.Semin.Leningrad Otdel. Mat.Inst.Steklov.(LOMI), 113(1981), 41-75. (Russian). English transl. in J.Soviet. Math.
10. Nikolskii, N.K. and Vasyunin, V.I.: Control Subspaces of minimal dimension and rootvectors, Integral Equations and Operator Theory, 6, N 2 (1983), 274-311.
11. Nikolskii, N.K. and Vasyunin, V.I.: Control subspaces of minimal dimension, unitary and model operators, J.Operator Theory, 10 (1983), 307-330.
12. Glimm, J. and Jaffe, A.: Quantum Physics, Springer-Verlag, 1981.
13. Vasyunin, V.I. and Karayev, M.T.: The multiplicity of some contractions, Zap.Naucn.Sem. Leningrad. Otdel.Mat.Inst. Steklov. (LOMI), 157 (1987), 23-29 (Russian).
14. Nikol'skii, N.K.: Treatise on the Shift Operator, Springer-Verlag, 1986.
15. Sz.-Nagy, B. and Foias, C.: Harmonic analysis of operators on Hilbert space, North Holland-Akadémiai Kiado, Amsterdam-Budapest, 1970.
16. Nikolskii, N.K.: The present state of the spectral analysis -synthesis problem. I, Operator Theory in Function Spaces (Proc.School, Novosibirsk, 1975), 1977, 240-282. AMS Transl. ser.2, 124 (1984), 97-129.
17. Nikolskii, N.K.: Invariant subspaces in operator theory and function theory, Itogi Nauki: Mat.Anal., 12, VINITI, Moscow,

1974, 199-412. (Russian). English transl. in Soviet Math. Surveys 5 (1976).
18. Herrero, D.A.: On multicyclic operators, Integral Equations and Operator Theory, 1 (1978), 57-102.
19. Volberg, A.L.: Asymptotically holomorphic functions and their applications, Dissertation, Steklov Math.Inst., Leningrad, 1989.
20. Markus, A.S.: The problem of spectral synthesis for operators with point spectrum, Izv.Akad.Nauk SSSR, Ser.Mat. 34 (1970), 662-688. (Russian). English Trans. in Math. USSR Izv. 4 (1970), 670-696.
21. Wolff, J.: Sur les séries $\sum_1^\infty \frac{A_k}{z-\alpha_k}$, C.r. Acad.sci., 173 (1921), 1327-1328.
22. Denjoy, A.: Sur les séries de fractions rationnelles, Bull. soc.math. France, 52 (1924), 418-434.
23. Gončar, A.A.: On quasianalytic continuation of analytic functions across a Jordan arc, Dorkl.Akad.Nauk SSSR, 166 (1966), 1028-1031. (Russian). English transl. in Soviet Math.Dokl. 7 (1966), 213-216.
24. Bourdon, P.S. and Shapiro, J.H.: Spectral synthesis and common cyclic vectors (to appear).
25. Bourdon, P.S. and Shapiro, J.H.: Cyclic Composition Operators on H^2 (to appear).
26. Solomyak, B.M.: Multiplicity, calculi and multiplication operators, Sibirskii Matem.Zhurn. 29, N 2 (1988), 167-175. (Russian).
27. Nikol'skii, N.K.: Ha-plitz operators: a survey of some recent results, Proc.Conf.Operators and Function Theory, Lancaster 1984, 87-137, Reidel, 1984.
28. Nikol'skii, N.K.: Designs for calculating the spectral multiplicity of orthogonal sums, Zapiski Nauchn.Semin. LOMI, 126 (1983), 150-158; English transl. in J.Soviet Math. 27, N 1 (1984).

Operator Theory:
Advances and Applications, Vol. 42

WHEN IS A FUNCTION OF A TOEPLITZ OPERATOR CLOSE TO A TOEPLITZ OPERATOR?

V.V.Peller

CONTENTS.

1. Introduction. 60
2. Preliminaries. 63
3. The case of real functions φ 66
4. The case $\|\varphi\|_{L^\infty} \leq 1$ 82
References. 84

1. INTRODUCTION

In this paper we consider conditions under which the operator $f(T_\varphi) - T_{f\circ\varphi}$ belongs to the Schatten - von Neumann class $\mathfrak{S}_p$ and in particular conditions when $f(T_\varphi) - T_{f\circ\varphi}$ is of trace class. Here T_φ is a Toeplitz operator which is defined for bounded φ on the Hardy class H^2 by

$$T_\varphi f = \mathbb{P}_+ \varphi f, \tag{1}$$

where $\mathbb{P}_+$ is the orthogonal projection from L^2 onto H^2. We shall consider Toeplitz operators T_φ for unbounded functions too. For $\varphi \in L^2$ the operator T_φ is defined by (1) on a dense subset of H^2.

Nuclearity conditions for $f(T_\varphi) - T_{f\circ\varphi}$ and corresponding trace formulae are important for asymptotic trace formulae for pseudodifferential operators (see [13], [14], [15]).

Besides Toeplitz operators we shall deal with Wiener - Hopf operators W_φ which are defined on the Hardy class $H^2(\mathbb{C}_+)$ of functions in the upper half-plane by

$$W_\varphi f = \mathcal{P}_+ \varphi f,$$

where $\mathcal{P}_+$ is the orthogonal projection from $L^2(\mathbb{R})$ onto $H^2(\mathbb{C}_+)$ under the natural identification of $H^2(\mathbb{C}_+)$ with a subspace of $L^2(\mathbb{R})$ (here $\mathbb{C}_+ = \{z \in \mathbb{C} : \operatorname{Im} z > 0\}$).

The question of nuclearity of $f(W_\varphi) - W_{f\circ\varphi}$ was studied by Widom in [13]. He considered functions φ of class $B_2^{1/2}(\mathbb{R})$, that is functions φ satisfying

$$\int_{-\infty}^{\infty}\int_{-\infty}^{\infty} \frac{|\varphi(x) - \varphi(y)|^2}{|x-y|^2}\, dx\, dy < \infty .$$

For bounded φ in $B_2^{1/2}(\mathbb{R})$ the nuclearity of $f(W_\varphi) - W_{f\circ\varphi}$ was obtained in the two following cases:

a) f is analytic in a neighbourhood of the closed convex hull of the essential range of φ ;

b) φ is real, $f \in L^1(\mathbb{R})$ and $\int_{\mathbb{R}} t^2 |(\mathcal{F}f)(t)|\,dt < \infty$, where $\mathcal{F}$ is Fourier transform.

In both cases he showed that the following trace formula is valid

$$\operatorname{trace}(f(W_\varphi) - W_{f\circ\varphi}) = \tag{2}$$

$$= \frac{1}{8\pi^2}\int_{-\infty}^{\infty}\int_{-\infty}^{\infty} \frac{(\varphi(x)-\varphi(y))^2}{(x-y)^2}\int_0^1 f''((1-\theta)\varphi(x)+\theta\varphi(y))(\theta\log\theta+(1-\theta)\log(1-\theta))\,d\theta\,dx\,dy.$$

In Section 3 of this paper in the case of real functions φ of class $B_2^{1/2}(\mathbb{R})$ (φ is not assumed to be bounded) we obtain the nuclearity of $f(W_\varphi) - W_{f\circ\varphi}$(or $f(T_\varphi) - T_{f\circ\varphi}$ which reduces to each other by conformal mapping of the unit disc onto the upper half-plane) and prove the trace formula (2) under weaker conditions on f .

In Section 4 we consider the case when $\varphi \in B_2^{1/2}$ and $\|\varphi\|_{L^\infty} \le 1$ and f is analytic in the unit disc $\mathbb{D}$. We shall find conditions on f under which $f(T_\varphi) - T_{f\circ\varphi}$ is of trace class and formula (2) holds.

We also consider other Schatten - von Neumann classes $\mathfrak{S}_p$. Note that in both cases considered in Section 3, 4 for bounded φ and continuous f the question of whether $f(T_\varphi) - T_{f\circ\varphi} \in \mathfrak{S}_p$, $p<\infty$, is equivalent to the question of whether $f(T_\varphi)$ differs from a Toeplitz operator by an operator of class $\mathfrak{S}_p$. Indeed, it is well known (see [6]) that if $f(T_\varphi) - T_g$ is compact then $g = f\circ\varphi$.

An essential role in the paper is played by Hankel operators. Let φ be a function of class BMO on the unit circle $\mathbb{T}$ (see Section 2). Then the **Hankel operator** H_φ **from** H^2 to $H^2_- \stackrel{def}{=} L^2 \ominus H^2$ **is defined by**

$$H_\varphi f = \mathbb{P}_- \varphi f ,$$

where $\mathbb{P}_- = I - \mathbb{P}_+$. By the Nehari-Fefferman theorem (see [5], [9]) H_φ is bounded for $\varphi \in BMO$.

For $\varphi \in BMO(\mathbb{R})$ the Hankel operator $\mathcal{H}_\varphi$ from $H^2(\mathbb{C}_+)$ to $L^2(\mathbb{R}) \ominus H^2(\mathbb{C}_+)$ is defined by

$$\mathcal{H}_\varphi f = \mathcal{P}_- \varphi f ,$$

where $\mathcal{P}_- = I - \mathcal{P}_+$.

The following formula for bounded φ, ψ on $\mathbb{T}$ is well-known (see e.g. [5]):

$$T_{\varphi\psi} - T_\varphi T_\psi = H^*_{\bar\varphi} H_\psi . \tag{3}$$

It is easy to see that it holds in the case $\varphi, \psi \in BMO$ too.

It is seen from this formula why in the study of nuclearity conditions for $f(T_\varphi) - T_{f\circ\varphi}$ it is natural to assume that $\varphi \in B_2^{1/2}$. Indeed, if φ is real and $f(x) = x^2$ then it follows from (3) that $f(T_\varphi) - T_{f\circ\varphi} \in \mathbf{S}_1$ if and only if $H_\varphi \in \mathbf{S}_2$ which is well-known to equivalent to the fact that $\varphi \in B_2^{1/2}$.

Analogously in the question of when $f(T_\varphi) - T_{f\circ\varphi} \in \mathbf{S}_p$ it is natural to assume that $\varphi \in B_{2p}^{1/2p}$ (see Section 2).

A question arises of to what extent the sufficiency conditions for the nuclearity of $f(T_\varphi) - T_{f\circ\varphi}$ obtained in Sections 3, 4 are close to necessary. Widom showed in [13] from the trace formula for $f(W_\varphi) - W_{f\circ\varphi}$ that if f satisfies the condition b) then

$$\sup\{|\operatorname{trace}(f(W_\varphi) - W_{f\circ\varphi}) : \|\varphi\|_{B_2^{1/2}(\mathbb{R})} \le 1\} \geqslant$$

$$\geqslant \sup\{|f''(x)| : x \in \mathbb{R}\} .$$

The sufficient condition found in Section 3 ($f \in B^2_{\infty 1}(\mathbb{R})$, see Section 2) is fairly close to the condition $f'' \in L^\infty(\mathbb{R})$. I do not know whether the condition $f \in B^2_{\infty 1}(\mathbb{R})$ is necessary for the nuclearity of $f(T_\varphi) - T_{f\circ\varphi}$ and I do not know whether the condition $f'' \in L^\infty(\mathbb{R})$ is sufficient (I suspect not).

In [13] Widom found asymptotic trace formulae for $f(W^{(\alpha)}_\varphi) - W^{(\alpha)}_{f\circ\varphi}$ for **finite Wiener - Hopf operators**. These operators are defined on the subspace $H^2_{(\alpha)}$, which consists of the functions whose Forier transforms are supported on $[0,\alpha]$, by

$$W^{(\alpha)}_\psi f = \mathcal{P}_\alpha \psi f$$

where $\mathcal{P}_\alpha$ is the orthogonal projection onto $H^2_{(\alpha)}$.

Widom showed in [13] that under the condition a) or b) the following asymptotic trace formula holds

$$\lim_{\alpha\to\infty} \operatorname{trace}(f(W^{(\alpha)}_\varphi) - W^{(\alpha)}_{f\circ\varphi}) = 2\operatorname{trace}(f(W_\varphi) - W_{f\circ\varphi}) . \tag{4}$$

In Sections 3, 4 we shall show that this formula is valid under the same conditions $\varphi \in B^{1/2}_2(\mathbb{R})$, $f \in B^2_{\infty 1}$. An analogous asymptotic formula holds for finite Toeplitz matrices too.

As shown in [13] formula (4) contains as a partial case the asymptotic Szegö formula for Toeplitz determinants. In Widom's paper [14] another expression for the trace of $f(W_\varphi) - W_{f\circ\varphi}$ was found. Analogous formulae were obtained by Budylin and Buslaev [1].

Note also that in [15] Widom used these trace formulae for obtaining asymptotic trace formulae for pseudodifferential operators.

2. PRELIMINARIES

In this section we present necessary information on the class BMO, Besov classes, Schatten - von Neumann classes and Hankel operators.

Function classes. The class BMO consists of the functions on the unit circle $\mathbb{T}$ satisfying

$$\|f\|_{BMO} \stackrel{def}{=} \sup_I \frac{1}{m(I)} \int_I |f - f_I| \, dm < \infty ,$$

where the supremum is taken over all subarcs I, m is normalized Lebesgue measure on $\mathbb{T}$, and $f_I \stackrel{def}{=} \frac{1}{m(I)} \int_I f \, dm$. Note that $\|\cdot\|_{BMO}$ vanishes on the constant functions.

By Fefferman's theorem $f \in BMO$ if and only if $f = g + \tilde{h}$ where $g, h \in L^\infty$ and $\tilde{h}$ is the harmonic conjugate of h.

The space BMO ($\mathbb{R}$) of functions on $\mathbb{R}$ is defined in a similar way. We refer the reader to [3] for more detailed information on BMO.

The Besov classes $B^s_{pq}(\mathbb{R})$, $0 < p \le \infty$, $0 < q \le \infty$, $s > 0$ (we shall consider the homogeneous Besov classes) can be defined in the following way. Let V_n, $n \in \mathbb{Z}$, be functions on $\mathbb{R}$ such that the Fourier transforms $\mathcal{F}V_n$ are C^∞-functions supported on $[2^{n-1}, 2^{n+1}]$, $(\mathcal{F}V_{n+1})(x) = (\mathcal{F}V_n)(2x)$, $n \in \mathbb{Z}$, and

$$\sum_{n \in \mathbb{Z}} (\mathcal{F}V_n)(x) = 1, \quad x > 0 .$$

Let V_n^*, $n \in \mathbb{Z}$, be the function on $\mathbb{R}$ such that $(\mathcal{F}V_n^*)(x) = (\mathcal{F}V_n)(-x)$.

Then the Besov class $B^s_{pq}(\mathbb{R})$ consists of the distributions f on $\mathbb{R}$ satisfying

$$\{2^{ns} \|f * V_n\|_{L^p(\mathbb{R})}\}_{n \in \mathbb{Z}} \in \ell^q, \quad \{2^{ns} \|f * V_n^*\|_{L^p(\mathbb{R})}\}_{n \in \mathbb{Z}} \in \ell^q .$$

We shall use the notation $B^s_p(\mathbb{R}) = B^s_{pp}(\mathbb{R})$.

Note that each $f \in B^s_{pq}(\mathbb{R})$ admits the representation

$$f = \sum_{n \in \mathbb{Z}} f * V_n^* + \sum_{n \in \mathbb{Z}} f * V_n + P , \tag{5}$$

where P is a polynomial. We shall say that the Fourier transform $\mathcal{F}f$ is supported on $(-\infty, 0)$ if $f = \sum_{n \in \mathbb{Z}} f * V_n^*$ and $\mathcal{F}f$ is supported on $(0,-\infty)$ if $f=\sum_{n\in\mathbb{Z}} f*V_n$.

For $s>1/p-1$ the space $B^s_{pq}(\mathbb{R})$ can be identified with the space of functions f on $\mathbb{R}$ such that

$$\frac{\|\Delta^n_t f\|_{L^p(\mathbb{R})}}{|t|^s}\in L^q(\mathbb{R},\frac{dt}{|t|}) \tag{6}$$

for some integer $n>s$, where $(\Delta_t f)(x)\overset{def}{=\!=}f(x+t)-f(x)$.

It is not always convenient to assume that the Besov classes contain the set of polynomials. We shall deal with the classes $B^{1/p}_p(\mathbb{R})$ and $B^2_{\infty 1}(\mathbb{R})$. In the first case we shall suppose that in representation (5) only constant functions P can occur, while in the second case we admit polynomials P of degree at most 2.

For $1<p<\infty$ it is convenient to endow the spaces $B^{1/p}_p(\mathbb{R})$ with the norm

$$\|f\|_{B^{1/p}_p(\mathbb{R})}\overset{def}{=\!=}\frac{1}{8\pi^2}\int_{-\infty}^{\infty}\int_{-\infty}^{\infty}\frac{|f(x)-f(y)|^p}{|x-y|^2}\,dx\,dy\ .$$

Actually $\|\cdot\|_{B^{1/p}_p(\mathbb{R})}$ is a quotient-norm since it annihilates the constant functions.

The Besov classes B^s_{pq} of functions on $\mathbb{T}$ are defined in a similar way. Their definition is simpler since there is no problem of which polynomials are admitted in representation (5). Let $\mathcal{U}_n$, $\mathcal{U}^*_n$, $n\geq 1$, be the trigonometric polynomials whose Fourier coefficients satisfy the conditions: $\hat{\mathcal{U}}_n(k)=(\mathcal{F}V_n)(k)$, $\hat{\mathcal{U}}^*_n(k)=(\mathcal{F}V^*_n)(k)$, $n\geq 1$, $k\in\mathbb{Z}$. Then f is said to belong to B^s_{pq} if

$$\{2^{ns}\|f*\mathcal{U}_n\|_{L^p}\}_{n\geq 1}\in\ell^q,\quad \{2^{ns}\|f*\mathcal{U}^*_n\|_{L^p}\}_{n\geq 1}\in\ell^q.$$

If $s>1/p-1$ then B^s_{pq} admits a description similar to (6). Moreover for any $n>s$ we obtain the same class of functions.

We refer the reader to [6] for more detailed information on Besov classes.

Schatten-von Neumann classes. Let $\mathcal{H}_1$, $\mathcal{H}_2$ be Hilbert spaces, T a bounded operator from $\mathcal{H}_1$ to $\mathcal{H}_2$. Then T is said to belong to the Schatten - von Neumann class $\mathcal{S}_p$, $0<p<\infty$, if

$$\{\inf \|T-R\| : \operatorname{rank} R \leq n\}_{n\geq 0} \in \ell^p .$$

For $p=1$ we obtain the class of nuclear operators (trace class) and for $p=2$ we obtain the Hilbert - Schmidt class. The functional **trace is defined on the nuclear operators on a Hilbert space $\mathcal{H}$ by**

$$\operatorname{trace} T = \sum_{n\geq 0} (Te_n, e_n), \quad T\in \mathcal{S}_1 ,$$

where $\{e_n\}_{n\geq 0}$ is an orthonormal basis in $\mathcal{H}$. The value of this functional does not depend on the choice of a basis. For more detailed information on the classes $\mathcal{S}_p$ we refer to [4].

Hankel operators. By the Nehari-Fefferman theorem a Hankel operator H_φ is bounded on H^2 if and only if $\mathbb{P}_-\varphi \in BMO$ and $\|H_\varphi\|$ is equivalent to $\|\mathbb{P}_-\varphi\|_{BMO}$ (see [5], [9]). The Hankel operators of class $\mathcal{S}_p$ are described as follows: $H_\varphi \in \mathcal{S}_p$, $p>0$, if and only if $\mathbb{P}_-\varphi \in B_p^{1/p}$ (see [7] for $p\geq 1$ and [8], [12] for $p<1$). A similar assertion holds for the operators $\mathcal{H}_\varphi$ too.

3. THE CASE OF REAL FUNCTIONS φ.

In this section we study the operators $f(T_\varphi)-T_{f\circ\varphi}$ for real functions φ. We show that if $\varphi\in BMO$ then T_φ is selfadjoint and so its functions are defined by the spectral theorem. If we are interested in the condition $f(T_\varphi)-T_{f\circ\varphi}\in \mathcal{S}_p$ then it is natural to assume that $\varphi\in B_{2p}^{1/2p}$ since it follows from (3) that for $f(x)=x^2$ the operator $f(T_\varphi)-T_{f\circ\varphi}$ belongs to $\mathcal{S}_p$ if and only if $H_\varphi\in \mathcal{S}_{2p}$, and the latter is equivalent to the fact that $\varphi\in B_{2p}^{1/2p}$ (see Section 2).

We show that under this assumption $f(T_\varphi)-T_{f\circ\varphi}\in \mathcal{S}_p$ for $f\in B_{\infty 1}^2(\mathbb{R})$. We also show that for $p=1$ the same assumptions guarantee the trace formula for $f(T_\varphi)-T_{f\circ\varphi}$ and the asymptotic trace formula for finite Toeplitz matrices which were obtained by Widom in 13 under more restrictive assumptions.

First of all let us show that the assumptions of Widom are more restrictive. Indeed, the condition $\int_{\mathbb{R}} t^2 |(\mathcal{F}f)(t)|\,dt < \infty$ is equivalent to the following one

$$\sum_{n\in\mathbb{Z}} 2^{2n} \int_{-\infty}^{\infty} |(\mathcal{F}(f*V_n))(t)|\,dt + \sum_{n\in\mathbb{Z}} 2^{2n} \int_{-\infty}^{\infty} |(\mathcal{F}(f*V_n^*))(t)|\,dt < \infty ,$$

while the condition $f \in B^2_{\infty 1}(\mathbb{R})$ means that

$$\sum_{n\in\mathbb{Z}} 2^{2n} \| f * V_n \|_{L^\infty} < \infty , \quad \sum_{n\in\mathbb{Z}} 2^{2n} \| f * V_n^* \|_{L^\infty} < \infty .$$

Let us show that the assumption $\varphi \in BMO$ implies that T_φ is a selfadjoint operator with domain $\{f \in H^2 : \mathbb{P}_+ \varphi f \in H^2\}$.

1. LEMMA. <u>Let φ be a real function in BMO. Then T_φ is a selfadjoint operator with domain</u> $\{f \in H^2 : \mathbb{P}_+ \varphi f \in H^2\}$.

PROOF. Let T be the operator defined on the set $\mathcal{D}(T) = \{f \in H^2 : \varphi f \in L^2\}$ by $Tf = \mathbb{P}_+ \varphi f$. Then T is a closed operator on H^2 with domain $\mathcal{D}(T)$. It can easily be verified that T^* is the operator defined on $\mathcal{D}(T^*) = \{f \in H^2 : \mathbb{P}_+ \varphi f \in H^2\}$ by $T^* f = \mathbb{P}_+ \varphi f$. Clearly the proof will be completed if we show that $\mathcal{D}(T) = \mathcal{D}(T^*)$. But this follows from the Nehari-Fefferman theorem. Indeed, let $f \in H^2$, then $H_\varphi f = \mathbb{P}_- \varphi f \in L^2$. So $f \in \mathcal{D}(T^*)$ if and only if $f \in \mathcal{D}(T)$. ●

To estimate the $\mathbf{S}_p$-norm of $f(T_\varphi) - T_{f\circ\varphi}$ we establish an integral formula for $f(T_\varphi) - T_{f\circ\varphi}$. This formula involves the backward translation operators which are defined as follows.

Let f be a function on $\mathbb{R}$ whose Fourier transform is supported on $[0, +\infty)$ and belongs to $L^1(\mathbb{R})$. Given $t > 0$, the function $S_t^* f$ is defined by

$$S_t^* f = \mathcal{P}_+ \mathcal{M}_t^* f ,$$

where $(\mathcal{M}_t^* f)(x) = e^{-itx} f(x)$ and $(\mathcal{P}_+ g)(x) = \int_0^\infty (\mathcal{F}g)(s)\, e^{isx}\, ds$.

We shall deal with bounded analytic function in $\mathbb{C}_+$. In this case the function $S_t^* f$ can be defined only up to a constant function, since for functions g in $L^\infty(\mathbb{R})$ the function $\mathcal{P}_+ g$ can be defined only up to a constant function. To be definite we put for $g \in L^\infty$

$$(\widetilde{\mathcal{P}}_+ g)(z) = \frac{1}{2\pi}\int_{-\infty}^{\infty}\left(\frac{1}{t-z} - \frac{t}{1+t^2}\right)\varphi(t)\,dt, \qquad \operatorname{Im} z > 0 .$$

It is easy to see that $(\widetilde{\mathcal{P}}_+ g)(i) = 0$ for any $g \in L^\infty$. If $\mathcal{F}g \in L^1(\mathbb{R})$ then $\mathcal{P}_+ g$ and $\widetilde{\mathcal{P}}_+ g$ differ by a constant function.

For $f \in H^\infty(G)$ we can define now

$$\widetilde{S}_t^* f = \widetilde{\mathcal{P}}_+ \mathcal{M}_t^* f .$$

If $f \in B^2_{\infty 1}(\mathbb{R})$, and $\mathcal{F}f$ is supported on $(0,+\infty)$ then

$$\widetilde{S}_t^* f \overset{def}{=\!=} \widetilde{\mathcal{P}}_+ \mathcal{M}_t^* \Big(\sum_{n \geqslant \frac{\log t}{\log 2}} f * V_n \Big) .$$

Note that $\sum\limits_{n \geqslant n_0} f * V_n \in H^\infty$ for any $n_0 \in \mathbb{Z}$.

The integral in the following theorem as well as all subsequent integrals of operator-valued functions should be understood in the sence of the strong operator topology.

2. THEOREM. Let φ be a real function in BMO and f be a function in $B^2_{\infty 1}(\mathbb{R})$ whose Fourier transform is supported on $(0,+\infty)$. Then

$$f(T_\varphi) - T_{f\circ\varphi} = -i\int_0^\infty H^*_{(\widetilde{S}_\tau^* f)\circ\varphi}\, H_\varphi\, e^{i\tau T_\varphi}\, d\tau . \tag{7}$$

Moreover, the integral on the right absolutely converges and so the operator on the right is bounded.

First we consider the partial case $f(x) = e^{itx}$. The following assertion for bounded φ is proved in [13].

3. LEMMA. Let φ be a real function in BMO. Then

$$e^{itT_\varphi} - T_{e^{it\varphi}} = -i\int_0^t H^*_{e^{-i(t-\tau)\varphi}} H_\varphi e^{i\tau T_\varphi}\, d\tau .$$

PROOF OF LEMMA 3. Put

$$\mathcal{D} = \{f \in H^2 : \mathbb{P}_+ \varphi f \in H^2\} .$$

Clearly, $\mathcal{D}$ is dense in H^2. It follows easily from the spectral theorem that $T_\varphi e^{-itT_\varphi} h \in H^2$ for $h \in \mathcal{D}$ and all $t \in \mathbb{R}$. It follows that $\varphi e^{-itT_\varphi} h \in L^2$ for $h \in \mathcal{D}$ and $t \in \mathbb{R}$ since

$$\varphi e^{-itT_\varphi} h = T_\varphi e^{-itT_\varphi} h + H_\varphi e^{-itT_\varphi} h$$

and H_φ is bounded from H^2 to L^2.

It is easy to see that for $h \in \mathcal{D}$

$$\lim_{t \to t_0} \frac{1}{t-t_0} (T_{e^{it\varphi}}\, e^{-itT_\varphi} - T_{e^{it_0\varphi}}\, e^{-it_0 T_\varphi})\, h =$$

$$= -iT_{e^{it_0\varphi}}\, T_\varphi e^{-it_0 T_\varphi} h + iT_{\varphi e^{it_0\varphi}}\, e^{-it_0 T_\varphi} h =$$

$$= iH^*_{e^{-it_0\varphi}} H_\varphi e^{-it_0 T_\varphi} h .$$

Therefore, for

$$T_{e^{it\varphi}}\, e^{-itT_\varphi} h = h + i\int_0^t H^*_{e^{-i\tau\varphi}} H_\varphi e^{-i\tau T_\varphi} h\, d\tau .$$

Since on both sides bounded operators apply to h, we have

$$T_{e^{it\varphi}}\, e^{-itT_\varphi} = I + i\int_0^t H^*_{e^{-i\tau\varphi}} H_\varphi e^{i(t-\tau)T_\varphi}\, d\tau,$$

which implies

$$e^{itT_\varphi} - T_{e^{it\varphi}} = -i\int_0^\infty H^*_{e^{-i\tau\varphi}} H_\varphi e^{i(t-\tau)T_\varphi}\,d\tau =$$

$$= -i\int_0^t H^*_{e^{-i(t-\tau)\varphi}} H_\varphi e^{i\tau T_\varphi}\,d\tau. \quad \bullet$$

In order to prove that the integral on the right-hand side of (7) converges absolutely we need the following assertion.

4. LEMMA. Let f be a function in $B^2_{\infty 1}(\mathbb{R})$ whose Fourier transform is supported on $(0,+\infty)$. Then

$$\int_0^\infty \|(\widetilde{S}^*_\tau f)'\|_{L^\infty(\mathbb{R})}\,d\tau < \infty.$$

PROOF OF LEMMA 4. We have $f = \sum_{n\in\mathbb{Z}} f * V_n$,

$$\sum_{n\in\mathbb{Z}} 2^{2n}\|f * V_n\|_{L^\infty(\mathbb{R})} < \infty,$$

and $\mathcal{F}(f * V_n)$ is supported on $[2^{n-1}, 2^{n+1}]$. Therefore it is sufficient to prove that

$$\int_0^\infty \|(\widetilde{S}^*_\tau g)'\|_{L^\infty}\,d\tau \le \text{const}\cdot 2^{2n}\|g\|_{L^\infty}$$

for $g\in L^\infty$ with $\operatorname{supp}(\mathcal{F}g)\subset[2^{n-1}, 2^{n+1}]$. For $0<\tau\le 2^{n+1}$ we consider the function W_τ such that $\mathcal{F}W_\tau$ vanishes outside $[\tau, 2^{n+2}+\tau]$, $(\mathcal{F}W_\tau)(2^{n+1}+\tau) = 1$, and $\mathcal{F}W_\tau$ is linear on $[\tau, 2^{n+1}+\tau]$ and $[2^{n+1}+\tau, 2^{n+2}+\tau]$.

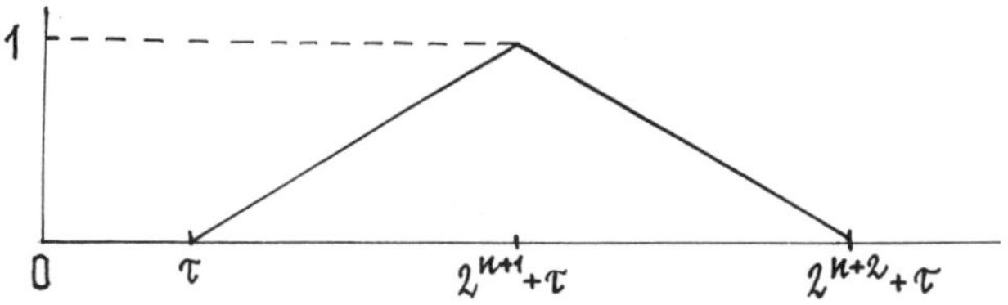

It is easy to see that

$$(\tilde{S}_\tau^* g)' = 2^{n+1} i \mathcal{M}_\tau^* (g * W_\tau) . \tag{8}$$

Therefore

$$\|(\tilde{S}_\tau^* g)'\|_{L^\infty} = 2^{n+1} \| g * W_\tau \|_{L^\infty} \le 2^{n+1} \|g\|_{L^\infty} ,$$

since $\|W_\tau\|_{L^1} = 1$.

It follows that

$$\int_0^{2^{n+1}} \|(\tilde{S}_\tau^* g)'\|_{L^\infty} d\tau \le 4 \cdot 2^{2n} \|g\|_{L^\infty} . \quad \bullet$$

PROOF OF THEOREM 2. Let us show first that the integral on the right-hand side of (7) converges absolutely. We have

$$\| H^*_{\overline{(\tilde{S}_\tau^* f)\circ\varphi}} H_\varphi e^{i\tau T_\varphi} \| \le \| H_{\overline{(\tilde{S}_\tau^* f)\circ\varphi}} \| \cdot \| H_\varphi \| \le$$

$$\le const \|(\tilde{S}_\tau f) \circ \varphi\|_{BMO} \cdot \|\varphi\|_{BMO} .$$

It follows from the definition of BMO that if g satisfies the Lipschitz condition and $\varphi \in BMO$ then $g \circ \varphi \in BMO$ and $\|g \circ \varphi\|_{BMO} \le \|g'\|_{L^\infty} \cdot \|\varphi\|_{BMO}$. Hence

$$\| H^*_{\overline{(\tilde{S}_\tau^* f)\circ\varphi}} H_\varphi e^{i\tau T_\varphi} \| \le const \|\varphi\|^2_{BMO} \cdot \|(\tilde{S}_\tau^* f)'\|_{L^\infty} .$$

Consequently,

$$\int_0^\infty \| H^*_{\overline{(\tilde{S}_\tau^* f)\circ\varphi}} H_\varphi e^{i\tau T_\varphi} \| d\tau \le const \|\varphi\|^2_{BMO} \int_0^\infty \|(\tilde{S}_\tau^* f)'\|_{L^\infty} d\tau \le$$

$$\leq \text{const}\,\|\varphi\|^2_{BMO}\cdot\|f\|_{B^2_{\infty 1}(\mathbb{R})}$$

by Lemma 4.

Let us now prove formula (7). Since $f=\sum_{n\in\mathbb{Z}} f*V_n$ and $\sum_{n\in\mathbb{Z}} 2^{2n}\|f*V_n\|_{L^\infty}<\infty$, it is sufficient to prove (7) in the case $\operatorname{supp}\mathcal{F}f\subset[2^{n-1},2^{n+1}]$, $n\in\mathbb{Z}$.

We consider first the case when $\mathcal{F}f\in L^1(\mathbb{R})$. We have

$$f(T_\varphi)-T_{f\circ\varphi}=\int_0^\infty(\mathcal{F}f)(t)\,e^{itT_\varphi}\,dt-\int_0^\infty(\mathcal{F}f)(t)\,T_{e^{it\varphi}}\,dt=$$

$$=\int_0^\infty(\mathcal{F}f)(t)\,(e^{itT_\varphi}-T_{e^{it\varphi}})\,dt=$$

$$=-i\int_0^\infty(\mathcal{F}f)(t)\int_0^t H^*_{e^{-i(t-\tau)\varphi}}H_\varphi e^{i\tau T_\varphi}\,d\tau\,dt$$

by Lemma 3.

Hence

$$f(T_\varphi)-T_{f\circ\varphi}=$$

$$=-i\int_0^\infty\Big(\int_\tau^\infty(\mathcal{F}f)(t)\,H^*_{e^{-i(t-\tau)\varphi}}\,dt\Big)H_\varphi e^{i\tau T_\varphi}\,d\tau=$$

$$=-i\int_0^\infty\Big(\int_0^\infty(\mathcal{F}f)(t+\tau)\,H^*_{e^{-it\varphi}}\,dt\Big)H_\varphi e^{i\tau T_\varphi}\,d\tau=$$

$$=-i\int_0^\infty H^*_{\overline{(S^*_\tau f)\circ\varphi}}H_\varphi e^{i\tau T_\varphi}\,d\tau=$$

$$=-i\int_0^\infty H^*_{\overline{(\tilde{S}^*_\tau f)\circ\varphi}}H_\varphi e^{i\tau T_\varphi}\,d\tau.$$

Since $(S^*_\tau f)\circ\varphi$ and $(\tilde{S}^*_\tau f)\circ\varphi$ differ by a constant function and the Hankel operator whose symbol is a constant function equals zero.

Suppose now that f is an arbitrary bounded function on $\mathbb{R}$ whose Fourier transform is supported on $[2^{n-1}, 2^{n+1}]$. Let τ be a smooth positive even function on $\mathbb{R}$ such that $\operatorname{supp}\tau\subset[-1,1]$, $\int_{-1}^{1}\tau(x)\,dx=1$. Put $\tau_\varepsilon(x)=\frac{1}{\varepsilon}\tau(x/\varepsilon)$, $\varepsilon>0$. Let now f_ε be the function such that $\mathcal{F}f_\varepsilon=(\mathcal{F}f)*\tau_\varepsilon$. Clearly, $\mathcal{F}f_\varepsilon\in L^1(\mathbb{R})$,

$$\lim_{\varepsilon\to 0}\|f_\varepsilon\|_{L^\infty(\mathbb{R})}=\|f\|_{L^\infty(\mathbb{R})}\quad\text{and}\quad\lim_{\varepsilon\to 0}f_\varepsilon(x)=f(x)$$

for any $x\in\mathbb{R}$.

Therefore (7) holds for f_ε. It is easy to see that

$$\lim_{\varepsilon\to 0}f_\varepsilon(T_\varphi)=f(T_\varphi),$$

$$\lim_{\varepsilon\to 0}T_{f_\varepsilon\circ\varphi}=T_{f\circ\varphi}$$

in the strong operator topology.

To prove that

$$\lim_{\varepsilon\to 0}H^*_{\overline{(\tilde S^*_\tau f_\varepsilon)\circ\varphi}}=H^*_{\overline{(\tilde S^*_\tau f)\circ\varphi}}\tag{9}$$

in the strong operator topology, we need the following assertion.

5. LEMMA. Let $\{\psi_m\}_{m\geq 1}$ be a sequence of functions on $\mathbb{T}$ such that $\|\psi_m\|_{BMO}\leq\mathrm{const}$ and $\lim_{m\to\infty}\psi_m(\zeta)=0$ for almost all $\zeta\in\mathbb{T}$. Then $\lim_{m\to\infty}H^*_{\psi_m}=0$ in the strong operator topology.

We postpone the proof of Lemma 5 and complete the proof of Theorem 2.

It is easy to see that $\lim_{\varepsilon\to 0}(\tilde S^*_\tau f_\varepsilon)(\varphi(\zeta))=(\tilde S^*_\tau f)(\varphi(\zeta))$ for almost all $\zeta\in\mathbb{T}$. We also have

$$\|(\tilde S^*_\tau f_\varepsilon)\circ\varphi\|_{BMO}\leq\|(S^*_\tau f_\varepsilon)'\|_{L^\infty}\cdot\|\varphi\|_{BMO}=$$

$$= 2^{n+1} \| W_\tau * f_\varepsilon \|_{L^\infty} \cdot \| \varphi \|_{BMO} \le 2^{n+1} \| f_\varepsilon \| \cdot \| \varphi \|_{BMO} .$$

Therefore by Lemma 5, (9) holds and so

$$\lim_{\varepsilon \to 0} \int_0^\infty H^*_{\overline{(\mathcal{S}_\tau^* f_\varepsilon)\circ\varphi}} H_\varphi e^{i\tau T_\varphi} d\tau = \int_0^\infty H^*_{\overline{(\mathcal{S}_\tau^* f)\circ\varphi}} H_\varphi e^{i\tau T_\varphi} d\tau$$

which proves (7). ●

PROOF OF LEMMA 3. Since $\| H^*_{\psi_m} \| \le const \| \psi_m \|_{BMO} \le const$, it follows that it is sufficient to prove that $\lim_{m\to\infty} \| H^*_{\psi_m} z^k \|_{H^2} = 0$ for all $k<0$. We have

$$\| H^*_{\psi_m} z^k \|_{H^2} = \| \mathbb{P}_+ z^k \psi_m \|_{L^2} \le \| \psi_m \|_{L^2} .$$

Since $BMO \subset L^4$ (see [3]), the result follows from the following easily verifiable assertion:

Suppose that $\psi_m \in L^4(\mathbb{T})$, $\| \psi_m \|_{L^4} \le const$, and $\psi_m \to \mathbb{O}$ in measure. Then $\lim_m \| \psi_m \|_{L^2} = 0$. ●

The following theorem establishes a sufficient condition for $f(T_\varphi) - T_{f\circ\varphi} \in \mathcal{S}_p$.

6. THEOREM. Let $f \in B^2_{\infty 1}(\mathbb{R})$. The following assertions hold.

1) If φ is a real function in BMO then $f(T_\varphi) - T_{f\circ\varphi}$ is bounded.

2) If $1 \le p < \infty$ and φ is a real function in $B^{1/2p}_{2p}$ then $f(T_\varphi) - T_{f\circ\varphi} \in \mathcal{S}_p$.

PROOF. 1) Let $f \in B^2_{\infty 1}(\mathbb{R})$. Then $f = f_- + f_0 + f_+$, where f_0 is a polynomial of degree at most 2, and f_- and f_+ are functions in $B^2_{\infty 1}(\mathbb{R})$ whose Fourier transforms are supported on $(-\infty, 0)$ and $(0, +\infty)$ respectively. The fact that $f_+(T_\varphi) - T_{f_+\circ\varphi}$ is bounded follows from Theorem 1. For f_- the result follows from $\bar{f}_-(T_\varphi) = (f_-(T_\varphi))^*$ and $T_{\bar{f}_-\circ\varphi} = T^*_{f_-\circ\varphi}$. Finally, the result for f_0 follows from the identity $T^2_\varphi - T_{\varphi^2} = -H^*_\varphi H_\varphi$.

2) As in 1) it is sufficient to consider the case when $\mathcal{F}f$

is supported on $(0,+\infty)$. Since $B_{2p}^{1/2p} \subset BMO$, we can use formula (7). We have

$$\|f(T_\varphi) - T_{f\circ\varphi}\|_{S_p} \le$$

$$\le \text{const} \int_0^\infty \| H^*_{\overline{(\tilde{S}^*_\tau f)\circ\varphi}} \|_{S_{2p}} \cdot \|H_\varphi\|_{S_{2p}} \cdot \|e^{i\tau T_\varphi}\| \, d\tau \le$$

$$\le \text{const} \int_0^\infty \|(\tilde{S}^*_\tau f)\circ\varphi\|_{B_{2p}^{1/2p}} \cdot \|\varphi\|_{B_{2p}^{1/2p}} \, d\tau .$$

It follows from the definition of $B_{2p}^{1/2p}$ in terms of the differences (it is here the condition $p>1/2$ is essential) that for $g'\in L^\infty$ and $\psi\in B_{2p}^{1/2p}$ the following inequality holds

$$\|g\circ\psi\|_{B_{2p}^{1/2p}} \le \|g'\|_{L^\infty} \cdot \|\psi\|_{B_{2p}^{1/2p}} .$$

Hence

$$\|f(T_\varphi) - T_{f\circ\varphi}\|_{S_p} \le$$

$$\le \text{const} \|\varphi\|^2_{B_{2p}^{1/2p}} \int_0^\infty \|(\tilde{S}^*_\tau f)'\|_{L^\infty} \, d\tau \le \text{const} \|\varphi\|^2_{B_{2p}^{1/2p}} \cdot \|f\|_{B^2_{\infty 1}(\mathbb{R})}$$

by Lemma 4. ●

Note that the sufficient condition $f\in B^2_{\infty 1}(\mathbb{R})$ for the boundedness of $f(T_\varphi) - T_{f\circ\varphi}$ is very far from being necessary. Indeed, if f is bounded and continuous on $\mathbb{R}$ then obviously $f(T_\varphi)-T_{f\circ\varphi}$ is bounded. But in the most interesting case $p=1$ the condition $f\in B^2_{\infty 1}(\mathbb{R})$ is fairly close to necessary. We shall discuss this in more detail after the proof of Theorem 7.

Similar assertions hold for Wiener-Hopf operators too. Moreover, they formally follow from the above assertions on Toeplitz operators, since under conformal mapping of the unit disc onto the upper half-plane Toeplitz operators are transformed into Wiener-Hopf operators (see [2]), Hankel operators on H^2 are transformed into Hankel operators on $H^2(\mathbb{C}_+)$ and the Besov classes $B_p^{1/p}$ are transformed into the Besov classes $B_p^{1/p}(\mathbb{R})$ (see [7]). Let us state the analogue of Theorem 6.

6'. THEOREM. Let $f \in B^2_{\infty 1}(\mathbb{R})$. The following assertions hold.

1) If φ is a real function in BMO $(\mathbb{R})$, then $f(W_\varphi) - W_{f\circ\varphi}$ is bounded.

2) If $1 \le p < \infty$ and φ is a real function in $B^{1/2p}_{2p}(\mathbb{R})$ then $f(W_\varphi) - W_{f\circ\varphi} \in \mathbf{S}_p$.

The analogue of formula (7) looks as follows. Suppose that $f \in B^2_{\infty 1}(\mathbb{R})$ and $\mathcal{F}f$ is supported on $(0,+\infty)$. Then

$$f(W_\varphi) - W_{f\circ\varphi} = \int_0^\infty \mathcal{H}^*_{\overline{(\tilde{S}^*_\tau f)\circ\varphi}} \mathcal{H}_\varphi e^{i\tau W_\varphi} \quad . \tag{10}$$

In [13] Widom proved that under the condition b) (see §1) formula (2) holds for the trace of $f(W_\varphi) - W_{f\circ\varphi}$. Let us show that formula (2) holds under the hypotheses of Theorem 6'.

7. THEOREM. Let $f \in B^2_{\infty 1}(\mathbb{R})$ and φ real function in $B_2^{1/2}(\mathbb{R})$. Then formula (2) holds.

PROOF. Let $f = f_- + f_0 + f_+$, where f_0 is a polynomial of degree at most 2, and f_- and f_+ are functions in $B^2_{\infty 1}(\mathbb{R})$ whose Fourier transforms are supported on $(-\infty, 0)$ and $(0,+\infty)$ respectively.

For polynomials formula (2) can be verified directly. It should be only noticed that

$$\operatorname{trace}(W_\varphi^2 - W_{\varphi^2}) = -\operatorname{trace} \mathcal{H}^*_\varphi \mathcal{H}_\varphi =$$

$$= -\frac{1}{8\pi^2}\int_{-\infty}^{\infty}\int_{-\infty}^{\infty} \frac{(\varphi(x)-\varphi(y))^2}{(x-y)^2}\,dx\,dy \, .$$

Therefore it is sufficient to consider f_+ since the corresponding assertion for f_- can be obtained by passing to complex conjugation. We have $f=\sum_{n\in\mathbb{Z}} f*V_n$ and $\sum_{n\in\mathbb{Z}} 2^{2n}\|f*V_n\|_{L^\infty} <$ $<\infty$. It follows that it is sufficient to prove (2) for those f in L^∞ whose Fourier transform is supported on $[2^{n-1}, 2^{n+1}]$. Assume first that $\mathcal{F}f\in L^1(\mathbb{R})$ and $\varphi\in L^\infty(\mathbb{R})$. In this case formula (2) was proved by Widom [13].

Suppose now that f is an arbitrary function in $L^\infty(\mathbb{R})$ whose Fourier transform is supported on $[2^{n-1}, 2^{n+1}]$ and $\varphi\in B_2^{1/2}(\mathbb{R})\cap L^\infty(\mathbb{R})$. **Consider the function f_ε defined in the proof of Theorem 2. Clearly**

$$\lim_{\varepsilon\to 0}\int_{-\infty}^{\infty}\int_{-\infty}^{\infty}\frac{(\varphi(x)-\varphi(y))^2}{(x-y)^2}\int_0^1 f_\varepsilon''((1-\theta)\varphi(x)+\theta\varphi(y))(\theta\log\theta+(1-\theta)\log(1-\theta))\,d\theta\,dx\,dy=$$

$$=\int_{-\infty}^{\infty}\int_{-\infty}^{\infty}\frac{(\varphi(x)-\varphi(y))^2}{(x-y)^2}\int_0^1 f''((1-\theta)\varphi(x)+\theta\varphi(y))(\theta\log\theta+(1-\theta)\log(1-\theta))\,d\theta\,dx\,dy\,.$$

Let us show that

$$\lim_{\varepsilon\to 0}(f_\varepsilon(W_\varphi)-W_{f_\varepsilon\circ\varphi})=f(W_\varphi)-W_{f\circ\varphi}$$

in the norm of $\boldsymbol{S}_1$. For this aim we apply formula (7). It follows that it is sufficient to show that for $0<\tau\le 2^{n+1}$ the following equality holds

$$\lim_{\varepsilon\to 0}(\tilde{S}_\tau^* f_\varepsilon)\circ\varphi=(\tilde{S}_\tau^* f)\circ\varphi$$

in the norm of $B_2^{1/2}(\mathbb{R})$. Suppose that $\varphi(\mathbb{R})\subset[-M,M]$. Then it is evident that

$$\|(\tilde{S}_\tau^* f_\varepsilon)\circ\varphi-(\tilde{S}_\tau^* f)\circ\varphi\|_{B_2^{1/2}(\mathbb{R})}\le\|\varphi\|_{B_2^{1/2}(\mathbb{R})}\sup_{t\in[-M,M]}|(\tilde{S}_\tau^*(f_\varepsilon-f))'(t)|$$

and the right-hand side tends to zero.

Suppose now that $\varphi\notin L^\infty(\mathbb{R})$. Pick a sequence $\{\varphi_n\}_{n\ge 0}$ such that $\lim_n\varphi_n(x)=\varphi(x)$ for almost all $x\in\mathbb{R}$, $\varphi_n\in L^\infty(\mathbb{R})$, and $\lim_n\|\varphi_n-\varphi\|_{B_2^{1/2}(\mathbb{R})}=0$. Then formula (2) holds for φ_n and it is easy to check that

$$\lim_{n\to\infty}\int_{-\infty}^{\infty}\int_{-\infty}^{\infty}\left(\frac{\varphi_n(x)-\varphi_n(y)}{x-y}\right)^2\int_0^1 f''(\theta\varphi(x)+(1-\theta)\varphi(y))(\theta\log\theta+(1-\theta)\log(1-\theta))d\theta\,dx\,dy =$$

$$=\int_{-\infty}^{\infty}\int_{-\infty}^{\infty}\left(\frac{\varphi(x)-\varphi(y)}{x-y}\right)^2\int_0^1 f''(\theta\varphi(x)+(1-\theta)\varphi(y))(\theta\log\theta+(1-\theta)\log(1-\theta))d\theta\,dx\,dy .$$

The proof will be completed if we show that

$$\lim_n (f(W_{\varphi_n}) - W_{f\circ\varphi_n}) = f(W_\varphi) - W_{f\circ\varphi} \tag{11}$$

in the norm of $\mathfrak{S}_1$. To prove this we need the following assertion.

8. LEMMA. Suppose that g is a C^1-function on $\mathbb{R}$ with bounded derivative, φ , $\varphi_n \in B_2^{1/2}(\mathbb{R})$, $\lim_n \|\varphi_n - \varphi\|_{B_2^{1/2}(\mathbb{R})} = 0$, and $\lim_n \varphi_n(x) = \varphi(x)$ for almost all $x \in \mathbb{R}$. Then $\lim_n \|g\circ\varphi_n - g\circ\varphi\|_{B_2^{1/2}(\mathbb{R})} = 0$.

PROOF OF LEMMA 8. We have to show that the sequence of functions in two variables $g(\varphi_n(x)) - g(\varphi_n(y))$ converges to the function $g(\varphi(x)) - g(\varphi(y))$ in $L^2(\mathbb{R}\times\mathbb{R}, \frac{dx}{(x-y)^2})$. We have

$$g(\varphi_n(x)) - g(\varphi_n(y)) = \frac{g(\varphi_n(x)) - g(\varphi_n(y))}{\varphi_n(x)-\varphi_n(y)}(\varphi_n(x)-\varphi_n(y)) .$$

By the assumptions $\lim_n (\varphi_n(x) - \varphi_n(y)) = \varphi(x) - \varphi(y)$ in $L^2(\mathbb{R}\times\mathbb{R}, \frac{dx}{(x-y)^2})$. It is also clear that the functions $\frac{g(\varphi_n(x)) - g(\varphi_n(y))}{\varphi_n(x)-\varphi_n(y)}$ are uniformly bounded and coverge to the function $\frac{g(\varphi(x)) - g(\varphi(y))}{\varphi(x)-\varphi(y)}$ almost everywhere which proves the lemma. ●

Let us complete the proof of Theorem 7. By (8) the function $\widetilde{S}_\tau^*(f)$ has bounded derivative and belongs to C^1 . Therefore by

Lemma 8 for all $\tau \leq 2^{n+1}$

$$\lim_{n\to\infty} (\tilde{S}_\tau^* f)\circ\varphi_n = (\tilde{S}_\tau^* f)\circ\varphi$$

in the norm of $B_2^{1/2}(\mathbb{R})$. This implies (11) which completes the proof. ●

REMARK. Let us compare the estimates from above for $\|f(W_\varphi)-W_{f\circ\varphi}\|_{S_1}$ obtained in Theorem 6' with those from below obtained by Widom in [13]. Using formula (2), Widom showed that for functions f satisfying the condition b) (see §1)

$$|||f||| \stackrel{\text{def}}{=} \{|\operatorname{trace}(f(W_\varphi)-W_{f\circ\varphi})| : \varphi\in \operatorname{Re} B_2^{1/2}(\mathbb{R});\ \|\varphi\|_{B_2^{1/2}(\mathbb{R})}\leq 1\} \tag{12}$$

can be estimated from below in terms of const $\cdot\|f''\|_{L^\infty(\mathbb{R})}$. This can be seen from formula (2) by taking

$$\varphi(t) = x + n^{-1/2}(\cos nt - \cos(n-1)t)$$

and making n tend to ∞ . It can be shown that if we consider the set X of functions f continuous on $\mathbb{R}$ such that $f(W_\varphi)-$ $-W_{f\circ\varphi}\in S_1$ for any real φ in $B_2^{1/2}(\mathbb{R})$ and $|||f|||<\infty$ then $X\subset\{f: f''\in L^\infty(\mathbb{R})\}$ and $|||f|||\geq \text{const}\,\|f''\|_{L^\infty}$.

The difference between $B^2_{\infty 1}(\mathbb{R})$ and $\{f: f''\in L^\infty(\mathbb{R})\}$ can be seen from the following equality

$$B^2_{\infty 1}(\mathbb{R}) = \{f: \sum_{n\in\mathbb{Z}}\|(f*V_n)''\|_{L^\infty(\mathbb{R})} + \sum_{n\in\mathbb{Z}}\|(f*V_n^*)''\|_{L^\infty(\mathbb{R})}<\infty\}.$$

In [13] Widom considered the asymptotic behaviour of

$$\operatorname{trace}(f(W_\varphi^{(\alpha)}) - W_{f\circ\varphi}^{(\alpha)})$$

for finite Wiener - Hopf operators (see Section 1). He showed that for real φ in $B_2^{1/2}(\mathbb{R})\cap L^\infty(\mathbb{R})$ and f satisfying the condition b) (see §1) the following asymptotic trace formula holds

$$\lim_{\alpha\to\infty} \operatorname{trace}(f(W_\varphi^{(\alpha)}) - W_{f\circ\varphi}^{(\alpha)}) =$$

$$= \frac{1}{4\pi^2}\int_{-\infty}^{\infty}\int_{-\infty}^{\infty}\frac{(\varphi(x)-\varphi(y))^2}{(x-y)^2}\int_0^1 f''(\theta\varphi(x)+(1-\theta)\varphi(y))(\theta\log\theta+(1-\theta)\log(1-\theta))d\theta\, dx\, dy, \tag{13}$$

that is

$$\lim_{\alpha\to\infty} \operatorname{trace}(f(W_\varphi^{(\alpha)}) - W_{f\circ\varphi}^{(\alpha)}) = 2\operatorname{trace}(f(W_\varphi) - W_{f\circ\varphi}).$$

Let us show that formula (13) holds under the hypotheses of Theorem 6'.

9. THEOREM. Let $f\in B^2_{\infty 1}(\mathbb{R})$ and φ real function in $B_2^{1/2}(\mathbb{R})$. Then formula (13) holds.

PROOF. We shall use the following well-known formula (see [13])

$$W_{\varphi_1\varphi_2}^{(\alpha)}h - W_{\varphi_1}^{(\alpha)}W_{\varphi_2}^{(\alpha)}h =$$

$$= \mathcal{P}_\alpha \mathcal{H}^*_{\bar\varphi_1}\mathcal{H}_{\varphi_2}h + \mathcal{P}_\alpha \mathcal{M}_\alpha \mathcal{H}_{\varphi_1}\mathcal{H}^*_{\bar\varphi_2}\mathcal{M}^*_\alpha h,$$

where $(\mathcal{M}_\alpha f)(t) = e^{i\alpha t}f(t)$.

As in Lemma 3 it can be proved that

$$e^{itW_\varphi^{(\alpha)}} - W^{(\alpha)}_{e^{it\varphi}} =$$

$$= -i\mathcal{P}_\alpha\int_0^t \mathcal{H}^*_{e^{-i(t-\tau)\varphi}}\mathcal{H}_\varphi e^{i\tau W_\varphi^{(\alpha)}}d\tau -$$

$$-i\mathcal{P}_\alpha\mathcal{M}_\alpha\int_0^t \mathcal{H}_{e^{i(t-\tau)\varphi}}\mathcal{H}^*_\varphi\mathcal{M}^*_\alpha e^{i\tau W_\varphi^{(\alpha)}}.$$

As in Theorem 2 it can be proved that for functions f in $B^2_{\infty 1}(\mathbb{R})$ whose Fourier transform is supported on $(0,+\infty)$ the following formula holds

$$f(W^{(\alpha)}_\varphi) - W^{(\alpha)}_{f\circ\varphi} =$$

$$= -i\mathcal{P}_\alpha \int_0^\infty \mathcal{H}^*_{\overline{(\tilde{S}^*_\tau f)\circ\varphi}} \mathcal{H}_\varphi e^{i\tau W^{(\alpha)}_\varphi} d\tau -$$

$$- i\mathcal{P}_\alpha \mathcal{M}_\alpha \int_0^\infty \mathcal{H}_{(\tilde{S}^*_\tau f)\circ\varphi} \mathcal{H}^*_\varphi \mathcal{M}^*_\alpha e^{i\tau W^{(\alpha)}_\varphi} d\tau .$$

It is easy to see that

$$\lim_{\alpha\to\infty} \operatorname{trace}(-i\mathcal{P}_\alpha \int_0^{+\infty} \mathcal{H}^*_{\overline{(\tilde{S}^*_\tau f)\circ\varphi}} \mathcal{H}_\varphi e^{i\tau W^{(\alpha)}_\varphi} d\tau) =$$

$$= \operatorname{trace}(-i\int_0^\infty \mathcal{H}^*_{\overline{(\tilde{S}^*_\tau f)\circ\varphi}} \mathcal{H}_\varphi e^{i\tau W_\varphi} d\tau) = \operatorname{trace}(f(W_\varphi)-W_{f\circ\varphi}) .$$

Next,

$$\lim_{\alpha\to\infty} \operatorname{trace}(-i\mathcal{P}_\alpha \mathcal{M}_\alpha \int_0^\infty \mathcal{H}_{(\tilde{S}^*_\tau f)\circ\varphi} \mathcal{H}^*_\varphi \mathcal{M}^*_\alpha e^{i\tau W^{(\alpha)}_\varphi} d\tau) =$$

$$= \operatorname{trace}(f(W_\varphi) - W_{f\circ\varphi}) .$$

In the case $\varphi \in B^{1/2}_2(\mathbb{R}) \cap L^\infty$ and $\mathcal{F}f'' \in L^1(\mathbb{R})$ this equality is proved in [13]. In the general case this equality can be established using approximation arguments as in Theorems 2 and 7. ●

A similar trace formula holds in the case of Toeplitz operators:

$$\operatorname{trace}(f(T_\varphi) - T_{f\circ\varphi}) =$$

$$= \frac{1}{2}\iint_{\mathbb{T}\mathbb{T}} \left(\frac{\varphi(\zeta_1)-\varphi(\zeta_2)}{\zeta_1-\zeta_2}\right)^2 \int_0^1 f''(\theta\varphi(\zeta_1)+(1-\theta)\varphi(\zeta_2))(\theta\log\theta+(1-\theta)\log(1-\theta))\,d\theta\, dm(\zeta_1)dm(\zeta_2) \quad (14)$$

(here f and φ satisfy the hypotheses of Theorem 2 with $p=1$). This formula can be proved in the same way as formula (2) but it can also be derived from formula (2) with the help of conformal mapping from the unit disc onto the upper half-plane.

It is also possible to state an analogue of Theorem 9 for finite Toeplitz matrices. Namely, let $T_\psi^{(N)} = \{\hat{\psi}(j-k)\}_{j,k=0}^{N}$. Then under the same assumptions on f and φ

$$\lim_{N\to\infty} \operatorname{trace}(f(T_\varphi^{(N)}) - T_{f\circ\varphi}^{(N)}) = 2\operatorname{trace}(f(T_\varphi) - T_{f\circ\varphi}). \tag{15}$$

Note that trace $(T_{f\circ\varphi}^{(N)}) = (N+1)\widehat{f\circ\varphi}(0)$. In the case of finite Wiener-Hopf operators, $W_{f\circ\varphi}^{(\alpha)}$ is not always nuclear. See [11], [10] where criteria for $W_g^{(\alpha)}$ to be in $\mathbf{S}_p$ are given. But if $W_{f\circ\varphi}^{(\alpha)} \in \mathbf{S}_1$ then we have trace $(W_{f\circ\varphi}^{(\alpha)}) = (\alpha/2\pi)(\mathcal{F}(f\circ\varphi))(0)$.

Note that in all results of this section in the case we deal with bounded φ we can replace the condition $f \in B_{\infty 1}^{2}(\mathbb{R})$ by the condition $f \,|\, [\operatorname{ess\,inf} \varphi, \operatorname{ess\,sup} \varphi] \in B_{\infty 1}^{2}(\mathbb{R}) \,|\, [\operatorname{ess\,inf} \varphi, \operatorname{ess\,sup} \varphi]$.

4. THE CASE $\|\varphi\|_{L^\infty} \leqslant 1$

In this section we consider the operators $f(T_\varphi) - T_{f\circ\varphi}$ for contractive T_φ, that is $\|\varphi\|_{L^\infty} \leqslant 1$. In this case the operators $f(T_\varphi)$ and $T_{f\circ\varphi}$ are defined for functions f analytic in $\operatorname{clos} \mathbb{D}$. We shall show that for $\varphi \in BMO$ the mapping $f \mapsto f(T_\varphi) - T_{f\circ\varphi}$ extends to $B_{\infty 1}^{2}$ and $f(T_\varphi) - T_{f\circ\varphi} \in \mathbf{S}_p$ for $\varphi \in B_{2p}^{1/2p}$, $p \geqslant 1$. Moreover for $p=1$ trace formula (14) and asymptotic trace formula (15) hold. The proofs of these assertions are analogous to those of the corresponding results of the preceding section but much simpler than they.

First of all we establish an analogue of formula (7).

10. THEOREM. <u>Suppose that</u> φ <u>is a function on</u> $\mathbb{T}$ <u>such that</u> $\|\varphi\|_{L^\infty} \leqslant 1$ <u>and</u> f <u>is a function analytic in</u> $\operatorname{clos} \mathbb{D}$. <u>Then the following formula holds</u>

$$f(T_\varphi) - T_{f\circ\varphi} = -\sum_{k\geqslant 1} H^*_{\overline{(S_*^k f)\circ\varphi}} H_\varphi T_\varphi^{k-1}, \tag{16}$$

where the operator S_* is defined by

$$S_* g = \sum_{k\geq 1} \hat{g}(k+1) z^k .$$

PROOF. Using formula (3) one can easily show by induction that

$$T_\varphi^n - T_{\varphi^n} = -\sum_{k=1}^{n-1} H^*_{\overline{\varphi^{n-k}}} H_\varphi T_\varphi^{k-1} .$$

It follows that

$$f(T_\varphi) - T_{f\circ\varphi} = -\sum_{n\geq 0} \sum_{k=1}^{n-1} \hat{f}(n) H^*_{\overline{\varphi^{n-k}}} T_\varphi^{k-1} =$$

$$= -\sum_{k\geq 1} \Big(\sum_{n\geq k+1} \hat{f}(n) H^*_{\overline{\varphi^{n-k}}} \Big) H_\varphi T_\varphi^{k-1} =$$

$$= -\sum_{k\geq 1} H^*_{\overline{\sum_{n\geq k+1} \hat{f}(n)\varphi^{n-k}}} H_\varphi T_\varphi^{k-1} =$$

$$= -\sum_{k\geq 1} H^*_{\overline{(S_*^k f)\circ\varphi}} H_\varphi T_\varphi^{k-1} . \bullet$$

11. THEOREM. Let $1 \leq p < \infty$, $\varphi \in B_{2p}^{1/2p}$, $\|\varphi\|_{L^\infty} \leq 1$, and f function in $B^2_{\infty 1}$, analytic in $\mathbb{D}$. Then the series on the right-hand side of (16) converges absolutely in the norm of S_p and so $f(T_\varphi) - T_{f\circ\varphi} \in S_p$.

12. THEOREM. Let $\varphi \in B_2^{1/2}$, $\|\varphi\|_{L^\infty} \leq 1$ and f function in $B^2_{\infty 1}$ analytic in $\mathbb{D}$. Then the trace formula (14) holds.

13. THEOREM. Under the hypotheses of Theorem 7 the asymptotic trace formula (15) holds.

Analogous results also hold for Wiener-Hopf operators.

As in the preceding section it can be shown that if

$$\sup\{|\operatorname{trace}(f(T_\varphi)-T_{f\circ\varphi})| : \|\varphi\|_{L^\infty}\le 1, \|\varphi\|_{B_2^{1/2}}\le 1\}<\infty$$

then $f''\in H^\infty$.

REFERENCES

1. Budylin A.M., Buslaev V.S., On the asymptotic behaviour of spectral characteristics of an integral operator with difference kernel on expanding domains (Russian), Doklady Akad. Nauk SSSR, 287, N 3 (1986), 529-532.
2. Douglas R.G., Banach algebra techniques in operator theory, Acad.Press, New York - London, 1972.
3. Garnett J.B., Bounded analytic functions, Acad.Press, New York - London - Toronto, 1981.
4. Gohberg I.Ts., Krein M.G., Introduction to the theory of linear nonselfadjoint operators in Hilbert space, "Nauka", Moscow, 1965. Translation: Amer.Math.Soc., Providence, R.I., 1969.
5. Nikolskii N.K. Treatise on the shift operator, Springer-Verlag, Berlin - Heidelberg - New York, 1985.
6. Peetre J., New Thoughts on Besov spaces, Duke Univ.Press, Durham, 1976.
7. Peller V.V., Hankel operators of class $\mathfrak{S}_p$ and their applications (rational approximation, (Russian), Matem.Sbornik, 113, N 4 (1980), 538-581. Translation: Math.USSR Sbornik, 41, N 4, 443-479.
8. Peller V.V., A description of Hankel operators of class $\mathfrak{S}_p$ for $p>0$, an investigation of the rate of rational approximation, and other applications, (Russian), Matem.Sbornik, 122, N 4 (1983), 481-510. Translation: Math.USSR Sbornik, 50, N 2, 465-494.
9. Peller V.V., Khrushchev S.V., Hankel operators, best approximations and stationary Gaussian processes, Uspekhi Mat.Nauk, 37, N 1 (1982), 53-124. Translation: Russian Math.Surveys, 37 (1982), 61-144.

10. Peller V.V., Wiener-Hopf operators on a finite interval and Schatten - von Neumann classes, Uppsala Univ.Dep.,Math., Report 186:9.
11. Rochberg R. Toeplitz and Hankel operators on the Paley - Wiener space, Integral Equat.Oper.Theory, 10 (1987), 187-235.
12. Semmes S., Trace ideal criteria for Hankel operators and applications to Besov spaces, Integral Equat.Operator Theory, 7 (1984), 241-281.
13. Widom H. A trace formula for Wiener-Hopf operators, J.Operator Theory, 8 (1982), 279-298.
14. Widom H., Trace formulas for Wiener-Hopf operators, Operator Theory: Advances and Appl., 24 (1987), 365-371, Birkhäuser Verlag, Basel.
15. Widom H., Asymptotic expansions for pseudodifferential operators on bounded domains, Lect.Notes Math., 1152 (1985), Springer-Verlag, Berlin - Heidelberg - New York.

Operator Theory:
Advances and Applications, Vol. 42

MULTIPLICITY OF ANALYTIC TOEPLITZ OPERATORS

B.M.Solomyak and A.L.Volberg

CONTENTS

Chapter 1. Statement of the results and discussion. 89
1. Introduction. 89
2. Main results and examples. 93
3. An outline of the proof. The cocycle problem. 101
Chapter 2. The beginning of the proof of the Main Theorem. 106
4. Technical Background I. 106
5. Estimates of multiplicity from below. 111
6. The construction of a cyclic set: simple cases. 117
Chapter 3. Conditions sufficient for the cyclicity. 124
7. Technical background II. 124
8. The case of essentially univalent φ 128
9. Reduction. 135
10. Conditions sufficient for the cyclicity. Statement and discussion. 139
11. Conditions sufficient for the cyclicity. Proof. 143
Chapter 4. "Vector bundle": the construction of a cyclic set. 153
12. Reduction to the problem on weighted cocycles. 155
13. The beginning of the proof of Theorem 12.3: the construction of a bundle. 163

14. Solution of the Riemann-Hilbert problem. 169
15. How to ensure the condition (10.4) of analytic independence? 179
Appendix. Proof of approximation lemmas on a Riemann surface. 183
Notation 186
References. 188

CHAPTER I. STATEMENT OF THE RESULTS AND DISCUSSION

1. INTRODUCTION

1.1. THE MULTIPLICITY $\mu(T)$ of an operator T acting on a linear topological space $\mathcal{X}$ is defined to be the least cardinal number of $\mathcal{Y}\subset\mathcal{X}$ such that

$$\mathrm{Span}(T^n\mathcal{Y}: n\geq 0)=\mathcal{X}. \tag{1.1}$$

Any $\mathcal{Y}$ satisfying (1.1) is called a cyclic set for T.

The multiplicity is an important characteristic of an operator. It is essential for example in the Sz.-Nagy-Foiaş theory of Jordan models for the class C_0 [45], [46]. There is vast literature devoted to operators with $\mu(T)=1$, i.e. those having a cyclic vector. The notion of multiplicity and related subjects were studied by N.Nikol'skii and V.Vasyunin [30], [33], [34], [53], A.Atzmon [3], and especially by D.Herrero and his co-authors [22], [23], [24], [25], [26]. In several papers the multiplicity of operators of various classes was determined. J.Bram [7] proved as a matter of fact that for a normal operator on a Hilbert space multiplicity is equal to the essential supremum of the local spectral multiplicity. In a paper of W.Wogen [61] it is established that Toeplitz operators with antianalytic symbols are cyclic. V.Vasyunin [52] found the multiplicity of a contraction with finite defect indices. In this paper we shall be concerned with the class of analytic Toeplitz operators.

The Analytic Toeplitz operator (AT operator) with symbol φ is the multiplication operator $T_\varphi: f\mapsto\varphi f$, acting on the Hardy space H^2, or, more general, on some space of functions analytic in the unit disc $\mathbb{D}$. It should be noted that AT operators are of some importance in the models up to a similarity developed by D.Clark [8], [9], D.Wang [58] and recently by D.Yakubovich [62], [63] for Toeplitz operators with rational (or smooth) symbols. This enlarges the scope of their applications.

Estimates of multiplicity of AT operators were initiated by N.Nikolskii [30]. The main result of the present paper is the computation of $\mu(T_\varphi)$ for φ analytic on the closed unit disc.

1.2. LOCAL MULTIPLICITY. In the finite-dimensional case and in the case of normal Hilbert space operators one is concerned mostly with the function of local spectral multiplicity. The multiplicity is then the maximum (or the essential supremum) of this function. In the general case one can consider the function (see [31])

$$\nu_T(\lambda) = \dim \operatorname{Ker}(T^* - \bar{\lambda} I), \ \lambda \in \mathbb{C},$$

as it were an analogue of "local multiplicity". Denote $n(T) = \sup\{\nu_T(\lambda) : \lambda \in \mathbb{C}\}$. The estimate $\mu(T) \geq n(T)$ is quite easy (see [22] and also 5.2 below). Sometimes this estimate can be sharpened. Consider the "set of maximal local multiplicity"

$$\Omega(T) = \{\lambda \in \mathbb{C} : \nu_T(\lambda) = n(T)\}.$$

It is shown in Corollary 5.4 that if $\mathbb{C} \setminus \operatorname{int} \Omega(T)$ is disconnected and a special basis of $\operatorname{Ker}(T^* - \bar{\lambda} I)$, $\lambda \in \operatorname{int} \Omega(T)$, can be chosen, then $\mu(T) \geq n(T) + 1$. (A related assertion was proved by D.Herrero [22], [24]).

Estimates of $\mu(T)$ from above in terms of $n(T)$ are impossible in the general case since T^* can have empty point spectrum along with $\mu(T) = \infty$.

1.3. GENERAL DESCRIPTION OF THE MAIN RESULT. Suppose now that $T = T_\varphi$ is an AT operator. Assume that φ is analytic on $\operatorname{clos} \mathbb{D}$ and $\varphi' | \partial\mathbb{D} \neq 0$ (the latter condition is imposed in the Introduction for the sake of simplicity). It was proved in [39] that $\mu(T) \leq n(T) + 1$. (Note that for a function univalent on $\mathbb{D}$ this estimate follows from [22, Prop.2]). The question of whether the multiplicity is equal to $n(T)$ or

$n(T)+1$ reduces to a problem of weighted polynomial approximation on the set $V(T) = \operatorname{clos} \Omega(T)$. The answer is, roughly speaking, the following: $\mu(T) = n(T)$ if $\mathbb{C} \setminus V(T)$ is connected (the simple case, [39]) or if it is disconnected but the "separators" between its components are "narrow". Otherwise $\mu(T) = n(T) + 1$. This result was stated in our paper [56] and the proof was outlined under several simplifying assumptions.

Note that for an AT operator $T = T_\varphi$ the value of $\nu_T(\lambda)$ is equal to $\operatorname{card}(\varphi^{-1}(\lambda) \cap \mathbb{D})$ counting multiplicities. So we see that $\mu(T)$ has a clear geometrical sense. The number

$$n(T_\varphi) = \max\{\operatorname{card}(\varphi^{-1}(\lambda) \cap \mathbb{D}) : \lambda \in \mathbb{C}\}$$

will be called the <u>maximal valency</u> of φ.

The study of symbols which are not so smooth is perhaps a difficult problem that will not be touched in the present paper. At least it is clear that the boundary behaviour of φ should be taken carefully into account.(See also Remark 2.10 about this).

1.4. OTHER PROBLEMS ON AT OPERATORS. The computation of multiplicity may be viewed as the first step in the study of invariant subspaces. The next step is to describe all cyclic sets. We present some necessary and, separately, sufficient conditions for the cyclicity. For some classes of AT operators similar methods provide a complete description of cyclic sets [39], [40] and even the invariant subspace lattice [63].

There are many papers devoted to other topics related to AT operators. Among them we mention the paper of E.Nordgen [35] on reducing subspaces and vast literature dealing with the commutant of an AT operator (see [4], [48], [10], [43] and the references given there). It is noteworthy that in those questions the leading part is played not by the maximal but by the minimal nonzero valency of the symbol.

1.5. A MORE GENERAL VIEW AND ITS APPLICATIONS. In fact it is more natural to consider a wider class than AT operators. We shall be concerned with multiplication operators acting on a space of analytic functions defined in an open domain $\mathcal{R}$ (not necessarily connected) in a Riemann surface. It will be convenient to take the algebra of bounded analytic functions $H^\infty(\mathcal{R})$ endowed with the weak-$*$ topology. The difficult part of the result on multiplicity for H^2 will follow easily. Multiplication operators acting on some other spaces (e.g., disc algebra, $A^k(\mathcal{R})$ etc.) can be treated similarly, see [41], [42]. Now we mention some important special cases and applications.

(1). The operator of multiplication by z on E^2 on a plane domain (In this connection note that the invariant subspace lattice for $f \mapsto zf$ on an annulus attracted attention for a long time, see [37], [27]).

(2) The orthogonal sum of several AT operators (Here $\mathcal{R}$ is the disjoint union of discs).

(3) The matrix case. Let Φ be an $n\times n$ matrix-valued function with entries in H^∞. Define a matrix analytic Toeplitz operator (MAT operator) acting on the vector-valued Hardy space H^2_n as follows: $T_\Phi : x \mapsto \Phi x$. It turns out that T_Φ (for Φ analytic on $\operatorname{clos}\mathbb{D}$) can be reduced in a sense to an operator of multiplication acting on E^2 on the Riemann surface of characteristic polynomial of Φ. This enables us to compute the multiplicity of T_Φ mainly in the same way as that of an AT operator (see [57] in the present volume).

(4) Let T be a completely non-unitary contraction on Hilbert space and let $\varphi \in H^\infty$. It is easy to see that $\mu(\varphi(T)) \leqslant \mu_\varphi\, \mu(T)$, where μ_φ is the multiplicity of $T_\varphi : f \mapsto \varphi f$ acting on H^∞. Similar considerations can be developed to obtain estimates of the multiplicity for other functional calculi [42].

Acknowledgement. We are deeply grateful to N.K.Nikol'skii for his permanent interest in our work and to I.M.Spitkovskii for an important consultation on the Riemann-Hilbert problem. We wish to thank A.B.Aleksandrov, E.M.Dyn'kin, A.E.Eremenko, M.Ju.Ljubich, D.V.Yakubovich for helpful discussions and V.V.Peller for language consultations.

2. MAIN RESULTS AND EXAMPLES.

2.1. DEFINITION. Let $\mathcal{K}$ be a compact Riemann surface with boundary. We shall say that an open set R contained in $\mathcal{K}$ is <u>admissible</u> if $\operatorname{clos} R \subset \operatorname{int} \mathcal{K}$, ∂R is a finite union of analytic Jordan arcs and $R = \operatorname{int}(\operatorname{clos} R)$. (The latter condition implies that incisions and punctured points are excluded.) Suppose that R is an admissible set and φ is analytic on $\operatorname{clos} R$ (i.e. in a neighbourhood of $\operatorname{clos} R$). Then we say that φ and the pair (R, φ) are admissible.

Let (R, φ) be an admissible pair. We consider the algebra $H^\infty(R)$ of bounded analytic functions in R, equipped with the weak-$*$ topology. The topology is defined by the seminorms $f \mapsto |\int_{\partial R} f\, d\omega|$, where $d\omega$ is a differential which can be written locally in the form $d\omega = g\, dz$, $g \in L^1(\partial R)$. The collection of such differentials will be denoted by $\mathcal{L}^1(\partial R)$, the measure is always assumed to be the arc length unless otherwise specified.

We consider the operator $T_\varphi : f \mapsto \varphi f$ on $H^\infty(R)$. The question is to find its multiplicity $\mu_\varphi \overset{def}{=\!=} \mu(T_\varphi)$.

Let us note that μ_φ has the following interpretation: it is the minimal number of $f_1, \dots, f_n \in H^\infty(R)$ such that the sums $\sum_{i \le n} p_i(\varphi) f_i$ are dense in $H^\infty(R)$, the p_i being arbitrary polynomials.

To state the results we have to introduce the notion of "essential maximal valency".

2.2. DEFINITION. Suppose (R, φ) is an admissible pair. Put

$$\Pi^k(\varphi) = \{ z \in \mathbb{C} : \operatorname{card}(\varphi^{-1}(z) \cap \operatorname{clos} R) = k \}.$$

The cardinal number $n(\varphi)$,

$$n(\varphi) \overset{def}{=\!=} \max\{ k : \operatorname{card} \Pi^k(\varphi) = \infty \}$$

is called the <u>essential maximal valency</u> of φ (with respect to $\mathcal{R}$). The set

$$V(\varphi) = \operatorname{clos} \Pi^{n(\varphi)}(\varphi)$$

will be called the <u>set of essential maximal valency</u>.

Note that $n(\varphi)$ and $V(\varphi)$ do not depend upon whether we count the pre-images with multiplicities or not. This follows from the definition of an admissible pair.

2.3. COMPARISON OF $n(\varphi)$ AND $n(T_\varphi)$. It is not hard to see that $n(\varphi)$ is greater than the maximal valency $n(T_\varphi) = \max\{\operatorname{card}(\varphi^{-1}(z) \cap \mathcal{R})\}$ if and only if φ maps two disjoint arcs of $\partial\mathbb{D}$ homeomorphically onto a single arc in $\mathbb{C}$ with opposite orientations. Thus $n(\varphi) = n(T_\varphi)$ in the following cases: (a) φ is one-to-one on $\partial\mathcal{R}$ except for a finite set; (b) $\mathcal{R}$ is simply connected, $\partial\mathcal{R}$ is analytic (e.g., $\mathcal{R} = \mathbb{D}$) and $\varphi' | \partial\mathcal{R} \neq 0$. It is somewhat surprising that there exists a function φ analytic on $\operatorname{clos}\mathbb{D}$ such that $n(\varphi) > n(T_\varphi)$.

2.4. DEFINITION. The <u>belt</u> is a closed connected set $B \subset \mathbb{C}$ such that $B = \operatorname{clos} V_B \setminus G_B$ where $G_B \subset V_B$ and V_B, G_B are both simply connected admissible domains. We shall say that the belt B is <u>wide</u> if $\partial G_B \subset \partial(\operatorname{int} B)$ and for a neighbourhood U of the set $\partial G_B \cap \partial V_B$ the following holds

$$\omega_{G_B} \leq \operatorname{const} \omega_B \qquad \text{on } \partial G_B \cap U. \qquad (2.1)$$

Here ω_{G_B} is the harmonic measure for G_B and ω_B is the sum of harmonic measures for the components of $\operatorname{int} B$. We call the belt <u>narrow</u> if it is not wide.

2.5. DEFINITION. Let $\Phi \subset \mathbb{C}$ be the closure of an admissible plane domain. We shall say that a belt B is <u>subordinate</u> to Φ if (a) $B \subset \Phi$, $G_B \not\subset \Phi$; (b) there exists a neighbourhood U of the set $\partial G_B \cap \partial V_B$ such that

$$\partial B \cap U \subset \partial\Phi \cap U. \qquad (2.2)$$

2.6. MAIN THEOREM. *Let* (R,φ) *be an admissible pair. Then*

$$n(\varphi) \leq \mu_\varphi \leq n(\varphi)+1,$$

and $\mu_\varphi = n(\varphi)$ *if and only if every belt subordinate to* $V(\varphi)$ *is narrow.*

2.7. COROLLARY. *If* $\mathbb{C}\setminus V(\varphi)$ *is connected then* $\mu_\varphi = n(\varphi)$.

2.8. COROLLARY [39]. *Let* φ *be a function analytic on* $\operatorname{clos}\mathbb{D}$ *such that all the points of self-intersection of the curve* $t \mapsto \varphi(e^{it})$ *are simple and transversal* (such a function will be called *generic*.) *Then* $\mathbb{C}\setminus V(\varphi)$ *is connected and therefore* $\mu_\varphi = n(\varphi)$.

2.8.1. COROLLARY. *Let* (R,φ) *be an admissible pair. Then* $\mu_\varphi = n(\varphi)$ *if and only if the operator* $f \mapsto zf$ *on* $H^\infty(\operatorname{int} V(\varphi))$ *is cyclic* (*or* $\operatorname{int} V(\varphi)=\emptyset$).

Note that from the results of D.Herrero [22], it follows that the operator of multiplication by z on $H^\infty(\Omega)$ has multiplicity at most 2 and if $\mathbb{C}\setminus\Omega$ is disconnected, its multiplicity is 2 .

2.9. REMARK. Theorem 2.6 for $R=\mathbb{D}$ was established by the authors in [56]. Unfortunately the statement of Theorems 1, 2 in this paper needs a correction. Firstly, we have overseen the possibility $n(\varphi) > n(T_\varphi)$ mentioned in 2.3. So $n(\varphi)$ in [56] should be defined as in the present paper (or [39]). Secondly, there is an inaccuracy in the definition of a belt. To correct it one should allow the belt to have any finite number of complementary components. However this does not influence the proof in [56] since it was given under the assumptions which imply that $n(\varphi)=n(T_\varphi)$ and that all belts have two components of the complement.

2.10. REMARK. Under the hypotheses of the Main Theorem the AT operator T_φ has the same multiplicity in all spaces $E^P(R)$,

$1 \leq p < \infty$, and $H^{\infty}(R)$ (for the definition see 4.1, 4.5 below). The smoothness requirements imposed on (R, φ) in Theorem 2.6 can be somewhat relaxed if we impose additional conditions on the set $\Gamma = \varphi(\partial R)$. For instance, the result remains true if ∂R is piecewise C^2-smooth, φ is analytic on $\operatorname{clos} R$ as above, and no two ares of Γ are tangent. If ∂R is piece - wise C^r-smooth, then the tangency of order not exceeding $r-2$ can be allowed.

2.11. REMARK. The statement of Theorem 2.6 involves an infinite number of belts. In fact for any admissible pair (R, φ) it suffices to test finitely many belts whether they are wide. These belts can be described as follows.

Suppose $\mathbb{C} \setminus V(\varphi)$ is disconnected. The boundary $\partial V(\varphi)$ is a finite union of analytic arcs. Choose a finite set $\mathcal{A} \subset \partial V(\varphi)$ such that all components of $\partial V(\varphi) \setminus \mathcal{A}$ are Jordan arcs. (The points in $\mathcal{A}$ will be called the vertices of $\partial V(\varphi)$). Choose $\varepsilon > 0$ so that the discs $\{|z-a| \leq \varepsilon\}$, $a \in \mathcal{A}$, are disjoint and let $\mathcal{A}_\varepsilon$ denote their union. In Theorem 2.6 it is sufficient to consider only those belts whose boundary is contained in $\partial V(\varphi) \cup \cup \partial \mathcal{A}_\varepsilon$.

Now we present several **examples** illustrating the Main Theorem.

2.12. EXAMPLE. Suppose $R = \mathbb{D}$. The function φ is indicated in Fig.1 by the image $\varphi(\partial \mathbb{D})$. **The set shaded in the Figure** B **is**

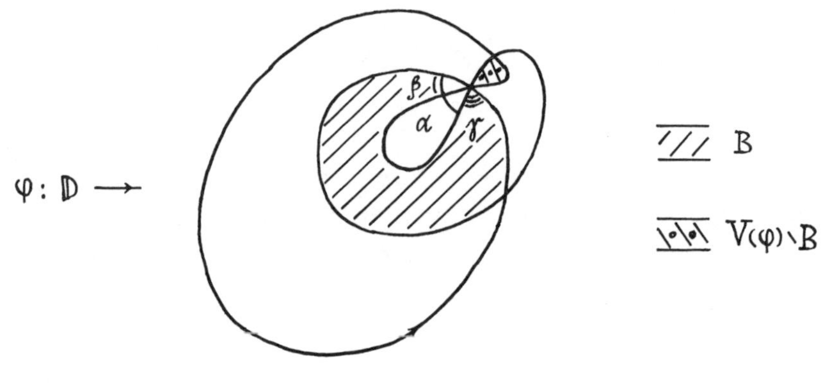

Figure 1.

the only belt that should be tested. The type of the belt is determined by the magnitude of α, β, γ. If $0<\alpha \leq \min(\beta,\gamma)$ then the belt B is wide and $\mu_\varphi = n(\varphi)+1=3$. If $\alpha > \min(\beta,\gamma)$ then the belt B is narrow and $\mu_\varphi = 2$. Note that the dependence of μ_φ from angles was discovered by N.Nikolskii in [30] for a special univalent φ (see Example 2.16 below).

2.13. EXAMPLE. Here we show why it is not sufficient to test the belts B such that $\partial B \subset \partial V(\varphi)$. In Fig.2 the set R is indicated and $\varphi(z)=z$.

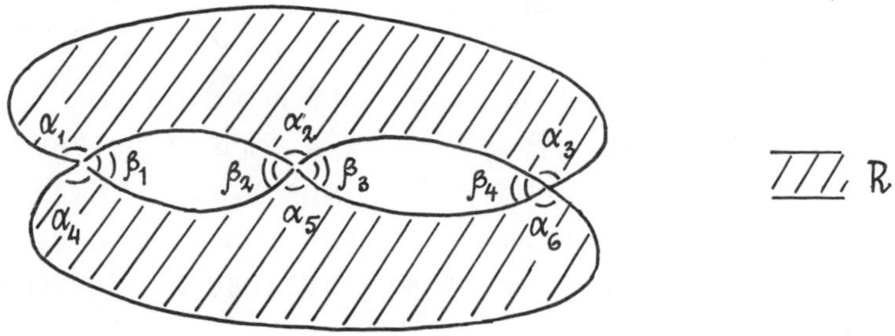

Figure 2.

Clearly $n(\varphi)=1$ so $1 \leq \mu_\varphi \leq 2$. To find μ_φ one should consider three belts obtained from $V(\varphi)=\operatorname{clos} R$ by removing a small neighbourhood of one of the three vertices. Theorem 2.6 implies that, provided all α_i, β_i are nonzero,

$$\mu_\varphi = 2 \Longleftrightarrow \begin{cases} \beta_1 \leq \min(\alpha_1,\alpha_4) \\ \beta_2 \leq \min(\alpha_2,\alpha_5) \end{cases} \vee \begin{cases} \beta_1 \leq \min(\alpha_1,\alpha_4) \\ \beta_4 \leq \min(\alpha_3,\alpha_6) \end{cases} \vee \begin{cases} \beta_3 \leq \min(\alpha_2,\alpha_5) \\ \beta_4 \leq \min(\alpha_3,\alpha_6) \end{cases}.$$

2.14. EXAMPLE. It is demonstrated that sometimes it is necessary to consider belts having more than two components of the complement (see Remark 2.9). Again let $\varphi(z) = z$ and let R be the union of two "crescent domains" R_1 and R_2 shaded in Fig.3a. In this case five belts **subordinate to the set** R

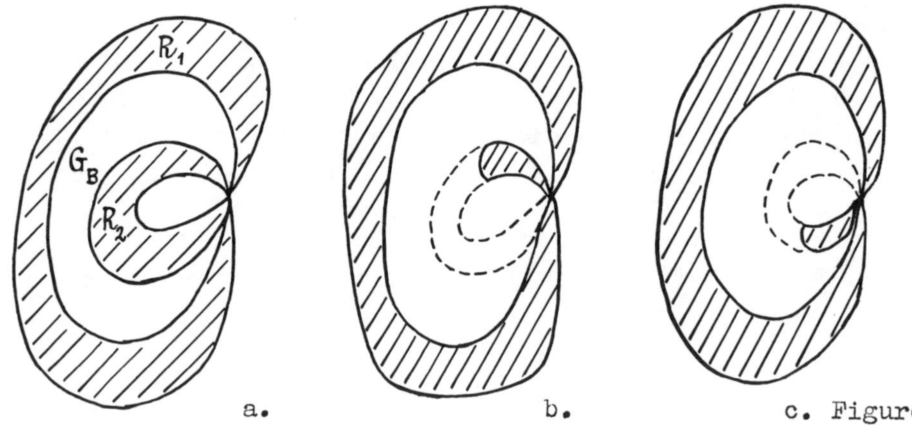

c. Figure 3

should be tested: 1) $B = \operatorname{clos} R$ (G_B is indicated in Fig.3a), 2) $B = \operatorname{clos} R_1$, 3) $B = \operatorname{clos} R_2$. The belts 4), 5) are presented in Fig.3 b, c. According to Remark 2.11 one should take into consideration the sixth belt, equal to the union of the belts 4) and 5). But if it is wide then so is the belt 1) hence we can skip the sixth belt.

2.15. EXAMPLES. Here a distinction between the maximal valency $n(T_\varphi)$ and the essential maximal valency $n(\varphi)$ is shown. In Fig.4 $n(\varphi) = 2$, $V(\varphi)$ is a line segnment while $n(T_\varphi) = 1$. So $\mu_\varphi = 2$.

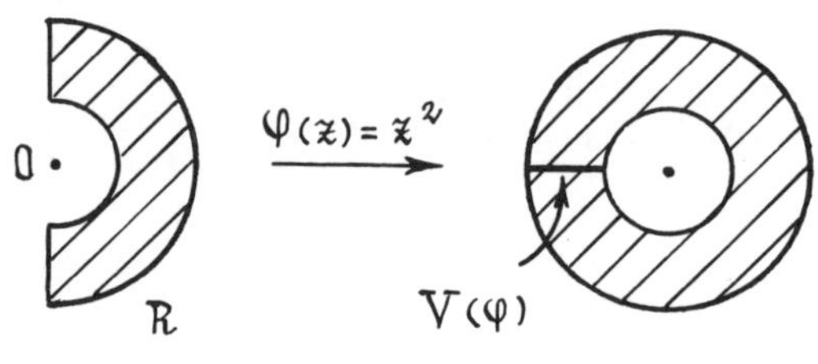

Figure 4

A more complicated picture is shown in Fig.5. Here $n(\varphi) = 3$, $V(\varphi) = [-2, 2]$, $n(T_\varphi) = 2$, $\mu_\varphi = 3$. It turns out that a similar situation can occur on $\mathbb{D}$ (with φ analytic on $\operatorname{clos} \mathbb{D}$)which provides the example mentioned in Remarks 2.3 and 2.9.

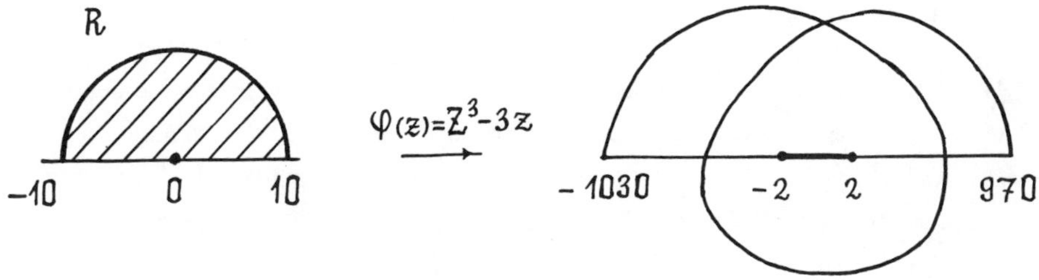

Figure 5

Now we are going to present some more examples in which cyclic sets are indicated. The proof of the cyclicity is omitted but can be extracted from Sect.11 without difficulty.

2.16. CRESCENT DOMAIN [30]. Let φ be the conformal mapping of the unit disc $\mathbb{D}$ onto the domain R in Fig.6. Suppose that R is bounded by analytic arcs and the angles α, β, γ are nonzero. Let z be the "vertex" of ∂R. Denote by Γ_1, Γ_2, Γ_3, Γ_4 small arcs ending at z as indicated in the Figure. Let α_{ij} be the angle between Γ_i and Γ_j.

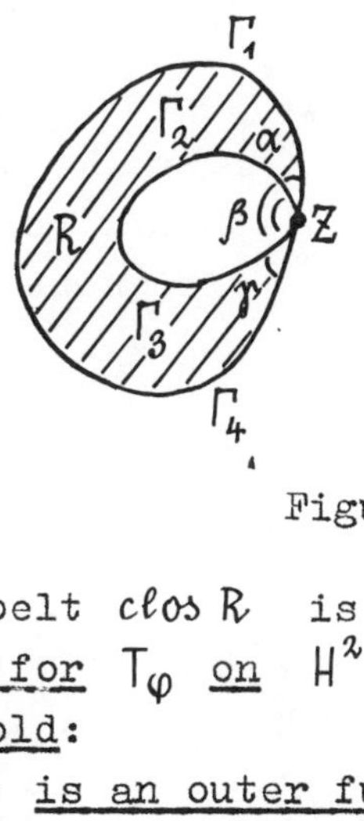

Figure 6.

Case 1. Suppose $\beta > \min(\alpha, \gamma)$. Then the belt $\operatorname{clos} R$ is narrow and $\mu_\varphi = 1$. A function $u \in H^2$ is cyclic for T_φ on H^2 if and only if the following two conditions hold:

(i) u is an outer function (in the sense of Beurling);

(ii) for all pairs i, j such that $1 \le i \le 2$, $3 \le j \le 4$,

$$\int_{\Gamma_i \cup \Gamma_j} \log |(u \circ \varphi^{-1})(\zeta)|\, |\zeta - z|^{\pi/\alpha_{ij}} |d\zeta| = -\infty .$$

REMARK. Recently P.Bourdon [6] mentioned that in a preprint by J.Akeroyd the cyclicity of T_φ is proved in the case of a crescent bounded by two internally tangent circles. It is to be noted that this result follows from [30].

Case 2. Suppose $\beta \leq min(\alpha, \gamma)$. Then the belt $clos\,R$ is wide and $\mu_\varphi = 2$. Let G be the bounded component of $\mathbb{C} \setminus clos\,R$. Denote by $GCD(f_i)$ the greatest common inner divisor of $\{f_i\}$. $N(G)$ is the Nevanlinna class of functions representable as a ratio of two functions from $H^\infty(G)$.

The set $\{u, v\} \subset H^2$ is cyclic for T_φ on H^2 if and only if (compare with [39], [40])

(i) $GCD(u,v) = \mathbb{1}$;

(ii) $\frac{u \circ \varphi^{-1}}{v \circ \varphi^{-1}}\Big|_{\partial G} \notin N(G)$, or in other words, the ratio $\frac{u\circ\varphi^{-1}}{v\circ\varphi^{-1}}$ has no pseudocontinuation to G.

For example any pair $\{\mathbb{1}, w \circ \varphi^{-1}\}$ with w analytic on $clos\,R$ and having an essential singularity in G, is cyclic.

2.17. GENERIC FUNCTION. Let φ be a generic function on the unit disc $\mathbb{D}$ as in Corollary 2.8. Let $n = n(\varphi)$. Choose Ω to be a Jordan domain with nowhere differentiable boundary and let $\omega : \mathbb{D} \to \Omega$ be the conformal mapping. Then for some constant $r(\varphi) < 1$ the set $U = \{\mathbb{1}, u, \dots, u^{n-1}\}$ is cyclic for T_φ where $u(z) = \omega(rz)$, $r \in [r(\varphi), 1)$.

2.18. A SPECIAL CASE. Let φ be a generic function having the property:

for every $k = 1, \dots, n(\varphi)$, the set $\mathbb{C} \setminus clos\,\Pi^k(\varphi)$ is connected. (2.3)

The condition (2.3) can be stated in another, equivalent form: For any component G of $\mathbb{C} \setminus \varphi(\partial R)$ there exists a chain of adjacent components $G_0, \dots, G_k$ of $\mathbb{C} \setminus \varphi(\partial R)$ such that $G_k = G$, G_0 is the unbounded component and the number of φ pre-images in G_0, $G_1, \dots, G_k$ increases monotonically.

For an AT operator with a symbol satisfying (2.3), in [40] a complete description of cyclic sets is given, and in [63] the invariant subspace lattice is found to some extent. For a cyclic set one can take $U = \{ 1, u, \dots, u^{n-1} \}, n = n(\varphi)$, where u is analytic and univalent on $\operatorname{clos} \mathbb{D}$.

3. OUTLINE OF THE PROOF. COCYCLE PROBLEM.

3.1. PLAN OF THE PROOF. 1) Estimates of the multiplicity μ_φ from below are comparatively easy. They are obtained in Sect.5. The estimate $\mu_\varphi \geqslant n(\varphi)$ is implicitly contained in [30]. The estimate $\mu_\varphi \geqslant n(\varphi)+1$ for an appropriate φ is based on a proposition dealing with an operator which has a wide belt subordinate to the set of maximal local multiplicity (see Corollary 5.5).

2) Estimates of μ_φ from above are proved by a more or less explicit presentation of the cyclic set. First an admissible domain is slightly enlarged to obtain some geometrical properties of the pair (R, φ) (Sect.9). The cyclic set is constructed on this larger domain. In Sect.10 we state and discuss the conditions sufficient for the cyclicity.

3) To prove that the set is cyclic we take a functional orthogonal to the invariant subspace it generates. The functional can be written in the form $f \mapsto \int_{\partial R} f\, d\omega$. We want to show that $d\omega$ extends to an analytic differential in R of the appropriate Hardy class. We transfer $d\omega$ to the set $\Gamma = \varphi(\partial R)$ yielding a differential $g\,dz$ and try to extend the function g . This extension is constructed in the components of $\mathbb{C} \setminus \Gamma$ separately and then is assembled from those pieces (Sect.11). A similar approach was used by J.Wermer in his works on generators of the disc algebra [60].

4) The main obstacle for the extension of an orthogonal differential mentioned above appears when one crosses a belt. It is overcome with the help of a theorem due to A.L.Volberg related with weighted approximation on a system of rays (Theorem 8.10). This part is little dependent on the valency. Thus we start with the important case of univalent φ to determine when T_φ is cyclic (Sect.8).

5) The second and rather independent part of the proof is

the construction of a set satisfying all the conditions stated in Sect.10. These conditions can be rewritten in terms of the differences $f_{ij}=u\circ\varphi^{-1}_{(i)}-u\circ\varphi^{-1}_{(j)}$, where $\varphi^{-1}_{(i)}$, $\varphi^{-1}_{(j)}$ are branches of φ^{-1}. Every branch is defined in its own domain. When f_{ij} are known u can be recovered by solving the First Cousin Problem. Thus we arrive at a problem of finding an additive cocycle with some additional properties. This problem is treated in more detail below. Here we only note that these cocycles are found as sections of a special holomorphic vector bundle (Sec.13-14).

6) In the case of a generic function (as in Corollary 2.8) there is a more direct approach to the computation of the multiplicity, presented in Sect.6. However, this approach gives a considerably smaller supply of cyclic sets.

3.2. THE COCYCLE PROBLEM. To provide the needed properties of cocycles, it is natural to represent them in the form $f_{ij}=w_{ij}g_{ij}$ where w_{ij} are certain weight functions. Thus we arrive at the following problem.

Let G_i, $i=1,\dots,N$ be plane domains, $G_{ij}=G_i\cap G_j$, $G_{ijk}=G_i\cap G_j\cap\cap G_k$. Suppose the weight functions $w_{ij}\in H^\infty(G_{ij})$ are given. **Can we find** $g_{ij}\not\equiv 0$ belonging to some class of functions analytic in G_{ij} such that for all i,j,k the cocycle condition (3.1) holds?

$$w_{ij}g_{ij}+w_{jk}g_{jk}+w_{ki}g_{ki}\equiv 0 \quad \text{in } G_{ijk}. \tag{3.1}$$

The family $\{g_{ij}\}$ satisfying (3.1) will be called a weighted cocycle. For our purpose the requirement $g_{ij}\not\equiv 0$ is not enough, however, and should be replaced with a stronger condition. Now we state results obtained in this direction.

By $A(\Omega)$, $A^1(\Omega)$, $\Lambda^\alpha(\Omega)$ denote the spaces of functions analytic in Ω and respectively continuous, having continuous derivative and satisfying the Lipschitz condition of order α in $\operatorname{clos}\Omega$. Let $\mathcal{Y}^\alpha(\Omega)$ denote the class of functions $f(z)$ representable as $f(z)=g(z)\prod_{i=1}^{m}\log^{k_i}(z-z_i)$, $k_i\geqslant 0$, where

$g \in \Lambda^{\alpha}(\Omega)$ and z_i are arbitrary points on $\partial\Omega$. The branches of the logarithm are supposed to be analytic single-valued in Ω. By a "nearly outer" function we mean the product of an outer function (see 4.4 below for the definition) and a polynomial.

3.3. THEOREM [56]. Let G_i be Jordan domains such that ∂G_i is C^2-smooth for all i. Suppose that the weight functions $w_{ij} \in A^1(G_{ij})$ satisfy

$$|w_{ij}(\zeta)| + |w_{jk}(\zeta)| > \delta > 0 \qquad \text{in } G_{ijk},\ \forall i \neq j \neq k. \tag{3.2}$$

Then there exist an $\alpha > 0$ and nearly outer functions $g_{ij} \in \Lambda^{\alpha}(G_{ij})$ such that the cocycle condition (3.1) holds.

Question 1. Can the functions g_{ij} in the preceding theorem be found outer?

If the answer is affirmative then in Theorem 2.6 one can find the cyclic set in the form $U = \{\mathbb{1}, u, \ldots, u^{n-1}\}$. Theorem 3.3 enables us to choose such a set to be "nearly cyclic".

3.4. DEFINITION. Suppose T acts on a space $\mathfrak{X}$. A set $U \subset \mathfrak{X}$ is called nearly cyclic for T if $\operatorname{span}(T^n U : n \geq 0)$ has finite codimension in $\mathfrak{X}$.

Now we state another question which looks elementary but nevertheless we have been unable to solve it. An affirmative answer to Question 1 seems to imply that the answer to Question 2 is also yes.

Question 2. Let (R, φ) be an admissible pair (on a Riemann surface). Is there a function $\psi \in A(R)$ such that $\{\varphi, \psi\}$ generate $A(R)$ as a Banach algebra? (A related question: is there a ψ such that the map $(\varphi, \psi) : R \to \mathbb{C}^2$ is an imbedding?)

Since we have not been able to answer Question 1, to prove the Main Theorem, we have constructed a family of cocycles for a given weight satisfying a condition of common nondegeneracy.

3.5. THEOREM. Let G_i, $i = 1, \ldots, N$, be as in Theorem 3.3 and let the functions $w_{ij} \in A^1(G_{ij})$ satisfy condition (3.2). Denote by n the maximal number of domains G_i having nonempty joint intersection. Then for some $\alpha > 0$ there exist functions $g_{ij}^{[r]} \in \mathcal{J}^{\alpha}(G_{ij})$, $r = 1, \dots, n-1$, having the cocycle property (3.1) for every r. Moreover, for any subset $\beta \subset \{1, \dots, N\}$, $\operatorname{card} \beta = k$, such that $G_\beta \overset{def}{=} \bigcap_{i \in \beta} G_i \neq \emptyset$, the matrix-valued function

$$\zeta \mapsto \left[g_{ij}^{[r]}(\zeta) \right]_{i<j;\, i,j \in \beta}^{1 \leq r \leq n-1}$$

is outer in G_β (which implies that its rank is maximal possible in G_β).

In fact we need a more precise result (Theorem 12.3) which is too long to be stated here. The main instrument in the proof of all these theorems is a construction of an analytic vector bundle and its trivialization. The latter reduces to a special Riemann-Hilbert matrix problem. We don't know to what extent the condition (3.2) is necessary. Note only that equations (3.1) under condition (3.2) have some resemblance with the Corona problem (see [19]).

Studying the matrix generalization we have been somewhat surprised to discover the following result

3.6. LEMMA. Let T_φ be an AT operator on $\mathcal{X} = E^p(\mathcal{R})$ or $H^\infty(\mathcal{R})$ for an admissible pair $(\mathcal{R}, \varphi)$. Suppose $\mathcal{Y} \subset \mathcal{X}$ is a T_φ-invariant subspace of finite codimension. Then $\mu(T_\varphi) = \mu(T_\varphi | \mathcal{Y})$.

The lemma is proved in [57] in the present volume.

Thus if we have a nearly cyclic set for T_φ consisting of n elements, then $\mu_\varphi \leq n$. Theorem 3.3 implies the result on multiplicity. However we have preserved the scheme of our paper because it has some advantages. First it may be useful to have conditions which guarantee that a set is cyclic and not only nearly cyclic. Also we produce a cyclic set more explicitly than it is done with the help of Lemma 3.6. As for Theorem 3.5 it has a more ultimate form than Theorem 3.3 and seems to be interesting in itself. One of its possible applications concerns the description of the invariant subspace lattice of an AT operator [63]. For the reader interested only in the computation of μ_φ we indicate the parts which are not necessary for the proof of the Main Theorem.

CHAPTER II. THE BEGINNING OF THE PROOF OF THE MAIN THEOREM.

4. TECHNICAL BACKGROUND I.

Here we recall well-known facts on Smirnov and Nevanlinna classes, harmonic measure and inner-outer factorization. We consider their Riemann surface analogues in a form convenient to us. Then some terminology is introduced and the structure of belts is discussed.

4.1. SMIRNOV CLASSES E^p. Let Ω be a finitely connected plane domain. A function f analytic in Ω is said to belong to the class $E^p(\Omega)$, $p>0$, (Smirnov class), if there exists a sequence of domains $\{\Omega_i\}$ bounded by a finite number of rectifiable curves such that $\Omega_i \uparrow \Omega$ and

$$\overline{\lim_{i\to\infty}} \int_{\partial\Omega_i} |f(z)|^p \, |dz| < +\infty \ . \tag{4.1}$$

Smirnov classes are treated in [20], [14]. The theory for multiply connected domains was developed by G.Tumarkin and S.Khavinson [50].

If $\partial\Omega$ is rectifiable, then angular boundary values exist a.e. on $\partial\Omega$ for any function $f\in E^p(\Omega)$, $p>0$. The boundary function, also denoted by f, belongs to $L^p(\partial\Omega)$. The structure of a Banach space for $p\geqslant 1$, a Hilbert space for $p=2$, and a linear topological space for $p<1$ is induced on $E^p(\Omega)$ from $L^p(\partial\Omega)$.

The algebra $H^\infty(\Omega)$ consists of all bounded analytic functions in Ω. We shall endow this space with the weak-$*$ topology, i.e. the topology determined by the seminorms

$f\mapsto \left| \int_{\partial\Omega} fg|dz| \right|$, where $g\in L^1(\partial\Omega)$ ($\partial\Omega$ is supposed to be rectifiable).

4.2. THE NEVANLINNA CLASS $N(\Omega)$ can be defined as follows:

$$N(\Omega)=\{g_1/g_2 : g_i\in H^\infty(\Omega)\}.$$

It is known [50] that

$$N(\Omega)\supset E^p(\Omega),\ p>0. \tag{4.2}$$

4.3. HARMONIC MEASURE. Let G be a simply connected plane domain with rectifiable boundary and $\theta:\mathbb{D}\to G$ the conformal mapping. By ω_G we denote the harmonic measure for G :

$$\omega_G(e)=\ell(\theta^{-1}e),\quad e\subset\partial G,$$

where $\ell(\cdot)$ denotes the arc length. Recall that [51] for $h\in L^p(\partial G)$

$$\exists f\in E^p(G): |f| = |g| \quad \text{a.e. on } \partial G \iff \int_{\partial G}|\log|h||\,d\omega_G<+\infty. \tag{4.3}$$

4.4. INNER-OUTER FACTORIZATION. Let G be a simply connected domain and let $\theta:\mathbb{D}\to G$ be the conformal mapping. A function f is said to be outer in G if $f\circ\theta$ is outer in the unit disc. If G belongs to the class (S) , i.e. if θ' is outer, then every function $f\in E^p(G)$, $p>0$, can be written in the form $f=f^{in}f^{out}$. Here f^{out} is outer in G and f^{in} is inner in G , which means that $f^{in}\in H^\infty(G)$, $|f^{in}|=1$ a.e. on ∂G . The inner functions arithmetic is transferred from the unit disc by the conformal mapping. It follows from [36, III.15] that a domain with piecewise-smooth boundary, so an admissible simply connected domain, belongs to the class (S) . Other conditions which ensure that $G\in(S)$ can be found in [11], [64].

4.5. E^p SPACES ON A RIEMANN SURFACE. Let R be an admissible domain in a compact Riemann surface with boundary $\mathcal{K}$ (see 2.1). Consider $L^p(\partial R)$, $0<p<\infty$, bearing in mind that $L^\infty(\partial R)$ is equipped with the weak-$*$ topology. One can define the following spaces (as usual, we do not distinguish between a function analytic in R and its boundary function):

1) $E^p(R)$, $0<p<\infty$, is the set of functions f analytic

in R such that for any $z \in \partial R$ there exists a neighbourhood U_z and a conformal mapping $\theta : \mathrm{clos}\, U_z \to \mathbb{C}$ onto a plane domain, having the property

$$f \circ \theta^{-1} \in E^p(\theta(R \cap U_z)).$$

2) $E^p_{(1)}(R)$ is the closure in $L^p(\partial R)$, $1 \le p \le \infty$, of the set of functions analytic on $\mathrm{clos}\, R$.

3) $E^p_{(2)}(R)$, $1 \le p \le \infty$, is the subspace of $L^p(\partial R)$ consisting of functions f such that $\int_{\partial R} f\, d\omega = 0$ for any differential analytic on $\mathrm{clos}\, R$.

4.6. LEMMA. $E^p(R) = E^p_{(1)}(R) = E^p_{(2)}(R)$, $1 \le p \le \infty$.

The proof is outlined in Appendix.

By a theorem of Behnke and Stein (see [16, § 25]) the following result is an immediate consequence of Lemma 4.6.

4.7. LEMMA. <u>Let</u> R <u>be an admissible domain in a compact Riemann surface with boundary</u> $\mathcal{K}$. <u>Fix a point in each component of</u> $\mathrm{int}\,\mathcal{K} \setminus \mathrm{clos}\, R$. <u>Then the set of all functions analytic in</u> $\mathrm{int}\,\mathcal{K} \setminus \{\text{fixed points}\}$ <u>is dense in</u> $E^p(R)$, $1 \le p \le \infty$.

4.8. LEMMA. <u>Let</u> R <u>be an admissible domain and let</u> τ <u>be a function analytic on</u> $\mathrm{clos}\, R$ <u>such that</u> $\tau | R \ne 0$. <u>Then</u> $\tau E^p(R)$ <u>is dense in</u> $E^p(R)$, $1 \le p \le \infty$, <u>i.e.</u> τ <u>is "outer"</u>.

The proof is given in Appendix.

4.9. ONE-DIMENSIONAL ANALYTIC SETS. TERMINOLOGY. Throughout the paper we shall deal with plane sets which are finite unions of analytic arcs. Let Γ be of this type. By an arc we always mean a Jordan arc i.e. the image of a closed interval under a one-to-one continuous mapping. The corresponding image of an open interval will also be called an arc.

The complement $\mathbb{C} \setminus \Gamma$ consists of a finite number of components (here the fact is used that Γ is piecewise analytic). These components are simply connected if Γ is connected. One can choose a finite subset $\mathcal{A} \subset \Gamma$ such that all components of

$\Gamma \setminus \mathcal{A}$ are analytic Jordan arcs. These arcs will be referred to as the <u>arcs of</u> Γ and the points in $\mathcal{A}$ will be called the <u>vertices of</u> Γ . Naturally the choice of $\mathcal{A}$ is not unique and one can always add a finite set to $\mathcal{A}$.

Let $z \in \mathcal{A}$ and let U_z be a disc centered at z . If U_z is small enough, it is divided by Γ into curvilinear sectors. Such a disc will be called a <u>regular neighbourhood of</u> z . By a <u>curvilinear sector</u> (or just by a sector) we mean a Jordan domain bounded by an arc of a circle and two arcs ending at its center. The latter will be called the <u>radial arcs</u>. We shall not distinguish between sectors arising from different regular neighbourhood of a vertex.

Now we pass from familiar subjects to our belts (see Definition 2.4), discuss their structure and give an equivalent statement of the "width" property.

4.10. THE STRUCTURE OF BELTS. TERMINOLOGY. Let B be a belt. Recall that $B = \operatorname{clos} V_B \setminus G_B$ where $V_B \supset G_B$ and V_B, G_B are simply connected admissible plane domains. Since the curves $\partial V_B, \partial G_B$ are piecewise analytic, $\partial V_B \cap \partial G_B$ consists of arcs and isolated points. Endpoints of these arcs and these isolated points will be called the <u>vertices of the belt</u>. If the intersection $\partial V_B \cap \partial G_B$ contains an arc, the belt will be called supernarrow (then $\partial G_B \not\subset \partial(\operatorname{int} B)$, see 2.4).

Evidently the set $B \setminus \operatorname{clos}(\operatorname{int} B)$ is the union of open arcs of $\partial V_B \cap \partial G_B$. These arcs and also the closures of components of $\operatorname{int} B$ will be referred to as <u>links of the belt</u>. Let W be a link of the belt B . Apparently W and G_B have a common boundary arc. If $\ell(\partial W \cap \partial V_B) > 0$ the link will be called <u>principal</u> and otherwise <u>additional</u>. Removing an additional link from the belt we get another belt with the same exterior $\mathbb{C} \setminus \operatorname{clos} V_B$. These notions are illustrated in Fig.7. The depicted belt has vertices z_1, z_2, z_3, z_4 . It has four principal links and two additional links K, L . This belt is supernarrow due to the link (z_2, z_3) .

Let z be a vertex of the belt B and let U_z be its re-

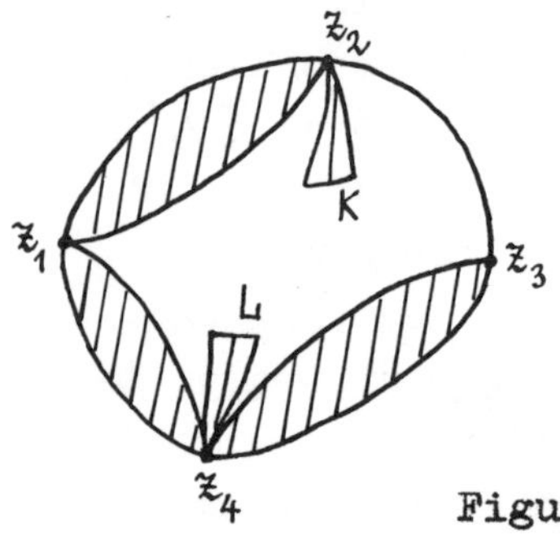

Figure 7

gular neighbourhood, i.e. a disc centered at z divided into curvilinear sectors by arcs of ∂B. The sectors lying in B will be referred to as <u>belt sectors</u>.

Now we want to state the "width" property of the belt in terms of its sectors. To do this we have to compare sectors with analytic radial arcs. The following two lemmas follow easily from Warschawskii's asymptotics [59].

4.11. LEMMA. <u>Let S be a curvilinear sector with vertex 0_S whose boundary arcs are analytic. Then there exists an increasing smooth function defined on $(0,1]$ such that</u>

$$\omega_S([0_S,\zeta]) \sim \Psi_S(|\zeta - 0_S|), \quad \zeta \in \partial S, \quad \zeta \to 0_S ,$$

<u>where ω_S is the harmonic measure for S</u>.

4.12. DEFINITION. We say that two curvilinear sectors S, T with analytic arcs are equivalent if $\Psi_S \sim \Psi_T$ at the zero. On the set of all such sectors we introduce the following ordering: $S \prec T$ if $\Psi_S = o(\Psi_T)$ as $\zeta \to 0$. A simple analytic arc γ will be regarded as a degenerate sector, and we assume that $\gamma \prec S$ for any ordinary sector. We write $S \precsim T$ if $S \prec T$ or S is equivalent to T.

Note that if at least one of the sectors has nonzero angle, the ordering introduced above coincides with the ordering by the value of their angles.

4.13. LEMMA. <u>The functions Ψ_S possess the following properties</u>:

1) <u>if</u> $S \prec T$ <u>then</u> $\Psi_S'(r) = o(\Psi_T'(r)), \quad r \to 0$;
2) <u>for any two sectors</u> S, T <u>either</u> $S \precsim T$ <u>or</u> $T \precsim S$.

Now Definition 2.4 readily implies the following lemma.

4.14. LEMMA. <u>A belt B is wide if and only if for any vertex</u> z

of B and for any two neighbouring belt sectors S_1, S_2 at this vertex, the sector $T \subset G_B$ which lies between them, satisfies $T \not\leq S_1$, $T \not\leq S_2$.

5. ESTIMATES OF MULTIPLICITY FROM BELOW.

5.0. Propositions 5.1 and 5.6 below contain the estimates of μ_φ from below in the Main Theorem 2.6. To prove them we present some more or less general conditions necessary for the cyclicity.

The estimate $\mu(T) \geq n(T)$ is well-known (see e.g., [22], [30]). Corollary 5.5. claims that under certain conditions $\mu(T) \geq n(T)+1$, if the set $\operatorname{clos} \Omega(T)$ contains a wide belt. The methods employed resemble those of the papers by D.Herrero [22] and D.Herrero, L.Rodman [25].

The following result for the case $R = \mathbb{D}$ is contained implicitly in the paper by N.Nikol'skii [30].

5.1. PROPOSITION. Let (R, φ) be an admissible pair. Then the multiplicity of the operator $T_\varphi : f \mapsto \varphi f$ acting on $E^2(R)$ is at least the maximal essential valency $n(\varphi)$.

REMARK. Since $H^\infty(R)$ is densely imbedded into $E^2(R)$, we obtain that $\mu_\varphi \geq n(\varphi)$. It is not hard to prove a similar estimate for $E^p(R)$, $p \geq 1$.

PROOF. Put $n = n(\varphi)$. By the definitions of the maximal essential valency and the admissible pair we see that there exists a Jordan arc γ whose pre-image under φ in $\operatorname{clos} R$ is a union of n arcs. Passing to a subarc one can assume that $\varphi^{-1}(\gamma) = K = \bigcup_{i=1}^{n} e_i$ where the arcs e_i are pairwise disjoint and either $e_i \subset \partial R$ or $e_i \subset \operatorname{int} R$. Clearly $E^2(R)$ is imbedded into $L^2(K)$. Let us show that $E^2(R)$ is dense in $L^2(K)$. To this end choose small topological discs U_i such that $U_i \supset e_i$, $U_i \setminus e_i$ is connected for all i , and the closures $\operatorname{clos} U_i$ are pairwise disjoint. Functions analytic in a neighbourhood of $\operatorname{clos} R$ are dense in $H^\infty(\bigcup_{i=1}^{n} U_i)$ by Lemma 4.7. Next $H^\infty(U_i)$ is dense in $C(K)$ by

Mergelyan's (even Lavrentiev's) Theorem. Thus $E^2(R)$ is densely imbedded in $L^2(K)$ and $\mu(T_\varphi)$ is at least the multiplicity of the operator $f \mapsto \varphi f$ acting on $L^2(K)$. But the last operator is normal; its local spectral multiplicity function is constant on K and equal to n. So $\mu(T_\varphi) \geq n$, and the proposition is proved. ▨

Throughout the paper we shall make use of the following well-known condition necessary for the cyclicity. Let $\mathfrak{X}$ be a locally convex space and let $\mathfrak{X}^*$ be its dual. For a subset $U \subset \mathfrak{X}$ let $U^\perp$ be its annihilator $\{ y \in \mathfrak{X}^* : y(x) = 0,\ \forall x \in U \}$.

5.2. THE LOCAL NECESSARY CONDITION FOR THE CYCLICITY (THE LOCAL CONDITION). Let T be a continuous operator on a locally convex space $\mathfrak{X}$, and let $U \subset \mathfrak{X}$ be a cyclic set for T. Then for all $\lambda \in \mathbb{C}$,

$$\operatorname{Ker}(T^* - \lambda I) \cap U^\perp = \{0\}. \tag{5.1}$$

PROOF. If y belongs to the left-hand side of (5.1), then $y(T^n U) = 0$, $n \geq 0$, and the cyclic property implies $y = 0$. ▨

REMARK. If $\mathfrak{X}$ is a finite-dimensional space, the local condition is also sufficient for the cyclicity. Below we need this and a slightly more general statement, see Lemma 11.7.

Mostly one uses the local condition in the following form. Let $x_\lambda^{(1)}, \dots, x_\lambda^{(n)}$, $n = \nu_T(\lambda)$, be a basis of the space $\operatorname{Ker}(T^* - \bar\lambda I)$. Then (5.1) holds if and only if

$$\operatorname{rank}\left[x_\lambda^{(1)}(u), \dots, x_\lambda^{(n)}(u) \right]_{u \in U} = n. \tag{5.2}$$

From (5.2) the estimate $\mu(T) \geq n(T) = \sup\{ \nu_T(\lambda) : \lambda \in \mathbb{C} \}$ is immediately seen.

Now let us specify the local condition for the AT operator T_φ acting on $\mathfrak{X}$ which is either $E^p(R)$ or $H^\infty(R)$ with the weak-* topology, for an admissible pair (R, φ). First observe that

$$\nu_{T_\varphi}(\lambda) = \dim \operatorname{Ker}(T_\varphi^* - \bar\lambda I) = \operatorname{codim}(\operatorname{clos}(\varphi - \lambda)\mathfrak{X}),$$

$$\nu_{T_\varphi}(\lambda) = \mathrm{card}(\varphi^{-1}(\lambda) \cap R) \tag{5.3}$$

counting multiplicities. The last equality follows from Lemma 4.8. Let $\varphi^{-1}(\lambda) \cap R = \{z_1, \dots, z_m\}$ and suppose that $d\varphi$ has a zero at z_j of multiplicity $k_j - 1$. Then fixing local coordinates at z_j, we obtain that the functionals $f \mapsto f^{(i)}(z_j)$, $i = 0, \dots, k_j - 1$, are linearly independent and belong to $\mathrm{Ker}(T_\varphi^* - \bar{\lambda} I)$. The equality (5.3) implies that they form a basis. Thus the condition (5.2) can be rewritten as follows: for all $\lambda \in \mathbb{C}$

$$\mathrm{rank}\left[u(z_1), \dots, u^{(k_1-1)}(z_1); \dots; u(z_m), \dots, u^{(k_m-1)}(z_m)\right]_{u \in U} = \sum_{i=1}^{m} k_i \,. \tag{5.4}$$

The following lemma also includes a general condition necessary for the cyclicity which sometimes enables one to prove that $\mu(T) \geq n(T) + 1$. Recall that $\Omega(T) = \{\lambda \in \mathbb{C} : \nu_T(\lambda) = n(T)\}$.

5.3. LEMMA. Let T be a bounded operator on a Banach space $\mathfrak{X}$ and suppose that $\mu(T) = n(T)$. Let $U = \{u_i\}_{i=1}^{n}$, $n = n(T)$, be a cyclic set for T. Let G be a simply connected domain with rectifiable boundary such that $G \not\subset \Omega(T)$ and $\partial G \setminus P \subset \Omega(T)$, the set P being of zero arc length. Suppose that $\varkappa_\lambda^{(1)}, \dots, \varkappa_\lambda^{(n)}$ is a basis of $\mathrm{Ker}(T^* - \bar{\lambda} I)$, $\lambda \in \partial G \setminus P$ such that the functions $\lambda \mapsto \varkappa_\lambda^{(j)}$ belong to the space $L^\infty(\partial G, \mathfrak{X}^*)$ of vector-valued functions. Then

$$\int_{\partial G} \log \left| \det \left[\varkappa_\lambda^{(j)}(u_i)\right]_{i \leq n}^{j \leq n} \right| d\omega_G = -\infty, \tag{5.5}$$

where ω_G is the harmonic measure for G.

PROOF. Set $\Psi(\lambda) = \det\left[\varkappa_\lambda^{(j)}(u_i)\right]_{i \leq n}^{j \leq n} \in L^\infty(\partial G)$. Consider the following functionals defined and bounded on $\mathfrak{X}$:

$$\tau_\lambda^{(s)}(f) = \det\Big[\varkappa_\lambda^{(j)}(u_1), \dots, \overbrace{\varkappa_\lambda^{(j)}(f)}^{s}, \dots, \varkappa_\lambda^{(j)}(u_n)\Big]^{j \leq n}, \quad \lambda \in \partial G \setminus P \,. \tag{5.6}$$

Then $\tau_\lambda^{(s)}(Tf) = \lambda\tau_\lambda^{(s)}$, $f \in \mathfrak{X}$; $\tau_\lambda^{(s)}(u_i) \neq 0$, $i \neq s$; $\tau_\lambda^{(s)}(u_s) = \Psi(\lambda)$.

If the condition (5.5) fails one can apply (4.3) and choose a function $h \in L^\infty(\partial G)$ such that the product $h\Psi$ can be extended to a bounded outer function in G. Fix a point $\mu \in G \setminus \Omega(T)$ and put

$$\chi^{(s)}(f) = \frac{1}{2\pi i}\int_{\partial G} \frac{\tau_\lambda^{(s)}(f)\, h(\lambda)}{\lambda - \mu}\, d\lambda .$$

Clearly $\chi^{(s)} \in \mathfrak{X}^*$, $\chi^{(s)}(T^n u_i) = 0$, $i \neq s$, and

$$\chi^{(s)}(T^n(u_s) = \frac{1}{2\pi i}\int_{\partial G} \frac{\lambda^n \Psi(\lambda)\, h(\lambda)}{\lambda - \mu}\, d\lambda = \mu^n (h\Psi)(\mu) = \mu^n \chi^{(s)}(u_s) .$$

Since U is cyclic, $\chi^{(s)} \in Ker(T^* - \bar{\mu} I)$. The functionals $\chi^{(1)}, \dots, \chi^{(n)}$ are linearly independent because $(h\Psi)(\mu) \neq 0$. Thus $\dim Ker(T^* - \bar{\mu} I) \geqslant n$ which leads to a contradiction. ▨

The following result seems to bear resemblance with a theorem of D.Herrero [22], [24], which claims that if $\mu(T) = n$, then all components of the set $\{\lambda : T - \lambda I$ is Fredholm, $ind(T - \lambda I) = -n\}$ are simply connected.

5.4. COROLLARY. <u>Let</u> T <u>be a bounded operator on a Banach space. Suppose that</u> γ, $\gamma \subset \Omega(T)$, <u>is a rectifiable Jordan contour non-contractible in</u> $\Omega(T)$, <u>and there exists a basis of</u> $Ker(T^* - \bar{\lambda} I)$, $\lambda \in \gamma$, <u>piecewise continuously depending on</u> λ. <u>Then</u> $\mu(T) \geqslant n(T) + 1$.

PROOF. Let G be the domain interior to γ. Suppose that $\mu(T) = n(T) = n$. If the set $U = \{u_i\}_{i=1}^n$ is cyclic, the determinant $\Psi(\lambda)$ in (5.5) does not vanish on $\gamma = \partial G$ by the condition (5.2). Since $\Psi(\lambda)$ is piecewise continuous, this function is bounded away from zero, so (5.5) fails. According to Lemma 5.3, $G \subset \Omega(T)$, which implies that γ is contractible in $\Omega(T)$. The contradiction concludes the proof. ▨

5.5. COROLLARY. <u>Let</u> T <u>be a bounded operator on a Banach space</u>

$\mathfrak{X}$. <u>Suppose that there exists a wide belt</u> B <u>such that</u> $int\,B \subset \Omega(T)$ <u>and the "interior component of its complement"</u> $G_B : G_B \not\subset \Omega(T)$. <u>Suppose also that there is a basis</u> $x_\lambda^{(1)}, \dots, x_\lambda^{(n)}$, $n = n(T)$, <u>of the space</u> $Ker(T^* - \bar{\lambda} I)$, $\lambda \in int\,B$, <u>such that the vector-valued functions</u> $\lambda \mapsto x_\lambda^{(j)}(f)$ <u>belong to</u> $H^\infty(int\,B, \mathfrak{X}^*)$. <u>Then</u> $\mu(T) \geq n(T) + 1$.

PROOF. Let $B = clos\,V_B \setminus G_B$ and let $\mathcal{A}$ be the set of all vertices of the belt B . We slightly enlarge G_B so that the resulting domain G is simply connected, $\mathcal{A} \subset \partial G$, $\partial G \setminus \mathcal{A} \subset int\,B$, and all arcs of ∂G are tangent with corresponding arcs of ∂G_B at the vertices (see Fig.8). The set $B' = clos\,V_B \setminus G$ is also a belt. Clearly the corresponding sectors of B and B' are equivalent in the sense of Definition 4.12 if the order of tangency is high enough. Suppose that $\mu(T) = n(T)$ and let $U = \{u_i\}_{i=1}^n$, $n = n(T)$, be a cyclic set. The function

$$\Psi(\lambda) = det\,[x_\lambda^{(j)}(u_i)]_{i,j \leq n}$$

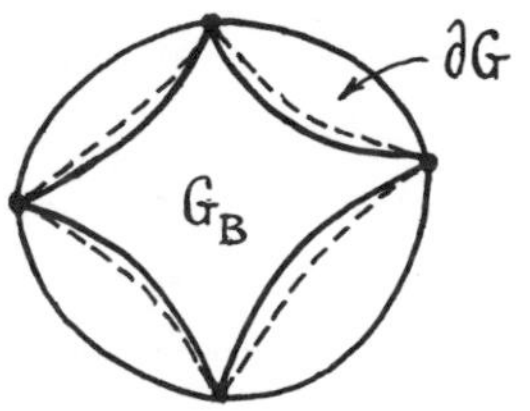

Figure 8

belongs to $H^\infty(int\,B')$ and is bounded away from zero outside any neighbourhood of $\mathcal{A}$ according to (5.2). In a neighbourhood of $\mathcal{A}$, $log|\Psi(\lambda)|$ is integrable with respect to the harmonic measure for the components of $int\,B'$. By the property (2.1) it is also integrable with respect to the measure ω_G . But it contradicts Lemma 5.3 and the proof is complete. ▨

Now let us return to analytic Toeplitz operators.

5.6. PROPOSITION. <u>Let</u> (R, φ) <u>be an admissible pair and suppose that there exists a wide belt</u> B <u>subordinate to the set of maximal essential valency</u> $V(\varphi)$. <u>Then for each</u> p , $1 \leq p \leq \infty$, <u>the multiplicity of</u> $T_\varphi : f \mapsto \varphi f$ <u>acting on</u> $E^p(R)$, <u>is at least</u> $n(\varphi) + 1$.

PROOF. Put $n = n(\varphi)$, $T = T_\varphi$. Since a wide belt has nonempty interior we have $n(\varphi) = n(T)$, $int\,V(\varphi) = \Omega(T)$. Let $\mathcal{A}$ be the set of vertices of the belt B. It is not hard to construct a new belt B' coinciding with B in a neighbourhood of $\mathcal{A}$, having no vertices except $\mathcal{A}$, and such that $\operatorname{int} B' \cap Z = \emptyset$, $Z \stackrel{def}{=} \varphi(\{\zeta \in \operatorname{clos} R : d\varphi(\zeta) = 0\})$. Clearly B' is also a wide belt subordinate to $V(\varphi)$. The belt B' must be further diminished in the same manner as in the proof of Corollary 5.5. The resulting belt B'' should have the same set of vertices $\mathcal{A}$, and satisfy $\partial B'' \setminus \mathcal{A} \subset \operatorname{int} B$. The corresponding arcs of $\partial B'$ and $\partial B''$ should have tangency of sufficiently high order so that B'' would also be wide. We are going to apply Corollary 5.5. The condition $G_{B''} \not\subset \Omega(T)$ follows from the definition of a subordinate belt (see 2.5). It remains to choose an appropriate basis of $\operatorname{Ker}(T^* - \bar{\lambda} I)$.

It is sufficient to consider the operator T_φ on $E^1(R)$, because all $E^p(R)$, $p > 1$, are densely contained in $E^1(R)$. Assume first that $\operatorname{int} B'$ is not simply connected. Then choose a Jordan contour $\gamma \subset \operatorname{int} B'$ noncontractible in $\operatorname{int} B$. By the definition of a belt the interior of γ intersects $\mathbb{C} \setminus V(\varphi)$ One can choose piecewise continuous branches $\varphi^{-1}_{(1)}, \dots, \varphi^{-1}_{(n)}$ of φ^{-1} on γ. Now we can put $\varkappa^{(j)}_\lambda(f) = (f \circ \varphi^{-1}_{(j)})(\lambda)$, $\lambda \in \gamma$. Clearly this is a suitable basis of $\operatorname{Ker}(T^* - \bar{\lambda} I)$, and the desired estimate follows from Corollary 5.4.

Suppose now that all components of $\operatorname{int} B'$ are simply connected. Since $\operatorname{int} B' \cap Z = \emptyset$, we may choose analytic single-valued branches $\varphi^{-1}_{(1)}, \dots, \varphi^{-1}_{(n)}$ of φ^{-1} defined in $\operatorname{int} B'$. Let m be the maximal order of tangency of $\partial B''$ and $\partial B'$. Put $q(\lambda) = \prod_{z_k \in \mathcal{A}} (\lambda - z_k)^m$ and set

$$\varkappa^{(j)}_\lambda(f) = (f \circ \varphi^{-1}_{(j)})(\lambda)\, q(\lambda).$$

Clearly, these functionals form a basis of $\operatorname{Ker}(T^* - \bar{\lambda} I)$, $\lambda \in \operatorname{int} B''$. In order to apply Corollary 5.5 we shall verify that

$$|(f \circ \varphi^{-1}_{(j)})(\lambda)| \leq c\, |q(\lambda)|^{-1}, \quad \lambda \in \operatorname{int} B'', \ f \in E^1(R), \ \|f\|_1 = 1.$$

Indeed, fix $z \in \mathcal{A}$, a regular neighbourhood $U_z \ni z$, and a local coordinate at the point $\varphi^{-1}_{(j)}(z)$. We have

$$(f\circ\varphi_{(j)}^{-1})(\varphi_{(j)}^{-1})'\in E^1(\mathrm{int}\,B'\cap U_z).$$

For $g\in E^1(\mathrm{int}\,B'\cap U_z)$, $\|g\|_1=1$, the following estimate holds

$$|g(\lambda)|\leqslant c_1|\,\mathrm{dist}(\lambda,\partial B')|^{-1}\leqslant c_2|\lambda-z|^{-m},\ \lambda\in\mathrm{int}\,B''\cap U_z\,.$$

On the other hand $|(\varphi_{(i)}^{-1})'(\lambda)|\geqslant \mathrm{const}$ since $d\varphi$ is analytic in a neighbourhood of $\mathrm{clos}\,R$. This implies the desired estimate. ▨

REMARKS. 1) If $R\subset\mathbb{C}$, the requirement that φ should be analytic on $\mathrm{clos}\,R$ (see the definition of an admissible pair, 2.1) can be weakened. Namely, in Proposition 5.6 it is sufficient that $\varphi\in A(R)$ and the image $\varphi(\partial G)$ is piecewise analytic. Thus we cover Example 2.16.

2) It is plausible that Corollary 5.5 may be applied to some nonanalytic Toeplitz operators.

6. THE CONSTRUCTION OF A CYCLIC SET: SIMPLE CASES

Here we produce a cyclic set for T_φ having $n(\varphi)$ elements in the case of a generic function and thus provide an independent proof of Corollary 2.8. Besides we indicate how to choose a cyclic set in the case when $\mathbb{C}\setminus V(\varphi)$ is connected (Corollary 2.7), though the proof of its cyclicity follows only from the considerations of Sect.11. To do this we construct a special imbedding of the set R into a Riemann surface that will also be useful later.

6.1. LEMMA. (see [39] for the case of the disc). Let (R,φ) be an admissible pair, $n=n(\varphi)$ and let B be the disc such that $\varphi(\mathrm{clos}\,R)\subset B=\mathrm{int}\,B$. Then there exist a compact Riemann surface $S\cup\partial S$, $\partial S\neq\emptyset$, a univalent analytic mapping $\omega:R\to S$, and an analytic n-sheeted branched covering $\pi:S\to B$ such that $\varphi=\pi\circ\omega$, and the set $\omega(R)$ is admissible.

6.2. DEFINITION. We shall say that an admissible pair (R,φ) is a <u>generic pair</u> if the boundary ∂R is the union of disjoint

Jordan contours, and the set $\varphi(\partial R)$ has a finite number of simple transversal self-intersection points. The latter means that for some finite set $\mathcal{P} \subset \partial R$ the restriction $\varphi | \partial R \setminus \mathcal{P}$ is one-to-one. Moreover if $z \in \varphi(\mathcal{P})$ then $\varphi^{-1}(z) = \{\zeta_1, \zeta_2\}$, $d\varphi(\zeta_1) \neq 0$, $d\varphi(\zeta_2) \neq 0$, and the angle between two arcs of $\varphi(\partial R)$ at the point z is nonzero.

6.3. A SUPPLEMENT TO LEMMA 6.1. <u>If</u> (R,φ) <u>is a generic pair, then the imbedding</u> ω <u>is Lemma 6.1 can be chosen to have a continuous and one-to-one extension to the closure of</u> R.

Before we prove these statements let us give their application. The following trivial result will be needed.

6.4. LEMMA. <u>Let</u> R, R_1 <u>be open sets and let</u> $\omega: R_1 \to R$ <u>be the conformal mapping</u>, $\varphi \in H^\infty(R)$. <u>Then</u> $\mu_\varphi = \mu_{\varphi \circ \omega}$. <u>The set</u> $X \subset H^\infty(R)$ <u>is cyclic for</u> T_φ <u>if and only if</u> $X \circ \omega$ <u>is cyclic for</u> $T_{\varphi \circ \omega}$ <u>acting on</u> $H^\infty(R_1)$.

The assertion follows immediately from the conformal invariance of $H^\infty(R)$. ▨

6.5. A CANDIDATE FOR A CYCLIC SET. Let (R,φ) be an admissible pair, $n = n(\varphi)$. Lemma 6.1 is applied. By replacing the pair with a conformally equivalent one (see Lemma 6.4), we can assume that $R \subset S$, $\varphi = \pi | R$ and $\pi : S \to \mathbb{D}$ is an n-sheeted branched covering. The candidate for a cyclic set for T_φ acting on $H^\infty(R)$ is any cyclic set for T_π on $E^2(S)$ restricted to R. Let us show that the last operator has multiplicity n.

We have $|\pi | \partial S| = 1$, so the operator T_π on $E^2(S)$ is a Hilbert space isometry. It is easy to see that T_π is completely nonunitary. Indeed $\pi(z) = 0$ for some $z \in \operatorname{int} S$. If $f \in \bigcap_{n \geq 0} T_\pi^n E^2(S)$ then f should have a zero of infinite order at z. Finally, $\dim \operatorname{Ker} T_\pi^* = n$ (see (5.3)). According to Wold's Lemma (see [47]), T_π is unitarily equivalent to the n-dimensional unilateral shift, so its multiplicity is equal to n.

Looking at (5.4) we see that the Local Necessary Condition 5.2 preserves under the restriction to a subdomain. So it is satisfied by the restriction of a T_π-cyclic set to R. That was the (only) reason to call it the candidate for a cyclic set for T_φ.

6.6. THE GENERIC CASE, COMPONENTS OF $\mathcal{R}$ SIMPLY CONNECTED. Applying 6.3 and Lemma 6.4, one can assume that $\mathcal{R} \subset \mathcal{S}$ and all components of $\operatorname{clos} \mathcal{R}$ are simply connected. Then, according to Lemma 4.7 the space $E^2(\mathcal{S})$ is dense in $H^\infty(\mathcal{R})$. So the cyclic set for T_π restricted to $\mathcal{R}$ is cyclic for T_φ as well. Thus $\mu_\varphi = n(\varphi)$ and Corollary 2.8 is proved.

6.7. CYCLIC SETS IN THE CASE WHEN THERE ARE NO BELTS. Let $\{v_1, \dots, v_n\}$ be a cyclic set for T_π on $E^2(\mathcal{S})$. Choose $g_1, \dots, g_n$ to be analytic on $\operatorname{clos} \mathcal{R} \subset \mathcal{S}$ and having pairly disjoint essential singularities in each component of $\mathcal{S} \setminus \operatorname{clos} \mathcal{R}$. Then we can put $u_i = v_i + \varepsilon g_i$ for ε sufficiently small. The cyclicity of $\{u_i\}$ can be derived by the arguments of Sect.11.

6.8. THE GENERIC CASE. ANOTHER APPROACH. Applying Lemma 6.2 with its supplement, one can construct a cyclic set more explicitly. Recall that $A(\mathcal{R})$ denotes the algebra of functions analytic in $\mathcal{R}$ and continuous on $\operatorname{clos} \mathcal{R}$. The following proposition employs the method which was used in the proof of [39, Theorem 3].

6.9. PROPOSITION. <u>Let</u> $(\mathcal{R}, \varphi)$ <u>be a generic pair and let components of</u> $\mathcal{R}$ <u>be simply connected. Then there exists</u> $v \in A(\mathcal{R})$ <u>such that the set</u> $\{\mathbb{1}, v, \dots, v^{n-1}\}$, $n = n(\varphi)$, <u>is cyclic for the operator</u> $T_\varphi : f \mapsto \varphi f$ <u>on</u> $A(\mathcal{R})$.

PROOF. As in 6.6, we can assume that $\mathcal{R} \subset \mathcal{S}$, all components of $\operatorname{clos} \mathcal{R}$ are simply connected, and $\varphi = \pi | \mathcal{R}$. Let us show first that there exists an analytic function $v : \mathcal{S} \to \mathbb{C}$ such that $v | \operatorname{clos} \mathcal{R}$ is one-to-one.

Indeed, choose open sets G_1, G_2 with the properties: $\operatorname{clos} \mathcal{R} \subset G_1$, $\operatorname{clos} G_1 \subset G_2$, $\operatorname{clos} G_2 \subset \operatorname{int} \mathcal{S}$, and all components of G_1, $\operatorname{clos} G_2$ are simply connected. By the Riemann Theorem there exists h mapping G_2 conformally onto the disjoint union of discs in the plane $\mathbb{C}$. By the Behnke-Stein Theorem (see [16,

§ 25]), h can be approximated by functions analytic in S uniformly on compact sets in G_2. Let $\varepsilon < dist(h(\partial R), h(\partial G_1))$. It is easy to see that a function v analytic in S and such that $|v-h| \mid G_1 < \varepsilon$, is univalent on $clos R$. (We are grateful to A.E.Eremenko for the construction).

Let us now prove that $U = \{ 1, v, \dots, v^{n-1} \}$ is cyclic. Since all components of the set $W = v(clos R)$ are simply connected, the polynomials are dense in $A(int W)$ and hence, the set $\{v^i\}_{i=0}^{\infty}$ is dense in $A(R)$. Thus it is sufficient to verify that $v^n \in Span\{U\varphi^k : k \geq 0\}$.

Denote by $\sigma_i(\Lambda)$ the symmetric polynomial in $\Lambda = \{\lambda_1, \dots, \lambda_n\}$: $\sigma_i(\Lambda) = (-1)^i \sum_{k_j \uparrow} \lambda_{k_1} \cdot \ldots \cdot \lambda_{k_i}$. Put $h_i = \sigma_i(v \circ \pi^{-1} \circ \varphi)$. The functions h_i are well-defined on the set where $card\, \pi^{-1}(\varphi(z)) = n$, i.e. outside a finite set. Since the h_i are bounded, by the Removing Singularity Theorem one has $h_i \in A(R)$. Recalling that $\varphi = \pi | R$ we obtain:

$$v^n(z) = -\sum_{i=1}^{n} v^{n-i}(z) h_i(z), \quad z \in clos R .$$

It remains to observe that $h_i = (\sigma_i(v \circ \pi^{-1})) \circ \varphi$, where the functions $\sigma_i(v \circ \pi^{-1})$ belong to $A(B)$, $B = \varphi(S) \supset \varphi(clos R)$. Thus, $h_i \in Span\{\varphi^k : k \geq 0\}$, and the proof is complete. ▨

6.10. THE BEGINNING OF THE PROOF OF LEMMA 6.1. The key point of the proof is a special triangulation with an additional structure which will be used further as well.

6.11. TRIANGULATION. The set $\Gamma = \varphi(\partial R)$ divides the complex plane into a finite number of components. Let $\mathcal{A}$ be the set of vertices, i.e. all components of $\Gamma \setminus \mathcal{A}$ are Jordan arcs (arcs of Γ). Set $Z = \varphi(\{\zeta \in clos R : d\varphi(\zeta) = 0\})$. It is not hard to find a set $S\Gamma \supset \Gamma$ with the following properties:

$S\Gamma$ is the union of a finite number of analytic Jordan arcs and their endpoints (vertices of $S\Gamma$); (6.1)
each vertex is the endpoint of at least three arcs;

$$\mathcal{S}\Gamma \supset \Gamma \cup \partial B \cup Z \ ; \qquad \text{all points of } Z \text{ are vertices of } \mathcal{S}\Gamma; \tag{6.2}$$

if z is a vertex of Γ which is an endpoint of at least three arcs of Γ, then for some neighbourhood $U_z \ni z$, holds: $U_z \cap \Gamma = U_z \cap \mathcal{S}\Gamma$. (6.3)

The bounded components of $\mathbb{C} \setminus \mathcal{S}\Gamma$ will be called "cells". The following conditions must also be satisfied:

each cell is a Jordan domain whose boundary consists of exactly three arcs (curvilinear triangle); (6.4)

on the boundary of a cell no more than one vertex of Γ can lie. (6.5)

The conditions (6.1) and (6.5) imply that two cells can have at most one common boundary arc: such cells will be referred to as <u>adjacent</u>. For each cell Ω_t we define its multiplicity n_t to be equal to the number of φ-preimages inside Ω_t. We have $n = n(\varphi) \geq \max_t n_t$. In the cell Ω_t there exist n_t single-valued analytic branches $\varphi^{-1}_{(1)}, \dots, \varphi^{-1}_{(n_t)}$ of φ^{-1}. Their enumeration will be fixed. Put $\mathcal{N} = \{1, \dots, n\}$.

6.12. CORRESPONDENCE FUNCTIONS. Let Ω_s, Ω_t be adjacent cells. A permutation $\sigma_{st} : \mathcal{N} \to \mathcal{N}$ will be called a <u>correspondence function</u> if

for $i \leq n_s, j \leq n_t$, the equality $\sigma_{st} i = j$ holds if and only if the branch $\varphi^{-1}_{(i)} \mid \Omega_s$ extends analytically across $\partial\Omega_s \cap \partial\Omega_t$ to $\varphi^{-1}_{(j)} \mid \Omega_t$. (6.6)

We shall assume that $\sigma_{st} = \sigma_{ts}^{-1}$. Since the number of branches of φ^{-1} on an arc is at most the essential maximal valency n, the correspondence functions can be defined for all pairs of adjacent cells. In what follows we assume that the correspondence functions σ_{st} are fixed.

6.13. CONSTRUCTION OF THE SURFACE $\mathcal{S}$. Consider the disjoint union of all cells, n copies each. Then let us identify the arcs according to the correspondence functions δ_{st}. Namely, paste $\Omega_s^{(i)}$ and $\Omega_t^{(j)}$ along their common boundary if Ω_s, Ω_t are adjacent and $\delta_{st}\, i=j$. At first we delete the vertices of cells. Apparently we obtain a surface with the natural projection mapping π. It is an n-sheeted unbranched covering of the set $B\setminus\mathcal{M}$ where $\mathcal{M}$ is the set of vertices of $S\Gamma$. For $z\in\mathcal{M}$, let U_z be a small neighbourhood of z. We add a point projecting into z to each component of $\pi^{-1}(U_z\setminus\{z\})$. Denote by $\mathcal{S}$ the resulting surface. Clearly $\pi:\mathcal{S}\to B$ is an n-sheeted branched covering. The conformal structure is defined in $\mathcal{S}$ in a standard fashion so that π becomes analytic.

6.14. THE IMBEDDING ω. THE END OF THE PROOF OF LEMMA 6.1. The mapping $\omega:\mathcal{R}\to\mathcal{S}$ is defined as follows: if $\varphi(z)\in\Omega_t$, $\varphi_{(i)}^{-1}(\varphi(z))= = z$, put $\omega(z)$ to be the point $\varphi(z)$ in the copy $\Omega_t^{(i)}$. By the property (6.6) ω is well-defined on the pre-images of arcs and vertices. Clearly, ω is one-to-one in $\mathcal{R}$. To complete the proof it remains to check that $\omega(\mathcal{R})$ is an admissible set. It follows from the "only if" part of the condition (6.6) that no two arcs of $\partial\mathcal{R}$ can be mapped onto one arc by ω. Hence $\omega(\mathcal{R})=\operatorname{int}(\operatorname{clos}\omega(\mathcal{R}))$ and the remaining properties of an admissible set (see 2.1) are obvious.

6.15. PROOF OF THE SUPPLEMENT TO LEMMA 6.1. It is seen from the construction of the imbedding that a discontinuity or a violation of univalence of ω can occur only at pre-images of a vertex of Γ. If $(\mathcal{R},\varphi)$ is a generic pair, then $\partial\mathcal{R}$ locally is a Jordan arc (see 6.2) and ω extends to $\partial\mathcal{R}$ continuously. The equality $\omega(\zeta_1)=\omega(\zeta_2)$, $\zeta_1,\zeta_2\in\partial\mathcal{R}$, can occur only if $\varphi(\zeta_1)= =\varphi(\zeta_2)=z$ is a self-intersection point of Γ. By Definition 6.2 a small neighbourhood of z is of the following kind (see Fig. 9). The point z is the vertex of four cells: Ω_1, Ω_2, Ω_3, Ω_4 (see (6.3)) and $n_1=k$, $n_2=k-1$, $n_3=k$, $n_4=k+1$. By (6.5) the functions δ_{ts} on the arcs ending at z can be choosen independently of the other vertices of Γ. Put $\gamma_{ij}=\partial\Omega_i\cap\partial\Omega_j$.

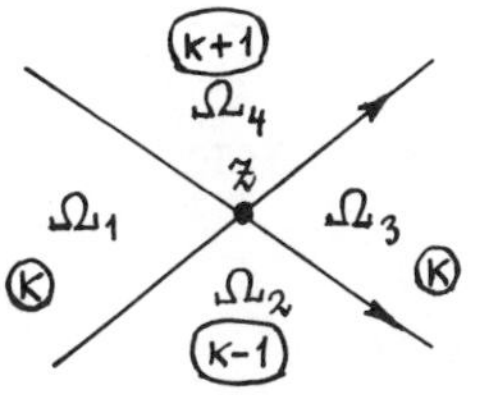

Figure 9

To define the correspondence functions on γ_{ij} we must fix the enumeration of the branches of φ^{-1} in $\Omega_1,\dots,\Omega_4$. Clearly, $\varphi^{-1}_{(1)},\dots,\varphi^{-1}_{(K-1)}$ can be chosen to agree on γ_{ij} and be analytic in $clos(\Omega_1\cup\Omega_2\cup\Omega_3\cup\Omega_4)$. Fix $\varphi^{-1}_{(K)}|\Omega_3$ coinciding with $\varphi^{-1}_{(K)}|\Omega_4$ on γ_{34}. Then $\varphi^{-1}_{(K)}|\Omega_1$ coincides automatically with $\varphi^{-1}_{(K+1)}|\Omega_4$ on γ_{14}. The conditions of 6.12 determine all $\sigma_{\alpha\beta}(i)$ uniquely except for $\sigma_{32}(K)$, $\sigma_{32}(K+1)$, $\sigma_{12}(K)$, $\sigma_{12}(K+1)$. If one puts $\sigma_{32}(K)=K$, $\sigma_{32}(K+1)=K+1$, $\sigma_{12}(K)=K+1$, $\sigma_{12}(K+1)=K$, then it is not hard to see that $\omega(\zeta_1)\neq\omega(\zeta_2)$. The proof is complete. Note that if the two last relations are replaced with $\sigma_{12}(K)=K$, $\sigma_{12}(K+1)=K+1$, then $\omega(\zeta_1)=\omega(\zeta_2)$ is a branch point of the surface S. ▨

CHAPTER III. SUFFICIENT CONDITIONS FOR THE CYCLICITY.

7. TECHNICAL BACKGROUND II.

7.1. CAUCHY INTEGRAL. Let Ω be a plane domain with rectifiable boundary, $f \in L^1(\partial\Omega)$. Consider the Cauchy integral

$$K_f(z) = \frac{1}{2\pi i}\int_{\partial\Omega} \frac{f(\zeta)\,d\zeta}{\zeta - z}, \qquad \zeta \in \mathbb{C} \setminus \partial\Omega .$$

By a well-known formula [36]

$$f = K_{f,\Omega} - K_{f,\mathbb{C}\setminus\Omega}, \tag{7.1}$$

where $K_{f,\Omega}$ and $K_{f,\mathbb{C}\setminus\Omega}$ stand for the angular boundary values of K_f (they exist almost everywhere on $\partial\Omega$ [65]). We shall refer to (7.1) as the "jump formula".

Suppose $\partial\Omega$ is piecewise analytic. Then comparing [36] and [21] (we do not need the recent results due to Calderon, David et al.) we have

$$f \in L^p(\partial\Omega) \Rightarrow K_f \in E^p(\Omega), \quad p > 1 .$$

If $f \in L^1(\partial\Omega)$ then K_f belongs to the weak-type space $E_0^{1,+\infty}(\Omega)$. We are not going to give the definition since for our purposes it is more convenient to use another class of functions.

NOTATION. Put

$$E^{1-0}(\Omega) = \{ f \in \bigcap_{p<1} E^p(\Omega) : \|f\|_p \cdot (1-p) \to 0,\ p \to 1- \}, \tag{7.2}$$

$$L^{1-0}(\gamma) = \{ f \in \bigcap_{p<1} L^p(\gamma) : \|f\|_p \cdot (1-p) \to 0,\ p \to 1- \}, \tag{7.3}$$

for a rectifiable arc γ.

We note that Lemmas 7.2-7.3 and Theorem 7.4 below remain true if one replaces piecewise analyticity (of γ, $\partial\Omega$) by piecewise C^2-smoothness and the class $E^{1-0}(\Omega)$ by $E_0^{1,+\infty}(\Omega)$ that is contained in $E^{1-0}(\Omega)$ (see [63]).

Suppose Ω is an admissible domain in a Riemann surface (see 2.1 for the definition). Recall that Hardy-Smirnov classes $E^p(\Omega)$, $p>0$, were defined in 4.5. The class $E^{1-0}(\Omega)$ is defined by (7.2).

7.2. LEMMA. Let Ω be an admissible domain in a Riemann surface. Suppose γ is an analytic arc contained in $\operatorname{clos}\Omega$ and Ω' is an admissible subdomain. Then $E^p(\Omega)\subset L^p(\gamma)$, $E^p(\Omega)\subset E^p(\Omega')$, $p>0$, $E^{1-0}(\Omega)\subset L^{1-0}(\gamma)$, $E^{1-0}(\Omega)\subset E^{1-0}(\Omega')$ which means that the corresponding restriction operators are continuous.

PROOF. Evidently the claim concerning γ implies the one concerning a subdomain. It is easy to see from 4.5 that the lemma may be verified for a plane domain. If Ω is a plane domain and simply connected, the claim follows easily from the Carleson Imbedding Theorem and the fact the imbedding constant is independent of p. For an arbitrary admissible $\Omega\subset\mathbb{C}$ it can be derived from the simply connected case by a theorem of G.Tumarkin and S.Khavinson [50]. ▨

7.3. LEMMA. Suppose Ω is a plane admissible domain. Then $f\in L^1(\partial\Omega)$ implies $K_f\in E^{1-0}(\Omega)$.

The assertion follows from [36] and Lemma 7.2.

7.4. THEOREM. (A.B.Aleksandrov). Let $\Omega,\Omega_1,\dots,\Omega_n$ be admissible domains in a Riemann surface such that $\Omega_i\cap\Omega_j=\emptyset$ for $i\neq j$ and $\operatorname{clos}\Omega=\bigcup_{i=1}^{n}\operatorname{clos}\Omega_i$. Suppose $f_i\in E^{1-0}(\Omega_i)$ and $f_i=f_j$ a.e. on $\partial\Omega_i\cap\partial\Omega_j$. Then there exists $f\in E^{1-0}(\Omega)$ such that $f\mid\Omega_i=f_i$.

PROOF. The theorem was proved by A.Aleksandrov [2] (see also [1]) for $\Omega=\mathbb{D}$, $n=2$, Ω_1 and Ω_2 being the upper and lower half-discs. However this implies the general case easily. Observe first that f_i and f_j are analytic continuations of each other on every analytic arc of $\partial\Omega_i \cap \partial\Omega_j$. This is a local fact and follows from Aleksandrov's case by a conformal change of variable straightening the arc. Thus we obtain f analytic in $\Omega \setminus \mathcal{B}$ for some finite set $\mathcal{B}$. It is easy to see from Lemma 7.2 that $f \in E^{1-0}(\Omega \setminus \mathcal{B})$ and hence $f \in E^{1-0}(\Omega)$. ▨

It is known (see [32, Lecture 1], in fact the idea goes back to V.I.Smirnov [38]) that the standard Phragmén-Lindelöf principle follows from the assertion: if $g \in L^2(\partial\mathbb{D})$, f is outer and $fg \in H^2$, then $g \in H^2$. We shall need a generalization of this result. First we introduce the notion of an outer function for domains in a Riemann surface.

7.5. DEFINITION. Let Ω be an admissible domain in a Riemann surface. Suppose first that Ω is simply connected and $\operatorname{clos}\Omega$ is contained in a topological disc U with conformal $\theta: U \to \mathbb{D}$. Then a function f analytic in Ω is called outer if $f \circ \theta^{-1}$ is outer in $\mathbb{D}$. Suppose Ω is arbitrary. A function $f, f \neq 0$, is said to be outer in Ω, if for any $z \in \partial\Omega$ there exists a small admissible neighbourhood U_z such that all components of $U_z \cap \Omega$ are simply connected, and the restriction of f to each of them is outer.

7.6. LEMMA (abstract Phragmén-Lindelöf principle). Let Ω be an admissible domain in a Riemann surface. Suppose $0 < r \leq p$, $q < \infty$. If $g \mid \partial\Omega \in L^p(\partial\Omega)$, f is outer in Ω, $f \in E^q(\Omega)$, and $fg \in E^r(\Omega)$, then $g \in E^p(\Omega)$.

PROOF. If Ω is simply connected and small enough, the assertion follows immediately from the well-known disc version of the lemma. Suppose Ω is arbitrary. Fix $z \in \partial\Omega$ and a small admissible parametric disc centered at z such that $\partial\Omega$ divides $U = U_z$ into curvilinear sectors. Choose the discs U_1, U_2 centered at z such that $\operatorname{clos} U_1 \subset U_2$, $\operatorname{clos} U_2 \subset U$. Denote $G = U \cap \Omega$, $G_1 = U_1 \cap \Omega$, $G_2 = U_2 \cap \Omega$. To prove that $g \in E^p(\Omega)$ it is enough to check that $g \mid G_1 \in E^p(G_1)$. By Definition 7.5 we may assume that $f|G$ is outer. According to (4.3) one can find an outer function h in G such that $|h|=1$ on $\partial G_1 \cap \partial G$ and $|h| = \min(1,|f|)$ on $\partial G \setminus \partial G_1$ (see Fig.10). The function h has the properties: (i) $h \in H^\infty(G)$; (ii) h is bounded away from zero on G; (iii) $|h/f| \le 1$ on $\partial G \setminus \partial G_1$. We have $fgh|G \in E^r(G)$ by Lemma 7.2 and $f|G$ is outer. By the choice of h, $gh|\partial\Omega \cap \partial G \in L^p(\partial\Omega \cap \partial G)$, and

$$gh|\partial G \setminus \partial\Omega = fg \cdot \frac{h}{f}|\partial G \setminus \partial\Omega \in L^r(\partial G \setminus \partial\Omega),$$

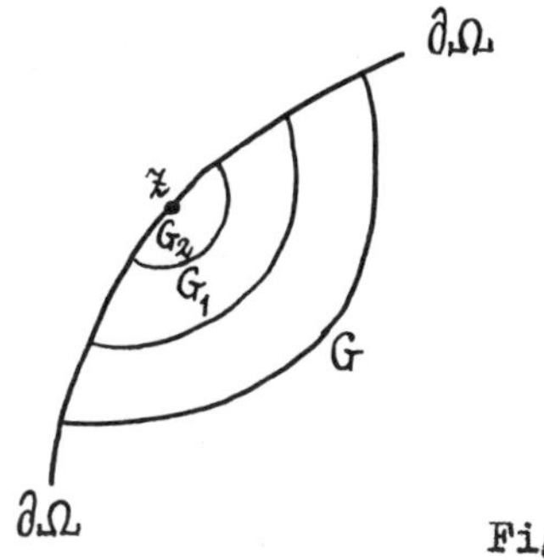

Figure 10

hence $gh|\partial G \in L^r(\partial G)$. Thus by the case already treated we obtain that $gh|G \in E^r(G)$. Property (ii) implies that $g|G_2 \in E^r(G)$. Since z has been chosen arbitrary we have proved that $g \in E^r(\Omega)$. To verify that $g \in E^p(\Omega)$ a similar trick can be applied. Let z, G, G_1, G_2 be as before. Choose h_1 to be outer in G and such that $|h_1|=1$ on $\partial G \cap \partial G_1$, $|h_1| = \min(1,|g|^{-1})$ on $\partial G \setminus \partial G_1$. Then $gh_1|\partial G \in L^p(\partial G)$ and $gh_1 \in E^r(G)$ hence $gh_1 \in E^p(G)$. Again we have $|h_1| > \delta > 0$ on G_2 which implies that $g \in E^p(G)$. The proof is complete. ▨

Now we make some preparations for the statement of cyclicity conditions.

7.7. FUNCTIONS ASSOCIATED WITH SECTORS. Let (R,φ) be an admissible pair. Recall that $V(\varphi)$ is the set of maximal essential valency (see 2.2). Let us fix a regular neighbourhood for each vertex of $\partial V(\varphi)$ (see 4.9). Let W be the union of these neighbourhoods. Define $\mathfrak{A}$ to be the set of all sectors constituting $V(\varphi) \cap W$. Clearly $\mathfrak{A}$ is finite. Recall that in 4.12 $\mathfrak{A}$ has been endowed with the ordering $\prec$. According to Lemma 4.13, one can choose for each sector $S \in \mathfrak{A}$ a smooth positive function on $(0,1]$ such that $h_S(r) \to +\infty$, $r \to +0$, and

$$\int_0 h_S(\tau)\,\psi_S'(\tau)\,d\tau < +\infty,$$

$$\int_0 h_S(\tau)\psi_T'(\tau)d\tau = +\infty, \quad \forall T\in\mathfrak{A},\ T\succ S.$$

Combining this with Lemma 4.11 we obtain

$$\int_{\partial S} h_S(|\zeta - 0_S|)\,d\omega_S(\zeta) < +\infty,$$
$$\int_{\partial T} h_S(|\zeta - 0_T|)\,d\omega_T(\zeta) = +\infty, \quad T\succ S, \tag{7.4}$$

where 0_S and 0_T stand for the vertices of sectors S,T and ω_S, ω_T denote their harmonic measures. We fix h_S and refer to it as the <u>function associated with the sector</u> S.

8. THE CASE OF ESSENTIALLY UNIVALENT φ.

Here we prove the Main Theorem in the case $n(\varphi)=1$ which needs a separate treatment. The argument will be useful for the general case and some parts of the proof will be employed directly. Moreover, the case of univalent φ is related to polynomial approximation which is illustrated by the following lemma.

8.1. LEMMA. (see [31]). <u>Let</u> R <u>be a plane domain</u>, $\varphi\in H^\infty(R)$ <u>The following statements are equivalent</u>:

(i) $\mu_\varphi = 1$

(ii) φ <u>is univalent and there exists a function</u> $u\in H^\infty(\Omega)$, $\Omega = \varphi(R)$, <u>such that the set</u> $\{up : p$ <u>polynomial</u>$\}$ <u>is weak</u>-$*$ <u>dense in</u> $H^\infty(\Omega)$.

The proof is trivial.

8.2. THE BEGINNING OF THE PROOF OF THEOREM 2.6. Since $n(\varphi)=1$, the domain $\Omega = \varphi(R)$ is admissible and the problem reduces to computing the multiplicity of the operator $T : f\mapsto zf$ acting on $H^\infty(\Omega)$ with the weak-$*$ topology. Estimates from below have been obtained in Sect.5. If the set $\mathbb{C}\setminus \operatorname{clos}\Omega$ is connected, the polynomials are weak-$*$ dense in $H^\infty(\Omega)$. So a nonzero constant function can be taken for a cyclic vector, and $\mu(T)=1$. Suppose $\mathbb{C}\setminus \operatorname{clos}\Omega$ is disconnected. The estimate $\mu(T)\leq 2$ follows from a result of D.Herrero [22, Prop.2]. In the following proposition we obtain a large supply of cyclic sets consisting of two elements. Its proof employs some ideas relevant to the general case.

The complex plane is divided by $\partial\Omega$ into open components, analytic Jordan arcs (they will be referred to as "the arcsof $\partial\Omega$ ") and isolated points, or "vertices of $\partial\Omega$ ".

8.3. PROPOSITION. Suppose that $w\in H^{\infty}(\Omega)$ and that w has the following property: For each component G of $\mathbb{C}\setminus \operatorname{clos}\Omega$ and for each arc $\gamma\subset\partial G$, the function w has no pseudocontinuation to G across γ, or more precisely,

$$w|\gamma\notin N(G)|\gamma ,$$

where $N(G)$ is the Nevanlinna class. Then the pair $\{\mathbb{1}, w\}$ is cyclic for T.

REMARKS. 1) For w one can take any function analytic (single-valued) on $\operatorname{clos}\Omega$ and having, say, a logarithmic branch point in each component of $\mathbb{C}\setminus \operatorname{clos}\Omega$.

2) The pseudocontinuation condition arises in the description of cyclic vectors of the backward shift operator [13]. In these terms it is possible to describe all cyclic sets for the operator $f\mapsto zf$ in a finitely connected domain with smooth boundary [39].

8.4. THE CASE OF NARROW BELTS. Suppose now that all belts subordinate to $\operatorname{clos}\Omega$ are narrow. Recall that in 7.7 the functions h_S associated with sectors of $V(\varphi)=\operatorname{clos}\Omega$ have been introduced. Note that all components of Ω are simply connected; otherwise there would exist a "superwide" belt without any vertices. By an outer function in Ω we mean an analytic function which is outer in each component of Ω.

8.5. PROPOSITION. Let $\Omega\subset\mathbb{C}$ be an admissible domain such that

<u>all belts subordinate to</u> $\operatorname{clos}\Omega$ <u>are narrow</u>. <u>Suppose</u> $v \in H^{\infty}(\Omega)$ <u>is outer in</u> Ω <u>and satisfies the estimate</u>

$$|v(\zeta)| \leq C \cdot \exp\{-h_S(|\zeta - O_S|)\}, \quad \zeta \in \partial S, \tag{8.1}$$

<u>for every sector</u> S <u>of</u> $\operatorname{clos}\Omega$ <u>with vertex</u> O_S. <u>Then</u> v <u>is a cyclic vector for</u> T.

REMARK. By (7.4) and (4.3) such a function v exists and hence, under the hypotheses of Proposition 8.5, we have $\mu(T) = 1$.

8.6. PROOF OF PROPOSITION 8.3. Let $\mathrm{U} \subset H^{\infty}(\Omega)$. To prove the cyclicity we take an "orthogonal" function $g \in L^1(\partial\Omega)$, i.e. satisfies

$$\int_{\partial\Omega} u z^n g \, dz = 0, \quad \forall u \in \mathrm{U}, \ n \geq 0. \tag{8.2}$$

We must verify that g determines the zero functional on $H^{\infty}(\Omega)$. The condition (8.2) is equivalent to

$$K_{ug}(z) \equiv 0, \quad \forall u \in \mathrm{U}, \tag{8.3}$$

in the unbounded component of $\mathbb{C} \setminus \operatorname{clos}\Omega$. Here K_{ug} is the Cauchy integral (see 7.1). Let us show that if the pair $\mathrm{U} = \{\mathbb{1}, w\}$ is chosen as stated in Proposition 8.3, the condition (8.3) holds <u>everywhere</u> in $\mathbb{C} \setminus \operatorname{clos}\Omega$.

Let A be a component of $\operatorname{int}\Omega$ and let B be a component of $\mathbb{C} \setminus \operatorname{clos}\Omega$ having a common boundary arc with A. Combining the "jump" formula (7.1) with the Uniqueness Theorem we see that if (8.3) holds in B, then

$$K_{wg}(z) = w(z) K_g(z), \quad z \in A. \tag{8.4}$$

Conversely, (7.1) shows that if (8.4) holds, then

$$K_{wg,B} = w K_{g,B} \qquad \text{a.e. on } \partial A \cap \partial B,$$

where $K_{wg,B}$, $K_{g,B}$ are boundary functions of the Cauchy integrals K_{wg}, K_g in B. Since the Cauchy integrals represent functions of the class $E^p(B)$, $p < 1$, we conclude from (4.2) that either $K_{g,B} \equiv K_{wg,B} \equiv 0$ or $w|_{\partial A \cap \partial B} \in N(B)$. The latter is excluded by the choice of w.

Thus proceeding from a component of $\mathbb{C} \setminus \partial\Omega$ to an adjacent one, we obtain that (8.4) holds everywhere in Ω and (8.3) holds everywhere in $\mathbb{C} \setminus \operatorname{clos}\Omega$. In particular, we have

$$K_g(\lambda) = \int_{\partial\Omega} \frac{g(\zeta)\,d\zeta}{\zeta - \lambda} = 0, \quad \lambda \notin \operatorname{clos}\Omega .$$

Since the rational fractions $\{(\zeta-\lambda)^{-1} : \lambda \notin \operatorname{clos}\Omega\}$ form a complete family in $H^\infty(\Omega)$, it follows that g corresponds to the zero functional. The proof is complete. ▨

8.7. THE BEGINNING OF THE PROOF OF PROPOSITION 8.5 repeats that of Proposition 8.3. Under the assumption that $g \in L^1(\partial\Omega)$ and

$$K_{vg}(z) \equiv 0 \tag{8.5}$$

in the unbounded component of $\mathbb{C} \setminus \operatorname{clos}\Omega$ we wish to prove that g determines the zero functional on $H^\infty(\Omega)$, i.e. that $g \in E^1(\Omega)$. Again it is crucial to establish that (8.5) holds in the whole of $\mathbb{C} \setminus \operatorname{clos}\Omega$.

Denote by $\mathcal{P}$ the union of components of $\mathbb{C} \setminus \operatorname{clos}\Omega$ in which (8.5) fails. Suppose $\mathcal{P} \neq \emptyset$. Let $\mathcal{T}$ be the unbounded component of $(\mathbb{C} \setminus \Omega) \setminus \mathcal{P}$. The components of $\mathbb{C} \setminus \mathcal{T}$ are admissible simply connected domains. Among them choose a domain Ψ such that $\Psi \cap \mathcal{P} \neq \emptyset$. Since $\operatorname{int}\mathcal{T} \subset \mathbb{C} \setminus \operatorname{clos}\Omega$ and $\mathcal{P} \subset \mathbb{C} \setminus \operatorname{clos}\Omega$ the set $\partial\Psi \cap \partial\mathcal{P}$ is finite. Put $\{z_1, \dots, z_\ell\} = \partial\Psi \cap \partial\mathcal{P}$, $\ell \geq 0$. Let $\varepsilon > 0$ be so small that the discs $|\zeta - z_i| \leq 2\varepsilon$ are pairwise disjoint and regular with respect to $\partial\Omega$ (see 4.9). Put

$$\Psi_\varepsilon = \{ z \in \Psi : \operatorname{dist}(z, \partial\Psi) > \varepsilon \} .$$

It is not hard to see that $B = (\operatorname{clos}\Psi \setminus \Psi_\varepsilon) \cap \operatorname{clos}\Omega$ is a belt subordinate to $\operatorname{clos}\Omega$ and $\{z_1, \dots, z_\ell\}$ is the family of its vertices. To each vertex z_i there correspond one or several sectors lying in Ψ (they will be referred to as "sectors of Ψ"). A sector of Ψ in turn is divided into several sectors by the set

$\partial\Omega$. Their number is necessarily odd; these sectors enumerated clockwise starting from the sector, having a common arc with $\partial\Psi$, will be denoted $S_1, \dots, S_{2K+1}$ (see Fig.11). Clearly $S_{2i} \subset \mathbb{C} \setminus \Omega$, $S_{2i+1} \subset \operatorname{clos}\Omega$.

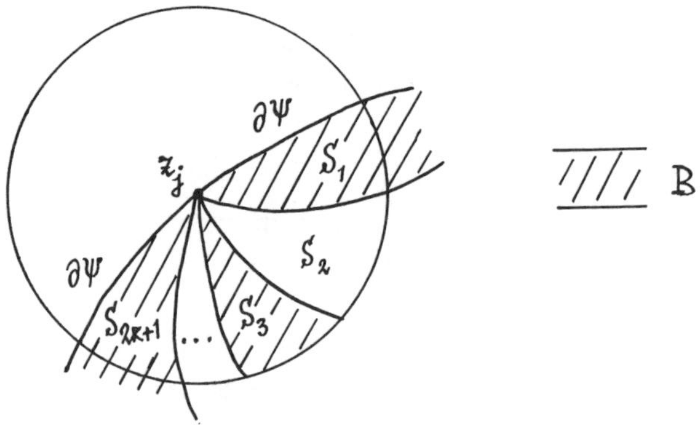

Figure 11

All sectors S_{2i+1} lie in the belt B , but the links containing S_1, S_{2K+1} are principal while all others are additional (see 4.10). By removing from B some of the additional links, one can obtain a family of belts with the same exterior. All of them are narrow by the hypothesis. This implies that $\partial\Psi \cap \cap \partial\mathcal{P} \neq \emptyset$ and there exists a sector of Ψ in which (2.1) fails for all belts that can be obtained from B . Fix such a sector Ψ_z with vertex z . We shall say that Ψ_z is a "<u>distinguished sector</u>".

At this stage we need some elementary combinatorics. Consider the sectors $S_1, \dots, S_{2K+1}$ comprising Ψ_z . The sectors lying in Ω , that is, S_{2i+1} , will be called "<u>black</u>" (they correspond to the shaded sectors in Fig.11). By a <u>configuration</u> we shall mean any collection of "black" sector including the sectors S_1 and S_{2K+1} (i.e. the principal links). A sector which consists of several S_i and lies between two neighbouring black sectors of a configuration will be called a "<u>white sector</u> of the configuration" or simply a "white" sector. A "white" sector between "black" sectors S_{2i+1} and S_{2j+1} will be denoted $\mathcal{D}_{ij}$. Configurations precisely correspond to belts which can be obtained from $\mathcal{B}$ as mentioned above. Recall that Ψ_z is a distinguished sector. By Lemma 4.14 this means that <u>for every configuration there exists a "white" sector which is greater than at least one of the adjacent "black" sectors</u>. (Here "greater" is in the sense of Definition 4.12).

Let us call a sector S_{2i} accessible if for any configuration the "white" sector containing S_{2i} is greater than at least one of the adjacent "black" sectors.

8.8. LEMMA. <u>There exists an accessible sector</u>.

PROOF. A "white" sector which is greater than at least one of the adjacent "black" sectors will be called a $\mathcal{B}$-sector. Other "white" sectors will be called $\mathcal{S}$-sectors.

Let us show first that if S_{2p}, S_{2r}, $p<r$, are both inaccessible, they are contained either in one $\mathcal{S}$-sector or in two disjoint $\mathcal{S}$-sectors. Indeed, S_{2p} and S_{2r} lie in some $\mathcal{S}$-sectors $\mathcal{D}_{ij}$, $\mathcal{D}_{i'j'}$ respectively. Let $2i+1<2p<2i'+1<2j+1<2r<2j'+1$ (otherwise the claim is obvious). Then $\mathcal{D}_{ii'}$ is an $\mathcal{S}$-sector too. In fact, $\mathcal{D}_{ii'} \precsim \mathcal{D}_{ij} \precsim S_{2i+1}$; $\mathcal{D}_{ii'} \preccurlyeq \mathcal{D}_{ij} \preccurlyeq S_{2j+1} \preccurlyeq \mathcal{D}_{i'j'} \preccurlyeq$ $\preccurlyeq S_{2i'+1}$ (we used the fact that $S_{2j+1} \subset \mathcal{D}_{i'j'}$). The claim is proved.

Suppose all the sectors S_{2p} are inaccessible. Then each of them is contained in an $\mathcal{S}$-sector. Using the assertion proved above we obtain a collection of disjoint $\mathcal{S}$-sectors which contain all sectors S_{2p}. But such a collection clearly corresponds to a configuration. This contradicts the assumption that Ψ_z is distinguished. ▨

8.9. CONTINUATION OF THE PROOF OF PROPOSITION 8.5. Consider the function $h(\zeta)=h_{\mathcal{S}}(|\zeta-z|)$ on $\partial\Omega \cap \partial\Psi_z$ where $\mathcal{S}$ is the black sector such that $\zeta\in\partial\mathcal{S}$. Let S_{2p} be the accessible sector from the preceding lemma. It follows from (7.4) that

$$\int_{\partial\mathcal{D}} h(\zeta)\,d\omega_{\mathcal{D}}(\zeta)=+\infty\,,\quad \mathcal{D}\supset S_{2p}\,, \tag{8.6}$$

where $\mathcal{D}$ is an arbitrary sector consisting of the sectors S_i.

Now we wish to show that the condition (8.5) holds in the

accessible sector and thus to arrive at a contradiction with the choice of Ψ . To this end the following theorem is applied.

8.10. THEOREM (A.L.Volberg, [55]). Suppose that the disc $\mathbb{D}$ is divided into sectors $S_i = 0A_iA_{i+1}$, $i=0,\dots,m;\ A_{m+1}=A_0$ by C^2-smooth disjoint Jordan arcs $(0,A_i]$. Let f_i be a function analytic in S_i such that f_i is a linear combination of Cauchy integrals of some functions in $L^1(0A_j)$, $j\le m$ with coefficients in $H^\infty(S_i)$. Mark the sectors S_0 and S_ℓ . Suppose next that for any i,j such that $0<i\le\ell$, $\ell+1\le j\le m+1$, the following property holds:

$$\int_{[0A_i]} \log|f_i - f_{i-1}|\,d\omega_{0A_iA_j} + \int_{[0A_j]} \log|f_j - f_{j-1}|\,d\omega_{0A_iA_j} = -\infty \tag{8.7}$$

where $\omega_{0A_iA_j}$ is the harmonic measure for the sector $0A_iA_j$ containing S_ℓ by not S_0 . Then

$$f_0 \equiv 0 \Longrightarrow f_\ell \equiv 0 .$$

REMARK. The cumbersome definition of the class to which f_i belongs is needed for the general case $n(\varphi)>1$.

8.11. COMPLETION OF THE PFOOF OF PROPOSITION 8.5. We shall apply the preceding theorem to the disc $U=\{\zeta : |\zeta - z| < \varepsilon\}$ divided into sectors $S_0, \dots, S_{2k+1}$, where $S_0 = U \setminus \operatorname{clos}\Psi_z$. Put $f_0\equiv 0$, $f_i = K_{vg} \mid S_i$. Observe that the components of $\mathbb{C}\setminus\operatorname{clos}\Omega$ bordering on the sectors S_1, S_{2k+1} from the exterior of the belt B , do not lie in Ψ and hence (8.5) holds in them. Therefore, all "jumps" $f_{i+1} - f_i$ on the arcs $\partial S_i \cap \partial S_{i+1}$ are equal to $\pm vg$. The conditions (7.4) and (8.6) imply the validity of all hypotheses of Theorem 8.10 for the marked sector S_{2p}. By the theorem, $K_{vg} \mid S_{2p} \equiv 0$, which contradicts the choice of Ψ .

Thus we have proved that (8.5) holds in the whole of $\mathbb{C}\setminus\operatorname{clos}\Omega$. By the jump formula (7.1) we have a.e. on $\partial\Omega$

$$vg = K_{vg,\Omega} .$$

The Cauchy integral K_{vg} belongs to $E^p(\Omega)$, $p<1$. Since $g\in L^1(\partial\Omega)$ and v is outer, using the Abstract Phragmen -Lindelöf principle (Lemma 7.6), we conclude that $g \in E^1(\Omega)$. The proof is complete. ▨

9. REDUCTION.

We come at last to the construction of a cyclic set in the main case $n(\varphi)>1$. In the next section, the list of conditions sufficient for cyclicity is given under some additional geometrical requirements on the pair (R,φ). Here we show that one can find a pair (R_1,φ_1), such that $R_1 \supset R$, $\varphi=\varphi_1 | R$, for which all these requirements are satisfied. The cyclic set for T_φ will be obtained as a restriction of a cyclic set for T_{φ_1} to R.

In the present section we pass from the space $H^\infty(R)$ to a space of functions possibly having weak (logarithmic) growth at several fixed points. This is stipulated by the method of constructing a cyclic set.

9.1. HOW TO OBTAIN THE ESTIMATE $\mu_\varphi \leqslant n(\varphi)+1$? We shall not prove it separately but deduce from the other claim of Theorem 2.6.

Let (R,φ) be an admissible pair, $n=n(\varphi)$. One can find an arc γ of $\varphi(\partial R)$ whose φ-inverse image consists of exactly n disjoint arcs in $\operatorname{clos} R$. Let $\widetilde{R}$ be the disjoint union of R and the unit disc $\mathbb{D}$. Define a function $\widetilde{\varphi}$ on $\mathbb{D}$ to be linear and such that the image $\widetilde{\varphi}(\mathbb{D})$ would be a sufficiently small disc intersecting γ. Set $\widetilde{\varphi}|R=\varphi$. The new pair $(\widetilde{R},\widetilde{\varphi})$ is clearly admissible, $n(\widetilde{\varphi})=n(\varphi)+1$, and the set of maximal essential valency $V(\widetilde{\varphi})$ (which is equal to an arc or a distorted half-disc) does not divide the complex plane. If we knew already that in such circumstances the multiplicity is equal to the maximal essential valency, we should have $\mu_{\widetilde{\varphi}}=n(\widetilde{\varphi})$. The restriction of a cyclic set for $T_{\widetilde{\varphi}}$ to R is apparently cyclic for T_φ and hence the estimate $\mu_\varphi \leqslant n(\varphi)+1$ would be proved.

Thus further we can confine ourselves to the pairs (R,φ) for which the Main Theorem claims the equality $\mu_\varphi=n(\varphi)$.

9.2. DEFINITION. We say that an admissible pair (R,φ) is

<u>strictly admissible</u> if for some finite set $P \subset \partial R$ the restriction $\varphi | \partial R \setminus P$ is one-to-one.

Let (R, φ) be an admissible pair. We shall say that an admissible pair (R_1, φ_1) is a <u>"good" extension</u> of (R, φ) if $R_1 \supset R$, $\varphi = \varphi_1 | R$, $H^\infty(R_1)$ is weak-* dense in $H^\infty(R)$, and moreover, the following condition is satisfied:

> $n(\varphi_1) = n(\varphi)$ and if there exists a wide belt subordinate to $V(\varphi_1)$ then there exists a wide belt subordinate to $V(\varphi)$.

It is clear that if we have a "good" extension, then the estimate from above in Theorem 2.6 for the pair (R, φ) follows from that for the pair (R_1, φ_1).

9.3. LEMMA. <u>Let</u> (R, φ) <u>be an admissible pair. Then there exists a strictly admissible pair</u> (R_1, φ_1) <u>which is a "good" extension of</u> (R, φ).

9.3.1. AGREEMENT. If a pair (R, φ) is admissible, then the curve $\Gamma = \varphi(\partial R)$ divides the complex plane into components, arcs and vertices (see (4.9)). Put $Z = \varphi(\{\zeta \in \operatorname{clos} R : d\varphi(\zeta) = 0\})$. We shall always add the finite set $Z \cap \Gamma$ to the collection of vertices.

9.3.2. PROOF OF LEMMA 9.3. Suppose that the pair (R, φ) is admissible but not strictly admissible. This means that one can find an arc β of $\Gamma = \varphi(\partial R)$ which is the image of at least two arcs $\alpha_1, \alpha_2 \subset \partial R$. The domain R_1 will be constructed in several steps. In Fig.12 an example of a one-step extension is given.

Let A, B denote the components of $\mathbb{C} \setminus \Gamma$ with the boundary arc β ($A = B$ is possible, as in Fig.12). By 9.3.1 $d\varphi$ does not vanish on α_1, so a small subdomain of R adjoining to α_1 is mapped homeomorphically, say, into A. Choose an analytic Jordan arc β' lying in B and tangent with β at both endpoints. Let $\varphi^{-1}_{(1)}$ be the branch of φ^{-1} such that $\varphi^{-1}_{(1)}(\beta) = \alpha_1$. Since φ is analytic in a neighbourhood of $\operatorname{clos} R$, the arc β' can be chosen to lie in the domain of $\varphi^{-1}_{(1)}$. The order of tan-

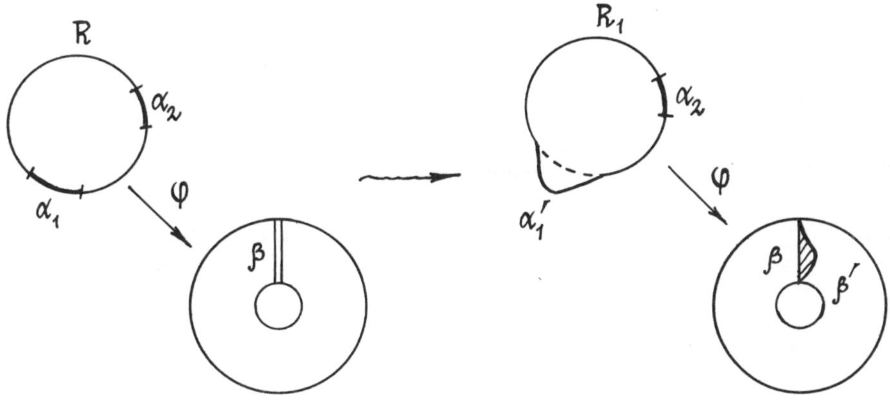

Figure 12

gency of β and β' can be made high enough to ensure that the two sectors between these arcs with the vertices at their common endpoints would be less than any sector formed by Γ (in the sense of Definition 4.12). Replace now the arc α_1 by $\alpha_1' = \varphi_{(1)}^{-1}(\beta')$. The arc α_1' has the same endpoints and lies outside $\mathcal{R}$. As a result we obtain an admissible domain $\mathcal{R}'$, $\mathcal{R}' \supset \mathcal{R}$. Clearly, the essential maximal valency does not increase (unlike the ordinary maximal valency, see Fig.12). To each belt subordinate to $V(\varphi|\mathcal{R})$ there corresponds naturally a belt subordinate to $V(\varphi|\mathcal{R}')$. According to Lemma 4.14 a narrow belt cannot become wide because of the high order of tangency of β and β'. Would a new belt appear, it would have a link including a sector between β and β', and by the same lemma it would be narrow. Finally, $H^\infty(\mathcal{R}')$ is weak-$*$ dense in $H^\infty(\mathcal{R})$ by Lemma 4.7, since each component of the complement of $\operatorname{clos}\mathcal{R}$ intersects the complement of $\operatorname{clos}\mathcal{R}'$. So we have a "good" extension. It is readily seen that by a finite number of steps just described, we arrive at a strictly admissible pair $(\mathcal{R}_1, \varphi_1)$. The lemma is proved. ▨

9.4. CONDITION AT THE BELTS' VERTICES. Put

$$\Omega^k(\varphi) \stackrel{def}{=} \{ z \in \mathbb{C} \setminus \Gamma : \operatorname{card}(\varphi^{-1}(z) \cap \mathcal{R}) = k \}$$

counting multiplicities. For a strictly admissible pair it is convenient to use $\Omega^k(\varphi)$ in place of $\Pi^k(\varphi)$. Clearly if (R,φ) is strictly admissible, then $\operatorname{clos}\Pi^k(\varphi)=\operatorname{clos}\Omega^k(\varphi)$, $V(\varphi) = \operatorname{clos}\Omega^{n(\varphi)}(\varphi)$.

We shall need the following condition:

for any belt subordinate to the set $V(\varphi)$ each of its vertices z has a neighbourhood U_z such that (9.1)

$$U_z \subset \operatorname{clos}(\Omega^{n-1}(\varphi) \cup \Omega^n(\varphi)) .$$

9.5. LEMMA. Let (R,φ) be a strictly admissible pair. Then there exists a strictly admissible pair $(\tilde{R},\tilde{\varphi})$ satisfying (9.1), which is a "good" extension of (R,φ).

PROOF. Let z be a vertex of some belt subordinate to $V(\varphi)$ and let $U_z = \{\zeta : |\zeta - z| < 2\varepsilon\}$ be its regular neighbourhood (see 4.9). Put $\Gamma = \varphi(\partial R)$, $n = n(\varphi)$, $\Omega^i = \Omega^i(\varphi)$. If all sectors constituting $U_z \setminus \Gamma$ lie in $\Omega^{n-1} \cup \Omega^n$, then the condition (9.1) is satisfied locally. If the contrary holds, choose a sector C having minimal multiplicity (i.e. the minimal value of the valence function) $k < n-1$. Let $\mathcal{K}$ be the Riemann surface in which R is contained (see 2.1). Put $\mathcal{K}_1 = \mathcal{K} \cup \operatorname{clos} U_z$, $C_\varepsilon = C \cap \{\zeta : |\zeta - z| < \varepsilon\}$, $R_1 = R \cup C_\varepsilon$. Define φ_1 to be identical on $\operatorname{clos} U_z$ and equal to φ on R. It is clear that (R_1, φ_1) is admissible. Since (R,φ) is strictly admissible the sectors adjacent to C have multiplicity $(k+1) < n$. Thus $n(\varphi_1) = n(\varphi)$ and the set $V(\varphi_1) \setminus V(\varphi)$ can include at most two radial arcs of $\partial C_\varepsilon \cap \Gamma$. Evidently (R_1, φ_1) is a "good" extension of (R,φ). The new pair is not strictly admissible, so we apply Lemma 9.3 (or rather, the construction in its proof) and obtain a good extension (R_2, φ_2) which is strictly admissible. It is easy to see that after finitely many steps just described we satisfy the condition (9.1). ▨

We are thankful to D.V.Yakubovich for simplifying the proof of Lemma 9.5.

9.6. WEIGHTED SPACES. In Sects.12-15 we shall suggest a construction of a cyclic set. Unfortunately this construction does not allow us to choose all functions of this cyclic set to be bounded. But these functions can be chosen to belong to a certain weighted space which we now introduce.

Let (R,φ) be an admissible pair. Set

$$H^{\infty}(R,w)=\{f\in E^1(R): fw^{-1}|\partial R\in L^{\infty}(\partial R)\},$$

where w is a positive function on ∂R which is finite outside a finite set S_w and satisfies the estimate $w(t)\leqslant \leqslant |\log|t-z||^k$ in a small parametric neighbourhood of an arbitrary point $z\in S_w$. The topology on $H^{\infty}(R,w)$ is determined by seminorms $f\mapsto|\int f\,d\omega|$, where $d\omega$ is a differential on ∂R and $d\omega$ is locally representable as $w^{-1}g\,dz$, $g\in L^1(\partial R)$. The collection of all such differentials will be denoted by $\mathcal{Z}^1(\partial R,w)$.

9.7. LEMMA. The multiplicity of T_φ on $H^{\infty}(R)$ does not exceed the multiplicity of T_φ on $H^{\infty}(R,w)$.

PROOF. Let $U\subset H^{\infty}(R,w)$ be a cyclic set. Consider a function τ_w analytic on $\operatorname{clos}R$, nonvanishing in R, and having zeros on S_w. The existence of such a function follows from the Generalized Weierstrass Theorem (see [16, § 26]). Clearly $w\tau_w\in L^{\infty}(\partial R)$. We claim that $\tau_w U$ is the cyclic set for T_φ on $H^{\infty}(R)$. Indeed, let $d\omega\in\mathcal{Z}^1(\partial R)$,

$$\int_{\partial R}\tau_w u\,\varphi^n d\omega=0,\quad u\in U,\ n\geqslant 0.$$

We have $\tau_w d\omega\in\mathcal{Z}^1(\partial R,w)$. The cyclic property of U implies that $\int_{\partial R} f\tau_w d\omega=0$ for all $f\in H^{\infty}(R,w)\supset H^{\infty}(R)$. By Lemma 4.8, $\tau_w H^{\infty}(R)$ is dense in $H^{\infty}(R)$ and thus $d\omega$ determines the zero functional in $H^{\infty}(R)$. The proof is complete. ▨

10. CONDITIONS SUFFICIENT FOR THE CYCLICITY. STATEMENT AND DISCUSSION.

It is shown in Sect.9 that when proving the Main Theorem one can assume that $(\mathcal{R},\varphi)$ is strictly admissible, that all belts subordinate to $V(\varphi)$ are narrow and that (9.1) holds. We suppose also that $n=n(\varphi)\geq 2$. In this setting we state conditions sufficient for the set $U=\{u^{[0]}, u^{[1]},\dots,u^{[n-1]}\}$ to be cyclic for T_φ on the space $H^\infty(\mathcal{R},w)$ supplied with the weak-$*$ topology.

10.1. PREPARATION. The set $\Gamma=\varphi(\partial\mathcal{R})$ divides $\mathbb{C}$ into components (not necessarily simply connected), arcs and vertices, By Ω^k, $0\leqslant k\leqslant n$, we denote the union of components in which the number of φ-inverse images is equal to k counting multiplicities (see 9.4). Since $(\mathcal{R},\varphi)$ is strictly admissible, a component of Ω^k can have a common boundary arc only with a component of Ω^{k-1} or Ω^{k+1}. Since all belts are assumed to be narrow, all components of Ω^n are simply connected. Put $Z=\varphi(\{z\in \operatorname{clos}\mathcal{R}: d\varphi(z)=0\}$. The intersection $Z\cap\Gamma$ is added to the set of vertices, so inside an arc we have $d\varphi(z)\neq 0$. Clearly in $\Omega^k\setminus Z$ one can consider k (locally)-analytic branches of φ^{-1} which are denoted by $\varphi^{-1}_{(1)},\dots,\varphi^{-1}_{(k)}$. We need the matrix-valued functions

$$H^{(k)}(\zeta)\overset{def}{=\!=}\left[(u^{[r]}\circ\varphi^{-1}_{(s)})(\zeta)\right]^{0\leqslant r\leqslant n-1}_{1\leqslant s\leqslant k},\quad \zeta\in\Omega^k\setminus Z,\ k=1,\dots,n\ .$$

Now we are in a position to state our sufficient conditions.

10.2. OUTER-TYPE CONDITIONS.

The function $\zeta\mapsto(\det H^{(n)}(\zeta))^2$ is a product of polynomial and an outer function in Ω^n. (10.1)

There exists a finite set $Y\supset Z$ and $\ell>0$ such that for all $k<n$ the following estimate holds:

$$\sum|\operatorname{Minor}(H^{(k)}(\zeta),k)|\geqslant c\prod_{z\in Y}|\zeta-z|^\ell,\quad \zeta\in\Omega^k\setminus Y\ . \tag{10.2}$$

Let us discuss these conditions. It is not hard to see that (10.1) is necessary for the cyclicity. The condition (10.2) can be weakened considerably but this leads to some complications in the proof. The necessary condition to be compared with (10.2), is the requirement for $\zeta \mapsto H^{(k)}(\zeta)$, $k < n$, to be "locally-outer" in $\Omega^k \setminus Z$. We do not know whether this is sufficient in the general case.

10.3. BELT CONDITION. (From now on by a belt we always mean a belt subordinate to $V(\varphi)$). Recall that for each sector S of the set of maximal essential valency $V(\varphi)$ in 7.7 we have fixed a function h_S satisfying (7.4). Besides that fix in S two branches of φ^{-1}. Namely, let α, β be the radial arcs of S. The index a is chosen so that $\varphi^{-1}_{(a)}(\alpha) \subset \partial R$. If $\varphi^{-1}_{(a)}(\beta) \subset \partial R$, we take for $\varphi^{-1}_{(b)}$ any other branch in S (it is used here that $n > 1$). Otherwise $\varphi^{-1}_{(b)}$ is chosen so that $\varphi^{-1}_{(b)}(\beta) \subset \partial R$. Now we can state the belt condition:

For each sector S (with vertex z) of $V(\varphi)$, which is contained in some belt, there exists a function $G \in L^1(\partial S)$, $G \geq 0$, such that (10.3)

$$|(u^{[r]} \circ \varphi^{-1}_{(a)})(\zeta) - (u^{[r]} \circ \varphi^{-1}_{(b)})(\zeta)| \leq e^{-h_S(|\zeta - z|)} \cdot G(\zeta),$$

$$\zeta \in \partial S \cap \Gamma, \quad r = 0, \dots, n-1 .$$

<u>Discussion</u>. It suffices to require the estimate not for all sectors but only for 'distinguished' ones for each belt (see 8.7). The condition (10.3) is not necessary for the cyclicity. It is to be compared (keeping (7.4) in mind) with the following necessary condition implied by Lemma 5.3. Namely if U is cyclic, then for any belt $B = \operatorname{clos} V_B \setminus G_B$, the following holds

$$\int_{\partial G_B} \log |\det H^{(n)}(\zeta)| \, d\omega = -\infty ,$$

where ω is the harmonic measure for G_B.

The condition (8.1) is the analogue of (10.3) in the case $n = 1$.

10.4. THE CONDITION OF ANALYTIC INDEPENDENCE.

For any arc α of Γ_{k+1} separating components $A \subset \Omega^k$ and $B \subset \Omega^{k+1}$, there exists an admissible Jordan domain $\tilde{A}_\alpha \subset A \setminus Z$ such that $\beta = \partial \tilde{A}_\alpha \cap \alpha$ is a Jordan arc and the family $\{u^{[r]} \circ \vartheta^{-1} | \beta, r = 0, \dots, n-1\}$, $\vartheta \overset{def}{=} \varphi | \partial R$ is linearly independent over the algebra (of restrictions to β of) $H^\infty(\tilde{A}_\alpha)$. (10.4)

This condition is related to the condition of nonexistence of a pseudocontinuation which figures in the description of cyclic vectors for the backward shift operator and also in Proposition 8.3. Surely (10.4) is far from being necessary. Instead of (10.4) it can be required for each component $\mathcal{O}$ of $\mathbb{C} \setminus \Gamma$ the existence of components $B_0, \dots, B_m$ such that:

$B_m = \mathcal{O}$, B_0 is unbounded; for $i = 0, 1, \dots, m-1$, $B_i \not\subset V(\varphi)$, B_i borders on B_{i+1}, and if the multiplicity of B_i is greater than that of B_{i+1} then the linear independence in (10.4) holds on $\partial B_i \cap \partial B_{i+1}$.

For example, under the hypothesis (2.3), the condition (10.4) can be omitted completely.

10.5. PROPOSITION. <u>Suppose</u> (R, φ) <u>is a strictly admissible pair satisfying (9.1), and all belts are narrow. If the set</u> $\mathcal{U} = \{u^{[r]}\}_{r=0}^{n-1} \subset H^\infty(R, w)$, $n = n(\varphi)$, <u>satisfies (10.1), (10.2), (10.3), (10.4), then</u> $\mathcal{U}$ <u>is nearly cyclic for</u> T_φ (<u>see Definition 3.4). If in addition (5.4) holds, then</u> $\mathcal{U}$ <u>is cyclic</u>.

REMARK. Regardless the presence of belts, (5.4), (10.2), (10.4) with n replaced by $n-1$ guarantee the cyclicity of $\tilde{\mathcal{U}} = \{u^{[r]}\}_{r=0}^{n}$.

10.6. SKETCH OF THE PROOF. THE RÔLE OF DIFFERENT CONDITIONS. Let $E(\mathcal{U})$ be the T_φ-invariant subspace spanned by $\mathcal{U} \subset H^\infty(R, w)$ and let $d\omega \in \mathcal{L}^1(\partial R, w)$ be a differential orthogonal to $E(\mathcal{U})$. Our aim is to show that $d\omega$ determines the zero functional on

$H^\infty(R,w)$. This means that $d\omega$ extends to R to a holomorphic differential belonging to an appropriate Hardy-Smirnov class.

The first step of the proof will be called "breaking through Γ". It will be shown that $d\omega$ extends to a meromorphic differential in $R\setminus\varphi^{-1}(\Gamma)$, agreeing a.e. on $\varphi^{-1}(\Gamma)$. This step suggests an interesting interpretation: "breaking through Γ" implies that the essential spectrum of $T_\varphi^* \mid E(U)^\perp$ is contained in Γ.

At this stage it will be convenient to use the following condition.

There exists a finite set $\mathcal{Y}\supset Z$ such that for $k=1,\dots,n$. (10.5)

$$rank\, H^{(k)}(\zeta) = rank\left[(u^{[\nu]}\circ\varphi_{(s)}^{-1})(\zeta)\right]_{1\le s\le k}^{0\le\nu\le n-1} = k,\ \zeta\in\Omega^k\setminus\mathcal{Y}.$$

It is evident that (10.1), (10.2) $\Longrightarrow$ (10.5) and (5.4)$\Rightarrow$(10.5).

The belt condition (10.3) and the condition of analytic independence (10.4) will also be used at this stage.

The second step is called "sticking". It will be proved that parts of the extension of $d\omega$ assemble in a single meromorphic differential (of an appropriate class) in R. To do this we shall make use of the conditions (10.1), (10.2) only.

These two steps imply that $E(U)$ has finite codimension, i.e. that U is nearly cyclic. Finally, the completion of the proof will be based on a general result concerning a set satisfying the Local Necessary Condition (5.1) (see Lemma 11.7 below).

11. CONDITIONS SUFFICIENT FOR THE CYCLICITY. PROOF.

11.1. THE BEGINNING OF THE PROOF OF PROPOSITION 10.5. Throughout the section we assume that all preliminary hypotheses of Proposition 10.5 hold. Namely, (R,φ) is strictly admissible, (9.1) is satisfied and all belts, if any, are narrow. To prove the cyclicity of $U=\{u^{[\nu]}\}_{\nu=0}^{n-1}$ we must verify that any differential $d\omega\in\mathcal{L}^1(\partial R,w)$ such that

$$\int_{\partial R} u^{[r]} \varphi^k \, d\omega = 0, \quad k \geq 0, \quad 0 \leq r \leq n-1, \tag{11.1}$$

determines the zero functional in $H^\infty(R, W)$. First we change the variable in (11.1). To this end denote by ϑ the restriction $\varphi | \partial R$. Since the pair (R, φ) is strictly admissible, ϑ^{-1} is single-valued a.e. on $\Gamma = \varphi(\partial R)$. Set $v^{[r]} = u^{[r]} \circ \vartheta^{-1}$, $W_\Gamma = W \circ \vartheta^{-1}$. The differential $d\omega$ can be written locally in the form $d\omega = \lambda(z)\, dz$, $\lambda \in L^1(\partial R, W)$. The function $g[d\omega] = \lambda(z \circ \vartheta^{-1}) \cdot (z \circ \vartheta^{-1})'$ does not depend on the local coordinate z by the definition of a differential, hence $g = g[d\omega] \in L^1(\Gamma, W)$, $v^{[r]} g \in L^1(\Gamma)$. After the change of variable (11.1) turns into $\int_\Gamma v^{[r]} \zeta^k d\zeta = 0$. Consider the Cauchy integrals

$$K_{v^{[r]} g}(z) = \frac{1}{2\pi i} \int_\Gamma \frac{v^{[r]} g \, d\zeta}{\zeta - z}.$$

Clearly (11.1) is equivalent to

$$K_{v^{[r]} g}(z) \equiv 0, \quad z \in \Omega_1^0, \quad r = 0, \ldots, n-1, \tag{11.2}$$

where Ω_1^0 is the unbounded component of $\mathbb{C} \setminus \Gamma$.

Now we start to carry out the plan presented in 10.6. At the stage of "breaking through Γ" we want to extend $d\omega$ to a meromorphic differential in $R \setminus \varphi^{-1}(\Gamma)$. The following lemma is the key point for this. We shall need the locally-analytic functions $v_{(j)}^{[r]} \overset{def}{=\!=} u^{[r]} \circ \varphi_{(j)}^{-1}$, $j \leq k$, defined on $\Omega^k \setminus Z$.

11.2. LEMMA. <u>Let</u> $U, d\omega, g = g[d\omega]$ <u>have the same meaning as before, in particular, let</u> U <u>and</u> $d\omega$ <u>satisfy (11.1). Assume also that (10.3), (10.4), (10.5) hold. Then, for all</u> $k \geq 0$,

$$\operatorname{rank}\left[K_{v^{[r]} g}(\zeta), v_{(1)}^{[r]}(\zeta), \ldots, v_{(k)}^{[r]}(\zeta)\right]^{0 \leq r \leq n-1} = k, \quad \zeta \in \Omega^k \setminus \mathcal{Y}, \tag{11.3}$$

$\mathcal{Y}$ <u>being the finite set in (10.5)</u>, $\mathcal{Y} \supset Z$.

PROOF. Since $v^{[r]}g \in L^1(\Gamma)$, the Cauchy integrals $K_{v^{[r]}g}$ determine functions belonging to E^p, $p<1$, is each component of $\mathbb{C}\setminus\Gamma$, and the "jump formula" (7.1) holds. For a component $A\subset\Omega^k$ consider the matrix-valued function T_A,

$$T_A(\zeta)=\left[K_{v^{[r]}g}(\zeta),\, v^{[r]}_{(1)}(\zeta),\dots v^{[r]}_{(k)}(\zeta)\right]^{0\le r\le n-1},\quad \zeta\in A\setminus Y .$$

It follows from (10.5) that $\operatorname{rank} T_A(\zeta)\ge \operatorname{rank} H^{(k)}(\zeta) = k$. Let Φ denote the closure of the union of components in which (11.3) is true. The equality (11.2) means that $\Omega_1^\circ\subset\Phi$.

Suppose that components $A\subset\Omega^k$, $B\subset\Omega^{k+1}$ have a common boundary arc α. We claim that $(A\subset\Phi)\Longrightarrow(B\subset\Phi)$, and if $k\le n-2$, then $(B\subset\Phi)\Longrightarrow(A\subset\Phi)$.

Assume that the enumeration of branches $\varphi^{-1}_{(i)}$ of φ^{-1} is concordant in A and B in a neighbourhood of α. More precisely, $\varphi^{-1}_{(i)}\,|\,A$ should extend across α to $\varphi^{-1}_{(i)}\,|\,B$ for $i\le k$. Then necessarily $\varphi^{-1}_{(k+1)}|\alpha = v^{-1}|\alpha$. From (7.1) we have

$$K_{v^{[r]}g,B}=K_{v^{[r]}g,A}+v^{[r]}g \qquad \text{a.e. on } \alpha, \qquad (11.4)$$

where $K_{v^{[r]}g,\Omega}$ stands for the boundary function of the Cauchy integral in Ω. Hence, a.e. on α

$$\operatorname{rank} T_B=\operatorname{rank}\left[K_{v^{[r]}g,A},\, v^{[r]}_{(1)},\dots, v^{[r]}_{(k)},\, v^{[r]}\right]^{0\le r\le n-1}. \qquad (11.5)$$

Clearly, if we know that $\operatorname{rank} T_A=k$, then $\operatorname{rank} T_B\le k+1$. By the Uniqueness Theorem we have the required equality (11.3) in $B\setminus Z$. Conversely, let $\operatorname{rank} T_B\le k+1$, $k\le n-2$. Then all senior minors of the matrix on the right of (11.5) are zero a.e. on α. Expanding such a minor in the last column we obtain that $\sum_{\ell=1}^{k+2} v^{[r_\ell]}h_\ell=0$ a.e. on α, where h_ℓ are senior minors of T_A. The functions $h_\ell(\zeta)$ belong to $E^p(\widetilde{A}_\alpha)$, $p<1$, since the domain $\widetilde{A}_\alpha$ in (10.4) does not contain images of branch points. Recalling (4.2), we conclude from (10.4) that $h_\ell\equiv 0$ for all ℓ. But it just amounts to the estimate $\operatorname{rank} T_A\le k$, hence $A \subset \Phi$. Our claim is proved.

It follows that if a component lies in $\mathbb{C} \setminus V(\varphi)$ and can be connected with Ω_1°, by a chain of adjacent components lying in $\mathbb{C} \setminus V(\varphi)$, then (11.3) holds in it. Moreover, $\partial\Phi \subset \partial V(\varphi)$. The inclusion $V(\varphi) = clos\, \Omega^n \subset \Phi$ is obvious. Thus, if $\mathbb{C} \setminus V(\varphi)$ is connected, we have $\Phi \subset \mathbb{C}$. It remains to consider the main case when $\mathbb{C} \setminus V(\varphi)$ is disconnected but all belts are narrow.

Now setting $\Omega = int\, V(\varphi)$, we shall proceed along the same lines as in the proof of Proposition 8.5. Put $\mathcal{P} = \mathbb{C} \setminus \Phi$. As in 8.7, let $\mathcal{T}$ be the unbounded component of $(\mathbb{C} \setminus \Omega) \setminus \mathcal{P}$. There exists a component Ψ of $\mathbb{C} \setminus \mathcal{T}$ such that $\Psi \cap \mathcal{P} \neq \emptyset$. Suppose $\partial\Psi \cap \partial\mathcal{P} = \{z_1, \dots, z_\ell\}$ (this set is clearly finite and it is nonempty because otherwise there would exist a wide belt). For z_i we consider a regular neighbourhood, whose intersection with Ψ consists of what we call "the sectors of Ψ with the vertex at z_i". Each of these sectors in turn is divided into sectors $S_1, \dots, S_{2k+1}$ by the arcs of $\partial V(\varphi)$ as in Fig.11 (see 8.7). The sectors S_{2j+1} lie in $V(\varphi)$ and S_{2j} lie in $\mathbb{C} \setminus V(\varphi)$. Among them S_1 and S_{2k+1} are the only ones having a common boundary arc with Ψ. As in the proof of Proposition 8.5 consider the belt $B = \Psi \setminus G_B$ where G_B is a union of $\{z \in \Psi : dist(z, \partial\Psi) > \varepsilon\}$ and the sectors S_{2j} corresponding to all the sectors of Ψ. There exists a distinguished sector of Ψ (see 8.7) which will be denoted by Ψ_z (z is its vertex). In view of the assumption (9.1) we have $S_{2j} \subset \Omega^{n-1}$ for all i. By Lemma 8.8 one can choose an accessible sector S_{2p}. We want to show that contradicting the choice of Ψ, (11.3) holds in this sector.

We are going to apply Theorem 8.10 in the disc $B = \{\zeta : |\zeta - z| < \varepsilon\}$ divided into sectors $S_0, \dots, S_{2k+1}$ where $S_0 \overset{def}{=} B \setminus clos\, \Psi_z$. Denote $\gamma_i = \partial S_i \cap \partial S_{i+1}$, $\gamma_{2k+1} = \partial S_{2k+1} \cap \partial S_0$. In the sectors S_{2i}, $i > 0$, set

$$f_{2i} = \pm \det \left[K_{v^{[r]}} g,\ v^{[r]}_{(1)}, \dots, v^{[r]}_{(n-1)} \right]^{0 \le r \le n-1}. \tag{11.6}$$

The sign will be determined a bit later. The enumeration of

branches $\varphi_{(j)}^{-1}$ and hence of $v_{(j)}^{[r]}$ can be fixed in S_{2i} since one can assume that $(B\cap Z)\setminus\{z\}=\emptyset$. All functions $v_{(j)}^{[r]}$ are analytic single-valued in S_{2i}. The equality (11.3) in S_{2i} is equivalent to $f_{2i}\equiv 0$. Put $f_0\equiv 0$. By the choice of Ψ, the condition (11.3) and hence (11.6) holds in the sectors of $B\setminus\Gamma$ adjacent to Ψ_z.

Now we define the functions in the sectors S_{2i+1}. Recall that for each sector $S=S_{2i+1}\subset\Omega^n$ in (10.1) we have fixed the indices $a=a(s)$, $b=b(s)$ such that both radial arcs of S are mapped into ∂R by one of the branches $\varphi_{(a)}^{-1}$, $\varphi_{(b)}^{-1}$. Set

$$f_{2i+1}=\pm\det\left[K_{v^{[r]}}g,\ v_{(1)}^{[r]},\dots,\check{a},\dots,v_{(n)}^{[r]}\right]^{0\leq r\leq n-1}. \tag{11.7}$$

The symbol $\check{a}$ means that the column with this index is skipped. The enumeration of the branches is chosen arbitrarily and the sign will be determined further.

The functions f_i "nearly" belong to the class, required in Theorem 8.10. The functions $v_{(j)}^{[r]}$ can have only finitely many logarithmic singularities because they arise from the elements of $H^\infty(R,w)$. Therefore, multiplying all f_i by a common polynomial we can ensure that the resulting functions be linear combinations of Cauchy integrals with bounded coefficients. Thus to apply Theorem 8.10 it remains to verify (8.7).

Consider the function $h(\zeta)=h_S(|\zeta-z|)$ on $\gamma_0\cup\dots\cup\gamma_{2K+1}$, where S is the sector contained in $V(\varphi)$ such that $\zeta\in\partial S$. Recall that h_S satisfies (7.4) and (10.3). The fact that S_{2p} is accessible means that

$$\int_{\partial\mathcal{D}} h(s)\,d\omega_{\mathcal{D}}(\zeta)=+\infty\ ,$$

where $\mathcal{D}$ is any sector consisting of S_i and such that $\mathcal{D}\supset S_{2p}$, $\mathcal{D}\not\supset S_0$. The condition (8.7) will thus follow from the estimate

$$|f_{i+1}-f_i|\,|\,\gamma_i\leq e^{-h}|F|\ , \tag{11.8}$$

for some $F\in L^p(\Gamma)$, $p>0$. Let us estimate the "jump" $|f_{2i+1}-f_{2i}|$ (the "jump" $|f_{2i}-f_{2i-1}|$ is treated similarly).

Fix the enumeration of branches $\varphi^{-1}_{(j)}$ of φ^{-1} agreeing in the sectors S_{2i} and S_{2i+1}. Then $\varphi^{-1}_{(n)}|\gamma_{2i} = \vartheta^{-1}|\gamma_{2i}$, $v^{[r]}_{(n)}|\gamma_{2i} = v^{[r]}|\gamma_{2i}$ (recall that $\vartheta = \varphi|\partial\mathcal{R}$). Two cases can occur.

<u>Case</u> I: $a(S_{2i+1}) = n$. Then by the "jump formula" (7.1) for a proper choice of signs in (11.6), (11.7) we obtain a.e. on γ_{2i}

$$f_{2i+1} - f_{2i} = \pm \det\left[v^{[r]}g,\ v^{[r]}_{(1)}, \dots, v^{[r]}_{(n-1)}\right].$$

Taking $g \in L^p(\Gamma)$ out of the first column, remembering that $v^{[r]}|\gamma_{2i} = v^{[r]}_{(n)}|\gamma_{2i}$ and subtracting the column with the index $b(S_{2i+1})$ from the first column, we deduce from (10.3) the estimate (11.8).

<u>Case</u> II: $b(S_{2i+1}) = n$. It can be assumed that $a(S_{2i+1}) = n-1$. Below by $K_{v^{[r]}g}$ we mean the boundary values of the Cauchy integral of $v^{[r]}g$ in S_{2i}. For a proper choice of signs in (11.6), (11.7) we have a.e. on γ_{2i} :

$$\pm(f_{2i+1} - f_{2i}) = \det\left[K_{v^{[r]}g} + v^{[r]}g,\ v^{[r]}_{(1)}, \dots, v^{[r]}_{(n-2)},\ v^{[r]}_{(n)}\right] -$$

$$- \det\left[K_{v^{[r]}g},\ v^{[r]}_{(1)}, \dots, v^{[r]}_{(n-2)},\ v^{[r]}_{(n-1)}\right] =$$

$$= \det\left[K_{v^{[r]}g},\ v^{[r]}_{(1)}, \dots, v^{[r]}_{(n-2)},\ v^{[r]}_{(n)} - v^{[r]}_{(n-1)}\right].$$

Now (11.8) follows from (10.3).

It remains to look after the choice of signs. Fix the sign for f_1 arbitrarily, then choose the sign for f_2 properly and so on. The jumps on γ_0 and γ_{2K+1} will be as desired, because $f_0 \equiv 0$.

Thus we may apply Theorem 8.10 and arrive at a contradiction. The proof of Lemma 11.2 is complete. ▨

11.3. CONTINUATION OF THE PROOF OF PROPOSITION 10.5. We shall deduce from (11.3) that the differential $d\omega$ satisfying (11.1) can be extended to a meromorphic differential in $\mathcal{R}\setminus\varphi^{-1}(\Gamma)$.

Fix a component $A\subset\Omega^{K}$. The equality (11.3) implies the existence of a unique set of functions $\{g^{A}_{(s)}\}^{K}_{s=1}$ such that

$$K_{v^{[\nu]}}g=\sum_{s\leqslant K}v^{[\nu]}_{(s)}\,g^{A}_{(s)}\ ,\quad \text{in}\quad A\setminus\mathcal{Y},\ \nu=0,\dots,n-1\,. \tag{11.9}$$

The uniqueness follows from (10.5). Recall that $\mathcal{Y}$ is the finite set figuring in (10.5) (or in (11.3)) which contains all images of branch points. Explicit determinant formulae show that the functions $g^{A}_{(s)}$ are locally meromorphic in $A\setminus\mathcal{Y}$ and permute when continued about the points of $\mathcal{Y}$ in the same way as the branches $\varphi^{-1}_{(s)}$ do. The angular boundary values of $g^{A}_{(s)}$ exist a.e. on ∂A .

11.4. LEMMA. <u>Suppose that the components</u> $A\subset\Omega^{K}$, $B\subset\Omega^{K+1}$ <u>have a common boundary arc</u> α . <u>Then a.e. on</u> α <u>in the concordant enumeration of</u> $\varphi^{-1}_{(s)}$,

$$g^{A}_{(s)}=g^{B}_{(s)}\ ,\quad s=1,\dots,K\,;\qquad g=g^{B}_{(K+1)}\,. \tag{11.10}$$

PROOF. The functions $g^{A}_{(s)}$, $g^{B}_{(s)}$ are determined uniquely a.e. on α . The "jump formula" (11.4) implies the desired result. ▨

Now consider the pre-image $\varphi^{-1}(\Gamma)$ which is a finite union of analytic arcs in the Riemann surface. This set divides the domain $\mathcal{R}$ into finitely many components. Put $\Xi=\varphi^{-1}(\mathcal{Y})$. Fix a component $\mathcal{O}$ of the set $\mathcal{R}\setminus\varphi^{-1}(\Gamma)$. Clearly $\mathcal{O}$ is mapped onto a component of $\mathbb{C}\setminus\Gamma$. Let $\varphi(\mathcal{O})\subset\Omega^{K}$ and let s be the index, $1\leqslant s\leqslant K$, such that $\varphi^{-1}_{(s)}\circ\varphi|\mathcal{O}=id$. Put

$$F_{\mathcal{O}}\overset{def}{=\!=}g^{\varphi(\mathcal{O})}_{(s)}\circ\varphi\,.$$

11.5. LEMMA. 1) <u>The function</u> $F_{\mathcal{O}}$ <u>is meromorphic (single-valued) of the class</u> $N(\mathcal{O})$.

2) <u>If</u> $\mathcal{O},G$ <u>are two adjacent components of</u> $\mathcal{R}\setminus\varphi^{-1}(\Gamma)$ <u>then</u>

$F_{\mathcal{O}} = F_G$ <u>a.e. on</u> $\partial\mathcal{O} \cap \partial G$.
3) $F_{\mathcal{O}} d\varphi = d\omega$ <u>a.e. on</u> $\partial\mathcal{O} \cap \partial R$.

PROOF. The formula (11.9) implies that in $\mathcal{O} \setminus \Xi$

$$F_{\mathcal{O}} = \pm \frac{\det\left[K_{v^{[r]}g} \circ \varphi,\ v^{[r]}_{(j)} \circ \varphi\right]^{r \in \mathfrak{A}}_{j \le k,\, j \ne s}}{\det\left[v^{[r]}_{(j)} \circ \varphi\right]^{r \in \mathfrak{A}}_{j \le k}} , \tag{11.11}$$

where $\mathfrak{A}$ stands for an arbitrary subset of $\{0, \dots, n-1\}$ such that $\operatorname{card} \mathfrak{A} = k$ and the denominator does not vanish identically. This formula yields the first claim and the coincidence of boundary values follows readily from Lemma 11.4. ▨

Thus we have extended an arbitrary "orthogonal" differential $d\omega$ to a meromorphic differential in $R \setminus \varphi^{-1}(\Gamma)$ having concordant boundary values on $\varphi^{-1}(\Gamma) \cap R$. Now with the help of the outer-type conditions (10.1), (10.2) we are going to show that "pieces" of the extension are in fact assembled to form a "decent" analytic differential of an appropriate class in R .

11.6. LEMMA. <u>Suppose that the equality (11.3) holds for any</u> $g = g[d\omega]$, <u>where</u> $d\omega$ <u>satisfies (11.1)</u>. <u>Let also (10.1), (10.2) be satisfied</u>. <u>Then</u> U <u>is nearly cyclic for</u> T_φ . <u>If in addition (5.4) holds, then</u> U <u>is cyclic</u>.

PROOF. Call a function τ R-entire if it is analytic in a neighbourhood of $\operatorname{clos} R$. By the Generalized Weierstrass Theorem (see [16, § 26]) one can find an R-entire function with an arbitrary finite set of zeros in $\operatorname{clos} R$.

We claim that for an appropriate choice of R-entire τ the function $\tau F_{\mathcal{O}}$ belongs to $E^{1-0}(\mathcal{O})$ for any component $\mathcal{O}$ of $R \setminus \varphi^{-1}(\Gamma)$ (see 7.2 for the definition of E^{1-0}). Look at the formula (11.11). First we see that $F_{\mathcal{O}}$ can have poles only at the points of Ξ and their order has a bound which does not depend on the "orthogonal" differential $d\omega$. So a proper R-entire function will take care of them. Consider next the numerator $N_{\mathcal{O}}$ in (11.11). It follows from (7.3) that $K_{v}[r]_q \in E^{1-0}(\varphi(\mathcal{O}))$ and since $d\varphi$ is analytic on $\operatorname{clos} R$, we have $(K_{v}[r]_q \circ \varphi)\tau_1 \in E^{1-0}(\mathcal{O})$ for an R-entire τ_1 having zeros of sufficiently high order at the zeros of $d\varphi$. The functions $v_{(j)}^{[r]} \circ \varphi$ can have at most finitely many weak singularities, so they can be made bounded if multiplied by some R-entire function. Thus we can find an R-entire function τ_2 independent of W, such that $\tau_2 N_{\mathcal{O}} | \mathcal{O}' \in E^{1-0}(\mathcal{O}')$ for any admissible simply connected subregion $\mathcal{O}' \subset \mathcal{O} \setminus \Xi$.

Suppose $\varphi(\mathcal{O}) \subset \Omega^{\kappa}$, $\kappa < n$. The denominator $\mathcal{D}_{\mathcal{O}}$ in (11.11) is equal to $\pm \operatorname{Minor}(H^{(\kappa)}(\varphi(\zeta)), \kappa)$, so applying (10.2) and the preceding observations we see that $\tau_3 F_{\mathcal{O}} \in E^{1-0}(\mathcal{O})$ for a proper R-entire function τ_3 independent of $\mathcal{O}$.

Now let $\varphi(\mathcal{O}) \subset \Omega^{n}$. The denominator $\mathcal{D}_{\mathcal{O}}$ is equal to $\pm \det H^{(n)}(\varphi(\zeta))$ (see 10.1) and (10.1) implies that $\mathcal{D}_{\mathcal{O}}^2$ is the product of an outer function (see Definition 7.5) and an R-entire function τ_4.

From Lemma 11.5 it is readily seen that $\tau_3 F_{\mathcal{O}} | \partial\mathcal{O} \in L^{1-0}(\partial\mathcal{O})$. At the same time fixing p, $p < 1/2$, we conclude that $(\tau_2 N_{\mathcal{O}})^2 \in E^{p}(\mathcal{O})$. Now we have

$$\tau_2^2 \tau_3^2 \tau_4 F_{\mathcal{O}}^2 | \partial\mathcal{O} \in L^{p}(\partial\mathcal{O}), \quad \tau_2^2 \tau_3^2 \tau_4 F_{\mathcal{O}}^2 (\mathcal{D}_{\mathcal{O}}^2 / \tau_4) =$$

$$= \tau_3^2 (\tau_2 N_{\mathcal{O}})^2 \in E^{p}(\mathcal{O}),$$

and $\mathcal{D}_{\mathcal{O}}^2 / \tau_4$ is an outer function.

So, by the Abstract Phragmen-Lindelöf principle (Lemma 7.6), $\tau_2^2 \tau_3^2 \tau_4 F_{\mathcal{O}}^2 \in E^{p}(\mathcal{O})$ and hence $\tau_2 \tau_3 \tau_4 F_{\mathcal{O}} \in E^{2p}(\mathcal{O})$. The boundary values of this function belong to $L^{1-0}(\partial\mathcal{O})$ and so $\tau_2 \tau_3 \tau_4 F_{\mathcal{O}} \in E^{1-0}(\mathcal{O})$. Thus the claim is proved.

Now we can make use of Theorem 7.4 to conclude that the functions $G_{\mathcal{O}} \overset{def}{=} \tau F_{\mathcal{O}}$ (where τ stands for $\tau_2 \tau_3 \tau_4$) form a single function $G \in E^{1-0}(R)$. By the choice of τ_2 we can have $G | \partial R \in L^{1}(\partial R)$ and so $G \in E^{1-0}(R)$. Consider the function $F = G/\tau$. It follows that $F d\varphi$ is a meromorphic differential in R with a finite number of poles of order not higher than n_j at fixed points λ_j and n_j, λ_j are independent of $d\omega$. Moreover,

$$F\,d\varphi \,|\, \partial R = d\omega .$$

Observe that for a small neighbourhood V of any point in ∂R the restriction $\tau | R \cap V$ is outer. Moreover, τ is bounded away from zero outside a neighbourhood of a finite set. Fixing the local coordinate in the Riemann surface we see by Lemma 7.6 that $F d\varphi = h\,dz$ in $R \cap V$ for some $h \in E^1(R \cap V)$.

Let χ be an R-entire function having zeros at λ_j of multiplicity n_j. It is not hard to see that the functional $h \mapsto \int_{\partial R} h\,d\omega$ vanishes on $\chi H^\infty(R)$, and hence on $\chi H^\infty(R, w)$ (since $H^\infty(R)$ is weak-* dense in $H^\infty(R, w)$). To prove this it is enough to decompose the domain R into a union of sufficiently small domains and compute the integral by the residue formula. Taking into consideration that $d\omega$ is an arbitrary differential satisfying (11.1) and χ does not depend on $d\omega$, we conclude that $span(T_\varphi^n U : n \geqslant 0) \supset BH^\infty(R, w)$. Therefore the T_φ-invariant subspace has finite codimension in $H^\infty(R, w)$ and hence U is nearly cyclic. To prove that under the condition (5.4) the set U is cyclic we use the following lemma.

11.7. LEMMA. <u>Let</u> T <u>be a continuous operator on a locally convex space</u> $\mathcal{H}$ <u>and let</u> $U \subset \mathcal{H}$. <u>Suppose</u> $E_T(U) \overset{def}{=} span(T^n U : n \geqslant 0)$ <u>has finite codimension. If also</u>

$$Ker(T^* \bar{\lambda} I) \cap U^{\perp} = \{0\}, \quad \forall \lambda \in \mathbb{C}, \tag{11.12}$$

<u>then</u> $E_T(U) = \mathcal{H}$, <u>i.e.</u> U <u>is cyclic for</u> T.

PROOF. The subspace $X = E_T(U)^{\perp} \subset \mathcal{H}^*$ has finite dimension and is invariant under T^*. If $X \neq \{0\}$, it must contain a non-zero eigen-vector which contradicts (11.12). ▨

To complete the proof of Lemma 11.6 it remains to observe that the conditions (5.4) and (11.12) (=(5.1)) are equivalent. This is proved in 5.2 for $\mathcal{H} = E^p(R)$ but the case of $H^\infty(R, w)$ is treated in a similar way.

Thus Lemma 11.6 is proved, as well as Proposition 10.5.

CHAPTER IV. VECTOR BUNDLE: THE CONSTRUCTION OF A CYCLIC SET

INTRODUCTION

In view of Proposition 10.5 for the proof of the Main Theorem it remains to produce a set $U=\{u^{[r]}\}_{r=0}^{n-1}$ satisfying (5.4), (10.1)-(10.4). In fact the condition (5.4) can be skipped. In this case we obtain a nearly cyclic set and the multiplicity is computed by Lemma 3.6. However it seems worthwhile to construct a cyclic set explicitly. When doing this we shall always indicate all parts of the proof that can be skipped if one is interested only in multiplicity.

The condition (10.4) unlike the others is of "generic quality". So first we find functions satisfying all the conditions except for (10.4). Then in Sect.15 a "small perturbation" of the construction will be used to ensure (10.4) without violation of (10.1)-(10.3).

GENERAL SCHEME OF THE CONSTRUCTION OF $u^{[r]}$. We put $u^{[0]}\equiv 1$. Then an easy computation involving determinants shows that all conditions sufficient for the cyclicity can be rewritten in terms of the differences $f_{ij}^{[r]}=u^{[r]}\circ\varphi_{(i)}^{-1}-u^{[r]}\circ\varphi_{(j)}^{-1}$, $r\geq 1$, where $\varphi_{(i)}^{-1}$, $\varphi_{(j)}^{-1}$ are some branches of φ^{-1}. Thus we begin by constructing $f_{ij}^{[r]}$ and then find $u^{[1]},\dots,u^{[n-1]}$.

The functions $f_{ij}^{[r]}$ must satisfy the additive cocycle condition: $f_{ij}^{[r]}+f_{jk}^{[r]}+f_{ki}^{[r]}\equiv 0$. To determine $u^{[r]}$, the cocycle $\{f_{ij}^{[r]}\}$ being known, it is essentially to solve the First Cousin Problem. The functions $\{f_{ij}^{[r]}\}$ must have several properties. Among them the most restrictive is the belt condition (10.3) which prescribes the rate of decrease of certain $f_{ij}^{[r]}$ in the belt sectors. The functions $f_{ij}^{[r]}$ will be searched for in the form $f_{ij}^{[r]}=w_{ij}g_{ij}^{[r]}$, where the "weights" w_{ij} are outer functions and have a required analytical behaviour. Thus our problem reduces to solving the system of equations

$$w_{ij}(\zeta)g_{ij}^{[\nu]}(\zeta)+w_{jk}(\zeta)g_{jk}^{[\nu]}(\zeta)+w_{ki}(\zeta)g_{ki}^{[\nu]}(\zeta)\equiv 0,$$

$$\zeta\in \mathrm{Dom}\,\varphi_{(i)}^{-1}\cap \mathrm{Dom}\,\varphi_{(j)}^{-1}\cap \mathrm{Dom}\,\varphi_{(k)}^{-1}. \qquad (*)$$

It is readily verified that for $\zeta\in\Omega^p$ the system (*) has $(p-1)$ linearly independent solutions. The condition (10.5) (which is a weaker form of (5.4), (10.1), (10.2)) just requires that the family $g_{ij}^{[\nu]}(\zeta)$, $\nu=1,\dots,n-1$, have the maximal possible rank $(p-1)$.

How do we solve (*)? To begin with, the image $\varphi(R)$ is divided into "cells" with a constant number of pre-images in each. (The cells are in general smaller than the components of $\mathbb{C}\setminus\Gamma$) In every cell the Oka principle enables us to choose an analytic basis in the space of solutions of (*). We must ensure only the proper "agreement" in adjacent cells. This problem may be interpreted as the problem of trivialization of a special analytic vector bundle.

The "analytic vector bundle" is meant in a broad sense. Suppose that in each cell we have an m-dimensional "subspace-valued" function and on its boundary we have a transition matrix mapping the subspaces of the cell onto the subspaces of the adjacent cell. Then we say that an "analytic vector bundle" is given.

To construct the "bundle" we extend the pair (R,φ) to form an n-sheeted covering (S,π) , as in Sect.6. As a matter of fact the elements of the cyclic set will be defined in S but they will be analytic only in $S\setminus\partial R$. Such an extension will add new equations to the system (*), so that their number will be the same in all cells. The dimension of the solution space will be just $(n-1)$. Then transition matrices should be defined in such a way that analytic agreement in adjoining cells would be guaranteed.

To "trivialize the bundle" or to "find a basis of sections"

means to find matrix-valued functions in each cell, whose columns form a basis of the solution space for (*). Mereover these matrices must agree on the common boundary of two adjacent cells in accordance with the transition matrix. This is a sort of the Riemann-Hilbert matrix problem.

However our circumstances are different from those familiar in the classical theory of analytic vector bundles. The sections under consideration must have some properties of smoothness up to the boundary. These properties are needed both for the proper analytic "sticking" and for the subsequent solving of the First Cousin problem. Specifically we shall deal with the Lipschitz class $Lip\,\alpha$, $\alpha<1$, for which the theory of the Riemann-Hilbert problem was studied extensively.

12. REDUCTION TO THE PROBLEM ON WEIGHTED COCYCLES.

In Theorem 3.3 the weighted cocycle problem is treated (see [56] for its proof). Here we state a more complicated assertion which allows us to construct a set of functions satisfying (5.4), (10.1)-(10.3).

12.1. PREPARATION. We shall apply the construction of Sect.6. Consider a triangulation possessing the properties (6.1)-(6.5). Recall that "curvilinear triangles" are called "cells" and with each cell Ω_t we associate its multiplicity n_t.

Let Ω_s and Ω_t have a common boundary arc. We shall say that they are <u>adjacent cells of the first type</u> if $n_s = n_t$, and <u>adjacent cells of the second type</u> otherwise. Since the pair (R,φ) is assumed to be strictly admissible (see Sect.9), no two arcs of ∂R have the same image. Thus adjacent cells are of the second type if and only if $\partial\Omega_s \cap \partial\Omega_t \subset \Gamma$. In this case it is clear that $|n_s - n_t| = 1$.

For each cell Ω_t consider the branches of φ^{-1}: $\varphi^{-1}_{(1)},\dots,\varphi^{-1}_{(n_t)}$. They are single-valued since all images of branch points (the set of images of branch points will be denoted by $\mathcal{Z}$) are vertices of the triangulation.

As in 6.12 we define correspondence functions δ_{st} for any two adjacent cells Ω_s, Ω_t. These functions are one-to-one, $\delta_{st}: \mathcal{N} \to \mathcal{N}$, $\mathcal{N} \overset{def}{=} \{1, \dots, n\}$, and satisfy (6.6). The indices $i, \tilde{i}$ such that $\delta_{st} i = \tilde{i}$, will be called associated (for fixed Ω_s, Ω_t). The domain R can be assumed to lie in the covering surface S as in 6.13, 6.14. Then for each cell Ω_t there are n sheets of the surface S over it, n_t of which, with numbers $1, \dots, n_t$, lie in R. The indices $1, \dots, n_t$ will be called "true indices of the cell Ω_t" and those exceeding n_t will be called "false indices".

12.2. CONDITIONS ON WEIGHT FUNCTIONS. Suppose that for each cell Ω_t we have functions $w_{ij}^{(t)} \in A^1(\Omega_t)$, $1 \le i < j \le n_t$. These functions should possess the following properties:

$$|w_{ij}^{(t)}(\zeta)| + |w_{jk}^{(t)}(\zeta)| \ge \delta > 0, \quad \zeta \in \Omega_t, \quad i < j < k; \tag{12.1}$$

if Ω_t, Ω_s are adjacent cells and the indices $i, j \le n_t$ are such that $\delta_{ts} i \le n_s$, $\delta_{ts} j \le n_s$, then (12.2)

$$|w_{ij}^{(t)}(\zeta)| \ge \delta > 0, \quad \zeta \in \partial\Omega_t \cap \partial\Omega_s.$$

In what follows we shall consider the functions $w_{ij}^{(t)}$ for $i, j \le n$, $i \ne j$, assuming that $w_{ji}^{(t)} = w_{ij}^{(t)}$ and $w_{ij}^{(t)} \equiv 1$ if either i or j exceeds n_t i.e. is a "false" index.

In the next theorem the existence of a family of weighted cocycles $\{g_{ij,(t)}^{[r]}\}$, $r = 1, \dots, n-1$, for each cell Ω_t is claimed. Consider the $\frac{n(n-1)}{2} \times (n-1)$ matrix-valued function:

$$G_t(\zeta) \overset{def}{=} \left[g_{ij,(t)}^{[r]}(\zeta) \right]_{1 \le i < j \le n}^{1 \le r \le n-1}. \tag{12.3}$$

We shall also need certain submatrices of G_t.

Denote by $Vert(S\Gamma)$ the set of all vertices of $S\Gamma$ (see 6.11 for the definition of $S\Gamma$). Put $Vert(\Omega_t) = clos\,\Omega_t \cap Vert(S\Gamma)$. Let $c \in Vert(\Omega_t)$ and let $\mathcal{K}$ be the set of indices $i \le n_t$ such that $\varphi_{(i)}^{-1}(c) \in R$. Suppose $\nu = card\,\mathcal{K} \ge 2$ and consider the $\frac{\nu(\nu-1)}{2} \times (n-1)$ matrix-valued function

$$G_{t,c}(\zeta) \stackrel{def}{=\!=} \left[g_{ij,(t)}^{[r]}(\zeta) \right]_{i<j;\, i,j\in \mathcal{K}}^{1\le r\le n-1} . \qquad (12.4)$$

By the symbol $\mathcal{J}^{\varepsilon}(\Omega_t)$ we shall denote the class of functions of the form $\sum_{k\le N} \tau_k(\zeta) P_k(\zeta)$, where $\tau_k \in \lambda^{\varepsilon}(\Omega_t)$ and the P_k are polynomials in $\log(\zeta - c_k)$, $c_k \in Vert(\Omega_t)$ (see 3.2). Now we can state the theorem.

12.3. THEOREM. Let the weight functions $w_{ij}^{(t)} \in A^1(\Omega_t)$ satisfy conditions (12.1), (12.2). Then there exist $\varepsilon > 0$, $\ell > 0$, and the functions $g_{ij,(t)}^{[r]} \in \mathcal{J}^{\varepsilon}(\Omega_t)$, $r = 1, \ldots, n-1;\ 1 \le i < j \le n$, such that the following conditions hold.

The cocycle condition:

$$\forall i,j,k,r,\quad w_{ij}^{(t)} g_{ij,(t)}^{[r]} + w_{jk}^{(t)} g_{jk,(t)}^{[r]} - w_{ik}^{(t)} g_{ik,(t)}^{[r]} \equiv 0 \quad \text{in } \Omega_t \qquad (12.5)$$

The agreement condition:

if Ω_t, Ω_s are adjacent cells and if $i, \tilde{i}$ and j $\tilde{j}$ are associated pairs of "true" indices, then on the arc $\partial\Omega_t \cap \partial\Omega_s$ perhaps except for the ends: (12.6)

$$g_{ij,(t)}^{[r]} w_{ij}^{(t)} = g_{\tilde{i}\tilde{j}}^{[r]} w_{\tilde{i}\tilde{j}}^{(s)} .$$

The nondegeneracy conditions: for each cell Ω_t

$$\sum |\operatorname{Minor}(G_t(\zeta), n-1)| \ge const \prod_{c\in Vert(\Omega_t)} |\zeta - c|^{\ell} ; \qquad (12.7)$$

for any vertex c of the cell Ω_t, in a neighbourhood of c

$$\sum |\operatorname{Minor}(G_{t,c}(\zeta), \nu - 1)| \ge const\, |\zeta - c|^{(\nu - \varkappa)/2} . \qquad (12.8)$$

Here G_t, $G_{t,c}$ are defined by (12.3), (12.4), $\nu = card\, \mathcal{K}$

(see 12.2), and $\varkappa$ is the number of distinct pre-images among $\{\varphi_{(i)}^{-1}(c)\}$, $i \in \mathcal{K}$. The sums are taken over all minors of indicated order.

REMARKS. 1) Theorem 3.3 follows from Theorem 12.3 by making a triangulation and passing to cells as in 12.1.

2) If we confine ourselves to constructing a nearly cyclic set, the condition (12.8) can be omitted.

Theorem 12.3 will be proved in sections 13, 14, and now we are going to apply it to the construction of a cyclic set $\{u^{[r]}\}$. Note right now that (10.1), (10.2) will follow from (12.7) and that the condition (5.4) at the vertices will be implied by (12.8).

12.4. THE CONSTRUCTION OF $u^{[r]}$. First the weight functions $W_{ij}^{(t)}$ will be chosen. If a cell Ω_t contains a sector S of some belt (see 4.10), consider the indices $a=a(S)$, $b=b(S)$, $a,b \leq n_t$, in (10.3). Recall that h_S is a smooth positive function integrable with respect to the harmonic measure for S (see (7.4)). For the function $W_{ab}^{(t)}(\zeta)$ we take the product of $(\zeta-z)^2$ (where z denotes the vertex of S) and of an outer function in Ω_t with modulus equal to $exp\{-h_S(|\zeta-z|)\}$ on $\partial\Omega_t$. Since h_S is smooth, clearly $W_{ab}^{(t)} \in A^1(\Omega_t)$. In all the other cases we set $W_{ij}^{(t)} \equiv 1$. The conditions (12.1), (12.2) clearly hold. We use Theorem 12.3 to obtain the functions $g_{ij,(t)}^{[r]}$. The next step is the solution of the First Cousin Problem in $\mathcal{Y}^{\varepsilon}$.

Put $f_{ij,(t)}^{[r]} = W_{ij}^{(t)} \cdot g_{ij,(t)}^{[r]}$, and assume that $f_{ji,(t)}^{[r]} = -f_{ij,(t)}^{[r]}$. We wish to construct functions $u^{[r]}, r=1,\ldots,n-1$, with the following properties:

(I) $u^{[r]}$ are analytic in R and bounded outside a neighbourhood of a finite set where they may have only logarithmic rate of growth;

(II) if $\zeta \in \Omega_t, i,j \leq n_t$, then

$$(u^{[r]}\circ\varphi_{(j)}^{-1})(\zeta) - (u^{[r]}\circ\varphi_{(i)}^{-1})(\zeta) = f_{ij,(t)}^{[r]}(\zeta), \quad r \leq n-1 . \tag{12.9}$$

The construction will proceed in two steps.

STEP 1. For each cell $\Omega_t \subset \varphi(R)$ consider the functions

$$y_{j,(t)}^{[r]} \stackrel{def}{=} f_{1j,(t)}^{[r]}, \quad j \leq n_t .$$

Recall that $S\Gamma = \bigcup_t \partial\Omega_t$ and put

$$x^{[r]}(\zeta) = y_{j,(t)}^{[r]}(\varphi(\zeta)), \qquad \zeta \in R \setminus \varphi^{-1}(S\Gamma),$$

where t and $j \leq n_t$ are fixed by the requirements $\varphi(\zeta) \in \Omega_t$, $\varphi_{(j)}^{-1}(\varphi(\zeta)) = \zeta$. The functions $x^{[r]}$ are analytic in $R \setminus \varphi^{-1}(S\Gamma)$ and satisfy the property (II) by the cocycle condition (12.5).

STEP 2. To "improve" $x^{[r]}$ consider the following functions analytic in $\mathbb{C} \setminus S\Gamma$:

$$\mathcal{K}^{[r]}(z) \stackrel{def}{=} \sum_{\gamma \subset S\Gamma} \frac{1}{2\pi i} \int_\gamma \left(y_{k,(t)}^{[r]} - y_{\ell,(s)}^{[r]}\right)(\zeta)(\zeta - z)^{-1} d\zeta, \quad 1 \leq r \leq n-1 . \tag{12.10}$$

Here the sum is taken over all arcs $\gamma \subset S\Gamma \cap \varphi(R)$ (the orientation on γ is not fixed yet); Ω_t and Ω_s are cells separated by γ ; k, ℓ are arbitrary associated "true" indices for Ω_t, Ω_s. The definition (12.10) does not depend on the choice of the "true" index k since by (12.6) we have on γ :

$$y_{k,(t)}^{[r]} - y_{k',(t)}^{[r]} = f_{k'k,(t)}^{[r]} = f_{\ell'\ell,(t)}^{[r]} = y_{\ell,(s)}^{[r]} - y_{\ell',(s)}^{[r]} ,$$

for $\ell = \delta_{ts} k$, $\ell' = \delta_{ts} k'$. The functions $\mathcal{K}^{[r]}$ are easily seen to have only logarithmic singularities at the vertices of $S\Gamma$. (Indeed, let γ be an arc ending at the origin, $\tau \in Lip^{\varepsilon}(\gamma)$. We have

$$\tau(z) \log^m(z) = \tau(0) \log^m(z) + [\tau(z) - \tau(0)] \log^m(z) = \tau_1(z) + \tau_2(z).$$

The Cauchy integral of τ_1 is easily computed to produce a power of logarithm. $\tau_2(z) \in Lip^{\varepsilon/2}(\gamma)$. Extend γ to form a closed smooth Jordan contour C and continue τ_2 to C by linearity. Clearly $\tau_2 \in Lip^{\varepsilon/2}(C)$. The Cauchy integral operator act on $Lip^{\varepsilon}(C)$, $\varepsilon<1$, (see [15]). So we obtain the required.) Now put

$$u^{[0]} \equiv 1, \quad u^{[r]} = x^{[r]} + \mathcal{K}^{[r]} \circ \varphi, \quad r=1,\dots,n-1. \tag{12.11}$$

Clearly for an appropriate choice of orientation on the arcs γ in (12.10), by the "jump" formula (7.1) we conclude that each of the $u^{[r]}$ has concordant boundary values on the arcs of $R \cap \varphi^{-1}(S\Gamma)$. Hence $u^{[r]}$ are analytic outside the vertices of $\varphi^{-1}(S\Gamma)$. At the vertices they can have only logarithmic rate of growth, and therefore all $u^{[r]}$ are analytic in R. The properties (I), (II) apparently hold.

Note that (10.3) follows immediately from the construction. Now we are to verify (5.4), (10.1), (10.2). Recall that we do not take care of (10.4) for the time being.

We need the following lemma; its proof is obvious.

12.5. LEMMA. Suppose a numerical matrix $A=\{a_{ij}\}_{i,j=1}^{m}$ has all the entries of the first column equal to 1. Consider the $\frac{m(m-1)}{2} \times (m-1)$ matrix

$$B = \{b_{(k,j),i}\}_{1\le k<j\le m}^{1\le i\le m-1}, \quad b_{(k,j),i} = a_{j,i+1} - a_{k,i+1}.$$

It is claimed that $\det A \neq 0$ if and only if $\operatorname{rank} B = m-1$, and in this case $\det A$ and all nonzero minors of B of order $(m-1)$ are equal (modulo the sign). Moreover, consider a full graph with vertices labelled $1,\dots,m$ and arcs (k,j). Then an $(m-1)$-minor of B is nonzero if and only if the subgraph corresponding to the indices of its rows is a tree.

12.6. VERIFICATION OF (10.1), (10.2) AND (5.4) OUTSIDE THE VERTICES. Note first that for Ω^1 all these conditions follows from the fact that $u^{[0]} \equiv 1$. Fix a cell Ω_t with $n_t > 1$. Consider the following matrix-valued functions:

$$H_t = \left[u^{[r]} \circ \varphi_{(i)}^{-1}\right]_{1\le i\le n_t}^{0\le r\le n-1}, \quad F_t = \left[f_{1i}^{[r]}\right]_{1\le i\le n}^{0\le r\le n-1},$$

assuming $f_{1i}^{[0]} \equiv 1$. Fix a vertex c of Ω_t and a neighbourhood $U_c \ni c$ with the only condition that $clos\, U_c \cap Vert(\Omega_t) = \{c\}$. In view of (12.11), $u^{[r]} \circ \varphi_{(i)}^{-1} = f_{1i}^{[r]} + K^{[r]}$, $r \geq 1$, and $u^{[0]} \equiv 1$, which implies the following estimate in $\Omega_t \cap U_c$:

$$\Sigma |\, Minor(H_t(\zeta), n_t - 1)| \geq const\, |\det F_t(\zeta)| \cdot \ln^{-q} |\zeta - c| \, . \tag{12.12}$$

Here we make use of the fact that $f_{1i}^{[r]}$ can have only logarithmic singularities at the vertices. By Lemma 12.5,

$$|\det F_t(\zeta)| \geq const \prod_{1 \leq i < j \leq n} |\, W_{ij}^{(t)}(\zeta)| \cdot \Sigma |\, Minor(G_t(\zeta), n-1)| \, , \tag{12.13}$$

with G_t defined by (12.3). Now, combining the estimates (12.7), (12.12), (12.13), we obtain for some positive m :

$$\Sigma |\, Minor(H_t(\zeta), n_t - 1)| \geq const \prod |\, W_{ij}^{(t)}(\zeta)| \cdot |\zeta - c|^{m} \, . \tag{12.14}$$

Recalling the choice of the weight functions $W_{ij}^{(t)}$ (see 12.4), we see that $H_t(\zeta)$ has maximal possible rank in $clos(\Omega_t \cap U_c) \setminus \{c\}$. The neighbourhood U_c for all $c \in Vert(\Omega_t)$ can be chosen to form a cover of Ω_t . Thus we deduce that $rank\, H_t(\zeta) = n_t$ for $\zeta \in clos\, \Omega_t \setminus Vert(\Omega_t)$. This implies the condition (5.4) for $a \in \mathbb{C} \setminus Vert(S\,\Gamma)$. The estimate (12.14) yields also (10.2) with $Y = Vert(S\Gamma)$. The condition (10.1) follows as well since each small sector of a 'belt is contained in some cell Ω_t by (6.3) and the $W_{ij}^{(t)}$ are outer.

REMARK. We have checked all the conditions (except for (10.4) as was agreed) sufficient for the set U to be nearly cyclic. Note that (12.8) was not used. The rest of the section is devoted to the verification of (5.4). Remind that we do not need (5.4) if we confine ourselves to the computation of multiplicity.

In the next lemma the condition (5.4) is rewritten in an equivalent form. Its proof is a routine procedure involving determinants and will be omitted.

12.7. LEMMA. <u>Suppose</u> $c \in \varphi(R)$, $\varphi^{-1}(c) \cap R = \{z_1, \dots, z_{\varkappa}\}$, z_j <u>being</u>

<u>the pre-image of</u> c <u>of multiplicity</u> $k_j, \nu = \sum_{j=1}^{\varkappa} k_j$. <u>Let</u> $f^{[0]}$, $f^{[1]}, \dots, f^{[\nu-1]}$ <u>be arbitrary functions analytic in</u> $\mathcal{R}$ <u>and let</u> $\varphi^{-1}_{(1)}, \dots, \varphi^{-1}_{(\nu)}$ <u>be (local) branches of</u> φ^{-1} <u>defined in a neighbourhood of</u> c . <u>Then the following are equivalent</u>:

$$(i) \quad \det \left[\dots ; f^{[r]}(z_j), \dots, (f^{[r]})^{(k_j-1)}(z_j); \dots \right]_{j \le \varkappa}^{0 \le r \le \nu-1} \neq 0 ;$$

$$(ii) \quad \left| \det \left[(f^{[r]} \circ \varphi^{-1}_{(s)})(\zeta) \right]_{s \le \nu}^{0 \le r \le \nu-1} \right| \sim \mathrm{const} \, |\zeta - c|^{\frac{\nu - \varkappa}{2}}, \quad \zeta \to c .$$

12.8. THE VERIFICATION OF (5.4) AT THE VERTICES. Let c be a vertex of $S\Gamma$. We use the notation introduced in the preceding lemma. We are to verify that

$$\mathrm{rank} \left[\dots ; u^{[r]}(z_j), \dots, (u^{[r]})^{(k_j-1)}(z_j); \dots \right]_{j \le \varkappa}^{0 \le r \le n-1} = \nu . \qquad (12.15)$$

The branches $\varphi^{-1}_{(1)}, \dots, \varphi^{-1}_{(\nu)}$ of φ^{-1} can be chosen analytic in a neighbourhood U_c of the vertex c cut along one of the arcs ending at c . When continued across this arc the set of branches $\{\varphi^{-1}_{(j)}\}$ undergoes a permutation. Consider all cells having c as a vertex and choose a cell Ω_t of minimal multiplicity n_t among them. In this cell the branches $\varphi^{-1}_{(s)}$, $1 \le s \le n_t$, are defined, and $n_t \ge \nu$. Put

$$A(\zeta) \overset{\text{det}}{=\!=} \left[(u^{[r]} \circ \varphi^{-1}_{(s)})(\zeta) \right]_{1 \le s \le \nu}^{0 \le r \le n-1} .$$

According to Lemma 12.5,

$$\Sigma \left| \mathrm{Minor}(A(\zeta), \nu) \right| \ge \mathrm{const} \left| \prod_{1 \le i < j \le \nu} w_{ij}^{(t)}(\zeta) \right| \cdot \Sigma \left| \mathrm{Minor}(G_{t,c}(\zeta), \nu - 1) \right| ,$$

where the matrix-valued function $G_{t,c}$ is defined by (12.4). We have $w_{ij}^{(t)}(c) \neq 0$, $1 \le i < j \le \nu$, by the choice of $w_{ij}^{(t)}$ (see

12.4) and we obtain from the estimate (12.8) that in a neighbourhood of c

$$\Sigma |\operatorname{Minor}(A(\zeta),\nu)| \geq \operatorname{const} |\zeta - c|^{(\nu-\varkappa)/2}.$$

Taking into account that all functions $\zeta \mapsto [\operatorname{Minor}(A(\zeta),\nu)]^2$ are analytic and single-valued in a neighbourhood of c, we conclude that at least one of them has a zero of order not greater than $(\nu-\varkappa)$ at c. By Lemma 12.7 this implies (12.15). Thus to complete the work it remains only to prove Theorem 12.3.

13. THE BEGINNING OF THE PROOF OF THEOREM 12.3: THE CONSTRUCTION OF A BUNDLE.

In accordance with the scheme presented at the beginning of the Chapter, we construct an "analytic vector bundle" (in a broad sense), whose sections will turn out to be the required cocycles.

We shall employ the following notation: if $\mathcal{A}(\Omega)$ is a space consisting of functions defined on the set Ω, then $\mathcal{A}_{k\times \ell}(\Omega)$ denotes the space of $k\times\ell$ matrix-valued functions with entries from $\mathcal{A}(\Omega)$.

First we note that in the case $n=2$ Theorem 12.3 is trivial. Indeed, if $n=2$ the cocycle condition disappears. In any cell Ω_t for which $n_t=2$, one can take $g^{[1]}_{12,(t)} \in \Lambda^{\varepsilon}(\Omega_t)$ nonvanishing in $\operatorname{clos}\Omega_t$. In the sequel we assume that $n\geq 3$.

13.1. CELL MATRIX. Put $m=\frac{n(n-1)}{2}$, $p=\frac{n(n-1)(n-2)}{6}$. Consider the sets

$$\mathcal{H} \overset{def}{=} \{(i,j) : 1\leq i<j\leq n\},\quad \mathcal{M} \overset{def}{=} \{(i,j,k) : 1\leq i<j<k\leq n\}.$$

In the spaces $\mathbb{C}^n, \mathbb{C}^m, \mathbb{C}^p$ we fix bases which will be indexed by the elements of $\mathcal{N}=\{1,\dots,n\}$, $\mathcal{H}$, $\mathcal{M}$ respectively: $e(i)$, $e(i,j)$, $e(i,j,k)$. Unless otherwise specified, linear operators acting between these spaces will be identified with their matrices in these bases. Assume that $e(j,i)=-e(i,j)$ and $e(\sigma(i),\sigma(j),\sigma(k)) = \operatorname{sgn}(\sigma)\, e(i,j,k)$, where σ is a permutation of (i,j,k). Sometimes it will be convenient to use the bases $\{e(\tilde{i})\}$, $\{e(\tilde{i},\tilde{j})\}$,

$e(\tilde{i},\tilde{j},\tilde{k})$, where $i \mapsto \tilde{i}$ is a permutation of $\mathcal{N}$.

Consider the operator $M : \mathbb{C}^p \to \mathbb{C}^m$ defined by

$$Me(i,j,k) = e(i,j) + e(j,k) - e(i,k) \, .$$

It is convenient to employ the pairing $\langle x,y \rangle = \Sigma x_\nu y_\nu$. For an operator $A : X \to Y$, the operator $A^T : Y \to X$ is defined by the equality $\langle Ax,y \rangle = \langle x,A^T y \rangle$; naturally its matrix is the transpose of the matrix of A .

With every cell Ω_t we associate a diagonal matrix-valued function $W_t : \mathbb{C}^m \to \mathbb{C}^m$, acting as follows:

$$W_t(\zeta)\, e(i,j) = w_{ij}^{(t)}(\zeta)\, e(i,j) \, , \quad i,j \le n \, .$$

The matrix-valued function $R_t = M^T W_t : \mathbb{C}^m \to \mathbb{C}^p$ will be called the <u>cell matrix</u> for Ω_t . Clearly $R_t \in A^1_{p\times m}(\Omega_t)$. Further consider the "space-valued" function

$$L_t(\zeta) = \operatorname{Ker} R_t(\zeta) \, .$$

The requirement that the value of a vector function $\{g_{ij}(\zeta)\}$ be in $L_t(\zeta)$ for any $\zeta \in \Omega_t$ implies the cocycle condition (12.5).

13.2. LEMMA. $\forall z \in \operatorname{clos} \Omega_t \, , \ \dim L_t(z) = n-1$.

PROOF. Fix a point $z \in \operatorname{clos} \Omega_t$. The assumption that $\{d_{ij}\} \in L_t(z)$ means that $\{w_{ij}^{(t)}(z)\, d_{ij}\}$ is a numerical cocycle and hence $w_{ij}^{(t)}(z)\, d_{ij} = c_i - c_j$ for some $c_k \in \mathbb{C}$, $k=1,\dots,n$. Let $\mathcal{A}_z = \{(i,j) : w_{ij}^{(t)}(z) = 0\}$. By (12.1) two different pairs $(i,j), (i',j') \in \mathcal{A}_z$ cannot have a common index. So $\{c_i\}$ are arbitrary numbers, satisfying the property: $c_i = c_j$ whenever $(i,j) \in \mathcal{A}_z$. If we fix $c_1 = 0$ the correspondence between such $\{d_{ij}\}_{(i,j) \notin \mathcal{A}_z}$ and such $\{c_i\}_{i=1}^n$ will be one-to-one. One can also choose $\{d_{ij}\}$, $(i,j) \in \mathcal{A}_z$, to be arbitrary complex numbers. Thus we see that

$$\dim L_t(z) = n - \operatorname{card} \mathcal{A}_z - 1 + \operatorname{card} \mathcal{A}_z = n-1 \, .$$

For the construction of a bundle and a basis of its sections we shall need the following lemma. We shall say that a $k \times k'$ matrix-valued function $U\,(k > k')$ has a <u>uniform estimate from below</u> on a set Ω if it has a left inverse matrix-valued function with uniformly bounded norms on Ω, or equivalently, if the sums of moduli of all senior minors of $\mathbb{U}$ are bounded away from zero on Ω.

13.3. LEMMA. Let Ω be a Jordan domain with piecewise C^2-smooth boundary and let $\mathbb{U}$ be a matrix-valued function, $\mathbb{U}\in A^1_{\kappa\times\kappa'}(\Omega)$, $\kappa>\kappa'$, having a uniform estimate from below on Ω. Then

(i) $\mathbb{U}$ can be complemented to a square matrix-valued function invertible in $A^1_{\kappa\times\kappa}(\Omega)$;

(ii) $\mathbb{U}$ has a left inverse $\widetilde{\mathbb{U}}\in A^1_{\kappa'\times\kappa}(\Omega)$;

(iii) there exists $V\in A^1_{\kappa\times(\kappa-\kappa')}(\Omega)$ with a uniform estimate from below on Ω and such that $\mathrm{Im}\,V=(\mathrm{Im}\,\mathbb{U})^{\perp}$.

All statements of the lemma are well-known, see for example, [29], [49]. Surely they are not independent but we have preferred to state the needed facts explicitly.

13.4. TRANSITION MATRICES. In order to define a bundle we must produce transition matrices. Below we list the properties to be reconciled.

For any two adjacent cells Ω_t, Ω_s we wish to construct an operator-valued function $B_{ts}:\mathbb{C}^m\to\mathbb{C}^m$ on their common boundary. B_{ts} will be written as a matrix in the bases $\{e(i,j)\}$, $\{e(\tilde{i},\tilde{j})\}$ where the permutation $i\mapsto\tilde{i}$ is defined by the correspondence function δ_{ts}. Thus the columns of B_{ts} will be indexed by $(i,j)\in\mathcal{H}$, and the rows will be indexed by $(\tilde{i},\tilde{j})$. The diagonal entry of a column (i,j) will be considered to stand in the row $(\tilde{i},\tilde{j})$. (Remind that $i,\tilde{i}$ and $j,\tilde{j}$ are called associated pairs of indices). Assume that $w^{(t)}_{ji}=w^{(t)}_{ij}$. The transition matrices B_{ts} should have the following properties:

$$B^{\pm1}_{ts}\in C^1_{m\times m}(\partial\Omega_t\cap\partial\Omega_s),\quad B_{ts}=B^{-1}_{st}\;; \tag{13.1}$$

$$B_{ts}L_t\subset L_s \qquad \text{on } \partial\Omega_t\cap\partial\Omega_s; \tag{13.2}$$

If $i,\tilde{i}$ and $j,\tilde{j}$ are associated pairs of "true" indices (see (12.1), then the only nonzero entry in the union of the row $(\tilde{i},\tilde{j})$ and the column (i,j) of B_{ts} is the diagonal entry and it is equal to (13.3)

$w_{ij}^{(t)} / w_{\tilde{i}\tilde{j}}^{(s)}$.

REMARK. For any vertex a (a vertex is an endpoint of an arc of $S\Gamma$; see 12.1, 6.11), let $B_1^{(a)}, \dots, B_N^{(a)}$ be all the transition matrices defined at a enumerated clockwise, and let $\mathcal{D}^{(a)} = B_1^{(a)} \cdot \ldots \cdot B_N^{(a)}$. The conditions (13.1), (13.2) determine the "bundle" essentially. It would be nice if $\mathcal{D}^{(a)}$ were equal to the unit matrix. But this is, in general, impossible, and as a consequence we shall obtain sections having logarithmic growth at the vertices.

Now proceed to the construction of B_{ts} . Let Ω_t, Ω_s be adjacent cells, $\gamma = \partial\Omega_t \cap \partial\Omega_s$; $\tilde{k} = \sigma_{ts} k$, $k \leq n$. Let $\mathcal{E}_m : \mathbb{C}^m \to \mathbb{C}^m$, $\mathcal{E}_p : \mathbb{C}^p \to \mathbb{C}^p$ be linear operators acting as follows: $\mathcal{E}_m e(i,j) = e(\tilde{i}, \tilde{j})$, $\mathcal{E}_p e(i,j,k) = e(\tilde{i}, \tilde{j}, \tilde{k})$. The condition (13.2) is clearly implied by

$$\mathcal{E}_p R_t = R_s B_{ts} \tag{13.4}$$

(recall that $L_t = \operatorname{Ker} R_t$). If Ω_t, Ω_s are adjacent cells of the first type (see 12.1), one can set

$$B_{ts} = \operatorname{diag}\left[w_{ij}^{(t)} / w_{\tilde{i}\tilde{j}}^{(s)}\right] = W_s^{-1} \mathcal{E}_m W_t$$

Then (13.4) follows from the obvious equality $M\mathcal{E}_p^{\pm 1} = \mathcal{E}_m^{\pm 1} M$
The condition (13.1) follows from (12.2), so all the properties of a transition matrix are satisfied.

Let Ω_t, Ω_s be adjacent cells of the second type. To be definite let $n_t = n_s + 1$ and $\tilde{1} > n_s$. The matrix-valued function B_{ts} will be searched for in the form

$$B_{ts} = b_{ts} \oplus \operatorname{diag}\left[w_{ij}^{(t)} / w_{\tilde{i},\tilde{j}}^{(s)}\right]_{2 \leq i < j \leq n} \tag{13.5}$$

Then the equality (13.4) reduces to $Q_s b_{ts} = Q_t$, where $p \times (n-1)$ matrix-valued functions Q_t, Q_s consist of the columns of $\mathcal{E}_p R_t, R_s$ having indices $(1,i), (\tilde{1}, \tilde{i})$ correspondingly. These matrices are of the following form (keep in mind that $w_{\tilde{1}\tilde{i}}^{(s)} \equiv 1$ since $\tilde{1}$ is a "false" index for Ω_s):

$$Q_t = \begin{array}{c|cccccc|} & (12) & (13) & \cdots & & & (1n) \\ \hline (\tilde{1}\tilde{2}\tilde{3}) & w_{12}^{(t)} & -w_{13}^{(t)} & 0 & \cdots & \cdots & 0 \\ \vdots & w_{12}^{(t)} & 0 & -w_{14}^{(t)} & \cdots & \cdots & 0 \\ \vdots & \vdots & \vdots & \vdots & \ddots & & \vdots \\ & w_{12}^{(t)} & 0 & 0 & \cdots & \cdots & -w_{1n}^{(t)} \\ \hline & 0 & w_{13}^{(t)} & -w_{14}^{(t)} & 0 & \cdots & 0 \\ & \vdots & \vdots & & \ddots & & \vdots \\ & 0 & w_{13}^{(t)} & 0 & \cdots & \cdots & -w_{1n}^{(t)} \\ \hline & \cdots & \cdots & \cdots & \cdots & \cdots & \cdots \\ \hline & 0 & \cdots & \cdots & 0 & w_{1n-1}^{(t)} & -w_{1n}^{(t)} \\ \hline \end{array} \qquad Q_s = \begin{array}{c|cccccc|} & (\tilde{1}\tilde{2}) & (\tilde{1}\tilde{3}) & \cdots & & & (\tilde{1}\tilde{n}) \\ \hline (\tilde{1}\tilde{2}\tilde{3}) & 1 & -1 & 0 & \cdots & \cdots & 0 \\ \vdots & 1 & 0 & -1 & \cdots & \cdots & 0 \\ \vdots & \vdots & \vdots & \vdots & \ddots & & \vdots \\ & 1 & 0 & 0 & \cdots & \cdots & -1 \\ \hline & 0 & 1 & -1 & 0 & \cdots & 0 \\ & \vdots & \vdots & & \ddots & & \vdots \\ & 0 & 1 & 0 & \cdots & \cdots & 1 \\ \hline & \cdots & \cdots & \cdots & \cdots & \cdots & \cdots \\ \hline & 0 & \cdots & \cdots & 0 & 1 & -1 \\ \hline \end{array}$$

By q_t, q_s denote submatrices of Q_t, Q_s consisting of their first $(n-2)$ rows. Observe that the row of Q_t with index $(\tilde{1},\tilde{i},\tilde{j})$ is equal to the difference of the rows with indices $(\tilde{1},\tilde{2},\tilde{j})$ and $(\tilde{1},\tilde{2},\tilde{i})$ i.e. the rows of q_t. The same is true for the matrix Q_s. Thus we see that the equality (13.4) reduces to $q_s \mathscr{b}_{ts} = q_t$. To find such a matrix we complement q_s and q_t to invertible $(n-1)\times(n-1)$ matrix-valued functions B_s, B_t and put $\mathscr{b}_{ts} = B_s^{-1} B_t$. The matrix-valued function B_s can be obtained by adding to q_s the row with entries 1. In order to complement q_t we note that its senior minors are equal to $\pm \prod w_{1i}^{(t)}$ $(2 \leqslant i \leqslant n,\ i \neq j)$ for $j = 2, \dots, n$. Since by (12.1) two functions $w_{1i}^{(t)}, w_{1j}^{(t)}$, $i \neq j$ cannot have a common zero, the sums of moduli of all senior minors are bounded away from zero. By Lemma 13.3 (i) we can find the desired B_t, $B_t^{\pm 1} \in A^1_{(n-1)\times(n-1)}(\Omega_t)$. Thus the transition matrices satisfying (13.1)-(13.3) are constructed.

13.5. RIEMANN-HILBERT PROBLEM. Now we would like to find a "basis of sections". Strictly speaking for each cell Ω_t we need a matrix-valued function $G_t \in \mathcal{J}^{\varepsilon}_{m\times(n-1)}(\Omega_t)$ whose columns form a basis of the space $L_t(\zeta)$, $\zeta\in\Omega_t$, and such that for any two adjacent cells Ω_t, Ω_s on their common boundary (perhaps except for the vertices) the following equality holds

$$B_{ts}\, G_t = G_s\,. \tag{13.6}$$

Once these matrix-valued functions are found, we put

$$q^{[r]}_{ij,(t)} = G_t((i,j),r)\,, \quad 1\le i<j\le n, \quad r\le n-1\,.$$

Then the conditions (12.5), (12.6) of Theorem 12.3 will be satisfied. Remind that G_t must satisfy the estimates (12.7), (12.8).

In order to find G_t we use Lemmas 13.2 and 13.3 (iii) and choose a matrix-valued function $\mathcal{X}_t \in A^1_{m\times(n-1)}(\Omega_t)$ whose columns form a basis of $L_t(\zeta)$. According to Lemma 13.3 (ii) one can find a left inverse $\widetilde{\mathcal{X}}_t \in A^1_{(n-1)\times m}(\Omega_t)$ for $\mathcal{X}_t$. Let Ω_t, Ω_s be adjacent cells having a common boundary arc γ. We put

$$U_{ts} \overset{def}{=\!=} \widetilde{\mathcal{X}}_s B_{ts}\, \mathcal{X}_t\,. \tag{13.7}$$

Since $\mathcal{X}_s \widetilde{\mathcal{X}}_s = P_{L_s}$ is a projection onto the subspace L_s (not necessarily orthogonal) and $B_{ts} L_t = L_s$, we have $U_{st} = U^{-1}_{ts}$ and hence $U^{\pm1}_{ts} \in C^1_{(n-1)\times(n-1)}(\gamma)$. We wish to find Φ_t such that

$$\Phi_t \in \mathcal{J}^{\varepsilon}_{(n-1)\times(n-1)}(\Omega_t)\,, \tag{13.8}$$

$$\|\Phi_t^{-1}(\zeta)\| \le const \prod_i |\zeta - z_i|^{-\ell}\,, \tag{13.9}$$

for some positive ℓ, where the $\{z_i\}$ are the vertices of Ω_t, and

$$U_{ts}\,\Phi_t = \Phi_s\,, \quad \partial\Omega_t \cap \partial\Omega_s \setminus \{z_i\}\,. \tag{13.10}$$

After that we can put $G_t = \mathcal{X}_t \Phi_t$. Clearly then $G_t \in \mathcal{J}^{\varepsilon}_{m\times(n-1)}(\Omega_t)$

and the columns of $G_t(\zeta)$ form a basis of $L_t(\zeta)$ whenever $\zeta \in \operatorname{clos}\Omega_t \setminus \{z_i\}$. Let us verify (13.6):

$$\widetilde{\mathcal{Z}}_s B_{ts} G_t = \widetilde{\mathcal{Z}}_s B_{ts} \mathcal{Z}_t \Phi_t = U_{ts} \Phi_t = \Phi_s$$

hence

$$B_{ts} G_t = P_{L_s} B_{ts} G_t = \mathcal{Z}_s \widetilde{\mathcal{Z}}_s B_{ts} G_t = \mathcal{Z}_s \Phi_s = G_s .$$

The following lemma shows that (12.7) follows from (13.9) (taking into account the fact that $\Phi_t^{-1} \widetilde{\mathcal{Z}}_t$ is a left inverse for G_t).

13.6. LEMMA. Let B be a numerical $k \times p$ matrix, $k > p$ and let $\widetilde{B}$ be its left inverse. Then

$$\Sigma \, | \operatorname{Minor}(B, p) | \geq c_{k,p} \| \widetilde{B} \|^{-p} .$$

The proof is elementary.

Thus the proof of Theorem 12.3 has reduced to solving the system of vectorial Riemann-Hilbert equation (13.8)-(13.10) with an additional requirement arising from (12.8).

14. SOLUTION OF THE RIEMANN-HILBERT PROBLEM.

The main tool will be the following variant of A.Cartan's lemma on the factorization of matrix-valued functions of class $\operatorname{Lip}\alpha$, $\alpha < 1$. A short proof of the lemma can be found in [12], see also [28, § 2].

14.1. LEMMA. Let Ω_1, Ω_2 be Jordan domains with piecewise C^2-smooth boundaries. Suppose that $\Omega_1 \cap \Omega_2 = \emptyset$ and that $\Gamma = \partial\Omega_1 \cap \partial\Omega_2$ is a simple Jordan arc. Let $T(z)$ be a matrix-valued function on Γ such that $T^{\pm 1} \in \operatorname{Lip}^{\alpha}_{n \times n}(\Gamma)$, $\alpha < 1$. Then there exist matrix-valued functions F_1, F_2 having the following properties:

(i) $F_i^{\pm 1} \in \Lambda^{\alpha}_{n \times n}(\Omega_i)$, $i = 1, 2$; (ii) $T(z) F_1(z) = F_2(z)$, $z \in \Gamma$.

<u>The general form of matrix-valued functions</u> Φ_1, Φ_2 <u>with the properties (i), (ii) is as follows</u>:

$$\text{(iii)} \quad \Phi_1 = F_1 X, \quad \Phi_2 = F_2 X, \quad X \in \Lambda^{\alpha}_{n\times n}(\Omega_1 \cup \Gamma \cup \Omega_2).$$

14.2. FACTORIZATION OF DISCONTINUOUS MATRIX-VALUED FUNCTIONS. We shall need a result similar to that of Lemma 14.1, but with T having a jump discontinuity on the arc Γ. This case can be reduced to the continuous one.

Let Ω_1, Ω_2 and Γ have the same meaning as in Lemma 14.1. The point c is supposed to be not an endpoint of Γ and so it divides Γ into two arcs Γ_1, Γ_2. We shall say that a matrix-valued function G on Γ belongs to the class $Lip^{\alpha}_{n\times n}(\Gamma, c)$ if $G|\Gamma_i \in Lip^{\alpha}_{n\times n}(\Gamma_i)$, $i=1,2$. In this case one can consider the following matrices (see [17], [54]):

$$\Delta = G^{-1}(c+0)\,G(c-0), \quad \lambda = (2\pi i)^{-1} \log \Delta .$$

The matrix λ will be determined uniquely if we fix the branch of logarithm of each eigenvalue of Δ. In what follows these branches are chosen in such a way that the real parts of eigenvalues of λ lie in $[0,1)$. Fix a point $z_0 \in \Omega_1$ and introduce the following functions of λ:

$$Y_1(z) = (z-c)^{\lambda}, \quad Y_2(z) = G(c-0)\left(\frac{z-c}{z-z_0}\right)^{\lambda} .$$

The matrix-valued function $Y_1(z)$ is single-valued in the complex plane cut along a curve going from c to ∞ without intersecting Ω_1. The matrix-valued function Y_2 is single-valued in the plane cut along an arc connecting c and z_0 in Ω_1 (see Fig.13).

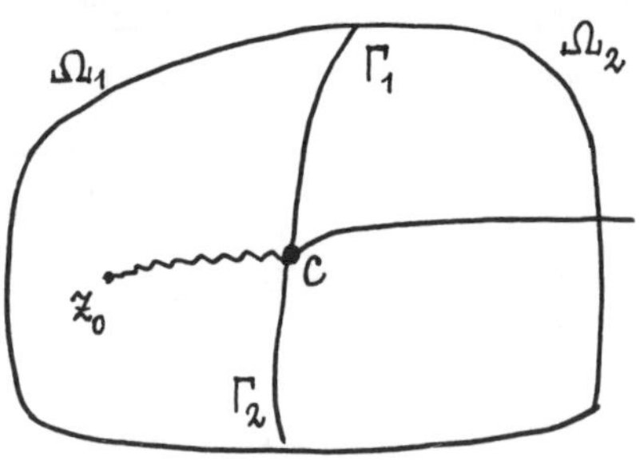

Figure 13

14.3. LEMMA. <u>Let</u> $\Omega_1, \Omega_2, \Gamma$ <u>and</u> G, Δ, Y_1, Y_2 <u>be as above. Let</u> $2\pi\alpha_0$ <u>be the maximum of the arguments of the eigenvalues of</u>

<u>the matrix</u> Δ (the arguments being chosen in $[0, 2\pi)$) <u>and let</u> $\alpha_0 < \alpha < 1$. <u>Then</u> $Y_2^{-1} G Y_1$ <u>belongs to</u> $Lip^{\alpha-\alpha_0}_{n\times n}(\Gamma)$ <u>and one can find matrix-valued functions</u> X_1, X_2 <u>such that</u>

(i) $X_i^{\pm 1} \in \Lambda^{\alpha-\alpha_0}_{n\times n}(\Omega_i)$, $i=1,2$;

(ii) <u>for any neighbourhood</u> $\mathcal{N}$ <u>of the point</u> c

$$X_i^{\pm 1} \in \Lambda^{\alpha}_{n\times n}(\Omega_i \setminus \mathcal{N}), \quad i=1,2;$$

(iii) $G Y_1 X_1 = Y_2 X_2$.

14.4. Now we return to our tringulation and put the cells in a suitable order. Enumerate the cells Ω_t by the indices $1,\dots, N$, so that $\Omega^{(k)} \overset{def}{=\!=} int(\bigcup_{i=1}^{k} clos\,\Omega_i)$ be a Jordan domain for every k, and $\Omega^{(k)}$, Ω_{k+1} have a common boundary arc. The solution of the problem (13.8)-(13.10) will be constructed by induction. On the kth step of the induction the required Φ_t (see (13.8)-(13.10)) will be constructed for all cells Ω_t such that $\Omega_t \subset \Omega^{(k)}$. To start the induction we put $\Phi_1^{(1)}$ to be the unit $((n-1)\times(n-1))$ matrix.

14.5. THE INDUCTION HYPOTHESIS. Suppose that we have already constructed matrix-valued functions $\Phi_i^{(k)}$, $i \le k$, satisfying (13.10) for adjoining cells Ω_i , Ω_j, $i,j \le k$. Suppose also that for some $\delta_k > 0$

$$\Phi_i^{(k)} \in \mathcal{J}^{\delta_k}_{(n-1)\times(n-1)}(\Omega_i),$$

the condition (13.9) holds, and also

$$\Phi_i^{(k)\pm 1} \mid \partial\Omega_i \cap \partial\Omega^{(k)} \in Lip^{1-\varepsilon}_{(n-1)\times(n-1)}(\partial\Omega_i \cap \partial\Omega^{(k)}). \qquad (14.1)$$

Here ε is a positive constant which will be chosen later. Assume also that (12.8) is true for the matrix-valued functions

$$G_i^{(k)} = \mathcal{Z}_i\, \Phi_i^{(k)} \qquad (14.2)$$

in all the vertices which lie in $int\,\Omega^{(k)}$.

14.6. THE STEP $k \to k+1$ OF THE INDUCTION. Since $\Omega^{(k)}$ is a

Jordan domain, the cell Ω_{k+1} (a curvilinear triangle) can have one or two common boundary arcs with $\Omega^{(k)}$. Consider these two cases.

CASE 1. The domains $\Omega^{(k)}$ and Ω_{k+1} have one common boundary arc $\gamma \in S\Gamma$. Let $\gamma = \partial\Omega_i \cap \partial\Omega_{k+1}$, $i \le k$. We apply Lemma 14.1 to the pair of domains $\Omega^{(k)}, \Omega_{k+1}$ and the matrix-valued function

$$U_{i,k+1}\Phi_i^{(k)} \in Lip^{1-\varepsilon}_{(n-1)\times(n-1)}(\gamma).$$

As a result we obtain Φ_k, Φ_{k+1} such that

$$\Phi_k^{\pm 1} \in \Lambda^{1-\varepsilon}_{(n-1)\times(n-1)}(\Omega^{(k)}), \qquad \Phi_{k+1}^{\pm 1} \in \Lambda^{1-\varepsilon}_{(n-1)\times(n-1)}(\Omega_{k+1}),$$

$$U_{i,k+1}\Phi_i^{(k)}\Phi_k = \Phi_{k+1} \quad \text{on} \quad \gamma .$$

Then we can put $\Phi_i^{(k+1)} = \Phi_i^{(k)}\Phi_k$, $i \le k$, $\Phi_{k+1}^{(k+1)} = \Phi_{k+1}$. Now the equality (13.10) holds for all pairs of adjacent cells among Ω_i, $i \le k+1$. It is readily verified that all other induction hypotheses (see 14.5) are also satisfied.

CASE 2. The domains $\Omega^{(k)}$ and Ω_{k+1} have two common boundary arcs γ_1, γ_2. Let $c = \gamma_1 \cap \gamma_2$. Consider a regular neighbourhood of the point c which is necessarily a vertex of $S\Gamma$ (see 6.11, 12.1). This neighbourhood is divided into curvilinear sectors which belong to different cells. Without loss of generality we can assume that these cells are $\Omega_1, \Omega_2, \dots, \Omega_r$, Ω_{k+1}, and that Ω_i, $2 \le i \le r-1$, lies between Ω_{i-1} and Ω_{i+1}, Ω_{k+1} lies between Ω_r and Ω_1 (see Fig.14).

Put $\gamma_1 = \partial\Omega_r \cap \partial\Omega_{k+1}$, $\gamma_2 = \partial\Omega_1 \cap \partial\Omega_{k+1}$. We have to factorize the discontinuous matrix-valued function G:

$$G|\gamma_1 = U_{r,k+1}\Phi_r^{(k)}, \qquad G|\gamma_2 = U_{1,k+1}\Phi_1^{(k)}.$$

Set $\Delta = G^{-1}(c+0)G(c-0)$ where $(c+0)$ corresponds to the arc γ_2.

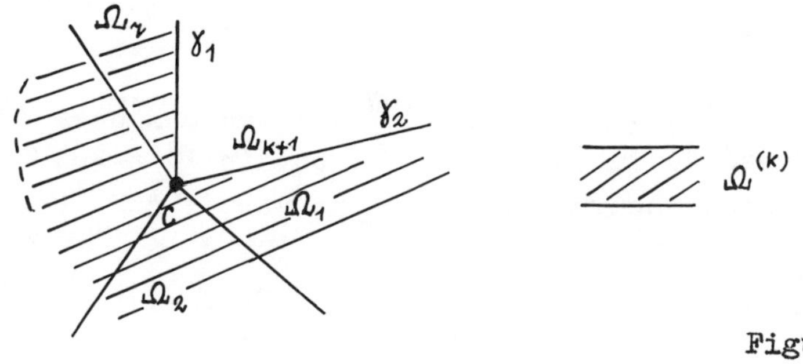

Figure 14

Let us compute this matrix using the invertibility of $\Phi_i^{(k)}(c)$, $i \leq 2$, implied by (14.1). We have

$$\Delta = ((\Phi_1^{(k)})^{-1} U_{1,k+1}^{-1} U_{2,k+1} \Phi_2^{(k)})(c) =$$

$$= ((\Phi_1^{(k)})^{-1} U_{1,k+1}^{-1} U_{2,k+1} U_{2-1,2} \Phi_{2-1}^{(k)})(c) = \dots =$$

$$= ((\Phi_1^{(k)})^{-1} U_{k+1,1} U_{2,k+1} U_{2-1,2} \cdot \ldots \cdot U_{12} \Phi_1^{(k)})(c).$$

Recalling that $B_{ts} L_t \subset L_s$ and $\mathcal{Z}_s \widetilde{\mathcal{Z}}_s = P_{L_s}$ and setting

$$B_c = (B_{k+1,1} \cdot B_{2,k+1} \cdot \ldots \cdot B_{12})(c) \tag{14.3}$$

we obtain from (13.7):

$$\Delta = (\Phi_1^{(k)}(c))^{-1} \widetilde{\mathcal{Z}}_1(c) B_c \mathcal{Z}_1(c) \Phi_1^{(k)}(c). \tag{14.4}$$

It is easy to see that $B_c L_1(c) \subset L_1(c)$ and the operator $\widetilde{\mathcal{Z}}_1(c) \cdot B_c \cdot \mathcal{Z}_1(c)$ is similar to the restriction $B_c \mid L_1(c)$. Hence all eigenvalues of the matrix Δ are among the eigenvalues of B_c. The constant $\varepsilon > 0$ in (14.1) must be chosen so small that for all vertices c the arguments of eigenvalues of B_c lie in $[0, 2\pi(1-2\varepsilon))$.

Now we apply Lemma 14.3 to $\Omega^{(k)}, \Omega_{k+1}$ and the matrix-valued function G. Put

$$\lambda=(2\pi i)^{-1}\log\Delta,\quad Y_1(z)=(z-c)^{\lambda},\quad Y_2(z)=G(c-0)\left(\frac{z-c}{z-z_0}\right)^{\lambda} \tag{14.5}$$

where z_0 is a fixed point in $\Omega^{(k)}$ and the branches of the powers are chosen as in 14.2. Thus we obtain Φ_k and Φ_{k+1} such that for any neighbourhood $\mathcal{N}$ of the vertex c

$$\Phi_k^{\pm 1}\in\Lambda^{1-\varepsilon}_{(n-1)\times(n-1)}(\Omega^{(k)}\setminus\mathcal{N}),\qquad \Phi_{k+1}^{\pm 1}\in\Lambda^{1-\varepsilon}_{(n-1)\times(n-1)}(\Omega_{k+1}\setminus\mathcal{N}),$$

$$\Phi_k^{\pm 1}\in\Lambda^{\varepsilon}_{(n-1)\times(n-1)}(\Omega^{(k)}),\qquad \Phi_{k+1}^{\pm 1}\in\Lambda^{\varepsilon}_{(n-1)\times(n-1)}(\Omega_{k+1}), \tag{14.6}$$

$$GY_1\Phi_k = Y_2\Phi_{k+1}\quad\text{on } \gamma_1\cup\gamma_2 .$$

Put

$$\Phi_i^{(k+1)}=\Phi_i^{(k)}Y_1\Phi_k,\quad i\leq k;\qquad \Phi_{k+1}^{(k+1)}=Y_2\Phi_{k+1}. \tag{14.7}$$

The condition (13.10) clearly holds. From the well-known explicit form of the matrix-valued function $\zeta\to\zeta^{\lambda}$ [18] it follows that

$$Y_1\in\mathcal{J}^{\delta}_{(n-1)\times(n-1)}(\Omega^{(k)}),\qquad Y_2\in\mathcal{J}^{\delta}_{(n-1)\times(n-1)}(\Omega_{k+1}),$$

for some $\delta>0$ which implies

$$\Phi_\ell^{(k+1)}\in\mathcal{J}^{\min(\delta,\delta_k)}_{(n-1)\times(n-1)}(\Omega_\ell),\quad \ell\leq k+1 .$$

Also (14.1) holds with the change k by $k+1$, since $c\in \operatorname{int}\Omega^{(k+1)}$. It remains to check (12.8) in the neighbourhood of c which appears to be not an easy task (the verification of (12.8) in all other vertices in $\operatorname{int}\Omega^{(k+1)}$ is immediate from the induction hypothesis and the formula for $\Phi_i^{(k+1)}$, $i\leq k$). Recall that (12.8) is not needed for the computation of the multiplicity.

14.7. VERIFICATION OF (12.8). Without loss of generality one can assume that $\varphi_{(i)}^{-1}(c)\in R$, $i\leq\nu$, $\varphi_{(i)}^{-1}(c)\in\partial R$, $i>\nu$. So

$\mathcal{K}$ in (12.4) is equal to $\{1,\dots,\nu\}$.

Let P_ν be the orthogonal projection an $\mathbb{C}^m$ onto the linear span of $e(i,j)$, $i,j\leq\nu$. We have $G_{t,c}=P_\nu G_t$ where $G_{t,c}$ is defined by (12.4). Let $P_\nu^{(0)}$ be the orthogonal projection on $\mathbb{C}^m$ onto the linear span of $e(1,j)$, $1<j\leq\nu$. To prove (12.8) it is sufficient to check that for $\ell=1,\dots,\kappa$, τ the following holds:

$$\sum\left|\operatorname{Minor}(P_\nu^{(0)}G_\ell^{(\kappa+1)}(z),\nu-1)\right|\geq \operatorname{const}|z-c|^{\frac{\nu-\varkappa}{2}},\quad z\in\Omega_\ell, \tag{14.8}$$

in a neighbourhood of c.

<u>Suppose first that</u> $\ell\leq\tau$ (see Fig.14). Recall that $\Phi_\kappa^{-1}(z)$ is continuous in $\operatorname{clos}\Omega_\ell$ (see (14.6)) and consider the matrix-valued function

$$A(z)=P_\nu^{(0)}G_\ell^{(\kappa+1)}\Phi_\kappa^{-1}(z).$$

We obtain from (14.2), (14.5), (14.7) that

$$A(z)=P_\nu^{(0)}\mathcal{Z}_\ell(z)\,\Phi_\ell^{(\kappa+1)}(z)\,\Phi_\kappa^{-1}(z)=P_\nu^{(0)}\mathcal{Z}_\ell(z)\,\Phi_\ell^{(\kappa)}(z)\,f_z(\Delta)$$

where $f_z(\Delta)=(z-c)^{\frac{\log\Delta}{2\pi i}}$ and Δ is defined by (14.4). By (14.4) we have

$$A(z)=P_\nu^{(0)}\mathcal{Z}_\ell(z)\Phi_\ell^{(\kappa)}(z)(\Phi_1^{(\kappa)}(c))^{-1}\widetilde{\mathcal{Z}}_1(c)f_z(B_c)\mathcal{Z}_1(c)\,\Phi_1^{(\kappa)}(c). \tag{14.9}$$

Put

$$B(z)\overset{\text{def}}{=\!=}P_\nu^{(0)}\mathcal{Z}_\ell(c)\,\Phi_\ell^{(\kappa)}(c)\,(\Phi_1^{(\kappa)}(c))^{-1}\widetilde{\mathcal{Z}}_1(c)f_z(B_c)\,\mathcal{Z}_1(c)\,\Phi_1^{(\kappa)}(c). \tag{14.10}$$

Using (13.10) and remembering that $B_{i-1,i}(c)L_{i-1}(c)\subset L_i(c)$, $\mathcal{Z}_i(c)\cdot\widetilde{\mathcal{Z}}_i(c)=P_{L_i(c)}$ one can write

$$\mathcal{Z}_\ell(c)\,\Phi_\ell^{(\kappa)}(c)(\Phi_1^{(\kappa)}(c))^{-1}\widetilde{\mathcal{Z}}_1(c)\;=\;B_{1,\ell}(c)P_{L_i(c)},$$

where $B_{1,\ell}(c)\overset{\text{def}}{=\!=}B_{\ell-1,\ell}(c)\,B_{\ell-2,\ell-1}(c)\cdot\ldots\cdot B_{12}(c)$. Next we use that all operators $B_{i-1,i}$ and hence $B_{1,\ell}$; B_c commute with P_ν by the property (13.3). Thus we have

$$B(z) = P_\nu^{(0)} P_\nu B_{1,\ell}(c) P_{L_1(c)} f_z(B_c) \mathcal{Z}_1(c) \Phi_1^{(k)}(c) =$$

$$= \underbrace{P_\nu^{(0)} B_{1,\ell}(c)}_{B_3} \cdot \underbrace{f_z(B_c)}_{B_2(z)} \cdot \underbrace{P_\nu \mathcal{Z}_1(c) \Phi_1^{(k)}(c)}_{B_1} . \tag{14.11}$$

Put $M_{p,\nu} \overset{def}{=\!=} P_\nu L_p(c)$, $p=1,\dots,k,r$. Since $P_\nu B_c = B_c P_\nu$ and $B_c L_1(c) \subset L_1(c)$, we see that $B_2(z) M_{1,\nu} \subset M_{1,\nu}$. Thus $B_2(z) | M_{1,\nu}$ is an isomorphism for $z \in \Omega_\ell$.

$$B_2(z) | M_{1,\nu} = (z-c)^{\frac{\log(B_c | M_{1,\nu})}{2\pi i}} .$$

Next we have to study $M_{p,\nu}$ and $B_c | M_{1,\nu}$ in detail.

Let us consider $M_{p,\nu}$ as a subspace of $\mathbb{C}^{\frac{\nu(\nu-1)}{2}}$. If $\{g_{ij}\}_{i,j\le\nu} \in M_{p,\nu}$, then $\{W_{ij}^{(p)}(c) g_{ij}\}$ is a cocycle. Thus $\dim M_{p,\nu} \le \nu-1$. Put $\tau(k) = \{g_{ij}\}_{i,j\le\nu}$ where $g_{1k}=1$; $g_{1j}=0$, $j \ne k$, and the remaining g_{ij} are determined by the cocycle condition. This can be done since $W_{ij}^{(\ell)}(c) \ne 0$, $i,j \le \nu$, by (12.2). Let us show that $\tau(k) \in M_{p,\nu}$. Indeed, by 12.2 there is at most one zero among $W_{ij}^{(p)}(c)$, $i,j \le n$. If $W_{\alpha\beta}^{(p)}(c)=0$ then either $\alpha > \nu$ or $\beta > \nu$. One can extend the cocycle $\{W_{ij}^{(p)}(c) g_{ij}\}_{i,j\le\nu}$ to a cocycle $\{W_{ij}^{(p)}(c) g_{ij}\}_{i,j\le n}$ by putting $g_{1\beta} = g_{1\alpha}$ and choosing g_{1j}, $j>\nu$, $j \ne \alpha, \beta$, arbitrary. Thus we have proved that $\tau(k) \in M_{p,\nu}$. It follows that $\dim M_{p,\nu} = \nu-1$, $\{\tau(2),\dots,\tau(\nu)\}$ is a basis of $M_{p,\nu}$ and $P_\nu^{(0)} | M_{p,\nu}$ is an isomorphism.

Now let us describe the action of $B_c | M_{1,\nu}$. Denote by δ the permutation of the set $\{1,\dots,\nu\}$ generated by the continuation of the branches $\varphi_{(1)}^{-1},\dots,\varphi_{(\nu)}^{-1}$ of φ^{-1} along a small circle centered at z. Clearly, $\delta = \delta_{k+1,1} \cdot \delta_{r-1,r} \cdot \ldots \cdot \delta_{12}$, where δ_{ts} are the correspondence functions. Let $\tilde{k} = \delta(k)$, $k \le \nu$. It follows from (13.3) that $B_c e(i,j) = W_{ij}^{(1)}(c) / W_{\tilde{i}\tilde{j}}^{(1)}(c)\, e(\tilde{i},\tilde{j})$. One verifies easily that

$$B_c \tau(K) = \frac{W^{(1)}_{1\widetilde{K}}(c)}{W^{(1)}_{1K}(c)} \cdot \tau(\widetilde{K}) , \qquad \text{if } \widetilde{K} \neq 1$$

$$B_c \tau(K) = -\frac{W^{(1)}_{1\widetilde{1}}(c)}{W^{(1)}_{1K}(c)} \cdot \tau(\widetilde{1}) , \qquad \text{if } \widetilde{K} = 1 .$$

Thus $B_c | M_{1,\nu}$ is a product of a diagonal operator and a permutation of the basis. The eigenvalues of $B_c | M_{1,\nu}$ can be computed easily.

The permutation δ of the set $\{1,\dots,\nu\}$ can be decomposed into a union of cycles. Clearly the number of cycles is equal to $\varkappa = card\,(\varphi^{-1}(c) \cap \mathbb{R})$. Without loss of generality we can assume that these cycles are $\{1,\dots,m_1\}$, $\{m_1+1,\dots,m_2\}$, ... $\{m_{\varkappa-1}+1,\dots,m_\varkappa\}$, $m_\varkappa = \nu$. Thus $\delta(1)=2,\dots,\delta(m_1)=1$, and so on. Then the permutation of $\{\tau(K): K=2,\dots,\nu\}$ induced by B_c is the union of cycles $\{\tau(2),\dots,\tau(m_1)\}$, $\{\tau(m_1+1),\dots$ $\dots,\tau(m_2)\},\dots,\{\tau(m_{\varkappa-1}+1),\dots,\tau(\nu)\}$. To each cycle corresponds a set of eigenvalues of $B_c | M_{1,\nu}$. Denote $n_j = m_{j+1} - m_j$.

$$exp\Big(\frac{\pi i}{n_1-1}\Big),\ exp\Big(\frac{3\pi i}{n_1-1}\Big), \dots, exp\Big(\frac{\pi i(2n_1-3)}{n_1-1}\Big)$$

are the eigenvalues corresponding to the first cycle, and

$$1,\ exp\Big(\frac{2\pi i}{n_p}\Big), \dots, exp\Big(\frac{2\pi i(n_p-1)}{n_p}\Big)$$

correspond to each of the remaining cycles, $p \geq 2$. Obvious changes must be made if $n_1 = 1$. Now it is easy to see that

$$|det(B_2(z)|M_{1,\nu})| = |det(z-c)^{\frac{log(B_c|M_{1,\nu})}{2\pi i}}| =$$

$$= |z-c|^{\frac{1}{2}\left(\frac{1}{n_1-1}+\dots+\frac{2n_1-3}{n_1-1}\right)+\left(\frac{1}{n_2}+\dots+\frac{n_2-1}{n_2}\right)+\dots} = |z-c|^{\frac{\nu-\varkappa}{2}} . \qquad (14.12)$$

Now we return to the formula (14.11). B_1 is a constant matrix mapping $\mathbb{C}^{n-1}$ onto $M_{1,\nu}$. One can find a right inverse matrix $\underset{\sim}{B}_1$, i.e. $B_1 \underset{\sim}{B}_1 = I_{\nu-1}$. Clearly $B_3 | M_{1,\nu}$ is an isomorphism onto $P_\nu^{(0)} M_{\ell,\nu}$. Thus

$$\underset{\sim}{B}(z) \overset{def}{=\!=} \underset{\sim}{B}_1 (z-c)^{-\frac{\log(B_c | M_{1,\nu})}{2\pi i}} (B_3 | M_{1,\nu})^{-1}$$

is a right inverse matrix-valued function for $B(z)$. It follows from (14.12) that

$$|\operatorname{Minor}(\underset{\sim}{B}(z), \nu-1)| \leq \operatorname{const} |z-c|^{-\frac{\nu-\varkappa}{2}} \tag{14.13}$$

for every minor of $\underset{\sim}{B}(z)$ of order $\nu-1$. All entries of $\underset{\sim}{B}(z)$ have modulus less than $\operatorname{const}|z-c|^{-(1-2\varepsilon)}$ by the choice of ε in 14.6. Since $\mathcal{Z}_\ell(z), \Phi_\ell^{(K)}(z)$ satisfy the Lipschitz condition of order $(1-\varepsilon)$, we obtain from (14.9), (14.10) that

$$\| A(z) - B(z) \| \leq \operatorname{const} \| \mathcal{Z}_\ell(z) \Phi_\ell^{(K)}(z) - \mathcal{Z}_\ell(c) \Phi_\ell^{(K)}(c) \| \leq \operatorname{const} |z-c|^{1-\varepsilon}.$$

Thus we have

$$A(z) \underset{\sim}{B}(z) = I_{\nu-1} + \varkappa_{\nu-1}(z), \quad \| \varkappa_{\nu-1}(z) \| \leq \operatorname{const} |z-c|^{\varepsilon}.$$

Hence $I_{\nu-1} + \varkappa_{\nu-1}(z)$ is invertible for z close to c. $\underset{\sim}{A}(z) = \underset{\sim}{B}(z)(I_{\nu-1} + \varkappa_{\nu-1}(z))^{-1}$ is a right inverse matrix-valued function for $A(z)$. From (14.13) we obtain the estimate

$$\Sigma | \operatorname{Minor}(A(z), \nu-1) | \geq \operatorname{const} |z-c|^{\frac{\nu-\varkappa}{2}}$$

which implies (14.8) since $\Phi_K^1(z)$ is continuous.

It remains to verify (12.8) in the cell Ω_{K+1}. Here we put $A(z) = P_\nu^{0} G_{K+1}^{(K+1)}(z) \Phi_{K+1}^{-1}(z)$. Proceeding as above we have

$$A(z) = P_\nu^{(0)} \mathcal{Z}_{K+1}(z) U_{\imath, K+1}(c) \Phi_\imath^{(K)}(c) (\Phi_1^{(K)}(c))^{-1} \widetilde{\mathcal{Z}}_1(c) g_z(B_c) \mathcal{Z}_1(c) \Phi_1^{(K)}(c)$$

where $g_z(B_c) = \left(\frac{z-c}{z-z_0}\right)^{\frac{\log B_c}{2\pi i}}$. We put

$$B(z) = P_\nu^{(0)} \mathcal{Z}_{k+1}(c)\, U_{\tau,k+1}(c)\, \Phi_\tau^{(k)}(c)(\Phi_1^{(k)}(c))^{-1} \widetilde{\mathcal{Z}}_1(c)\, q_z(B_c)\, \mathcal{Z}_1(c)\, \Phi_1^{(k)}(c) =$$

$$= P_\nu^{(0)} B_{1,k+1}(c)\, q_z(B_c)\, \mathcal{Z}_1(c)\, \Phi_1^{(k)}(c)\,.$$

Now the argument repeats that of the previous case.

Thus induction is completed along with the proof of Theorem 12.3.

15. HOW TO ENSURE THE CONDITION (10.4) (THE CONDITION OF ANALYTIC INDEPENDENCE).

Let us indicate the changes that must be made in order to ensure (10.4). We shall refer extensively to the previous sections.

We shall consider the condition (10.4) for the arcs of $S\Gamma \cap \Gamma$ which are, in general, smaller than the arcs of Γ.

Consider an arc α of $S\Gamma$ separating cells Ω_t, Ω_s for which $n_t = n_s + 1$. Fix a simply connected admissible region $\widetilde{\Omega}_\alpha \subset \Omega_s$ so that the intersection $\widetilde{\alpha} = \partial\widetilde{\Omega}_\alpha \cap \partial\Omega$ be an arc properly contained in α. Without loss of generality one can assume $\delta_{ts} 1 > n_s$. The region $\widetilde{\Omega}_\alpha$ is chosen sufficiently small so that the branch $\varphi^{-1}_{(1)}$ should continue analytically from Ω_t to the whole of $\widetilde{\Omega}_\alpha$ (this is possible since φ is analytic on $\operatorname{clos} R$ and there are no images of branch points on α by the definition of an arc).

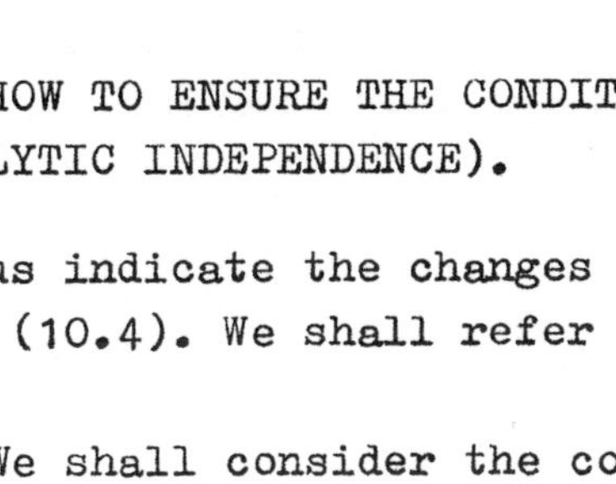

Figure 15

Put $\Omega_t' = \operatorname{int}(\operatorname{clos}(\Omega_t \cup \widetilde{\Omega}_\alpha))$.

First consider the case when $n_t > 1$. We can assume $\delta_{ts} 2 \le n_s$.

Choose a weight function $w_{12}^{(t)} \in A^1(\Omega_t')$ nonvanishing on $\operatorname{clos} \widetilde{\Omega}_\alpha$. When constructing B_{ts} we have defined Q_t, Q_s, q_t, q_s (see 13.4). The matrix-valued function q_t can be complemented to an invertible $((n-1)\times(n-1))$ matrix-valued function B_t in Ω'_t so that $B_t^{\pm 1}\in A^1_{(n-1)\times(n-1)}(\Omega'_t)$. Computing the minors of q_t we see that they are all nonvanishing outside a discrete set $\mathcal{D}$. Hence outside $\mathcal{D}$ all the elements of the last column of B_t^{-1} are also nonvanishing. The matrix q_s has been complemented to an invertible one (in 13.4) by adding a constant row with all the entries equal to 1. Now we replace the entries "1" by "$1+\varepsilon\psi_j$" where each of ψ_j is analytic and multi-valued on $clos(\Omega_s\cup\Omega_t)$ (having a (say) logarithmic branch point in $\widetilde{\Omega}_\alpha\setminus\mathcal{D}$). These branch points should be distinct for different ψ_j and ε should be so small that the resulting matrix-valued function B_s satisfy the condition:

$$B_s^{\pm 1}\in C^1_{(n-1)\times(n-1)}(\alpha).$$

The rest of the construction of the B_{ts} and all the further reasonings remain unchanged.

To verify the condition (10.4) on α we shall prove that if almost everywhere on $\widetilde{\alpha}$

$$\sum_{\nu=0}^{n_t-1} h_\nu u^{[\nu]}\circ\varphi_{(1)}^{-1}=0,\quad h_\nu\in H^\infty(\widetilde{\Omega}_\alpha) \tag{15.1}$$

then all $h_\nu\equiv 0$. From (15.1) we obtain that

$$\sum_{\nu=0}^{n_t-1} h_\nu(u^{[\nu]}\circ\varphi_{(1)}^{-1}-u^{[\nu]}\circ\varphi_{(2)}^{-1})=-\sum h_\nu(u^{[\nu]}\circ\varphi_{(2)}^{-1}),$$

hence by (12.9),

$$\sum_{\nu=1}^{n_t-1} h_\nu g^{[\nu]}_{12,(t)}=-(W_{12}^{(t)})^{-1}\sum_{\nu=0}^{n_t-1} h_\nu(u^{[\nu]}\circ\varphi_{(2)}^{-1}). \tag{15.2}$$

Observe that by the choice of $W_{12}^{(t)}$ the right-hand side of (15.2) is analytic in $\widetilde{\Omega}_\alpha$. From the equality $G_t=B_{st}G_s$ on α and the definition of B_{st} we obtain, denoting by $B_{st}((\kappa,\kappa'),(\ell,\ell'))$ the corresponding entry of B_{st} :

$$g^{[r]}_{12,(t)} = \sum_{i,j} B_{st}((1,2),(\tilde{i},\tilde{j}))\, g^{[r]}_{i,j,(s)} = \sum_{2\le j\le n} B_{st}((1,2),(\tilde{1},\tilde{j}))\, g^{[r]}_{\tilde{1},\tilde{j},(s)} .$$

Since $B_{st} = B_t^{-1} B_s$ one can write

$$B_{st}((1,2),(\tilde{1},\tilde{j})) = \sum_{\ell\le n-1} B_t^{-1}((1,2),\ell)\cdot B_s(\ell,(\tilde{1},\tilde{j})) .$$

All summands of the right-hand side of this formula except for the last one are analytic and single-valued in $\tilde{\Omega}_\alpha$. The last summand $(\ell = n-1)$ has a logarithmic branch point by the choice of B_s, B_t. Taking all this into account we conclude from (15.2) that

$$\sum_{r=1}^{n_t-1} h_r\, g^{[r]}_{\tilde{1},\tilde{j},(s)} \equiv 0, \quad 2\le j\le n .$$

But then if $h_r \not\equiv 0$ for some r we would obtain from the cocycle condition that the matrix-valued function $[g_{ij,(s)}]^{1\le r\le n_t-1}_{i,j\le n}$ would have in $\tilde{\Omega}_\alpha$ rank less than n_t-1. But this contradicts (12.7). So $h_r \equiv 0$, $r\ge 1$, and hence by (15.1) $h_0 \equiv 0$.

Thus we can ensure the condition of analytic independence on all arcs of $S\Gamma$ for which $n_t > 1$. It remains to study the case $n_t = 1$, i.e. the case $\Omega_s \not\subset \operatorname{clos}\varphi(R)$. Choose the functions $\psi^{[r]}$, $r = 1,\dots,n-1$, analytic on $\operatorname{clos}\varphi(R)$ and having pairwise disjoint essential singularities (e.g., logarithmic branch points) in $\tilde{\Omega}_\alpha$. For the cyclic set we take $u^{[0]} \equiv 1$, $u^{[r]} \equiv x^{[r]} + K^{[r]}\circ\varphi + \psi^{[r]}\circ\varphi$, $r = 1,\dots,n-1$, instead of (12.11). It is not hard to see that adding the last summand does not violate any conditions sufficient for cyclicity mentioned above (see Sect.10).

Let us verify (10.4) on the arc $\alpha = \partial\Omega_t \cap \partial\Omega_s$. Suppose (15.1) holds. According to the construction in Sect.12, $x^{[r]}|\Omega_t \equiv 0$, hence the functions $u^{[r]}\circ\varphi^{-1}_{(1)}$, $r\ge 1$, can be extended analytically to Ω_S and have distinct essential singularities there. Therefore all $h_\nu \equiv 0$ as required.

Thus we have constructed a set of n functions satisfying all the needed conditions. By Proposition 10.5 it is cyclic and hence we have completed the proof of the Main Theorem.

APPENDIX. PROOF OF APPROXIMATION LEMMAS ON A RIEMANN SURFACE.

PROOF OF LEMMA 4.6. The Carleson Imbedding Theorem implies that if $\mathcal{R}_1 \subset \mathcal{R}$ is an admissible subdomain then the restriction operator acts from $E^p(\mathcal{R})$ into $E^p(\mathcal{R}_1)$, $p>0$. Thus by the Cauchy Theorem making a fine partition we see that $E^p(\mathcal{R}) \subset \subset E^p_{(2)}(\mathcal{R})$ and evidently, $E^p_{(1)}(\mathcal{R}) \subset E^p_{(2)}(\mathcal{R})$. Let us prove that $E^p_{(2)}(\mathcal{R}) \subset E^p(\mathcal{R})$. We shall employ the following.

THEOREM (Behnke, Stein [5], see also [44]). Let the domain G have compact closure in a noncompact Riemann surface $\mathcal{K}$. Then for each $\zeta \in G$, there is a differential $d\omega(z,\zeta)$ which has a single pole and residue 1. This pole is of the first order, and it is located at ζ. If $z \in G$ and V is a parametric disc about z such that the differential $d\omega(z,\zeta)$ is represented in V by the function $A_V(z,\zeta)$, then $A_V(z,\zeta)$ is, as a function of ζ, a meromorphic function on G which has only one pole and residue -1. This pole is simple and is located at the point z.

Fix $f \in E^p_{(2)}(\mathcal{R})$ and consider the Behnke-Stein differential $d\omega(z,\zeta)$ for some G, $G \supset \operatorname{clos} \mathcal{R}$. Put

$$\tilde{f}(\zeta) = \frac{1}{2\pi i} \cdot \int_{\partial \mathcal{R}} f(z)\, d\omega(z,\zeta), \quad \zeta \in G \setminus \partial \mathcal{R}.$$

By the properties of $d\omega(z,\zeta)$ we see that $\tilde{f}(\zeta)$ is analytic in $\mathcal{R}$ and $\tilde{f}(\zeta) \equiv 0$ in $G \setminus \operatorname{clos} \mathcal{R}$. Let us prove that $\tilde{f} = f$ a.e. on $\partial \mathcal{R}$ and $\tilde{f} \in E^p(\mathcal{R})$. Let $z \in \partial \mathcal{R}$, and let U_1, U_2 be small discs about z such that $\operatorname{clos} U_1 \subset U_2$. Let $\theta : U_2 \to \mathbb{C}$ be the parametric conformal map. For $\zeta \in U_1 \setminus \partial \mathcal{R}$,

$$\tilde{f}(\zeta) = \frac{1}{2\pi i} \int_{\partial \mathcal{R} \setminus U_2} f(z)\, d\omega(z,\zeta) + \frac{1}{2\pi i} \int_{\partial \mathcal{R} \cap U_2} f(z)\, d\omega(z,\zeta) = f_1(\zeta) + f_2(\zeta).$$

Clearly $f_1(\zeta)$ is analytic in U_2. Let us show that $f_2 \circ \theta^{-1} \in$

$\in E^p(\theta(U_1 \cap R))$. One can write $f_2(\theta^{-1}\zeta) = \frac{1}{2\pi i}\int_{\theta(\partial R \cap U_2)} f(\theta^{-1}z)A(z,\zeta)\,dz$. by the preceding theorem,

$$A(z,\zeta) = \frac{1}{z-\zeta} + B(z,\zeta),$$

where $B(z,\zeta)$ is analytic in $\theta(U_2)\times\theta(U_2)$.

Denote $\gamma = \theta(\partial R \cap U_2)$. The function $\zeta \mapsto \frac{1}{2\pi i}\int_\gamma f(\theta^{-1}z)B(z,\zeta)dz$ is analytic in $\theta(U_2)$. The function

$$\zeta \longmapsto \frac{1}{2\pi i}\int_\gamma \frac{f(\theta^{-1}z)}{z-\zeta}\,dz$$

has a jump equal to $f(\theta^{-1}z)$ a.e. on γ and belongs to $E^q(\theta(U_1 \cap R))$, where $q=p$ if $1<p<\infty$, and $q<p$ otherwise. By the Phragmén-Lindelöf principle (Lemma 7.6) in case $p=1,\infty$ we conclude that $f\in E^p(R)$.

Now let us prove that $E^p(R)\subset E^p_{(1)}(R)$. We show that any functional on $L^p(\partial R)$ orthogonal to functions analytic on $\operatorname{clos} R$ is orthogonal to $E^p(R)$. Thus we have a differential $d\nu$, $d\nu \in \mathcal{X}^q(\partial R)$, $1/q + 1/p = 1$, such that

$$\int_{\partial R} f\,d\nu = 0, \qquad \text{for } f \text{ analytic on } \operatorname{clos} R.$$

It is well-known (see [16, §26]) that there exists a nonvanishing differential $d\lambda$ analytic on $\operatorname{clos} R$. Then clearly $d\nu/d\lambda$ is a function which belongs to $E^q_{(2)}(R)$. By what has just been proved $d\nu/d\lambda \in E^q(R)$. Making a fine partition and using the Cauchy Theorem we see that $\int_{\partial R} fg\,d\lambda = 0$ if $f\in E^p(R), g\in E^q(R)$. Thus $d\nu$ is orthogonal to $E^p(R)$, q.e.d.

PROOF OF LEMMA 4.8. Since $H^\infty(R)$ is dense in $E^p(R)$, it is sufficient to study the case $p=\infty$. By the Generalized Weierstrass Theorem (see [16, §26]) we can represent τ as a product of factors each having a single zero in a neighbourhood of $\operatorname{clos} R$. Thus it may be assumed that τ has a single zero

$z \in \partial R$. Then it is easy to see that $\tau(R)$ is an admissible domain in $\mathbb{C}$ such that the origin 0 is its boundary point. Suppose the lemma is proved for a plane domain. Then $\mathbb{1}$ belongs to the weak-$*$ closure of $\tau H^\infty(\tau(R))$, hence $\mathbb{1}$ belongs to the weak $*$-closure of $\tau H^\infty(R)$ which implies the desired result. Thus one can consider the case $R \subset \mathbb{C}$. Applying a transformation $\zeta \mapsto (\zeta - c)^{-1}$ for a proper c , we can assume that 0 lies in the closure of the unbounded component Ω of $\mathbb{C} \setminus \operatorname{clos} R$. Consider the hull $\hat{R} \overset{def}{=} \mathbb{C} \setminus \Omega$. Evidently $H^\infty(\hat{R}) \subset H^\infty(R)$. Thus it remains to prove the claim for a simply connected case. The latter follows, e.g., from the fact that the conformal mapping of the unit disc onto an admissible domain with 0 in the boundary cannot have a singular inner factor.

NOTATIONS

Here we list some notations. We give either an explanation, or a reference to the subsection where the symbol is defined, or both.

$Span$ - closed linear span
$\Sigma|Minor(H,\kappa)|$ - sum taken over all minors of H of order κ.
T_φ - analytic Toeplitz operator, 1.1, 2.1
$\mu(T)$ -multiplicity, 1.1
$\nu_T(\lambda)$ - 1.2
$\varkappa(T)$ - 1.2
(R,φ) - admissible pair, 2.1
$\mu_\varphi=\mu(T_\varphi)$, 2.1
$\Pi^{(\kappa)}(\varphi)$ - 2.2
$\varkappa(\varphi)$ - maximal essential valency, 2.2
$V(\varphi)$ - the set of maximal essential valency, 2.2, 9.4
$\Omega^\kappa(\varphi)$ - 9.4
K_f - Cauchy integral, 7.1

Classes of functions:

$A(\Omega)$ - 3.2
$A^1(\Omega)$ - 3.2
$\Lambda^\alpha(\Omega)$ - 3.2
$\mathcal{J}^\alpha(\Omega)$ - 3.2, 12.2
$H^\infty(R,w)$ - 9.6
$E^p(\Omega)$ - 4.1, 4.5
$N(\Omega)$ - 4.2
$E^{1-0}(\Omega)$ - 7.1
$L^{1-0}(\Omega)$ - 7.1
$Lip^\varepsilon(\gamma)$ - 12.4

Classes of differentials:

$\mathcal{Z}^1(\partial R)$ - 2.1
$\mathcal{Z}^1(\partial R,w)$ - 9.6

Functions:
$u^{[\nu]}$ - elements of a (hypothetically) cyclic set
$\varphi^{-1}_{(j)}$ - branches of φ^{-1}

$v_{(j)}^{[r]} = u^{[r]} \circ \varphi_{(j)}^{-1}$

$W_{ij}^{(t)}$ - weight function, 12.2

$\{g_{ij,(t)}^{[r]}\}$ - weighted cocycle, 12.2

Matrix-valued functions:

$H^{(\kappa)}(\zeta)$ - 10.2 H_t - 12.6

$G_t(\zeta)$ - 12.2 $G_{t,c}(\zeta)$ - 12.2

$L_t(\zeta)$ - "space-valued" function, 13.1

B_{ts} - transition matrix, 13.4

Other symbols:

$\mathrm{Vert}(\cdot)$ - the set of vertices, 12.2

δ_{ts} - correspondence functions, 6.12

$e(i)$, $e(i,j)$, $e(i,j,\kappa)$ - elements of the bases of $\mathbb{C}^n, \mathbb{C}^m, \mathbb{C}^p$ respectively, $m = C_n^2$, $p = C_n^3$, 13.1

REFERENCES

1. Aleksandrov, A.B.: Invariant subspaces of the shift operators. An axiomatic approach, Zapiski Nauch.Semin.L.O.M.I. 113 (1981), 7-26. (Russian)
2. Aleksandrov, A.B.: Essays on locally convex Hardy classes, Lecture Notes in Math. 864 (1981), 1-90.
3. Atzmon, A.: Multilinear mappings and estimates of multiplicity, Integral Equations and Operator Theory 10 (1987), 1-16.
4. Baker, I.N., Deddens, J.A. and Ullman, J.L.: A theorem on entire functions with applications to Toeplitz operators, Duke Math.J. 41 (1974), 739-745.
5. Behnke, H. and Sommer, F.: Theorie der analytischen Funktionen einer komplexen Veränderlichen, 3 Aufl., Springer, Berlin-Heidelberg-New York, 1965.
6. Bourdon, P.S.: Density of the polynomials in Bergman spaces, Pacific Math.J., 130, No.2 (1987), 215-222.
7. Bram, J.: Subnormal operators, Duke Math.J., 22 (1955), 75-94.
8. Clark, D.N.: On a similarity theory for rational Toeplitz operators, J.Reine Angew.Math. 320 (1980), 6-31.
9. Clark, D.N.: On Toeplitz operators with loops II, J.Operator Theory 7 (1982), 109-123.
10. Cowen, C.C.: The commutant of an analytic Toeplitz operator II, Indiana Univ.Math.J. 29, No.1 (1980), 1-12.
11. David, G.: Opérateurs intégraux singuliers sur certaines courbes du plan complexe, Ann.Sci.Ec.Norm.Sup., 4e série 17 (1984), 157-189.
12. Douady, A.: Le problème de modules pour sous-espaces analytiques compacts d'un espace analytique donné, Ann.Inst. Fourier 15, No.1 (1966), 1-94.
13. Douglas, R.G., Shapiro, H.S., Shields, A.L.: Cyclic vectors and invariant subspaces for the backward shift operator, Ann.Inst.Fourier 20, No.1 (1970), 37-76.
14. Duren, P.L: Theory of H^p spaces, Academic Press, New York-London, 1970.
15. Dyn'kin, E.M.: On the smoothness of Cauchy type integrals,

Zapiski Nauch.Semin.L.O.M.I. 92 (1979), 115-133.

16. Forster, O.: Riemannsche flächen, Springer, Berlin-- Heidelberg - New York, 1977.
17. Gahov, F.D.: A case of the Riemann Problem for a system of n function pairs, Izvestia Akad. Nauk SSSR 14, No.6 (1950), 549-568.
18. Gantmacher, F.R.: Matrizenrechung I, II, Deutscher Verlag der Wissenschaften, Berlin, 1958, 1966 (transl.from Russian).
19. Garnett, J.B.: Bounded analytic functions, Academic Press, New York, 1981.
20. Goluzin, G.M.: Geometric theory of functions of a complex variable, Amer.Math.Soc. R.I., Providence, 1969 (transl. from Russian).
21. Havin, V.P.: Boundary properties of the Cauchy integral and harmonic conjugate functions in the domains with rectifiable boundary, Mat.Sbornik 68,No.4 (1965), 499-517. (Russian).
22. Herrero D.: On multicyclic operators, Integral Equations and Operator Theory 1, No.1 (1978), 57-102.
23. Herrero D.: On the multiplicities of the powers of a Banach space operator, Proc.Amer.Math.Soc. 94, No. 2 (1985), 239-243.
24. Herrero D.: The Fredholm structure of an n-multicyclic operator, Indiana Univ.Math.J. 36, No.3 (1987), 549-566.
25. Herrero, D. and Rodman, L.: The multicyclic n-tuples of an n-multicyclic operator, and analytic structures on its spectrum, Indiana Univ.Math.J. 34, No.3 (1985), 619-629.
26. Herrero, D. and Wogen, W.: On the multiplicity of $T \oplus T \oplus \ldots \oplus T$ Preprint, 1987.
27. Hitt, D.: Invariant subspaces of H^2 of an annulus, Pacific J. of Math. 134, No.1 (1988) 101-120.
28. Leiterer, J.: Holomorphic vector bundles and the Oka-Grauert principle, Sovrem.Probl.Math., Fund.Napr., 10 (1986), 75-123, 'VINITI', Moscow. (Russian)
29. Lin, V.Ya.: Holomorphic bundles and multivalued functions of elements of a Banach algebra, Funkc.Anal.i ego Pril. 7 No.2 (1973), 43-51. (Russian).

30. Nikol'skii, N.K.: Designs for calculating the spectral multiplicity of orthogonal sums, Zapiski Nauchn.Semin.L.O.M.I. 126 (1983), 150-158; English transl. in J.Soviet Math. 27, No.1 (1984).
31. Nikol'skii, N.K.: Ha-plitz operators: a survey of some recent results, Proc.Conf.Operators and Function Theory, Lancaster 1984, 87-137, Reidel, 1984.
32. Nikol'skii, N.K.: Treatise on the shift operator, Springer-Verlag, 1986.
33. Nikol'skii, N.K. and Vasyunin, V.I.: Control subspaces of minimal dimension and root vectors, Integral Equations and Operator Theory 6 (1983), 274-311.
34. Nikol'skii, N.K. and Vasyunin,V.I.: Control subspaces of minimal dimension, unitary and model operators, J.Operator Theory 10 (1983), 307-330.
35. Nordgren, E.: Reducing subspaces of analytic Toeplitz operators, Duke Math.J. 34 (1967), 175-181.
36. Privalov, I.I.: Randeigenshhaften Aaalytishher Functionen, Deutscher Verlag der Wiss., East Berlin, 1956.
37. Sarason, D.: The H^p spaces of an annulus, Mem.Amer.Math. Soc. 56 (1965), 1-78.
38. Smirnov, V.I.: Sur les formules de Cauchy et de Green et quelques problèmes qui s'y rattachent, Izvestia Akad.Nauk S.S.S.R, ser.mat. 7(1932), 337-372.
39. Solomyak, B.M.: On the multiplicity of analytic Toeplitz operators, Dokl.Akad.Nauk S.S.S.R 286, No.6 (1986), 1308-1311; English transl.in Soviet Math.Dokl. 33 No.1 (1986).
40. Solomyak, B.M.: Cyclic sets for analytic Toeplitz operators, Zapiski Nauchn.Semin.L.O.M.I. 157 (1987), 88-102. (Russian)
41. Solomyak, B.M.: Calculi, invariant subspaces and multiplicity for some classes of operators, Dissertation, Leningrad State University, 1986. (Russian)
42. Solomyak, B.M.: Multiplicity, calculi and multiplication operators, Sibirskii Matem.Jurnal. 29, No.2 (1988), 167-175. (Russian)
43. Stephenson, K.: Analytic functions of finite valence, with

applications to Toeplitz operators, Michigan, Math.J. 32 No.1 (1985), 5-19.

44. Stout, E.L.: Bounded holomorphic functions on finite Riemann surfaces, Trans.Amer.Math.Soc. 120, No 2 (Nov.1965), 255-285.

45. Sz.-Nagy, B. and Foias, C.: Compléments à l'étude des opérateurs de classe Co, Acta Sci.Math. 31, No.3-4 (1970), 287-296.

46. Sz.-Nagy, B. and Foias, C.: Compléments à l'étude des opérateurs de classe Co II, Acta Sci.Math. 33, No.1-2 (1972), 113-116.

47. Sz.-Nagy, B. and Foias, C.: Harmonic Analysis of Operators on Hilbert space, North-Holland, 1970.

48. Thomson, J.E.: The commutant of a class of analytic Toeplitz operators, Amer.J.Math. 99 (1977), 522-529.

49. Tolokonnikov, V.A.: Estimates in the Carleson Corona Theorem ideals of the algebra , a problem of Sz.-Nagy, Zapiski Nauchn.Semin.L.O.M.I. 113 (1981), 178-199. (Russian)

50. Tumarkin, G.C. and Khavinson, S.Ya.: Classes of analytic functions on multiply-connected domains, Contemporary problems in the theory of functions of one complex variable, 45-77, Moscow, 1960, (Russian). French translation: Fonctions d'une variable complexe, Problémes contemporains, 37-71, Gauthiers-Villars, Paris, 1962.

51. Tumarkin, G.C. and Khavinson, S.Ya.: Existence of a single-valued function in a given class with a given modulus of its boundary values in multiply connected domains, Izvestia Akad.Nauk S.S.S.R. ser.Mat. 22 (1958, 543-562. (Russian)

52. Vasyunin, V.E.: Multiplicity of contractions with finite defect indices, present volume.

53. Vasyunin, V.I. and Karaev, M.T.: The multiplicity of some contractions, Zapiski Nauchn.Semin. L.O.M.I. 157 (1987), 23-29. (Russian)

54. Vekua, N.P.: The Riemann problem with discontinuous coefficients for several functions. Soobsch.Akad.Nauk GSSR 2 No.9 (1950), 533-538. (Russian)

55. Volberg, A.L.: How to break through a given contour, L.O.M.I. Preprints P-I-89, 1989 (Russian)

56. Volberg, A.L. and Solomyak, B.M.: The multiplicity of Toeplitz operators, weighted cocycles and the vector-valued Riemann problem, Funkc.Anal.i ego Pril. 21 No.3 (1987), 1-10. (Russian)

57. Volberg, A.L. and Solomyak, B.M.: Operator of multiplication by analytic matrix-valued function, present volume.

58. Wang, D.: Similarity of smooth Toeplitz operators, J.Operator Theory 12 (1984), 319-348.

59. Warschawskii, S.E.: On conformal mapping of regions bounded by smooth curves, Proc.Amer.Math.Soc. 2 (1951), 254-261.

60. Wermer, J.: Rings of analytic functions, Annals of Math. 67 No.3 (1958), 497-516.

61. Wogen, W.R.: On some operators with cyclic vectors, Indiana Univ.Math.J. 27 No.1 (1978), 163-171.

62. Yakubovich, D.V.: Similarity models of Toeplitz operators, Zapiski Nauchn.Semin.L.O.M.I. 157 (1987), 113-123. (Russian)

63. Yakubovich, D.V.: Riemann surface models of Toeplitz operators, present volume.

64. Zinsmeister, M.: Domain de Lavrentiev, Publication Math. d'ORSAY, 85-03, 1985.

65. Dyn'kin E.M. Methods of the theory of singular integrals. in Commutative Harmonic Analysis I. Encyclopaedia of Mathematical Sciences. Springer-Verlag, 1989.

Operator Theory:
Advances and Applications, Vol. 42

OPERATOR OF MULTIPLICATION BY AN ANALYTIC MATRIX-VALUED FUNCTION

B.M.Solomyak and A.L.Volberg

This work can be considered as a supplement to our paper 'Multiplicity of analytic Toeplitz operators' [2]. Our aim is to generalize the formula for multiplicity to the matrix case. However the main part of the present paper can be read independently of [2]. We prove that in a certain sense a matrix analytic Toeplitz operator reduces to a scalar operator of multiplication on a space of functions on a Riemann surface. In some cases this reduction provides model within to similarity which resemble that of D.Yakubovich [4]. This reduction, as we hope, may appear useful in other problems concerning matrix Toeplitz operators

1. STATEMENT OF RESULTS. SCHEME OF THE PROOF.

Consider the vector-valued Hardy space H^2_n in the unit disc $\mathbb{D}$ and an $n\times n$ matrix-valued function $\Phi(z)$ with entries in H^∞. The operator $T_\Phi : x \mapsto \Phi x$ acting on H^2_n will be called the <u>matrix analytic Toeplitz</u> (AT) <u>operator with symbol</u> Φ. This class of operators arises naturally, e.g., in the theory of Sz.-Nagy-Foias model [3]. In this paper we find the multiplicity of a matrix AT operator with symbol Φ analytic on $\operatorname{clos}\mathbb{D}$.

1.1. NOTATION AND TERMINOLOGY. An open set R is called <u>admissible</u> (see [2, 2.1]) if (a) $\mathrm{clos}\, R \subset \mathrm{int}\, \mathcal{K}$ where $\mathcal{K}$ is a compact Riemann surface with boundary, (b) $R = \mathrm{int}(\mathrm{clos}\, R)$, and (c) ∂R is a finite union of analytic Jordan arcs.

Let $\mathcal{X}(R)$ be a class of functions defined in R. Denote by $\mathcal{X}_n(R)$ ($\mathcal{X}_{n\times n}(R)$) the classes of vector-valued (respectively, matrix-valued) functions with entries in $\mathcal{X}(R)$. Let $\mathcal{A}^+(R)$ ($\mathcal{M}^+(R)$) be the sets of all functions analytic (meromorphic) on $\mathrm{clos}\, R$ (i.e. in a neighbourhood of $\mathrm{clos}\, R$). By I_n we denote the $n\times n$ unit matrix.

Let $\Phi(z)$ be a matrix-valued function, $\Phi(z) \in \mathcal{A}^+_{n\times n}(R)$. Put

$$\nu(\Phi,\lambda) = \sum_{z\in \mathrm{clos}\, R} \dim \mathrm{Ker}(\lambda I_n - \Phi(z))^n,$$

$$\Pi^K(\Phi) = \{\lambda \in \mathbb{C} : \nu(\Phi,\lambda) = K\},$$

$$n(\Phi) = \sup\{k : \mathrm{card}\, \Pi^K(\Phi) = \infty\},$$

$$V(\Phi) = \Pi^{n(\Phi)}(\Phi).$$

These characteristics are natural generalizations of those defined in [2, 2.2] for the scalar case $n=1$. For instance, if $n=1$ then $n(\Phi)$ turns into the essential maximal valency.

To state the main result we need also the notion of a wide belt. For the definition and relevant discussion the reader is referred to [2, Sect.2].

Out aim is to determine the multiplicity

$$\mu(T_\Phi) \overset{def}{=} \inf\{\dim \mathcal{Y} : \mathrm{span}(T_\Phi^n \mathcal{Y} : n \geq 0) = H^2_n\}.$$

1.2. THEOREM. <u>Let</u> $\Phi(z) \in \mathcal{A}^+_{n\times n}(\mathbb{D})$. <u>Then</u> $\mu(T_\Phi) = n(\Phi) + 1$, <u>if the set</u> $\mathbb{C} \setminus V(\Phi)$ <u>is disconnected and there exists a wide belt subordinate to</u> $V(\Phi)$. <u>Otherwise</u> $\mu(T_\Phi) = n(\Phi)$.

Now we give the scheme of the proof. We shall explain also what was meant by the reduction of a matrix AT operator to an AT operator on a space of functions on a Riemann surface.

Let $p(z,\lambda) = \det(\Phi(z)-\lambda I_n)$ be the characteristic polynomial One can assume that

$$p(z,\lambda) \not\equiv 0, \quad \forall \lambda \in \mathbb{C}. \tag{1.1}$$

It is quite easy that if (1.1) fails then both $n(\Phi)$ and $\mu(T_\Phi)$ are equal to infinity (see 2.2 below).

1.3. RIEMANN SURFACE. Fix $r>1$ such that $\Phi(z) \in \mathcal{A}^+_{n\times n}(r\mathbb{D})$. The polynomial $p(z,\lambda)$ can be decomposed into prime factors: $p = q_1^{m_1}\cdot \ldots \cdot q_\ell^{m_\ell}$ (see [1, Appendix 2]). The polynomials $q_i(z,\lambda)$ are irreducible and relatively prime with respect to the field of meromorphic functions in $r\mathbb{D}$. Let $K_i = \deg q_i$. Each of q_i defines a Riemann surface $\mathcal{K}_i$ which is a K_i-sheeted branched covering of $r\mathbb{D}$ (see [1, Appendix 5]). Roughly speaking $\mathcal{K}_i = \{(z,\lambda) \in \mathbb{C}^2 : q_i(z,\lambda) = 0, \ z \in r\mathbb{D}\}$. Let $\mathcal{K}$ be the disjoint union of $\mathcal{K}_i$'s, each $\mathcal{K}_i$ taken m_i times. Denote by π and Λ_Φ the projections $(z,\lambda) \mapsto z$ and $(z,\lambda) \mapsto \lambda$. It is clear that $\pi : \mathcal{K} \to r\mathbb{D}$ is an n-sheeted branched covering map and Λ_Φ is analytic in $\mathcal{K}$. Let $R = \pi^{-1}(\mathbb{D})$. It is readily seen that $n(\Phi) = n(\Lambda_\Phi | R)$, $\Pi^K(\Phi) = \Pi^K(\Lambda_\Phi | R)$. Define $E^2(R)$ as the closure of $\mathcal{A}^+(R)$ in $L^2(\partial R)$ (see [2, Section 4] for details on E^p classes).

1.4. PROPOSITION. <u>Suppose that</u> $\Phi(z) \in \mathcal{A}^+_{n\times n}(\mathbb{D})$ <u>and (1.1) is satisfied. Let</u> T_Λ <u>be the operator of multiplication by</u> Λ_Φ <u>on</u> $E^2(R)$. <u>Then</u> $\mu(T_\Phi) = \mu(T_\Lambda)$.

This proposition together with [2, Theorem 2.6] immediately implies Theorem 1.2. Proposition 1.4 is underlain by some pure algebra which is contained in the following proposition.

1.5. PROPOSITION. <u>In the setting of Proposition 1.4 the operators</u> T_Φ <u>on</u> $\mathcal{M}^+_n(\mathbb{D})$ <u>and</u> T_Λ <u>on</u> $\mathcal{M}^+(R)$ <u>are linearly similar (in a nontopological sense)</u>.

The crucial point in the proof of this proposition is the following lemma which claims that a Jordan block and the corresponding diagonal matrix-valued function are similar when regarded as multiplication operators on the space of meromorphic vector-valued functions. To be precise, let $\mathcal{R}$ be an admissible domain in a Riemann surface (see 1.1). Suppose $\psi \in \mathcal{A}^+(\mathcal{R})$, $\psi \not\equiv const$. Then the operator $d/d\psi$, locally acting as $f \mapsto f'/\psi'$, is well-defined on $\mathcal{M}^+(\mathcal{R})$. Denote by J_m the nilpotent $m \times m$ Jordan block.

1.6. LEMMA. Operators $T_{(\psi I_m + J_m)}$ and $T_{\psi I_m}$ are linearly similar on $\mathcal{M}^+_m(\mathcal{R})$. In fact the following operator relation holds:

$$exp(-\frac{\partial}{\partial\psi}\cdot J_m)\cdot T_{(\psi I_m + J_m)}\cdot exp(\frac{\partial}{\partial\psi}\cdot J_m) = T_{\psi I_m} .$$

Now we are going to pass from the spaces of meromorphic functions to Hardy-Smirnov classes. Recall the following (see [3]).

1.7. DEFINITION. Let $\mathfrak{X}_1, \mathfrak{X}_2$ be Banach spaces and let T_1, T_2 be bounded operators on $\mathfrak{X}_1, \mathfrak{X}_2$. T_1 is said to be a deformation of T_2 (notation: $T_1 \prec T_2$) if there exists $A: \mathfrak{X}_1 \to \mathfrak{X}_2$ with trivial kernel and dense range, such that $AT_1 = T_2 A$. The operators T_1, T_2 are called quasisimilar, if $T_1 \prec T_2$ and $T_2 \prec T_1$ (then we write $T_1 \sim T_2$). If $T_1 \prec T_2$ and A^{-1} is bounded, T_1 and T_2 are similar. Notation: $T_1 \simeq T_2$.

We shall say that T_1 and T_2 are weakly quasisimilar if there exist T_i-invariant subspaces $X_i \subset \mathfrak{X}_i$, $i = 1,2$, such that $codim\, X_i < \infty$ and $T_1 \prec T_2 | X_2$, $T_2 \prec T_1 | X_1$. If $T_1 \simeq T_2 | X_2$, $T_2 \simeq T_1 | X_1$ the operators T_1, T_2 will be called weakly similar.

It is well-known (and trivial) that $T_1 \prec T_2$ implies $\mu(T_2) \leqslant \mu(T_1)$, hence multiplicity is preserved under quasisimilarity. The following two lemmas justify the use of weak quasisimilarity.

1.8. LEMMA. The multiplicity of an operator on a linear topolo-

<u>gical space does not increase under a restriction of the operator to an invariant subspace of finite codimension.</u>

Clearly this is no longer true for subspaces of infinite codimension. For example, the operator $f \mapsto (z-\frac{1}{2})^n f$ on $L^2(\partial \mathbb{D})$ has multiplicity one, while its restriction to H^2 has multiplicity n.

1.9. LEMMA. <u>Suppose that</u> $R_1, \ldots, R_\ell$ <u>are admissible domains and the matrix-valued functions</u> $\Psi_i(z) \in \mathcal{A}^+_{n_i \times n_i}(R_i)$ <u>satisfy (1.1). Consider the operator</u> $T = \sum_{i=1}^{\ell} \oplus T_{\Psi_i}$ <u>on</u> $H = \sum_{i=1}^{\ell} \oplus E^2_{n_i}(R_i)$. <u>Then for any</u> $X \subset H$, $\operatorname{codim} X < \infty$, <u>such that</u> $TX \subset X$ <u>there exists</u> $\mathcal{Y} \subset X$ <u>such that</u> $\operatorname{codim} \mathcal{Y} < \infty$, $T\mathcal{Y} \subset \mathcal{Y}$ <u>and the restriction operator</u> $T \mid \mathcal{Y}$ <u>is similar to</u> T.

COROLLARY. <u>Under the hypotheses of Lemma 1.9</u> $\mu(T) = \mu(T \mid X)$.

Thus we see that multiplicity is preserved under weak quasisimilarity if we deal with orthogonal sums of matrix AT operators.

Consider once again the irreducible factorization of the characteristic polynomial: $p = q_1^{m_1} \cdot \ldots \cdot q_\ell^{m_\ell}$. There exists a matrix-valued function $W(z)$ such that $W(z)^{\pm 1} \in \mathcal{M}^+_{n \times n}(r\mathbb{D})$ and $\widetilde{\Phi}(z) = W(z)\Phi(z)W(z)^{-1}$ is block diagonal. Namely, $\widetilde{\Phi}(z) = \Psi_1(z) \oplus \ldots \Psi_\ell(z)$, where $\Psi_i(z) \in \mathcal{A}^+_{k_i \times k_i}(r\mathbb{D})$ and $q_i^{m_i}$ is the characteristic polynomial of $\Psi_i(z)$ (see [1, 4.6-4.7]). Put $n_i = m_i \deg q_i$.

1.10. LEMMA. <u>The operator</u> T_Φ <u>on</u> H^2_n <u>is weakly quasisimilar to the orthogonal sum</u> $\Sigma_{i=1}^{\ell} \oplus T_{\Psi_i}$ <u>on</u> $\Sigma_{i=1}^{\ell} \oplus H^2_{n_i}$.

From now on we shall restrict ourselves to the case of one irreducible factor. The general case follows easily by taking orthogonal sums. (Note that an orthogonal sum of AT operators on $E^2(R_i)$'s is unitarily equivalent to an AT operator on the disjoint union of R_i 's, and hence this case is covered by [2, Theorem 2.6].) Thus suppose that $p(z,\lambda) = q(z,\lambda)^m$, q is irreducible on $r\mathbb{D}$, $\deg q = \kappa$, and (1.1) holds (that is, $q(z,\lambda) \not\equiv \lambda - c$). Let $\mathcal{K}$ be the κ-sheeted branched covering over $r\mathbb{D}$, determined by q . Let π be the projection map and let Λ_Φ be the "eigenvalue" map on $\mathcal{K}$ as above. It is well-known (see [1, Chapter 5]) that outside a finite set Z the Jordan normal form of $\Phi(z)$ has a structure independent of z and consists of κ major $m\times m$ blocks corresponding to different eigenvalues. The nilpotent parts of all these blocks are the same. Let $\mathcal{D}$ be the $m\times m$ matrix-valued function on $\mathcal{K}$ in the Jordan normal form with this very nilpotent part and with the diagonal entries equal to Λ_Φ . Clearly $R = \pi^{-1}(\mathbb{D})$ is admissible.

1.11. LEMMA <u>The operator T_Φ on H^2_n is weakly quasisimilar to $T_{\mathcal{D}}: x \mapsto \mathcal{D}x$ on $E^2_m(R)$. If $Z \cap \partial\mathbb{D} = \emptyset$, these operators are weakly similar.</u>

Thus we can restrict ourselves to a single Jordan block.

1.12. LEMMA. <u>Suppose that R , R_1 , R_2 are admissible domains in a Riemann surface such that $R_1 \subset R \subset R_2$, both sets $\partial R_1 \cap \partial R$, $\partial R \cap \partial R_2$ are finite and both sets $R \setminus R_1$, $R_2 \setminus R$ have no compact components. Suppose $\psi \in \mathcal{A}^+(R_2)$, $\psi \not\equiv const$. Let $\Lambda = \psi I_m$ be the diagonal matrix-valued function on let $\mathcal{D} = \Lambda + J_m$ be the corresponding Jordan block. Then there exists a T_Λ-invariant subspace $X_1 \subset E^2_m(R_1)$ of finite codimension and a $T_{\mathcal{D}}$-invariant subspace $X \subset E^2_m(R)$ of finite codimension such that</u>

$$T_{\mathcal{D}} \mid E^2_m(R) \prec T_\Lambda \mid X_1 , \quad T_\Lambda \mid E^2_m(R_2) \prec T_{\mathcal{D}} \mid X .$$

To finish the proof we apply the following lemma which is the only part of the proof where the results of [2] are used explicitly.

1.13. LEMMA. <u>Suppose that R is an admissible domain in a Riemann surface and $\varphi \in \mathcal{A}^+(R)$. Then there exist admissible domains</u> R_1, R_2 such that $R_1 \subset R \subset R_2$, $\varphi \in \mathcal{A}^+(R_2)$, $\mathrm{card}(\partial R_1 \cap \partial R) < \infty$,

$\operatorname{card}(\partial R \cap \partial R_2) < \infty$, both sets $R \setminus R_1$, $R_2 \setminus R$ have no compact components and all the three operators of multiplication by φ on $E^2(R_1)$, $E^2(R)$, $E^2(R_2)$ have the same multiplicity.

The proof follows from [2, Theorem 2.6] and [2, 4.11]. This completes the proof of Proposition 1.4 and at the same time the proof of Theorem 1.2. It remains to prove Lemmas 1.8-1.12. Lemma 1.6 and Proposition 1.5 will be obtained as by-products . This will be done in Section 2.

Now let us discuss the 'models' for matrix AT operators in a special case.

1.14. SPECIAL CASE. Suppose that $\Phi(z) \in \mathcal{A}^+_{n\times n}(\mathbb{D})$ has no multiple eigenvalues for some $z \in \mathbb{D}$. Then there exists a finite set Ξ such that

$$\Phi(z) \text{ has no multiple eigenvalues for } z \in \operatorname{clos} \mathbb{D} \setminus \Xi \qquad (1.2)$$

The following proposition is implied by Lemmas 1.10, 1.11 and Remark 2.9 below.

1.15. PROPOSITION. Suppose that $\Phi(z) \in \mathcal{A}^+_{n\times n}(\mathbb{D})$ and (1.2) is satisfied. Let (R, Λ_Φ) be the admissible pair constructed in 1.3. Then T_Φ on H^2_n is quasisimiliar to $T_\Lambda : f \mapsto \Lambda_\Phi f$ on $E^2(R)$ restricted to an invariant subspace of finite codimension. If $\Xi \cap \partial\mathbb{D} = \emptyset$, then 'quasisimilarity' can be replaced by 'similarity'.

The subspace of finite codimension in Proposition 1.15 can be decribed in more detail. It consists of those functions in $E^2(R)$ whose Taylor coefficients at the points $\zeta = \pi^{-1}(z)$, $z \in \Xi \cap \mathbb{D}$, satisfy a certain system of $K_z \geq 0$ linear equations. Note that in the models developed by D.Yakubovich [4] certain restrictions of multiplication operators on $E^2(R)$ appear as well.

Sometimes a restriction of T_Λ to an invariant subspace is similar to T_Λ . For instance, if $\tau \in \mathcal{A}^+(R)$, $\tau | \partial R \neq 0$, then clearly $T_\Lambda \simeq T_\Lambda | \tau E^2(R)$. Now we present examples which show that, in general, (i) the restriction of T_Λ arising in Proposition 1.15 is not similar to T_Λ ; (ii) 'quasisimilarity' cannot be replaced by 'similarity'.

1.16. EXAMPLE. Let $n=2$, $\Phi(z)=\begin{bmatrix} 0 & 1 \\ z(z-a)^2 & 0 \end{bmatrix}$, $0<a<1$.

It is not hard to see that the pair $(\mathcal{R}, \Lambda_\Phi)$ is conformally equivalent to the pair $(\mathbb{D},\varphi)$, $\varphi(z)=z(z^2-a)$. The operator T_Φ on $H^2\oplus H^2$ is similar to the operator $T_\varphi : y\mapsto \varphi y$ on $\mathcal{Y} = \{y\in H^2 : y(-\sqrt{a}) = y(\sqrt{a})\}$. The similarity is provided by the intertwining operator $X: H^2\oplus H^2 \to H^2$, $X:\begin{bmatrix} f \\ g\end{bmatrix} \mapsto z(z^2-a)f(z^2)+g(z^2)$. One can show that $T_\varphi | H^2$ is not a deformation of $T_\varphi|\mathcal{Y}$.

1.17. EXAMPLE. Let $\Lambda(z)=\begin{bmatrix} z-1 & 0 \\ 0 & 1-z\end{bmatrix}$, $T(z)=\begin{bmatrix} 1 & 1 \\ 1 & z\end{bmatrix}$,

$M(z)=T(z)\Lambda(z)T(z)^{-1}=\begin{bmatrix} z+1 & -2 \\ 2z & -1-z\end{bmatrix}$. The operators

T_Λ and T_M on $H^2\oplus H^2$ are quasisimilar but not similar. The latter follows from the comparison of α_z , the angle between $\mathrm{Ker}(T_\Lambda^*-(\bar z\cdot 1)I)$, $\mathrm{Ker}(T_\Lambda^*-(1-\bar z)I)$, and β_z , the angle between $\mathrm{Ker}(T_M^*-(\bar z-1)I)$, $\mathrm{Ker}(T_M^*-(1-\bar z)I)$. It is easy to see that $\alpha_z=\pi/2$, while $\beta_z\to 0$, $z\to 1$.

2. PROOFS OF THE LEMMAS.

2.1. PROOF OF LEMMA 1.8. Let $\mathfrak{X}$ be a linear topological space, $\mathcal{Y}\subset\mathfrak{X}$, $\dim(\mathfrak{X}/\mathcal{Y})=k<\infty$. Suppose that T is an operator on $\mathfrak{X}$ such that $T\mathcal{Y}\subset\mathcal{Y}$. We wish to prove that $\mu(T/\mathcal{Y})\leqslant \mu(T)$.

Put $Z=\mathfrak{X}/\mathcal{Y}$ and consider the factor-operator $[T]$ on Z . There exists a chain $\{0\}\subset Z_1\subset\ldots\subset Z_n = Z$ of $[T]$-invariant subspaces such that $\dim Z_i = i$. Hence we obtain a chain $\mathcal{Y}\subset\mathcal{Y}_1\subset\ldots\subset\mathcal{Y}_n=\mathfrak{X}$ of T-invariant subspaces such that $\dim(\mathcal{Y}_i/\mathcal{Y}_{i-1})=1$. Thus one can assume $n=1$.

Take a subspace $\mathbb{U}$ cyclic for T , $\dim\mathbb{U}=\mu(T)$. Clearly $\mathbb{U}\not\subset\mathcal{Y}$. Fix an element $u\in\mathbb{U}\setminus\mathcal{Y}$. Since $n=1$ we have

$span(u, \mathcal{U} \cap \mathcal{Y}) = \mathcal{U}$. One can write $Tu = \lambda u + y$, $y \in \mathcal{Y}$. Evidently, $\mathcal{X} = span(T^n \mathcal{U} : n \geq 0) = span(u, T^n y, T^n(\mathcal{U} \cap \mathcal{Y}) : n \geq 0\}$, hence $\mathcal{Y} = span(T^n y, T^n(\mathcal{U} \cap \mathcal{Y}) : n \geq 0\}$, and the subspace $span(y, \mathcal{U} \cap \mathcal{Y})$ is cyclic for $T|\mathcal{Y}$. The lemma is proved. ▨

Before we proceed further let us show that $\mu(T_\Phi) < \infty$ implies (1.1).

2.2 LOWER ESTIMATE FOR $\mu(T_\Phi)$. Suppose that $\Phi(z) \in H^\infty_{n \times n}$. Fix $z \in \mathbb{D}$, an eigenvalue λ of $\Phi(z)$ and an eigenvector $h_\lambda \in \mathbb{C}^n$ of $\Phi(z)^*$ such that $\Phi(z)^* h_\lambda = \bar{\lambda} h_\lambda$. Put

$$\varkappa_{z,\lambda}(u) = \langle u(z), h_\lambda \rangle, \quad u \in H^2_n$$

where $\langle \cdot, \cdot \rangle$ stands for the scalar product in $\mathbb{C}^n$. Evidently, $\varkappa_{z,\lambda} \in Ker(T^*_\Phi - \bar{\lambda} I)$ and the functionals $\varkappa_{z,\lambda}$ corresponding to different z, are linearly independent. (Note that $\{\varkappa_{z,\lambda}\}$ may fail to span $Ker(T^*_\Phi - \bar{\lambda} I)$.) Thus, by a well-known estimate (see [2, Sect.5]).

$$\mu(T_\Phi) \geq dim\, Ker(T^*_\Phi - \bar{\lambda} I) \geq card\, \{z \in \mathbb{D} : \lambda \text{ is an eigenvalue of } \Phi(z)\}.$$

It follows that either $n(\Phi) = \mu(T_\Phi) = \infty$, or the condition (1.1) holds.

2.3. PROOF OF LEMMA 1.9. Let X be an invariant subspace of finite codimension of an orthogonal sum $T = T_1 \oplus T_2$ on $H = H_1 \oplus H_2$. Then $X^\perp = H \ominus X$ is T^*-invariant and finite-dimensional. The projections of $X^\perp$ onto H_i are T_i^*-invariant and of finite rank. Thus X contains a T-invariant subspace of the form $X_1 \oplus X_2$ where X_i is of finite codimension in H_i. This argument shows that the lemma can be proved for a single matrix AT operator T_Ψ on $E^2_n(R)$ where R is admissible and $\Psi \in \mathcal{A}^+_{n \times n}(R)$.

Suppose that $T_\Psi X \subset X$, $codim\, X = r < \infty$. The orthogonal complement $X^{\perp}$ is finite-dimensional and hence it is spanned by the eigenvectors and root vectors of T_{Ψ}^{*} . Put $S=\sigma(T_{\Psi}^{*} \mid X^{\perp})$. For $\lambda \in S$ put $q_{\lambda}(z)=\det(\Psi(z)-\lambda I_{n})$. It follows from (1.1) that $q_{\lambda} \not\equiv 0$ and clearly $q_{\lambda} \in \mathcal{A}^{+}(R)$. It is easy to see that

$$(\Psi(z)-\lambda I_{n})\mathcal{A}_{n}^{+}(R) \supset q_{\lambda}\mathcal{A}_{n}^{+}(R)$$

hence

$$\operatorname{clos}(T_{\Psi}-\lambda I)E_{n}^{2}(R) \supset \operatorname{clos} q_{\lambda}E_{n}^{2}(R).$$

It is readily seen that $X \supset \mathcal{Y} \overset{def}{=} \operatorname{clos}_{\lambda \in S} q_{\lambda}^{r} E_{n}^{2}(R)$. The last subspace can be written in the form $\mathcal{Y}=qE_{n}^{2}(R)$ where $q \in \mathcal{A}^{+}(R)$ has the same zeros as $\prod_{\lambda \in S} q_{\lambda}^{r}$ in the <u>interior</u> of R and $q \neq 0$ on ∂R . This follows from the fact that a function analytic on $\operatorname{clos} R$ which does not vanish in R is "outer" in R (see 2, 4.8).

Now observe that the multiplication operator $f \mapsto qf$ is an isomorphism of $E_{n}^{2}(R)$ onto $\mathcal{Y}$ commuting with T_{Ψ} . Hence T_{Ψ} is similar to $T_{\Psi} \mid \mathcal{Y}$ and the lemma is proved. ▨

2.4. PROOF OF LEMMA 1.10. Denote by $\mathcal{M}(\Omega)$ the field of meromorphic functions in Ω (not to be confused with $\mathcal{M}^{+}(\Omega)$). First we fix $r_1 > r > 1$ such that $\Phi(z) \in \mathcal{A}_{n\times n}^{+}(r_1 \mathbb{D})$. The characteristic polynomial $p(z,\lambda)$ can be decomposed into factors irreducible over the field $\mathcal{M}(r_1\mathbb{D}) : p=q_1^{m_1}\cdot \ldots \cdot q_{\ell}^{m_{\ell}}$. It follows from [1, 4.6-4.7] that there exists a matrix-valued function $W(z)$, $W(z)^{\pm 1} \in \mathcal{M}_{n\times n}(r_1\mathbb{D})$ such that $\widetilde{\Phi}(z)=W(z)\cdot \Phi(z)\cdot W(z)^{-1}$ is block diagonal: $\widetilde{\Phi}(z)=\Psi_1(z)\oplus \ldots \oplus \Psi_{\ell}(z)$, $\Psi_i(z) \in \mathcal{A}_{n_i\times n_i}^{+}(r\mathbb{D})$, and $q_i^{m_i}$ is the characteristic polynomial of $\Psi_i(z)$. Put $n_i=m_i \deg q_i$. We are to prove the weak quasisimilarity of T_{Φ} on H_{n}^{2} and $\sum_{i=1}^{\ell} \oplus T_{\Psi_i}$ on $\sum_{i=1}^{\ell} \oplus H_{n_i}^{2}$. Obviously $x \mapsto W(z)x$ provides an intertwining operator acting on $\mathcal{M}_{n}^{+}(\mathbb{D})$. The transition from $\mathcal{M}_{n}^{+}(\mathbb{D})$

to $H^2_n = E^2_n(\mathbb{D})$ goes along the same lines as in the proof of Lemma 1.11 below and is left to the reader. ▨

Now suppose that $p(z,\lambda) = \det(\Phi(z) - \lambda I_n) = q(z,\lambda)^m$, where q is irreducible over $\mathcal{M}(r_1\mathbb{D})$, $r_1 > 1$, $\deg q = k$. Before we turn to the proof of Lemma 1.11 let us discuss the definition of the Riemann surface under consideration in more detail. We shall need some known results partaining to the Jordan structure of an analytic matrix-valued function and its lifting to the Riemann surface. They are collected in 2.5-2.7 and can be found in the book by H.Baumgärtel [1].

2.5. EIGENVALUES AND JORDAN STRUCTURE OF AN ANALYTIC MATRIX-VALUED FUNCTION. Fix $r \in (1, r_1)$. An irreducible polynomial $q(z,\lambda)$ has no multiple zeros in $r\mathbb{D}$ outside a finite set Z_1. One can write $q(z,\lambda) = (\lambda - \lambda_1(z)) \cdot \ldots \cdot (\lambda - \lambda_k(z))$ where $\lambda_i(z)$ are all distinct for $z \in r\mathbb{D} \setminus Z_1$. The functions $\lambda_i(z)$ are analytic in any simply connected subdomain of $r\mathbb{D} \setminus Z_1$ and may undergo a permutation when continued along a path surrounding a point $z \in Z_1$. Evidently all $\lambda_i(z)$ have finite limits at the points of Z_1.

The structure of the Jordan normal form of $\Phi(z)$ does not depend on z outside a finite set $Z \supset Z_1$. We need the "transformation" matrix-valued function, i.e. a function $T(z)$, such that $T^{-1}(z)\,\Phi(z)\,T(z)$ is in the Jordan normal form. Fix the enumeration $\lambda_1(z), \ldots, \lambda_k(z)$ in a simply connected subdomain of $r\mathbb{D} \setminus Z$. Then $T(z)$ can be chosen in the form $[T_1(z), \ldots, T_k(z)]$ where $T_i(z)$ is an $n \times m$ analytic matrix-valued function whose columns are eigenvectors and root vectors corresponding to $\lambda_i(z)$ When continued along a path surrounding a point $z \in Z$, the matrices $T_i(z)$ permute agreeing with the permutation of $\lambda_i(z)$. Moreover, $T(z)$ can be chosen so that all entries of both $T(z)$ and $T(z)^{-1}$ have absolute valued not exceeding a negative power of $\operatorname{dist}(z, Z)$ times a constant.

2.6. RIEMANN SURFACE. Consider $\mathcal{K}^* = \{(z,\lambda) \in \mathbb{C}^2 : q(z,\lambda) = 0,\ z \in r\mathbb{D} \setminus Z\}$. Clearly $\pi : (z,\lambda) \mapsto z$ is a k-sheeted unbranched covering of $r\mathbb{D}\setminus \mathcal{Z}$ and $\mathcal{K}^*$ is a Riemann surface. For $\zeta\in\mathcal{Z}$ we add a point projecting into ζ to each component of $\pi^{-1}(B(\zeta)\setminus\{\zeta\})$, where $B(\zeta)$ is a small disc about ζ. We obtain a κ-sheeted branched covering of $r\mathbb{D}$. The conformal structure at the points of $\pi^{-1}(\mathcal{Z})$ is defined in a standard fashion to produce a Riemann surface $\mathcal{K}$. Define Λ_Φ as follows: $\Lambda_\Phi:(z,\lambda)\mapsto\lambda$. This function is analytic and bounded in $\mathcal{K}^*$, hence it extends to $\mathcal{K}$ analytically. Observe that $\mathcal{K}$ is connected since q is irreducible (this fact will not be used in the sequel). Clearly $R=\pi^{-1}(\mathbb{D})$ is admissible. Define T_Λ to be the operator $x\mapsto\Lambda x$ where Λ is an $m\times m$ diagonal matrix-valued function with the diagonal entries equal to Λ_Φ.

2.7. LIFTING OF THE JORDAN STRUCTURE. The Jordan normal form **can** also be lifted naturally to the Riemann surface. As in 2.5 we start with a simply connected subdomain of $r\mathbb{D}\setminus\mathcal{Z}$, where the transformation matrix-valued function $T(z)$ is single-valued. It is easily seen that $T(z)^{-1}\Phi(z)T(z) = \mathcal{D}_1(z)\oplus\ldots\oplus\mathcal{D}_k(z)$, where the $\mathcal{D}_i(z)$ are $m\times m$ analytic matrix-valued functions in the Jordan normal form. $\mathcal{D}_i(z)$ has $\lambda_i(z)$ as the diagonal entries and the nilpotent part is the same for all $i=1,\ldots,\kappa$. Consider in $\mathcal{K}$ a matrix-valued function $\mathcal{D}(\zeta)$ in the Jordan normal form with this very nilpotent part and the diagonal entries equal to Λ_Φ. Clearly $\mathcal{D}(\zeta)=\mathcal{D}_i(\pi(\zeta))$, where i is the index such that $\lambda_i(\pi(\zeta))=\zeta$. Evidently $\mathcal{D}(\zeta)$ is analytic in $\mathcal{K}$. Likewise we put $\widetilde{T}(z)=T_i(\pi(\zeta))$, $\zeta\in\mathcal{K}\setminus\pi^{-1}(\mathcal{Z})$, where $T_i(z)$ is the $n\times m$ submatrix of $T(z)$ defined in 2.5. It follows from 2.5 that $\widetilde{T}(\zeta)$ cannot grow too rapidly at the points of $\pi^{-1}(\mathcal{Z})$. Hence $\widetilde{T}(\zeta)$ is meromorphic in $\mathcal{K}$. The inverse matrix-valued function $T(z)^{-1}$, $z\in r\mathbb{D}\setminus\mathcal{Z}$ consists of κ submatrices $(T^{-1})_i(z)$ whose rows are eigenvectors and root vectors corresponding to $\lambda_i(z)$. They are lifted to $\mathcal{K}$, in a similar fashion, to a single meromorphic $m\times n$ matrix-valued function $(T^{-1})^\sim(\zeta)$.

2.8. PROOF OF LEMMA 1.11. Let $f\in\mathcal{M}_m^+(R)$. Put

$$(Vf)(z) = \sum_{\zeta \in \pi^{-1}(z)} \tilde{T}(\zeta) f(\zeta), \quad z \in r\mathbb{D} \setminus \mathcal{Z} .$$

It is clear that $Vf \in \mathcal{M}_n^+(\mathbb{D})$. The intertwining relation $T_\Phi V = VT_\mathcal{D}$ follows immediately from the fact that $\Phi(\pi(\zeta))\cdot \tilde{T}(\zeta) = \tilde{T}(\zeta)\mathcal{D}(\zeta)$, $\zeta \in \mathcal{K} \setminus \pi^{-1}(\mathcal{Z})$. Now define for $g \in \mathcal{M}_n^+(\mathbb{D})$

$$(Wg)(\zeta) = (T^{-1})^{\sim}(\zeta)\, g(\pi(\zeta)), \quad \zeta \in \mathcal{K} \setminus \pi^{-1}(\mathcal{Z}) .$$

It is easy to see that $Wg \in \mathcal{M}_m^+(\mathcal{R})$ and $W = V^{-1}$. Thus we have proved that T_Φ on $\mathcal{M}_n^+(\mathbb{D})$ and $T_\mathcal{D}$ on $\mathcal{M}_m^+(\mathcal{R})$ are similar (in a nontopological sense). To obtain the desired result on weak quasisimilarity we must look after the poles and zeros.

It is clear that $V\mathcal{A}_m^+(\mathcal{R}) \subset \frac{1}{f}\mathcal{A}_n^+(\mathbb{D})$ and $W\mathcal{A}_n^+(\mathbb{D}) \subset \subset \frac{1}{g\circ\pi}\mathcal{A}_m^+(\mathcal{R})$ where $f \in \mathcal{A}^+(\mathbb{D})$, $g \in \mathcal{A}^+(\mathbb{D})$ have zeros of sufficiently high order at the points of $\mathcal{Z}$. **Hence we have** $fV\mathcal{A}_m^+(\mathcal{R}) \supset fV\cdot(g\circ\pi)W\mathcal{A}_n^+(\mathbb{D}) = f\cdot(g\circ\pi)\,\mathcal{A}_n^+(\mathbb{D})$ and it follows that $\mathcal{Y} \overset{def}{=} fV\mathcal{A}_m^+(\mathcal{R})$ has finite codimension in $\mathcal{A}_n^+(\mathbb{D})$. The operator $V' = fV$ acts continuously from $E^2(\mathcal{R})$ to H_n^2 . Let X be the closure of $V'E_m^2(\mathcal{R})$ in H_n^2 . Clearly $X = clos\,\mathcal{Y}$, and hence $codim\, X < \infty$. Since $T_\Phi V' = V'T_\mathcal{D}$, the subspace X is T_Φ-invariant. Evidently $Ker\, V' = \{0\}$ and we conclude that $T_\mathcal{D} \mid E_m^2(\mathcal{R}) \prec T_\Phi \mid X$.

The second deformation is constructed in the same manner. It is easy to see that if $\mathcal{Z} \cap \partial\mathbb{D} = \emptyset$, $f \mid \partial\mathbb{D} \neq 0$ then $V'E_m^2(\mathcal{R})$ is closed. This observation implies the desired result on weak similarity. ▨

2.9. REMARK. It is not hard to show that $T_\mathcal{D}$ on $E_m^2(\mathcal{R})$ and $T_\Phi \mid X$ are actually quasisimilar, and thus we obtain a bit more than weak quasisimilarity.

2.10. PROOF OF PROPOSITION 1.5, LEMMA 1.6 AND LEMMA 1.12. Let R be an admissible domain in a Riemann surface, $\psi \in \mathcal{A}^+(R)$, $\psi \not\equiv const$. Let I_m be the unit $m \times m$ matrix and let J_m be the nilpotent $m \times m$ Jordan block. Put $\Lambda = \psi I_m$, $\mathcal{D} = \psi I_m + J_m$. By 2.8, to prove Proposition 1.5 it is sufficient to show that T_Λ and $T_{\mathcal{D}}$ are similar as linear operators on $\mathcal{M}_m^+(R)$. The operator $d/d\psi$ acting locally by the formula $f \mapsto f'/\psi'$ is well-defined on $\mathcal{M}^+(R)$. Put

$$B = exp\left(\frac{d}{d\psi} \cdot J_m\right) = \begin{bmatrix} 1, & \frac{d}{d\psi}, & \dots, & \frac{1}{m!} & \frac{d^m}{d\psi^m} \\ & \ddots & & & \\ 0 & & & \ddots & 1 \end{bmatrix}. \tag{2.1}$$

Clearly B acts on $\mathcal{M}_m^-(R)$. The relation $BT_\Lambda = T_{\mathcal{D}} B$ can be verified by a direct computation, but we give a short explanation which is due to M.Yu.Ljubich.

Let $[K_1, K_2] = K_1K_2 - K_2K_1$. The formulae below will be understood in the sense of operators on $\mathcal{M}_m^+(R)$. Put $C = \frac{d}{d\psi} \cdot J_m$. We have $[C, \Lambda] = J_m$. Since taking commutator is a differentiation of the operator algebra, one obtains $[C^n, \Lambda] = n J_m C^{n-1}$ and $[exp(C), \Lambda] = J_m \cdot exp(C)$. But this means that $exp(C) \cdot \Lambda = (\Lambda + J_m) \cdot exp(C)$. Clearly $B^{-1} = exp(-C)$ also acts on $\mathcal{M}_m^+(R)$. Thus we have proved Lemma 1.6 and completed the proof of Proposition 1.5. ▨

In the setting of Lemma 1.12 we have admissible domains R_1, R_2 such that $R_1 \subset R \subset R_2$, $\psi \in \mathcal{A}^+(R_2)$, $card(\partial R_1 \cap \partial R) < \infty$, $card(\partial R \cap \partial R_2) < \infty$, and both sets $R_2 \setminus R$, $R \setminus R_1$ have no compact components. Consider the operator B defined by (2.1) and let us show that for a proper function g, $g \in \mathcal{A}^+(R_2)$, the operator gB acts from $E_m^2(R_2)$ to $E_m^2(R)$. First we choose $g_1 \in \mathcal{A}^+(R_2)$ having zeros of sufficiently high order at the zeros of $d\psi$ so that the function $g \mid (\psi')^m$ (in local coordinates) be analytic. Then $g_1 B(\mathcal{A}_m^+(R_2)) \subset$

$\subset \mathcal{A}_m^+(R_2)$. Since R, R_2 are admissible, ∂R and ∂R_2 are finite unions of analytic arcts. The arcs of ∂R and ∂R_2 can have at most a tangency of finite order. It is well-known that the modulus of the ℓ-th derivative of a function in $E^2(\Omega)$ is not greater than $const \cdot [dist(z, \partial\Omega)]^{-\ell-1}$. Thus multiplying $g_1 \cdot d^\ell/d\psi^\ell(f)$ by a function $g_2 \in \mathcal{A}^+(R_2)$ having zeros of high order at the points of $\partial R_2 \cap \partial R$, we obtain a function belonging to $H^\infty(R)$, for any $f \in E^2(R_2)$, $\ell \leq m$. It follows that $g_1 g_2 B$ sends $E_m^2(R_2)$ to $H_m^\infty(R)$.

We have $g_1 g_2 B T_{\mathbb{D}} \mid E_m^2(R_2) = T_\Lambda g_1 g_2 B$ and $Ker(g_1 g_2 B) = \{0\}$. To obtain the desired deformation it remains to check that the closure of $(g_1 g_2 B)(E_m^2(R_2))$ has finite codimension in $E_m^2(R)$. Indeed, from the explicit form of B^{-1} it follows that $(g_1 B^{-1})(\mathcal{A}_m^+(R_2)) \subset \mathcal{A}_m^+(R_2)$. Hence $(g_1 g_2 B)(E_m^2(R_2)) \supset g_1^2 g_2 \mathcal{A}_m^+(R_2)$. By the Behnke-Stein Theorem, $\mathcal{A}^+(R_2)$ is dense in $\mathcal{A}^+(R)$. Thus the closed subspace in question contains $clos(g_1^2 g_2 E_m^2(R))$ which has finite codimension in $E_m^2(R)$.

The second deformation is constructed similarly. Thus we have completed the proof of Lemma 1.12, Proposition 1.4 and Theorem 1.2. ▨

In conclusion, we pose a question, related to Lemmas 1.8, 1.9.

QUESTION. Let $T_\varphi : f \mapsto \varphi f$ be an AT operator on H^2 and let X be a closed subspace invariant for T_φ. Is it true that $\mu(T_\varphi | X) \leq \mu(T_\varphi)$?

REFERENCES

1. Baumgärtel, H.: Analytic Perturbation Theory for Matrices and Operators, Birkhäuser, 1985.
2. Solomyak, B.M. and Volberg, A.L.: Multiplicity of analytic Toeplitz operators, present volume.
3. Sz.-Nagy, B. and Foias, C.: Harmonic analysis of Operators on Hilbert space, North-Holland, 1970.
4. Yakubovich, D.V.: Riemann surface models of Toeplitz operators, present volume.

Operator Theory:
Advances and Applications, Vol. 42

GEOMETRIC METHODS IN SPECTRAL THEORY OF VECTOR-VALUED FUNCTIONS: SOME RECENT RESULTS

S.R.Treil'

CONTENTS

1. Introduction. 211
2. Some well-known results. 213
Chapter 1. Szegö and Kolmogorov extremal problems. . . . 217
3. Statement of the problem. Discussions. 217
4. Constructive solution of the Szegö problem. 221
5. Szegö conditions. 225
6. Kolmogorov extremal problem. 228
Chapter 2. Extreme points. 232
7. Extreme points of the unit ball of the operator Hardy space $H^\infty(E \to E_*)$ 232
Chapter 3. Vector version of the Koosis theorem. 236
8. Statement of the problem and main results. 236
9. Proof of necessity in the main theorem 238
10. Proof of sufficiency in the main theorem. 240
11. How to get the operator $C^{1/2} W^{1/2}$ closed. 248
12. Proof of the simplest generalization of the Koosis theorem. 249
Chapter 4. Unconditional bases of eigenvectors of contraction. 251

13. Statement of the problem. 251
14. Statement of the main result. Discussions. 254
15. Proof of the main result. 258
Chapter 5. Angles between coinvariant subspaces and the Operator Corona Problem (the Sz.-Nagy problem). 261
16. Statement of the problem. 261
17. Connection of the Corona Problem with the angles between coinvariant subspaces. 263
18. Main results. 266
19. Proof of the main theorem. 267
20. Construction of the system of rational vector-valued functions. 271
21. Remark. 275
Appendix. . . . .
22. Some open problems. 276

1. INTRODUCTION

1.1 WHAT IS THIS PAPER ABOUT? The Spectral Theory of Functions (STF) is a branch of analysis pertaining to both function and operator theories. This theory deals with various "spectral" properties of functions by which we mean, speaking very generally, the properties of their singularities of no matter what nature. Many classical problems of function theory are spectral in essence and are connected with the classical problems of operator theory. This is a reason of the fact that the synthesis of "analytic" and "operator" methods (a characteristic feature of the spectral theory of functions) has been revealing itself so fruitful.

An important rôle in STF is played by questions connected with various function models of operators allowing to represent a given operator as the composition of an operator of multiplication by a function and a projection. This multiplication acts in a space of analytic functions (generally speaking, vector-valued). Various properties of operator we start with can be expressed, under the model representation, in terms of the so-called characteristic function (some bounded analytic operator-valued function), see, for example, [7], [15], [9], [10].

This theory is well-developed in the scalar case, when the characteristic function is a usual (scalar) bounded analytic function (especially, if it is an inner function), because in this case one can use the well-developed technique of Hardy classes theory (H^p-theory). In the general (matrix- or operator-valued) case the analogous theory has not been worked out in such details. The analytic difficulties that arise in this case are mainly due to the fact that multiplication of operators is non-commutative. But it often happens that simple geometric ideas and reasons help to overcome these difficulties. Geometric considerations constitute the core of this paper as is reflected in the title.

In general, geometric language seems to the author to

be most adequate and natural for the purposes of the Spectral Theory of Vector- and Operator-valued Functions.

In this paper solutions of some problems in the Spectral Theory of Vector-valued Functions (STVF) are given. We list the topics included: the vectorial Szegö and Kolmogorov extremal problems, a description of the Szegö weights, a description of the extreme points of the unit ball of the operator Hardy space H^∞, weighted norm inequalities for the Hilbert transform, the Sz.-Nagy problem (the Operator Corona Problem), bases of vector-valued rational functions.

These results were obtained by the author during the last few years.

1.2. BASIC DEFINITIONS AND DENOTATION.

$\mathbb{C}$ - the complex plane;

$\mathbb{D}$ - the unit disc of the complex plane, $\mathbb{D} \overset{def}{=} \{\xi \in \mathbb{C} : |\xi| < 1\}$;

$\mathbb{T}$ - the unit circle, $\mathbb{T} \overset{def}{=} \{\xi \in \mathbb{T} : |\xi| = 1\}$;

m - the normalized Lebesque measure on the circle $\mathbb{T}$, $m(\mathbb{T}) = 1$;

z - the identical mapping of a part of the complex plane onto itself (usually of the circle $\mathbb{T}$ or the disc $\mathbb{D}$), $z(\xi) = \xi$;

$\hat{f}(k)$ - the k-th Fourier coefficient of a summable function f on $\mathbb{T}$, i.e.

$$\hat{f}(k) \overset{def}{=} \int_{\mathbb{T}} f z^{-k} \, dm \; ;$$

$\mathcal{L}(\dots)$ - the linear hull of the set $(\dots)$;

$\operatorname{span}(\dots)$ - the closure of $\mathcal{L}(\dots)$.

Let E be a Hilbert space. (All Hilbert spaces in this paper are supposed to be separable). The symbol $L^2(E)$ stands for the space of all square-summable E-valued functions on $\mathbb{T}$ with the norm

$$\|f\|^2_{L^2(E)} = \int_{\mathbb{T}} \|f(\xi)\|^2_E \, dm(\xi) ,$$

and the symbol $H^2(E)$ denotes the Hardy space

$$H^2(E) \overset{def}{=} \{ f \in L^2(E) : \hat{f}(k) = 0 \text{ if } k < 0 \} .$$

The symbol $L^\infty(E \to E_*)$ (E, E_* are Hilbert spaces) denotes the space of all bounded measurable functions on the circle $\mathbb{T}$ taking values in the space of linear operators acting from E onto E_* , and $H^\infty(E \to E_*)$ is the corresponding Hardy class, i.e.

$$H^\infty(E \to E_*) \overset{def}{=} \{ F \in L^\infty(E \to E_*) : \hat{F}(k) = 0 \text{ if } k < 0 \} .$$

Note that the mapping $\Sigma \hat{f}(k)(z \mid \mathbb{T})^k \mapsto \Sigma \hat{f}(k)(z \mid \mathbb{D})^k$ enables us to identify Hardy classes with the spaces of analytic vector-valued and operator-valued functions.

The symbol S stands for the shift operator acting on $H^2(E)$, $Sf = zf$.

We will also need a strong operator-valued H^2-space $H^2(E \to E_*)$, which is the space of all analytic operator-valued functions F on the disc $\mathbb{D}$, such that the vector-valued function $F(\cdot)e$ belongs to $H^2(E_*)$ for every vector e from E . It is obvious that $H^\infty(E \to E_*) \subset H^2(E \to E_*)$.

Note that unlike the scalar (and even vectorial) H^2-spaces, it is impossible to define the operator Hardy class $H^2(E \to E_*)$ in terms of boundary values, because the functions in $H^2(E \to E_*)$ in general have no boundary values in any reasonable sense (see Section 3.3).

A function Θ in $H^\infty(E \to E_*)$ is said to be inner if the operators $\Theta(\xi)$ are isometries a.e. on $\mathbb{T}$. A function F in $H^2(E \to E_*)$ is said to be outer, if $clos_{H^2(E_*)}(F \mathcal{P}ol_A) = H^2$ where $\mathcal{P}ol_A$ is a set of analytic vector-valued polynomials in $L^2(E)$, i.e. $\mathcal{P}ol_A = \{ f = \Sigma_{k=0}^N \hat{f}(k) z^k : \hat{f}(k) \in E \}$.

2. SOME WELL-KNOWN RESULTS

2.1. THEOREM (Lax, Beurling, Halmos). Let $\mathcal{E}$ be a closed invariant subspace for the shift operator S , $Sf = z \cdot f$ on the vector Hardy space $H^2(E)$. Then there exist a unique, up to the constant unitary factor , inner function $\Theta \in H^\infty(E_* \to E)$ (E_* is some auxiliary Hilbert space) such that

$$\mathcal{E} = \Theta H^2(E).$$

This theorem is well known and can be found in monographs [3], [7], [15]. The theorem implies the following result on factorization of operator-valued functions.

2.2. THEOREM (see [3], [7], [15]). Let $F \in H^2(E \to E_*)$. There exist an auxiliary Hilbert space E_{**} , an inner function $\Theta \in H^\infty(E_{**} \to E_*)$ and an outer function $G \in H^2(E \to E_{**})$ such that

$$F = \Theta G \tag{2.1}$$

NOTE. If an operator-valued function F is in $H^\infty(E \to E_*)$ then equality (2.1) holds not only in the open disk $\mathbb{D}$ but also on its boundary $\mathbb{T}$.

2.3. NOTE ON BOUNDARY VALUES OF OPERATOR-VALUED FUNCTIONS. As it was referred to above, the operator Hardy space $H^\infty(E \to E_*)$ defined as a subspace of the operator space $L^\infty(E \to E_*)$ may be identified with the space of bounded analytic operator-valued functions on the disk $\mathbb{D}$. Such functions, like in the usual (scalar) Hardy spaces theory have a.e. on the circle $\mathbb{T}$ non-tangential boundary values (in uniform operator topology), and can be recovered from these boundary values by Poisson integral. Analogous proposition holds for the vector Hardy class $H^2(E)$ too. We do not present proofs of these facts since such proofs are analogous to the corresponding ones in the scalar theory.

The case of the space $H^2(E \to E_*)$ is opposite to the above two cases. At the first glance there seems to be no reasonable way to associate boundary values to an arbitrary function in $H^2(E \to E_*)$. Consider for example the function

$\mathbb{F} \in H^2(H^2 \to \mathbb{C})$ (H^2 is the usual scalar Hardy space)

$$\mathbb{F}(\lambda) f \overset{def}{=} f(\lambda), \ \lambda \in \mathbb{D}, \ f \in H^2 .$$

For the boundary values $\mathbb{F}(\xi)$, $\xi \in \mathbb{T}$, of this function and continuous f the following formula, of course, should be true:

$$\mathbb{F}(\xi) f = f(\xi), \quad \xi \in \mathbb{T} .$$

But the value of an H^2-function at a fixed point of the circle $\mathbb{T}$ is a very "bad" operator. Were the nonboundedness its unique shortcoming, the situation would had been not so much hopeless but the operator in question is non-closable. The work with non-closable operators is very difficult, and thus it makes no sense to attempt to define boundary values of functions in $H^2(E \to E_*)$ as usual operator-valued functions.

But one can still interprete boundary values of such functions as operator measures (complex operator measures). A E-valued operator measure (a semispectral measure in the terminology of [14]) is, by definition, a countably additive (in weak operator topology) function defined on Borel subsets of $\mathbb{T}$ with values in the set of bounded selfadjoint nonnegative operators on the Hilbert space E . The complex operator measure is defined as above, but its values may be non-selfadjoint operators.

Using the concept of operator measure one can define a "square of the modulus" $\mathbb{F}^* \mathbb{F}$ of a function $\mathbb{F}$ in $H^2(E \to E_*)$ on the circle $\mathbb{T}$ as an operator measure $\mathcal{W}$, such that

$$\| \mathbb{F} p \|_{H^2(E_*)} = \int_{\mathbb{T}} (d\mathcal{W} p, p)$$

for polynomials $p \in H^2(E)$.

Note in the end that in the finite dimensional case $(\dim E < \infty)$ the boundary values exist a.e., and we do not need the concept of operator measure for their interpretation.

2.4. OPERATOR WEIGHT can be interpreted as a special case of operator measure. By an operator weight we mean an operator-valued function W on the circle $\mathbb{T}$ taking values in the set of selfadjoint nonnegative (not necessarily bounded)operators, which is summable, i.e. for all vectors $e \in E$ the inclusion $e \in \mathcal{D}om\, W^{1/2}(\xi)$ holds a.e. on $\mathbb{T}$, and

$$\int_{\mathbb{T}} (W(\xi)e,e)\,dm(\xi) \overset{def}{=} \int_{\mathbb{T}} \| W^{1/2}(\xi)e \|^2 dm(\xi) < \infty .$$

The last condition means that the mean value $W(0)$ of the function W,

$$W(0) \overset{def}{=} \int_{\mathbb{T}} W(\xi)\,dm(\xi)$$

(the integral is interpreted in the weak sense) is a bounded operator.

1.3. ACKNOWLEDGEMENT. The author wishes to express his deep gratitude to N.K.Nikol'skii and V.I.Vasyunin for valuable discussions and advice. The author also thanks V.I.Vasyunin, V.V.Peller and S.V.Kislyakov for their help in translation and preparing the manuscript.

CHAPTER I. SZEGÖ AND KOLMOGOROV EXTREMAL PROBLEMS

The present chapter is devoted to a generalization to the operator weights of the well-known results by Szegö and Kolmogorov on approximation in the weighted L^2-norm. Recall these results. Let w be a nonnegative summable function (weight) on the circle $\mathbb{T}$ and let $\mathcal{P}ol^0$ and $\mathcal{P}ol^0_A$ denote the sets of triginometric polynomials and analytic trigonometric polynomials with vanishing Fourier coefficient $\hat{f}(0)$:

$$\mathcal{P}ol^0=\{f: f=\sum_{k\neq 0}\hat{f}(k)z^k\}, \quad \mathcal{P}ol^0_A=\{f: f=\sum_{k>0}\hat{f}(k)z^k\}.$$

The theorems by Szegö and Kolmogorov assert (see [2], [4]) that

$$dist_{L^2(w)}(\mathbb{1},\mathcal{P}ol^0_A)=exp(\int_{\mathbb{T}} log\, w^{1/2} dm) \qquad \text{(Szegö)},$$

and

$$dist_{L^2(w)}(\mathbb{1},\mathcal{P}ol^0)=(\int_{\mathbb{T}} w^{-1} dm)^{-1/2} \qquad \text{(Kolmogorov)};$$

here $\mathbb{1}$ is the function identically equal to 1 ; if integrals are divergent, the right sides of the above equalities are considered to be equal 0 .

These results play an important role in applications, for example to stationary random processes [12], [14], and it would be interesting for such applications to obtain a generalization of the above results to operator weights. Note that these problems are interesting even in a simpler case of matrix weights ($dim\, E<\infty$) .

3. STATEMENT OF THE PROBLEM. DISCUSSION

Let W be an operator weight, $W(\xi): E\to E$, $\xi\in\mathbb{T}$.

Denote by $\mathcal{P}ol$ the set of all vector-valued trigonometric polynomials of the form $f = \sum_k z^k \hat{f}(k)$ (the sums are finite). Put

$$\mathcal{P}ol_A \overset{def}{=} \{ f \in \mathcal{P}ol : \hat{f}(k) = 0 \quad \text{if } k < 0 \},$$

$$\mathcal{P}ol^\circ \overset{def}{=} \{ f \in \mathcal{P}ol : \hat{f}(0) = 0 \},$$

$$\mathcal{P}ol_A^\circ \overset{def}{=} \mathcal{P}ol_A \cap \mathcal{P}ol^\circ = \{ f \in \mathcal{P}ol : \hat{f}(k) = 0 \quad k \leqslant 0 \}.$$

Define on the set $\mathcal{P}ol$ the weighted L^2-norm $\|\cdot\|_W$:

$$\| f \|_W^2 \overset{def}{=} \int_{\mathbb{T}} (W(\xi) f(\xi), f(\xi))_E \, dm(\xi).$$

Fix some vector $e \in E$. By the vectorial extremal Szegö and Kolmogorov problems we understand the problems of finding the following distances:

$$\operatorname{dist}_{L^2(E,W)} (e, \mathcal{P}ol_A^\circ) \qquad \text{(Szegö problem)}$$

and

$$\operatorname{dist}_{L^2(E,W)} (e, \mathcal{P}ol^\circ) \qquad \text{(Kolmogorov problem)}$$

Note that one can consider analogous problems for the operator measures $\mathcal{W}$; corresponding weighted norm $\|\cdot\|_{\mathcal{W}}$ is defined by

$$\| f \|_{\mathcal{W}}^2 = \int_{\mathbb{T}} (d\mathcal{W} f, f) .$$

The following theorem solving Szegö extremal problem is well-known, but the author cannot be a judge in the question of priority. In the case of an operator weight in $L^\infty(E \to E)$ this theorem is contained, for example, in the books of Helson [3] and Sz.-Nagy - Foias [15]. The general case of operator measures has been investigated in the works of Roumanian mathematicains I.Suciu and I.Valusescu, see [14].

3.1 THEOREM. Let $\mathcal{W}$ be a E-valued operator measure on the circle $\mathbb{T}$. There exists a unique (up to a constant unitary factor on the left) outer operator-valued function $\mathbb{F} \in H^{\infty}(E \to E_*)$ (E_* is an auxiliary Hilbert space) such that

(i) $\mathbb{F}^*(\xi)\mathbb{F}(\xi) \leq \mathcal{W}(\xi)$.

(ii) $\mathbb{F}$ is maximal in the sense that for any analytic operator-valued function $\mathbb{G}$, satisfying (i), the inequality

$$\mathbb{G}^*\mathbb{G} \leq \mathbb{F}^*\mathbb{F}$$

holds.

Moreover, such $\mathbb{F}$ solves Szegö problem:

(iii) $dist_{L^2(E,W)}(e, \mathcal{P}ol_A^{\circ}) = \| \mathbb{F}(0)e \| \quad \forall e \in E$.

REMARK. Inequalities (i) and (ii) are inequalities for operator measures. If corresponding functions have boundary values, the above inequalities are interpreted to be the inequalities for boundary values and hold a.e. on $\mathbb{T}$.

3.2. PROOF OF THEOREM 3.1. Consider the weighted L^2-space $L^2(E, \mathcal{W})$, which is obtained from $\mathcal{P}ol$ by factorization and completion in the weighted norm $\|\cdot\|_{\mathcal{W}}$. Let $\mathcal{E} \stackrel{def}{=} clos_{L^2(E,W)} \mathcal{P}ol_A$. The operator of multiplication by z, defined on polynomials, is extended to a unitary operator S on $L^2(E, \mathcal{W})$. It is evident that $\mathcal{E}$ is an invariant subspace of S, i.e. $S\mathcal{E} \subset \mathcal{E}$. Let $\mathcal{E}'$ be the maximal completely nonreducing part of $\mathcal{E}$. In accordance with the Wold-Kolmogorov decomposition (see [3], [7], [15]) the space $\mathcal{E}'$ can be realized as a space of the form $H^2(E_*)$, i.e. there exists a unitary operator $V : \mathcal{E}' \to H^2(E_*)$ such that $VSf = zVf \quad \forall f \in H^2(E_*)$. Define the operator-valued function $\mathbb{F} \in H^2(E \to E_*)$ by

$$\mathbb{F}(\lambda)e = (VP_{\mathcal{E}'}e)(\lambda), \quad e \in E, \ \lambda \in \mathbb{D},$$

where $P_{\mathcal{E}'}$ stands for the orthogonal projection of $L^2(E, \mathcal{W})$ onto $\mathcal{E}'$. Then $\mathbb{F}f = VP_{\mathcal{E}'}f$ for all polynomials $f \in \mathcal{P}ol_A$.

Since the set $P_{\mathcal{E}'}\mathcal{P}ol_A$ is dense in $\mathcal{E}'$, the set $\mathbb{F}\mathcal{P}ol_A$ is dense in $H^2(E_*)$ and therefore $\mathbb{F}$ is an outer function. Note that $\|\mathbb{F}f\|_{H^2(E_*)}=\|P_{\mathcal{E}'}f\|_W \leq \|f\|_W$ for all $f\in\mathcal{P}ol_A$ and hence, by definition, $\mathbb{F}^*\mathbb{F}\leq W$. Check up (iii):

$$\mathrm{dist}_{L^2(E,W)}(e,\mathcal{P}ol_A^0)=\mathrm{dist}_{L^2(E,W)}(e,S\mathcal{E})=\|P_{\mathcal{E}\ominus S\mathcal{E}}e\|_W=$$

$$=\|P_{\mathcal{E}'\ominus S\mathcal{E}'}e\|_W=\|VP_{\mathcal{E}'\ominus S\mathcal{E}'}e\|_{H^2(E_*)}=\|P_{E_*}\mathbb{F}e\|_{H^2(E_*)}=\|\mathbb{F}(0)e\|_{E_*}.$$

To prove (ii) we note that if $G^*G\leq W$ for some $G\in H^2(E\to E')$, then

$$\mathrm{dist}_{L^2(E,W)}(e,\mathcal{P}ol_A^0)\geq\mathrm{dist}_{L^2(E,G^*G)}(e,\mathcal{P}ol_A^0)=$$

$$=\mathrm{dist}_{H^2(E')}(Ge,G\mathcal{P}ol_A^0)\geq\mathrm{dist}_{H^2(E')}(Ge,zH^2(E'))=\|G(0)e\|,$$

and therefore $\|G(0)e\|\leq\|\mathbb{F}(0)e\|$. Using conformal mappings of the disk $\mathbb{D}$ onto itself one can show that $\|G(\lambda)e\|\leq\|\mathbb{F}(\lambda)e\|$ for all $\lambda\in\mathbb{D}, e\in E$. The uniqueness of $\mathbb{F}$ is evident. ●

3.3. REMARK. It follows from the proof of Theorem 3.1 that an operator measure W admits a factorization of the form $W=\mathbb{F}^*\mathbb{F}$, $\mathbb{F}\in H^2(E\to E_*)$ iff the subspace $\mathcal{E}\overset{\mathrm{def}}{=\!=}\mathrm{clos}_{L^2(E,W)}\mathcal{P}ol_A$ is the completely nonreducing subspace of the multiplication by z (shift operator S) in $L^2(E,W)$.

3.4. EXAMPLE. Let w_1, w_2 be (scalar) weights such that $\int\log w_1\,dm>-\infty$, $\int\log w_2\,dm=-\infty$, and W be the 2x2 matrix weight

$$W=\begin{pmatrix} w_1 & 0\\ 0 & w_2\end{pmatrix}.$$

It is evident that for this weight W the operator-valued function $\mathbb{F}$ from Theorem 3.1 has the form

$$\mathbb{F} = (f_1, 0) \ , \quad \mathbb{F} \in H^2(\mathbb{C}^2 \to \mathbb{C})$$

where f_1 is the outer function in H^2 with $|f_1|^2 = w_1$. Note that $W \neq \mathbb{F}^* \mathbb{F}$.

This example shows that the absence of equality in Theorem 3.1, (i), is not connected with the specific nature of the infinite-dimensional case, or with the singular parts of the matrix measure (in the scalar case the inequality arises due to the singular part of the measure) but it is a characteristic feature of Szegö problem even in the matrix weight case.

4. CONSTRUCTIVE SOLUTION OF VECTORIAL SZEGO PROBLEM

The main shortcoming of the Szegö problem solution given by Theorem 3.1 is its nonconstructivity. The theorem below, obtained by the author in [22] is perhaps more constructive than Theorem 3.1 because it is possible to check up its condition, at least in matrix case (especially for small dimensions).

4.1. THEOREM. Let W be an operator weigh in $L^\infty(E \to E)$. Then, for any vector $e \in E$

$$\Delta_W[e] \overset{def}{=} dist_{L^2(E,W)}(e, \mathcal{P}ol_A^0) = exp\Big(-\frac{1}{2} \inf_{\substack{f \in H^2(E) \\ (f(0),e)=1}} \int_{\mathbb{T}} \log(W^{-1}(\xi)f(\xi), f(\xi))\, dm(\xi)\Big)$$

and the infimum is attained.

4.2. REMARK. In the integral we set $(W^{-1}f, f) \overset{def}{=} \| W^{-1/2} f \|^2$. We take $W^{-1/2} f$ to be the vector $g \in (\mathrm{Ker}\ W^{1/2})^\perp$ such that $W^{1/2} g = f$ if $f \in \mathrm{Range}\ W^{1/2}$ and put $\| W^{-1/2} f \| = \infty$ otherwise.

4.3. DISCUSSION. The author considers Theorem 4.1 as useful by two reasons. Firstly, using it, one can decide when the distance $\Delta_W[e]$ is not equal to zero. Secondly, one can obtain lower estimates for this distance. The definition of the distance gives an ability to obtain upper estimates, and therefore we can obtain double-sided estimations of the distance $\Delta_W[e]$ with an arbitrary hight precision.

To prove Theorem 4.1 we need the following lemma.

4.4. LEMMA. Suppose that E is a Hilbert space, W is a selfadjoint nonnegative operator on E, and f is a vector in E. Then the inequality

$$|(f,x)|^2 \leqslant (Wx,x)$$

holds for all $x\in E$ iff $(W^{-1}f,f)\leqslant 1$ ($(W^{-1}f,f)$ is understood as in Remark 4.2).

PROOF. Without loss of generality we may suppose that $x=f+y$ where $y\perp f$. Then the inequality will assume the form

$$\|f\|^4 \leqslant (W(f+y),f+y) \quad \forall y\perp f,$$

or equivalently

$$\|f\|^2 \leqslant \inf_{y\perp f}\|W^{1/2}(f+y)\| = \sup_{\substack{g\in(W^{1/2}f^\perp)^\perp\\ \|g\|=1}} (W^{1/2}f,g).$$

But $(W^{1/2}f^\perp)^\perp$ is nothing but the linear span of the vector $W^{-1/2}f$ (it is evident that $f\perp \operatorname{Ker} W$, $f\in \operatorname{Range} W^{1/2}$ and without loss of generality we may consider $\operatorname{Ker} W=\{0\}$). Therefore supremum on the right is equal to $\|f\|^2/\|W^{-1/2}f\|$ and thus $(W^{-1}f,f)\leqslant 1$.

To prove the converse implication it is sufficient to reverse the calculations. ●

4.5. PROOF OF THE THEOREM 4.1. Let $f\in H^2(E)$, $(f(0),e)=1$ and $\int_{\mathbb{T}}\log(W^{-1}(\xi)f(\bar\xi),f(\bar\xi))\,dm(\xi) > -\infty$. Let Φ be an outer function in (scalar) H^2 such that

$$|\Phi(\xi)| = \| W^{-1/2}(\bar{\xi}) f(\xi) \|^{-1}, \quad \xi \in \mathbb{T} .$$

Then for $g \in H^2(E)$, $g \overset{def}{=} \Phi f$, we have

$$|(g(0), e)| = |\Phi(0)| = exp(-\int_{\mathbb{T}} log \| W^{-1/2}(\bar{\xi}) f(\xi) \| dm(\xi)),$$

and $\| W^{-1/2}(\bar{\xi}) g(\xi) \| = 1$ a.e. on $\mathbb{T}$.

Hence, for the operator-valued function $\mathbb{F} \in H^2(E \to \mathbb{C})$ defined by the formula

$$\mathbb{F}(\xi) x = (x, g(\bar{\xi})), \quad x \in E, \quad \xi \in \mathbb{T}$$

the inequality $\mathbb{F}^* \mathbb{F} \leq W$ holds by Lemma 4.4. Consequently (see Theorem 3.1)

$$\Delta_W[e] \geq \| \mathbb{F}(0) e \| = |(e, g(0))| = exp(-\frac{1}{2} \int_{\mathbb{T}} log(W^{-1}(\xi) f(\bar{\xi}), f(\bar{\xi})) dm(\xi)) .$$

And so, to prove the theorem it is sufficient to present a function f for which the infimum is attained.

Let $\widetilde{\mathbb{F}}$ be a maximal outer operator function such that $\widetilde{\mathbb{F}}^* \widetilde{\mathbb{F}} \leq W$, existence of which is guaranteed by Theorem 3.1. Recall that $W \in L^\infty(E \to E)$ and hence $\widetilde{\mathbb{F}} \in H^\infty(E \to E_*)$ Set $\mathbb{F} = P \widetilde{\mathbb{F}}$ where P is the orthogonal projection of E_* onto $\mathcal{L}(\widetilde{\mathbb{F}}(0) e)$. It is evident that $\mathbb{F} \in H^\infty(E \to E_*)$ and $rank\ \mathbb{F}(\xi) = 1$ a.e. on $\mathbb{T}$. Therefore

$$\mathbb{F}(\xi) x = (x, f(\bar{\xi})) \widetilde{\mathbb{F}}(0) e, \quad x \in E, \quad \xi \in \mathbb{T},$$

where f is some function in $H^\infty(E) \subset H^2(E)$. It is also evident that $\mathbb{F}^* \mathbb{F} \leq W$ and $\mathbb{F}(0) e = \widetilde{\mathbb{F}}(0) e$.

By Lemma 4.4 the inequality $\mathbb{F}^* \mathbb{F} \leq W$ implies that $\| W^{-1/2}(\bar{\xi}) f(\xi) \| \leq \| \mathbb{F}(0) e \|^{-1}$.

Now we will prove that in fact the equality holds. Suppose to the contrary that $\| W^{-1/2}(\bar{\xi}) f(\xi) \| \cdot \| \mathbb{F}(0) e \| \leq 1, \ \xi \in \mathbb{T}$, and that on a set of nonzero Lebesque measure the strong inequality holds. Then for the outer function $\varphi \in H^\infty$,

$$|\varphi(\xi)| = max(\| W^{-1/2}(\bar\xi) f(\xi)\| \cdot \| F(0)e\|, 1/2), \quad \xi \in \mathbb{T},$$

we have $|\varphi(0)| < 1$. Set $g = f/\varphi$. It is evident that $\| W^{-1/2}(\bar\xi) g(\xi) \| \le \| F(0)e \|$, and $|(g(0), e)| > |(f(0), e)| = 1$. Hence for the outer operator function G, $G \in H^\infty(E \to E_*)$, $G(\xi)x = (x, f(\bar\xi)) F(0)e$, $x \in E$, the following inequalities will hold

$$G^*G \leqslant W \qquad \text{(by Lemma 4.4)},$$

$$\| G(0)e \| = |(e, g(0))| \cdot \| F(0)e \| > |(e, f(0)| \cdot \| \widetilde{F}(0)e \| = \| \widetilde{F}(0)e \| .$$

This contradicts the maximality of $\widetilde{F}$.

The equality $\| W^{-1/2}(\bar\xi) f(\xi) \| = \| F(0)e \|$ implies that

$$\Delta_W[e] = \| \widetilde{F}(0)e \| = \| F(0)e \| = exp(-\int_{\mathbb{T}} log \| W^{-1/2}(\xi) f(\bar\xi) \| dm(\xi)),$$

Q.E.D. ●

4.6. REMARK. The analysis of the proof of Theorem 4.1 shows that the condition of boundedness of W may be changed by a weaker one $\| W(\cdot) \| \in L^1$. This condition holds, for example, for any matrix weight W (i.e. if $dim\, E < \infty$).

The following lemma makes it possible to use the Theorem 4.1 to solve the extremal Szegö problem even for operator measures.

4.7. LEMMA. <u>Let $\mathcal{W}$ be an operator measure, W its Poisson integral, which is a bounded analytic function in the unit disk $\mathbb{D}$, and W_r, $0 < r < 1$, be the operator weights in $L^\infty(E \to E)$ defined by $W_r(\xi) = W(r\xi)$, $\xi \in \mathbb{T}$. Then $\Delta_{\mathcal{W}}[e] = \lim_{r \to 1^-} \Delta_{W_r}[e]$ for any vector $e \in E$.</u>

PROOF. Since for every trigonometric polynomial f the equality $\| f \|_{\mathcal{W}} = \lim_{r \to 1^-} \| f \|_{W_r}$ holds, it is evident that $\Delta_{\mathcal{W}}[e] \geqslant \geqslant \lim_{r \to 1^-} \Delta_{W_r}[e]$.

Let us prove the converse inequality. Let $F \in H^2(E \to E_*)$

be an operator function from Theorem 3.1. Set $F_r(\xi) = F(r\xi)$, $\xi \in \mathbb{T}$, $0 < r < 1$. It is easy to prove that $F_r^* F_r \leq W_r$, and therefore $\Delta_W[e] = \| F(0) e \| = \Delta_{F_r^* F_r}[e] \leq \Delta_{W_r}[e]$ ●

5. SZEGÖ CONDITIONS

Introduce some vector generalizations of the well known Szegö condition $\log w \in L^1$.

5.1. DEFINITION. We say that an operator weight W satisfies the (vector) Szegö condition (respectively the star Szegö condition) and write $W \in (S)$ (respectively $W \in (S)^*$) if the set of analytic polynomials $\mathcal{P}ol_A$ (respectively that of antyanalytic polynomials $\mathcal{P}ol_{\bar{A}} \overset{def}{=} \{ f : f(\bar{z}) \in \mathcal{P}ol \}$) is not dense in the weighted space $L^2(E,W)$. These conditions are equivalent to the existence of such a vector $e \in E$ that

$$dist_{L^2(E,W)}(e, \mathcal{P}ol_A^\circ) > 0 \qquad \text{(Szegö condition)},$$

$$dist_{L^2(E,W)}(e, \mathcal{P}ol_{\bar{A}}^\circ = \mathcal{P}ol_{\bar{A}} \cap \mathcal{P}ol^\circ) > 0 \qquad \text{(star Szegö condition)}.$$

Note that the conditions (S) and $(S)^*$ are not equivalent (see Example 5.9). But there is a simple connection between these two conditions. Namely, mapping $\#$, $f^\#(\xi) = f(\bar{\xi})$, $\xi \in \mathbb{T}$ maps $L^2(E,W)$ onto $L^2(E,W^\#)$ unitarily, and therefore

$$dist_{L^2(E,W)}(e, \mathcal{P}ol_A^\circ) = dist_{L^2(E,W^\#)}(e, \mathcal{P}ol_{\bar{A}}^\circ) .$$

Consequently, the Szegö condition for W is equivalent to the star Szegö condition for $W^\#$ and conversely. So $W \in (S) \iff W^\# \in (S)^*$.

Similarly, each of the conditions defined below is accompanied by the 'star' condition obtained by the substitution of $W^\#$ for W .

5.2. DEFINITION. A weight W is said to satisfy the strong

<u>Szegö condition if</u>

$$\Delta_W[e] = \mathrm{dist}_{L^2(E,W)}(e, \mathcal{P}ol_A^{\circ}) > 0 \qquad \forall e \in E \qquad (\mathcal{SS})$$

and the <u>uniform Szegö condition</u>, if

$$\inf_{\substack{e\in E\\ \|e\|=1}} \mathrm{dist}_{L^2(E,W)}(e, \mathcal{P}ol_A^{\circ}) > 0 \qquad (\mathcal{US})$$

Note that for matrix weights these two conditions are equivalent.

5.3. DEFINITION. A weight W will be called Szegö factorable and it will be written $W \in (\mathbb{F}\mathcal{S})$ if W can be represented in the form

$$W = \mathbb{F}^*\mathbb{F}, \quad \mathbb{F} \in H^2(E \to E_*).$$

It follows from the inner-outer factorization (see Theorem 2.2) that if $W \in (\mathbb{F}\mathcal{S})$ then the function $\mathbb{F}$ can be chosen to be outer.

5.4. REMARK. For the scalar weights w all conditions $(\mathcal{S})$, $(\mathcal{SS})$, $(\mathcal{US})$, $(\mathbb{F}\mathcal{S})$ and star ones coincide, and are equivalent to the condition $\log w \in L^1$. In the general (operator) case these conditions are different.

The following two well-known results describe matrix weights satisfying $(\mathbb{F}\mathcal{S})$ or $(\mathcal{SS})$.

5.5. THEOREM. Let W be an E-valued operator weight, and $\dim E < \infty$. The following are equivalent.

(i) $W \in (\mathbb{F}\mathcal{S})$

(ii) <u>The range function</u> $E(\xi) \overset{def}{=} W(\xi)E$ <u>is coanalytic, i.e.</u> $E(\xi) = \mathrm{span}\{f(\bar{\xi}) : f \in H^2(E),\ f(\bar{\zeta}) \in E(\zeta) \text{ a.e.}\}$ **<u>for almost all</u>** ξ, $\xi \in \mathbb{T}$, <u>and</u>

$$\int_{\mathbb{T}} \log\det(W(\xi)|\,E(\xi))\,dm(\xi) > -\infty .$$

(iii) rank $W(\xi)$ is a constant (say, r) a.e. If $e_1, \dots, e_n$ is an orthogonal basis in E, then there exists a principal minor $\delta = \det \{(W(\cdot) e_{i_p}, e_{i_q})\}_{1 \le p, q \le r}$ such that

$$\int_{\mathbb{T}} \log \delta \, dm > -\infty$$

and the quotients

$$\bar{\delta}_{k,p} \;/\; \bar{\delta}$$

are boundary values of some functions in Nev, the class of all Nevanlinna meromorphic functions (recall that $Nev = \{ f/g : f, g \in H^\infty \}$); here $1 \le p \le r$, $1 \le k \le n$ and $\delta_{k,p}$ stands for the minor of W which is obtained by replacing of the column $\{(W(\cdot) e_{i_p}, e_{i_q})\}_{1 \le q \le r}$ by $\{(W(\cdot) e_k, e_{i_q})\}_{1 \le q \le r}$ in δ.

The equivalence (i)$\Longleftrightarrow$(ii) of the theorem is by Helson-Lowdenslager (see [3]), and the one (i)$\Longleftrightarrow$(iii) is by Matveev - Rozanov.

5.6. COROLLARY. If $\dim E < \infty$ and W is an operator E-valued weight which is invertible a.e., then the following are equivalent.

(i) $W \in (S)$
(ii) $W \in (US)$
(iii) $W \in (SS)$
(iv) $\int_{\mathbb{T}} \log \det W \, dm > -\infty$.

Statements 5.5, 5.6 give us criteria for a weight W to satisfy conditions (FS) and (US) ((SS)). As far as the Szegö condition (S) is concerned the characterizations of weights satisfying these conditions besides the nonconstructive one, given by Theorem 3.1 are not known untill now. Using Theorem 4.1 one can obtain the following, more constructive reformulation of this condition.

5.7. PROPOSITION. An operator weight W satisfies the Szegö condition (S) iff there exists a function $f \in H^2(E)$ such that

$$\int_{\mathbb{T}} \log(W^{-1}(\xi) f(\bar{\xi}), f(\bar{\xi}))\, dm(\xi)$$

(the integral is defined as in Remark 4.2, i.e. we set $(W^{-1}f,f)=\infty$ if $f \notin \operatorname{Range} W^{1/2}$).

5.8. COROLLARY. If for an operator weight W the matrix of operator function W^{-1} in some orthonormal basis has at least one element with logarithm in L^1 on the main diagonal, then $W \in (\mathcal{S})$ (and also $W \in (\mathcal{S})^*$).

The following example is inspired by [9] and in essence is presented in that work. It is published also in [21].

5.9. EXAMPLE. Let $E = \mathbb{C}^2$, $f = \begin{pmatrix} f_1 \\ f_2 \end{pmatrix}$, $\log|f_{1,2}| \in L^1$ and $W = ff^* = \begin{pmatrix} f_1 \\ f_2 \end{pmatrix} \cdot (\bar{f}_1, \bar{f}_2)$. Using Theorem 5.5 one can easily show that $W \in (\mathcal{S})$ iff $\overline{W} \in (\mathbb{F}\mathcal{S})$, and, by Corollary 5.6, the latter holds iff $\bar{f}_2/\bar{f}_1 \in Nev$. Similarly $W \in (\mathcal{S})^* \Longleftrightarrow f_2/f_1 \in Nev$. Therefore choosing f_1 and f_2 for which $\bar{f}_2/\bar{f}_1 \in Nev$ but $f_2/f_1 \notin Nev$, we obtain a weight W such that $W \in (\mathcal{S})$ and $W \notin (\mathcal{S})^*$.

It is evident by Corollary 5.6 that in matrix case the conditions $(\mathcal{S}\mathcal{S})$ and $(\mathcal{U}\mathcal{S})$ are equivalent to the star ones $(\mathcal{S}\mathcal{S})^*$ and $(\mathcal{U}\mathcal{S})^*$. The author does not know if this is true in general operator case.

6. KOLMOGOROV EXTREMAL PROBLEM

This problem is simpler than the Szegö one. Its solution is given by the following simple theorem, proved in [24] (see Lemma 9 there; in this lemma a bit stronger statement than the theorem below is proved), see also [22] p.68.

6.1. THEOREM. Let W be an E-valued operator weight, and e be some vector in E . Then

$$dist_{L^2(E,W)}(e,\mathcal{P}ol^\circ)=\| [\int_{\mathbb{T}} W^{-1}dm]^{-1/2}e\| .$$

6.2. REMARK. A natural question arises about the meaning of the integral in Theorem 6.1., since the weight is not necessarily invertible a.e., and, moreover, even if it is invertible the operator-valued function W^{-1} needs not to be summable.

But one can give a sense to this expression. The integral

$$\int_{\mathbb{T}} (W^{-1}(\xi)e,e)\,dm(\xi)$$

(we assume as in Remark 4.2 that $(W^{-1}e,e)=\infty$ if $e\notin \mathrm{Range}\,W^{1/2}$) defines a nonnegative quadratic form $A[\cdot]$ in E, perhaps with values equal to ∞. One can show that its restriction $A_0[\cdot]$ to the set $\{f: A[f]<\infty\}$ is a closed quadratic form. Consequently it is a quadratic form of some selfadjoint (may be unbounded) operator A_0, acting on some subspace $E_0=\mathrm{clos}\,\mathcal{D}om\,A_0$, which may differ from E.

It is natural to define the operator $[\int W^{-1}dm]^{-1/2}$ as the operator whose matrix in the decomposition $E=E_0+E_0^\perp$ is

$$A^{-1/2}=\begin{pmatrix} A_0^{-1/2} & \mathbb{O} \\ \mathbb{O} & \mathbb{O}\end{pmatrix} .$$

This definition is really natural. It becomes clear if we note that the matrix of the "operator" A, which (formally) defines the quadratic form $A[\cdot]$ has in the same decomposition $E=E_0+E_0^\perp$ the following form

$$A=\begin{pmatrix} A_0 & \mathbb{O} \\ \mathbb{O} & \infty I\end{pmatrix}$$

There also exists another method of the interpretation of the integral which of course leads to the same result. Set $W_\varepsilon=W+\varepsilon I$, $\varepsilon>0$. The interpretation of the expression $A_\varepsilon^{-1/2}$ $[\int W_\varepsilon^{-1}dm]^{-1/2}$ is obvious, because $\|W_\varepsilon^{-1}\|\le 1/\varepsilon$.

We then define $A^{-1/2}$ to be the weak limit, $A^{-1/2} = w\text{-}\lim_{\varepsilon\to 0} A_\varepsilon^{-1/2}$. This weak limit exists because the operators $A_\varepsilon^{-1/2}$ decrease when ε goes down to 0 .

6.3. PROOF OF THEOREM 6.1. Let $e\in E$. Then

$$\mathrm{dist}_{L^2(E,W)}(e,\mathcal{P}ol^\circ) = \mathrm{dist}_{L^2(E)}(W^{1/2}e, W^{1/2}\mathcal{P}ol^\circ) =$$

$$= \sup\{|(W^{1/2}e,g)_{L^2(E)}| : g\in(W^{1/2}\mathcal{P}ol^\circ)^\perp,\ \|g\|_{L^2(E)}\le 1\}$$

($\perp$ denotes the orthogonal complement in $L^2(E)$). If the weight W is invertible a.e. and the weight W^{-1} is uniformly bounded, then it is easy to prove that

$$(W^{1/2}\mathcal{P}ol^\circ)^\perp = W^{-1/2}E .$$

Therefore the above chain of equalities can be continued

$$\mathrm{dist}_{L^2(E,W)}(e,\mathcal{P}ol^\circ) = \sup\{|(W^{1/2}e, W^{-1/2}f)_{L^2(E)}| : f\in E, \|W^{-1/2}f\|\le 1\} =$$

$$= \sup\{|(e,f)_E| : f\in E,\ ([\int_{\mathbb{T}} W^{-1}dm]f,f)\le 1\} =$$

$$= \sup\{|(e,[\int_{\mathbb{T}} W^{-1}dm]^{-1/2}g)_E| : g\in E, \|g\|\le 1\} =$$

$$= \|[\int_{\mathbb{T}} W^{-1}dm]^{-1/2}e\| .$$

Let now W be an aribtrary weight. Set $W_\varepsilon = W+\varepsilon I$, $\varepsilon>0$. As it was shown above

$$\mathrm{dist}_{L^2(E,W)}(e,\mathcal{P}ol^\circ) = \|[\int_{\mathbb{T}} W^{-1}dm]^{-1/2}e\| .$$

Since for any trigonometric polynomial f its weighted norm $\|f\|_{L^2(E,W_\varepsilon)}$ tends decreasing to $\|f\|_{L^2(E,W)}$ as

$\varepsilon \downarrow 0$, it follows that

$$\operatorname{dist}_{L^2(E,W)}(e, \mathcal{P}ol^0) = \lim_{\varepsilon \to 0^+} \operatorname{dist}_{L^2(E,W_\varepsilon)}(e, \mathcal{P}ol^0).$$

To find the limit on the right and to finish the proof we note that it follows from the spectral theorem that for any selfadjoint nonnegative operator W the following equality holds:

$$\lim_{\varepsilon \to 0} ((W^{-1} + \varepsilon I)^{-1} e, e) = \begin{cases} (W_0^{-1} e, e) & \text{if} \quad e \in \operatorname{Range} W^{1/2} \\ \infty & \text{if} \quad e \notin \operatorname{Range} W^{1/2} \end{cases}$$

here W_0 denotes the restriction of W to the space $(\operatorname{Ker} W)^\perp$. ∎

CHAPTER 2. EXTREME POINTS

7. EXTREME POINTS OF THE UNIT BALL OF THE OPERATOR HARDY SPACE

The following description of extreme points of the unit ball of the (scalar) Hardy space H^∞ is well known: a function f, $f \in H^\infty$, $\|f\|_\infty = 1$ is an extreme point iff $\log(1-|f|^2) \notin L^1$ see, for example [2]. This chapter is devoted to the generalization of this result on the operator case which was obtained in [21].

The problem of extreme points in the operator Hardy space $H^\infty(E \to E_*)$ arises in the study of the functional model (see [15], [9]), especially if one follows the coordinate free approach proposed by N.K.Nikol'skii and V.I.Vasyunin, see [10].

We have seen in the preceding chapter (Section 5) that the Szegö condition $\log w \in L^1$ is generalised to the vector case in different non equivalent forms. So the reader should not find the following statement unexpected.

7.1. THEOREM. <u>Let</u> $F \in H^\infty(E \to E_*)$, $\|F\|_\infty \leq 1$. <u>Then</u> F <u>is an extreme point of the unit ball of the operator Hardy space</u> $H^\infty(E \to E_*)$ <u>if and only if at least one of the two following condition holds</u>

a) $D_F^2 \stackrel{def}{=} (I - F^*F) \notin (S)$

b) $D_{F^*}^2 \stackrel{def}{=} (I - FF^*) \in (S)^*$.

The description of the weights satisfying the Szegö condition (S) was given in the above chapter, see Sections 5.1, 5.7, 5.8, 3.1, and we will not discuss it here. Note that in the vector case the theorem is more complicated than in the scalar one, because in the vector case conditions (S) and $(S)^*$ do not coinside, see example 5.9.

7.2. PROOF OF THE THEOREM. The sufficiency of condition a) (or b)) was proved in [9]. We present this proof for the sake of completeness. Let F, $\|F\|_\infty \leq 1$, be not an extreme point of the unit ball of $H^\infty(E \to E_*)$. Then there exists a function $\varepsilon \in H^\infty(E \to E_*)$, $\varepsilon \not\equiv 0$, such that $\|F \pm \varepsilon\|_\infty \leq 1$. Hence for all vectors $e \in E$ inequalities $\|(F \pm \varepsilon)e\| \leq \|e\|$ hold. By the parallelogramm identity we have $\|Fe\|^2 + \|\varepsilon e\|^2 = = \frac{1}{2}(\|(F+\varepsilon)e\|^2 + \|(F-\varepsilon)e\|^2) \leq \|e\|^2$. Consequently, $\varepsilon^*\varepsilon \leq I - F^*F = \mathcal{D}_F^2$ and by Theorem 3.1 $\mathcal{D}_F^2 \in (S)$. Similarly, using the inequalities $\|F^* \pm \varepsilon^*\| \leq 1$ we obtain $\varepsilon^*\varepsilon \leq \leq I - FF^* = \mathcal{D}_{F^*}^2$ and this implies $\mathcal{D}_{F^*}^2 \in (S)^*$.

Let now $\mathcal{D}_F^2 \in (S)$ and $\mathcal{D}_{F^*}^2 \in (S)^*$. Show that in this case the function F is not an extreme point. Assume that we can represent F in the form $F = F_2 F_1$, where $F_1 \in H^\infty(E \to E')$, $F_2 \in H^\infty(E' \to E_*)$ (E' is an auxiliary Hilbert space), and $|F_1| = |F|^{1/2}$, $|F_2^*| = |F^*|^{1/2}$; here and below the symbol $|A|$ denotes the modulus of the operator A, i.e. the nonnegative operator $(A^*A)^{1/2}$. Since $\mathcal{D}_{F_1}^2 = = I - |F| = \mathcal{D}_F^2 \cdot (I + |F|)^{-1}$, we have $\mathcal{D}_{F_1}^2 \in (S)$. Therefore (see Theorem 3.1) there exists a function $\varepsilon_1 \in H^\infty(E \to G)$, $\varepsilon_1 \not\equiv 0$, such that $\varepsilon_1^*\varepsilon_1 \leq \mathcal{D}_{F_1}^2$. Note that without loss of generality one can assume $\dim G = 1$. The inequality $\varepsilon_1^*\varepsilon_1 \leq \mathcal{D}_{F_1}^2$ means that the operator functions $\mathcal{F}_1^\pm$, $\mathcal{F}_1^\pm \in H^\infty(E \to E' \oplus G)$, $\mathcal{F}_1^\pm = \begin{pmatrix} F_1 \\ \pm\varepsilon_1 \end{pmatrix}$ are contractive ($\|\mathcal{F}_1^\pm\|_\infty \leq 1$). Similarly, $\mathcal{D}_{F_2^*}^2 \in (S)^*$ and hence there exists $\varepsilon_2 \in H^\infty(G_* \to E_*)$, $\varepsilon_2 \not\equiv 0$ such that $\varepsilon_2\varepsilon_2^* \leq \mathcal{D}_{F_2^*}^2$. Again, without loss of generality one may assume $\dim G_* = 1$, and below we will set $G_* = G$. The inequality $\varepsilon_2\varepsilon_2^* \leq \mathcal{D}_{F_2^*}^2$ means that the operator function $\mathcal{F}_2$, $\mathcal{F}_2 \in H^\infty(E' \oplus G \to E_*)$, $\mathcal{F}_2 = (F_2, \varepsilon_2)$ is also contractive. Therefore $1 \geq \|\mathcal{F}_2\mathcal{F}_1^\pm\|_\infty = \|F_2F_1 \pm \varepsilon_2\varepsilon_1\| = = \|F \pm \varepsilon_2\varepsilon_1\|$. Since $\dim G = 1$, the product $\varepsilon_2\varepsilon_1$ is not zero, and hence F is not an extreme point.

To finish the proof we need to represent F as a product $F = F_2F_1$. Choose an outer function $F_1 \in H^\infty(E \to E')$ such that $F_1^*F_1 = |F|$. Such a function F_1 exists because

(see Remark 3.3)

$$\mathrm{clos}_{L^2(E,|F|)} \mathcal{P}ol_A \subset \mathrm{clos}_{L^2(E,|F|^2)} \mathcal{P}ol_A$$

and the latter is a completely nonreducing subspace of multiplication by z. Let $F = U\cdot|F|$, $F_1 = V\cdot|F_1|$ be polar decompositions of F and F_1. Set $F_2 \overset{def}{=} U|F|^{1/2}V^*$. It is easy to prove that $F = F_2F_1$, and moreover $F_2 \in H^\infty(E' \to E)$. Indeed, since F_1 is an outer function, $F_1H^2(E) = H^2(E')$ and consequently $F_2H^2(E') \subset \mathrm{clos}\, FH^2(E) \subset H^2(E_*)$.
Calculate F_2^*:

$$F_2F_2^* = U|F|^{1/2}V^*V|F|^{1/2}U^* = U|F|U^* = |F^*|$$

i.e. $|F_2^*| = |F^*|^{1/2}$. ●

7.3. COROLLARY. (V.I.Vasyunin, private communication). <u>Let</u> $F \in H^\infty(\mathbb{C} \to E)$, $\|F\|_\infty \le 1$. <u>Then</u> F <u>is an extreme point iff</u> $\log \mathcal{D}_F^2 \notin L^1$.

PROOF. One can prove this proposition directly, but we deduce it from the theorem. It is sufficient to prove that if $\log \mathcal{D}_F^2 \in L^1$ then $\mathcal{D}_{F^*}^2 \in (S)^*$. Let $F = \Theta f$, where f is the outer part of F (in our case f is a scalar outer function in H^∞, $|f|^2 = F^*F$), and Θ is an inner function in $H^\infty(\mathbb{C} \to E)$. Besides, let h be a function in H^∞ such that $|h|^2 \le \mathcal{D}_F^2$. Such h exists because $\log \mathcal{D}_F^2 \in L^1$. Then

$$\mathcal{D}_{F^*}^2 = I - FF^* = I - \Theta|f|^2\Theta^* \ge \Theta(1-|f|^2)\Theta^* \ge (\Theta h)\cdot(\Theta h)^*$$

and therefore $\mathcal{D}_{F^*}^2 \in (S)^*$. ●

7.4. EXAMPLE ([9]). Let $f_{1,2} \in H^\infty$, $h_{1,2} \in H^\infty$, $|h_1|^2 + |h_2|^2 = 1$, $|f_1|^2 + |f_2|^2 = 1$. Put $h = \begin{pmatrix} h_1 \\ h_2 \end{pmatrix}$, $f_* = \begin{pmatrix} \bar f_1 \\ \bar f_2 \end{pmatrix}$, and $F = hf_*^* = \begin{pmatrix} h_1 \\ h_2 \end{pmatrix}(f_1, f_2)$. Obviously $F \in H^\infty(\mathbb{C}^2 \to \mathbb{C}^2)$ and $\|F\|_\infty \le 1$. It is easy to verify that $\mathcal{D}_F^2 = I - f_*f_*^* = ff^*$ and $\mathcal{D}_{F^*}^2 = I - hh^* = h_*h_*^*$, where $f = \begin{pmatrix} f_2 \\ -f_1 \end{pmatrix}$, $h_* = \begin{pmatrix} \bar h_2 \\ -\bar h_1 \end{pmatrix}$.
Consequently, choosing $f_{1,2}$ and $h_{1,2}$ so that $\bar f_1/\bar f_2 \in Nev$

$\bar{h}_1 / \bar{h}_2 \notin Nev$ (see Section 5.9) we have $\mathfrak{D}_F^2 \in (S)$, but $\mathfrak{D}_{F^*}^2 \notin (S)^*$.

This example shows that the case of $H^\infty(\mathbb{C} \to E)$ examined in Section 7.3 is an exceptional one.

CHAPTER 3. VECTOR VERSION OF THE KOOSIS THEOREM

The following result on weighted norm inequalities for the Riesz projection (or Hilbert transform) which was proved recently by P.Koosis [5] is now well known. Recall that the Riesz projection P_+ is defined by the formula $P_+ f = \sum_{k\geq 0} \hat{f}(k) z^k$

KOOSIS THEOREM. Let w be a weight on the unit circle $\mathbb{T}$. Then a nonzero weight v for which the inequality

$$\int_{\mathbb{T}} |P_+ f|^2 v\, dm \leq \int_{\mathbb{T}} |f|^2 w\, dm \qquad \forall f \in \mathcal{P}ol$$

holds, exists iff $\int_{\mathbb{T}} w^{-1} dm < \infty$.

In this chapter the generalization of this result to operator weights, obtained by the author in [24] (see also [22]), is presented. The proofs are based on the geometric (operator) approach to the weighted norm inequalities.

8. STATEMENT OF THE PROBLEM AND THE MAIN RESULTS

Let E be a separable Hilbert space and W be an E-valued operator weight. The problem is to find conditions of the existence of a nonzero operator weight V such that

$$\int_{\mathbb{T}} (VP_+ f, P_+ f)\, dm \leq \int_{\mathbb{T}} (Wf, f)\, dm \qquad \forall f \in \mathcal{P}ol \tag{8.1}$$

i.e. such that the Riesz projection P_+ is a bounded operator from $L^2(E,W)$ to $L^2(E,V)$. The Riesz projection in the vector case is defined as in the scalar one, $P_+ f = \sum_{k\geq 0} z^k \hat{f}(k)$ for any vector-valued polynomial f. It is required also to learn how "great" such weight V can be.

The following two results can be considered as a natural generalization of Koosis theorem.

8.1. THEOREM. A nonzero weight V, satisfying (8.1), exists if and only if $\int_{\mathbb{T}} (W^{-1}e, e)\, dm < \infty$ for some vector $e \in E$.

Moreover, the weight V can be chosen to satisfy the Szegö condition (S) (and the star one $(S)^*$).

This theorem can be proved in a simple way (see Section 12) by a reduction to the Koosis theorem. The following result seems to be more interesting.

8.2. THEOREM. Let W be an operator weight with the invertible mean value $W(0) = \int W\, dm$. Then a nonzero weight V with the invertible mean value $V(0)$ satisfying (8.1) exists iff the operator

$$W^{-1}(0) \overset{def}{=} \int W^{-1} dm$$

is bounded. Moreover, the weight V can be chosen to satisfy the uniform Szegö condition (US) (and also the star one $(US)^*$), and

$$\Delta_V[e] \overset{def}{=} dist_{L^2(E,V)}(e, \mathcal{P}ol_A^\circ) \geq \frac{1}{8} \left\| \left[\int_{\mathbb{T}} W^{-1} dm\right]^{-1/2} e \right\| \quad \forall e \in E,$$

$$\Delta_{V^\#}[e] \overset{def}{=} dist_{L^2(E,V^\#)}(e, \mathcal{P}ol_A^\circ) \geq \frac{1}{8} \left\| \left[\int_{\mathbb{T}} W^{-1} dm\right]^{-1/2} e \right\| \quad \forall e \in E.$$

(Recall, that $V^\#(\xi) = V(\bar{\xi})$, see Section 5.1).

Note that $\|[V(0)]^{1/2} e\| \geq \Delta_V[e]$ and therefore the above inequalities imply an estimate from below for the mean value $V(0)$. Recall also that $\|[\int_{\mathbb{T}} W^{-1} dm]^{-1/2} e\|$ is equal (see Section 6.1) to the distance $dist_{L^2(E,W)}(e, \mathcal{P}ol^\circ)$.

8.3. REMARK. In the matrix case ($\dim E < \infty$) the boundedness of $\int W^{-1} dm$ is equivalent to the convergence of the integral $\int \|W^{-1}\| dm$. If this condition holds, then there exists a scalar weight w, $w, \frac{1}{w} \in L^1$ such that

$$wI \leq W$$

and Theorem 8.2 (without estimates) follows from the Koosis theorem. But it is impossible to obtain in such a way the estimates of the distances $\Delta_V[e]$ and $\Delta_{V\#}[e]$ which do not depend on n. Hence from the point of such estimates Theorem 8.2 is of interest even in the matrix case.

Besides, in the case $\dim E = 1$ the proof of this theorem, presented below in Sections 9-11 can be considered as a new (operator) proof of the Koosis theorem.

8.4. REMARK. The condition of the invertibility of the mean value $W(0)$ is not a limitation but is rather some normalization condition. Firstly, if we substitute a space E for $(\operatorname{Ker} W(0))^{\perp}$ the weighted space $L^2(E,W)$ will not change. Secondly, the mapping $f \mapsto [W(0)]^{1/2} f$, $f \in L^2(E,W)$ isomorphically maps the weighted space $L^2(E,W)$ onto $L^2(E,W_1)$, where $W_1 \overset{def}{=} [W(0)]^{-1/2} W [W(0)]^{-1/2}$ (here we assume $\operatorname{Ker} W(0) = \{0\}$) and the mean $W_1(0) = \int W_1 dm$ is equal to the identical operator I. But W_1 is not necessarily a weight, because it is not evident that operators $[W(0)]^{-1/2} W(\xi) [W(0)]^{-1/2}$, $\xi \in \mathbb{T}$, are closed. Therefore, to avoid the technical difficulties we will simply assume that the weight W has the invertible mean value $W(0)$.

9. PROOF OF THE NECESSITY IN THEOREM 8.2

Denote by P_n the orthogonal projection in $L^2(E)$ acting on trigonometric polynomials f in the following way:

$$P_n f = \sum_{k=0}^{n} z^k \hat{f}(k) .$$

Since $P_n = P_+ - S^{n+1} P_+ S^{-n-1}$, where $Sf = zf$, $f \in \mathcal{P}ol$, and the operator P_+ is a strong limit of P_n, $n \to \infty$, the required boundedness of P_+ is equivalent to the uniform boundedness of the operators P_n acting from $L^2(E,W)$ to $L^2(E,V)$. To prove the necessity of the boundedness of $\int W^{-1}\, dm$ we calculate the norm of the operator P_0, $P_0 : L^2(E,W) \to L^2(E \to V)$:

$$\| P_0 \|_{L^2(E,W) \to L^2(E,V)} = \sup_{f \in \mathcal{P}ol} \frac{\| \hat{f}(0) \|_{L^2(E,V)}}{\| f \|_{L^2(E,W)}} =$$

$$= \sup_{e \in E} \| e \|_{L^2(E,V)} / \operatorname{dist}_{L^2(E,W)} (e, \mathcal{P}ol^0) .$$

The last equality holds because if we fix a vector $e \in E$, then

$$\sup_{\substack{f \in \mathcal{P}ol \\ \hat{f}(0) = e}} \frac{1}{\| f \|_{L^2(E,W)}} = 1 / \operatorname{dist}_{L^2(E,W)} (e, \mathcal{P}ol^0) .$$

But theorem 6.1 implies

$$\operatorname{dist}_{L^2(E,W)} (e, \mathcal{P}ol^0) = \| [\int W^{-1} dm]^{-1/2} e \| ,$$

and the norm we want to find is equal to

$$\sup_{e \in E} \frac{[\int (Ve, e)\, dm]^{1/2}}{\| [\int W^{-1} dm]^{-1/2} e \|} = \sup_{e \in E} \frac{\| [V(0)]^{1/2} e \|}{\| [W^{-1}(0)]^{-1/2} e \|} =$$

$$= \sup_{\substack{e \in E \\ \|e\| = 1}} \| [V(0)]^{1/2} [W^{-1}(0)]^{1/2} e \| = \| [V(0)]^{1/2} [W^{-1}(0)]^{1/2} \|_{E \to E} .$$

Since by the hypotheses the operator $V(0) = \int V\, dm$ is invertible, P_0 is a bounded operator from $L^2(E,W)$ to $L^2(E,V)$ if and only if the operator $W^{-1}(0) = \int W^{-1}\, dm$ is bounded. ●

9.1. REMARK. Together with the operator P_0 one can study the operators P^k, $P^k f = z^k \hat{f}(k)$, $f \in \mathcal{P}ol$ (note that $P^0 = P_0$). Since $P^k = S^k P_0 S^{-k}$, we have

$$\|P^k\|_{L^2(E,W)\to L^2(E,V)} = \|P_0\|_{L^2(E,W)\to L^2(E,V)}$$

for all operator weights V, W (in particular for $V = W$). Therefore the boundedness of $\int W^{-1} dm$ implies the uniform boundedness of the operators P^k, or, equivalently, it implies the so-called uniform minimality ($\mathcal{UM}$) of the system of subspaces $\{z^n E\}_{n\in\mathbb{Z}}$ in the weighted space $L^2(E,W)$. Recall that the uniform minimality of the system of subspaces $\{\mathcal{E}_n\}$ can be defined as the uniform boundedness of the projections $\mathcal{P}^n$, where $\mathcal{P}^n$ is the projection onto $\mathcal{E}_n$ with the kernel $\operatorname{span}(\mathcal{E}_k : k \neq n)$.

10. PROOF OF THE SUFFICIENCY IN THEOREM 8.2

As it was noted above, to prove theorem 8.2 it is sufficient to find a weight V such that the P_n are uniformly bounded operators from $L^2(E,W)$ to $L^2(E,V)$, or equivalently, the operators $V^{1/2} P_n W^{-1/2}$ are uniformly bounded in $L^2(E)$; here $V^{1/2}$ and $W^{-1/2}$ stands for the operators of multiplication by $V^{1/2}$ and $W^{-1/2}$ (with the natural domains) respectively. If we find a bounded operator-valued function C for which

$$\sup_n \|C\mathcal{P}_n\| < \infty ,$$

where $\mathcal{P}_n = W^{1/2} P_n W^{-1/2}$, then, setting $V = W^{1/2} C W^{1/2}$ we will prove the theorem. The consistency of the definition of V will be explained below. To explain it, it is sufficient to show that the operators $C^{1/2}(\xi) W^{1/2}(\xi)$ are closed (closable) a.e. on $\mathbb{T}$. For the moment we will assume the operators $W(\xi)$ to be bounded for a.e. $\xi \in \mathbb{T}$. This always takes place in the matrix case and the question about consistency does not arise in this case.

Before proving the theorem we introduce

10.1. SOME NOTATION. Denote

$\mathcal{P}_n \overset{def}{=} W^{1/2} P_n W^{-1/2}$, $\mathcal{P}^n \overset{def}{=} W^{1/2} P^n W^{-1/2}$ (recall that $P^n f = z^n \hat{f}(n)$).

Since $\mathcal{P}_n = \sum_{k=0}^n \mathcal{P}^k$ and the operators $\mathcal{P}^k$ are uniformly bounded (see Sections 9.1, 10), the operators $\mathcal{P}_n$ are well-defined on $L^2(E)$. Let $\mathcal{B}$ be the unit ball in the Hilbert space $L^2(E)$, $\mathcal{B} = \{f \in L^2(E) : \|f\| \le 1\}$ and $\mathcal{B}_n \overset{def}{=} \mathcal{P}_n \mathcal{B}$.

Denote also

$$\mathcal{X}_n = span(z^k W^{1/2} E : 0 \le k \le n)$$

and

$$\mathcal{X}_n' = span(z^k W^{-1/2} E : 0 \le k \le n)$$

(note that the space $\mathcal{X}_n'$ is the orthogonal complement in $L^2(E)$ to $span(z^k W^{1/2} E : k \ne 0, 1, \dots n)$).

10.2. SKETCH OF THE PROOF. To begin with, we will construct a non-negative bounded operator $\mathbb{D}$ on $L^2(E)$ such that

$$(\mathbb{D}\mathcal{P}_n f, \mathcal{P}_n f) \le \|f\|^2 \quad \forall f \in L^2(E), \ \forall n \tag{10.1}$$

and

$$\inf_{e \in E, \, n, \, \|e\|=1} dist_{\mathbb{D}}(z^n W^{1/2} e, \mathcal{X}_{n-1}) > 0 \tag{10.2}$$

here $dist$ denotes the distance in the metric $\|.\|_{\mathbb{D}}$, $\|f\|_{\mathbb{D}}^2 = (\mathbb{D}f, f)$.

Then we consider a weak limit point C of the averages

$$\frac{1}{n+1} \sum_{k=0} S^{-k} \mathbb{D} S^k,$$

where S is the operator of multiplication by z on $L^2(E)$. The above two estimates will be shown to hold with

$\mathcal{C}$ substituted for $\mathcal{D}$. The operator $\mathcal{C}$ commutes with $\mathcal{S}$. Therefore it is an operator of multiplication by some bounded operator weight C.

Estimate (10.1) with $\mathcal{C}$ substituted for $\mathcal{D}$ is equivalent to uniform boundedness of the operators $C^{1/2}\mathcal{P}_n$. Estimate (10.2) can be rewritten in this case (because $\mathcal{C}$ commutes with $\mathcal{S}$) in the following form

$$\inf_{\substack{e\in E\\ \|e\|=1}} \operatorname{dist}_{L^2(E,C)}(W^{1/2}e,\ \operatorname{span}(z^kW^{1/2}E : k<0)) > 0$$

or, if we set $V_1 = W^{1/2}CW^{1/2}$, in the form

$$\inf_{\substack{e\in E\\ \|e\|=1}} \operatorname{dist}_{L^2(E,V_1)}(e,\ \operatorname{span}(z^kE : k<0)) > 0 .$$

This means that the weight V_1 satisfies the uniform Szegö condition $(\mathcal{US})^*$.

Using the same construction for the weight $W^\#$ $(W^\#(\xi)=W(\bar\xi))$ we obtain a weight V_2 satisfies analogous inequalities with W replaced by $W^\#$ (i.e. operators $V_2^{1/2}P_nW^{\#-1/2}$ are uniformly bounded and $V_2\in(\mathcal{US})^*$). Therefore operators $V_2^{\#\,1/2}P_nW^{-1/2}$ are uniformly bounded also, and $V_2^\#\in(\mathcal{US})$. Hence the weight $V=V_1+V_2$ (divided by an appropriate constant, which equals 8 in our case) satisfies the condition of the theorem.

The operator $\mathcal{D}$ will be defined as the limit of some sequence of operators. To construct this sequence we need some simple lemmas. Here after symbol P_E denotes the orthogonal projection onto E, and $\mathcal{B}^A$ is the unit ball in the norm generated by a nonnegative bounded operator A: $\mathcal{B}^A \stackrel{def}{=} \{f\in \mathcal{D}om\,A : (Af,f)\le 1\}$.

10.3. LEMMA. <u>Let</u> $A_n = P_{\mathcal{X}_n}P_{\mathcal{X}'_n}\mid \mathcal{X}_n$ (see Section 10.1). <u>Then the operator</u> A_n <u>is invertible, and</u> $\mathcal{B}^{A_n}=\mathcal{B}_n\,(\stackrel{def}{=}\mathcal{P}_n\mathcal{B})$.

PROOF. The identity $\mathcal{B}^{A_n}=\mathcal{B}_n$ follows immediately from the equalities $(A_nx,x)=\|P_{\mathcal{X}'_n}x\|^2=\|P_{\mathcal{X}'_n}y\|^2$ and

$x = \mathcal{P}_n P_{\mathcal{X}'_n} x$ that hold for $x = \mathcal{P}_n y$ (see Fig.10.1). If A_n were noninvertible the corresponding "unit ball" $\mathcal{B}^{A_n}$ (which is equal to $\mathcal{B}_n$, as it has just been shown would be unbounded. But the operator $\mathcal{P}_n$ is bounded, because $\mathcal{P}_n = \Sigma_{k=0}^{n} \mathcal{P}^k$ and $\|\mathcal{P}^k\| = \|P_0\|_{L^2(E,W)} < \infty$ (see 9.1). Consequently the ball $\mathcal{B}_n = \mathcal{P}_n \mathcal{B}$ is bounded too. ◍

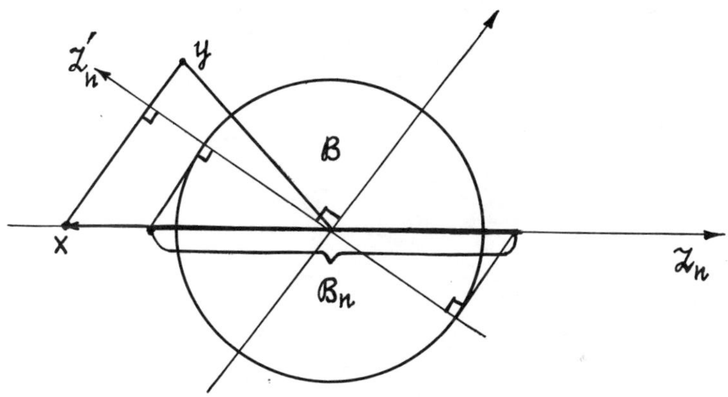

Fig.10.1

10.4. LEMMA. <u>Let</u> A_n <u>be the operator from the above lemma. Then</u>

(i) $A_n \le I$

(ii) $A_n \le P_{\mathcal{X}_n} A_{n+1} \mid \mathcal{X}_n$

<u>i.e.</u> $(A_n x, x) \le (A_{n+1} x, x) \quad \forall x \in \mathcal{X}_n$.

PROOF.(i) is evident because A is a product of orthogonal projections. Inequality (ii) follows from the inclusion $\mathcal{X}'_n \subset \mathcal{X}'_{n+1}$. ◍

10.5. LEMMA. <u>Suppose that</u> E <u>is a Hilbert space,</u> E_1 <u>is a closed subspace of</u> E, $E_1 \ne E$ <u>and</u> A <u>is nonnegative invertible operator on</u> E. <u>Then there exists a subspace</u> E_2 <u>such</u>

<u>that</u>

(i) <u>The angle between</u> E_1 <u>and</u> E_2 <u>is positive</u>.

(ii) $E = E_1 + E_2$.

(iii) $(A(x_1+x_2), x_1+x_2) = (Ax_1, x_1) + (Ax_2, x_2)$, $x_i \in E_i$, $i = 1, 2$

(<u>see Fig.10.2</u>).

PROOF. Let $E_2 = (AE_1)^{\perp}$. Then E_2 is the orthogonal complement of E_1 with respect to the scalar product $(\cdot, \cdot)_A$, $(f, g)_A \overset{def}{=\!=} (Af, g)$. This implies (iii). The first two statements follow from the equivalence of the norm $\|\cdot\|_A$, $\|f\|_A^2 = (Af, f)$ to the usual norm in E. ●

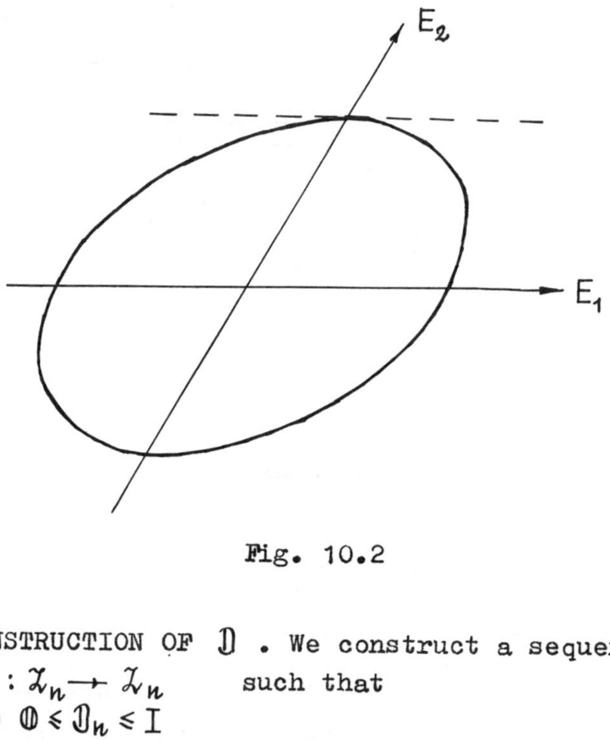

Fig. 10.2

10.6. THE CONSTRUCTION OF $\mathcal{D}$. We construct a sequence of operators $\mathcal{D}_n$, $\mathcal{D}_n : \mathcal{X}_n \to \mathcal{X}_n$ such that

(i) $\mathbb{O} \leqslant \mathcal{D}_n \leqslant I$

(ii) The operator $\mathcal{D}_n$ is a dilation of $\mathcal{D}_{n-1}$ i.e.

$\mathcal{D}_{n-1} = P_{\mathcal{X}_{n-1}} \mathcal{D}_n \mid \mathcal{X}_{n-1}$, or equivalently

$(\mathcal{D}_{n-1} x, x) = (\mathcal{D}_n x, x) \quad \forall x \in \mathcal{X}_{n-1}$

(iii) $\mathcal{P}^n \mathcal{B}^{\mathcal{D}_n} = \mathcal{P}^n \mathcal{B}^{A_n} \,(= \mathcal{P}^n \mathcal{B})$

(recall that $\mathcal{B}^A$ stands for the "unit ball" of the operator

A, $\mathcal{B}^A \overset{\text{def}}{=} \{ f \in \mathcal{D}om\, A : (Af, f) \leqslant 1 \}$, see Section 10.1, and note that the equality $\mathcal{P}^n \mathcal{B}^{A_n} = \mathcal{P}^n \mathcal{B}$ is valid by Lemma 10.3).

The operator $\mathcal{D}$ will be the strong limit of sequence $\{\mathcal{D}_n\}$.

To construct this sequence $\{\mathcal{D}_n\}$ we set $\mathcal{D}_0 = A_0$, and proceed by induction. Suppose that the operators $\mathcal{D}_k$, $k < n$ have already been constructed. Then apply Lemma 10.5 to the operator $A = A_n$ and the subspace $E_1 = \mathcal{Z}_{n-1}$. We obtain that there exists a subspace $E_2 \subset \mathcal{Z}_n$ such that

$$\mathcal{Z}_n = \mathcal{Z}_{n-1} \dotplus E_2$$

and

$$(A_n(x_1 + x_2), x_1 + x_2) = (A_n x_1, x_1) + (A_n x_2, x_2), \; x_1 \in E_1 = \mathcal{Z}_{n-1}, \; x_2 \in E_2 .$$

Define the operator $\mathcal{D}_n$ as follows:

$$(\mathcal{D}_n(x_1 + x_2), x_1 + x_2) = (\mathcal{D}_{n-1} x_1, x_1) + (A_n x_2, x_2), \; x_1 \in \mathcal{Z}_{n-1}, \; x_2 \in E_2 .$$

Check up the properties of $\mathcal{D}_n$. Property (ii) follows immediately from the construction of $\mathcal{D}_n$. To prove (i) we use the induction assumption $\mathcal{D}_{n-1} \leqslant A_{n-1}$ and take into account that by Lemma 10.4 $(A_{n-1} x, x) \leqslant (A_n x, x)$ for all $x \in \mathcal{Z}_{n-1}$ (see Fig. 10.3). Property (iii) is proved by the following chain of equalities (see also Fig. 10.3)

$$\mathcal{P}^n \mathcal{B}^{A_n} = \mathcal{P}^n (\mathcal{B}^{A_n} \cap E_2) = \mathcal{P}^n (\mathcal{B}^{\mathcal{D}_n} \cap E_2) = \mathcal{P}^n \mathcal{B}^{\mathcal{D}_n} .$$

The sequence $\{\mathcal{D}_n\}$ defines a bounded nonnegative operator $\mathcal{D}$ on $span(z^k W^{1/2} E : k \geqslant 0)$ such that

$$(\mathcal{D} \mathcal{P}_n f, \mathcal{P}_n f) \leqslant \| f \|^2 \quad \forall n \geqslant 0 \quad \forall f \in L^2(E) .$$

Show that the inequality (10.2) holds too. Let $e \in E$. Then

$$dist_{\mathcal{D}}(z^n W^{1/2} e, \mathcal{Z}_{n-1}) = \inf \{ (\mathcal{D} f, f)^{1/2} : f \in \mathcal{Z}_n, \; \mathcal{P}^n f = z^n W^{1/2} e \} =$$

$$= 1/\sup\{ |\lambda| : \lambda z^n W^{1/2} e \in \mathcal{P}^n \mathcal{B}^{\mathcal{D}_n} \; (= \mathcal{P}^n \mathcal{B}, \text{ see (iii) }) \} =$$

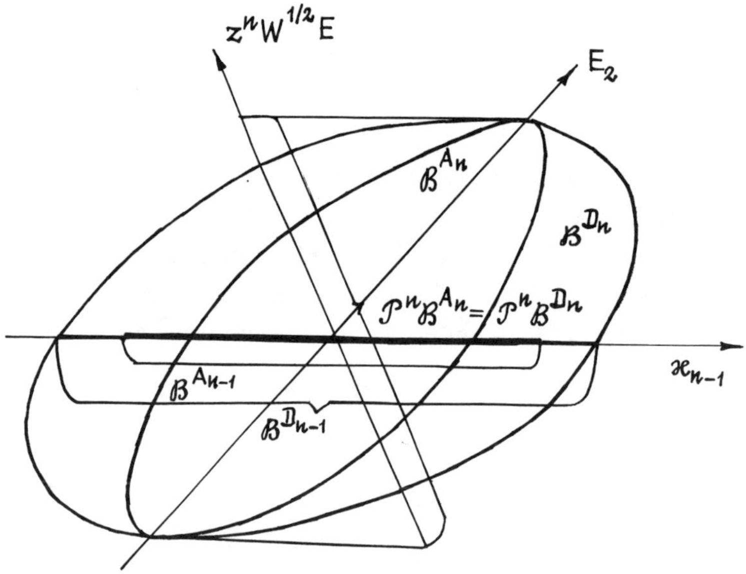

Fig. 10.3

$$= 1/\sup\{|\lambda| : \lambda W^{1/2} e \in \mathcal{P}_0\, \mathcal{B}\} =$$

(the last equality holds because the multiplication by the independent variable z is a unitary operator on $L^2(E)$)

$$= \operatorname{dist}_{L^2(E)}(W^{1/2}e, W^{1/2}\mathcal{P}ol^0) = \operatorname{dist}_{L^2(E,W)}(e, \mathcal{P}ol^0) = \| [\int_{\mathbb{T}} W^{-1} dm]^{-1/2} e \| .$$

Taking the infimum over all $e \in E$, $\|e\| = 1$, we find that $\mathfrak{D}$ satisfies the required property. Moreover we obtain the sharp equality

$$\operatorname{dist}_{\mathfrak{D}}(z^n W^{1/2} e, \mathcal{X}_{n-1}) = \| [\int_{\mathbb{T}} W^{-1} dm]^{-1/2} e \| . \tag{10.3}$$

Thus the operator $\mathfrak{D}$ is constructed and the following step of the proof is:

10.7. **AVERAGING.** Set $C_n = \frac{1}{n} \sum_{k=1}^{n} S^{-k} \mathfrak{D} S^k$ and let $C = C_I$

be a weak limit point of the sequence C_n. It is easy to show that C is a bounded operator commuting with the shift S. To prove the required estimates for the limit operator C we show that such estimates hold for the operators C_n uniformly in n. We have

$$(C_n \mathcal{P}_m f, \mathcal{P}_m f) = \frac{1}{n} \sum_{k=1}^{n} (\mathcal{D} S^k \mathcal{P}_m f, S^k \mathcal{P}_m f) =$$

$$= \frac{1}{n} \sum_{k=1}^{n} (\mathcal{D}(\mathcal{P}_{m+k} - \mathcal{P}_{k-1}) S^k f, (\mathcal{P}_{m+k} - \mathcal{P}_{k-1}) S^k f)$$

as $S^k \mathcal{P}_m = (\mathcal{P}_{m+k} - \mathcal{P}_{k-1}) S^k$. Since $\|S^k f\| = \|f\|$ and since it is evident that $(\mathcal{D}(\mathcal{P}_n - \mathcal{P}_m) f, (\mathcal{P}_n - \mathcal{P}_m) f) \leq 4\|f\|^2$ for all positive m, n, it is easy to conclude that

$$(C_n \mathcal{P}_m f, \mathcal{P}_m f) \leq 4\|f\|^2 \quad \forall m, n, \quad \forall f \in L^2(E).$$

Consequently

$$(C \mathcal{P}_n f, \mathcal{P}_n f) \leq 4\|f\|^2 \quad \forall n, \quad \forall f \in L^2(E).$$

It is also easy to check up the second inequality for C ((10.2) with $\mathcal{D}$ replaced by C):

$$\operatorname{dist}_{C_m}(z^n W^{1/2} e, \mathcal{X}_{n-1}) \geq$$

$$\geq \frac{1}{m} \sum_{j=1}^{m} \operatorname{dist}_{\mathcal{D}}(z^{n+j} W^{1/2} e, S^j \mathcal{X}_{n-1}) \geq$$

$$\geq \frac{1}{m} \sum_{j=1}^{m} \operatorname{dist}_{\mathcal{D}}(z^{n+j} W^{1/2} e, \mathcal{X}^{n+j-1}) =$$

$$= \left\| \left[\int_{\mathbb{T}} W^{-1} dm\right]^{-1/2} e \right\| \qquad \text{(see (10.3))},$$

and therefore for the limit operator C we have

$$\mathrm{dist}_C(z^n W^{1/2}e, \mathcal{X}_{n-1}) \geq \| [\int_{\mathbb{T}} W^{-1}dm]^{-1/2} e \| .$$

Define the weight $V_1 = W^{1/2} C W^{1/2}$. The weight $\frac{1}{4} V_1$ evidently satisfies (8.1) (see Section 10.1) and the uniform Szegö condition $(\mathcal{US})^*$.

10.8. FINISH OF THE PROOF. Applying the construction described in Sections 10.2-10.7 to the weight $W^\#$ ($W^\#(\xi) \overset{def}{=} W(\bar{\xi})$) we obtain another weight V_2 . It is evident that the weight $V = \frac{1}{8}(V_1 + V_2^\#)$ satisfies all requirements of the theorem. ●

11. HOW TO GET THE OPERATOR $C^{1/2}W^{1/2}$ CLOSED

Proving Theorem 8.2 we supposed the weight $W^{1/2}CW^{1/2}$ to be well-defined (see Section 10). This is the case if, for example, the operators $C^{1/2}(\xi)W^{1/2}(\xi)$ are closed (closable) a.e. on $\mathbb{T}$. To avoid this technical difficulty it was supposed (see Section 10.2) that the operators $W(\xi)$ are bounded a.e. on $\mathbb{T}$. We shall show now that we may assume this without loss of generality.

11.1. LEMMA. Let W be an operator weight with an invertible mean value $W(0)$ such that the operator $W^{-1}(0) = \int_{\mathbb{T}} W^{-1}dm$ is bounded, N be a positive number, $\varphi(t) \overset{def}{=} \min(t, N)$. Then the weight $\underline{W} \overset{def}{=} \varphi(W)$ is bounded a.e., its mean $\underline{W}(0)$ is invertible, and for the mean value $\underline{W}^{-1}(0) = \int_{\mathbb{T}} \underline{W}^{-1}dm$ the inequality

$$\| W^{-1}(0) - \underline{W}^{-1}(0) \| \leq \frac{1}{N}$$

holds.

PROOF. Consider the spectral representation of the operator of multiplication by W on $L^2(E)$, i.e. let this operator be realized as the multiplication by the independent variable in von Neumann integral $\int_0^\infty \oplus H(t)\,d\mu(t)$.

Fix a vector $e \in E$ (considered as an element of $L^2(E)$) and let $e(\cdot)$ be its spectral representation in the von Neumann integral. Choose ε , $0<\varepsilon<(4\|W^{-1}(0)\|)^{-1}$. Then

$$\int_0^\varepsilon \|e(t)\|^2 d\mu(t) \leq \varepsilon \int_0^\infty \frac{1}{t}\|e(t)\|^2 d\mu(t) = \varepsilon(W^{-1}e,e)_{L^2(E)} =$$

$$= \varepsilon(W^{-1}(0)e,e)_E \leq \frac{1}{4}\|e\|^2.$$

Hence

$$(\underline{W}(0)e,e)_E = (\underline{W}e,e)_{L^2(E)} = \int_0^\infty \varphi(t)\,\|e(t)\|^2 d\mu(t) \geq$$

$$\geq \varepsilon \int_\varepsilon^\infty \|e(t)\|^2 d\mu(t) \geq \frac{3}{4}\|e\|^2$$

if $N>\varepsilon$. This implies the invertibility of $\underline{W}(0)$. Similarly,

$$|((W^{-1}(0)-\underline{W}^{-1}(0))e,e)| = |((W^{-1}-\underline{W}^{-1})e,e)_{L^2(E)}| =$$

$$= \int_0^\infty \left|\frac{1}{\varphi(t)}-\frac{1}{t}\right|\cdot\|e(t)\|^2 d\mu(t) \leq \frac{1}{N}\int_N^\infty \|e(t)\|^2 d\mu(t) \leq \frac{1}{N}\|e\|^2.$$

and the required estimate is proved.

12. PROOF OF THEOREM 8.1

The necessity of the condition $\int (W^{-1}e,e)\,dm<\infty$ is simple: let e be a vector such that $(V(0)e,e)>0$. Then for inequality (8.1) to hold it is necessary (see Section 9) that

$$dist_{L^2(E,W)}(e,\mathcal{P}ol^0)>0 .$$

The latter is **equivalent** to the convergence of the integral

$\int (W^{-1}e,e)\,dm$ for some vector $e \in E$.

Let now $\int (W^{-1}e,e)\,dm < \infty$, $e \in E$, $\|e\| = 1$. It is easy to show that the weight w, $w(\xi) \stackrel{def}{=} (W^{-1}(\xi)e,e)^{-1}$ satisfies the conditions $w \in L^1$, $\frac{1}{w} \in L^1$. By Lemma 4.4 the inequality

$$w \cdot P_{\mathcal{X}(e)} \leqslant W$$

holds ($P_{\mathcal{X}(e)}$ denotes the orthogonal projection of E onto $\mathcal{X}(e)$), and the proof follows from the Koosis theorem. ●

CHAPTER IV. UNCONDITIONAL BASES OF EIGENVECTORS OF CONTRACTIONS

The present chapter is devoted to a problem concerning bases of vector-valued rational functions (eigenvectors of a contraction) posed by N.K.Nikol'skii, B.S.Pavlov and V.I.Vasyunin in problem book [6] and solved by the author in [20]. The questions considered in this and also in following chapters are not closely connected to the above three chapters; they concern another branch of the Spectral Theory of Vector-valued Functions (STVF). But nevertheless, there are certainly some connections. For example, the concept of uniform minimality, used in the previous chapter (see Section 9.1) is important in the present one. The tecnique used here is also similar to that used above. It is basdd on estimates of angles, distances, norms of skew projections and so on.

13. STATEMENT OF THE PROBLEM

Consider a contraction T (i.e. an operator T such that $\|T\| \leq 1$) on a Hilbert space. It is known that the linear span of the eigenvectors x_λ of the unitary spectrum ($Tx_\lambda = \lambda x_\lambda$, $|\lambda| = 1$) is orthogonal to the other eigenvectors of T and that there is an orthonormal basis of eigenvectors in this span. The action of T on the space H spanned by eigenvectors x_λ, $Tx_\lambda = \bar{\lambda} x_\lambda$ with $|\lambda| < 1$ is described by a unitary equivalent model; here vectors x_λ are carried into the rational functions. Namely, set

$$\mathcal{D}_T \overset{def}{=} (I - T^*T)^{1/2}, \quad E = clos\, \mathcal{D}_T H$$

and

$$Vx \overset{def}{=\!=} \sum_{k\geq 0} z^k \mathcal{D}_T T^k x, \quad x \in H.$$

Then V is a unitary mapping of H onto a subspace K of the vector Hardy space $H^2(E)$ while the action of the model operator $M = VTV^{-1}$ reduces to the inverse shift S^* :

$$S^* f = (f - f(0))/z$$

(see, for example [7], [15]). The normalised eigenvector $\|Vx_\lambda\|^{-1} Vx_\lambda$ has the form

$$\|Vx_\lambda\|^{-1} Vx_\lambda = \frac{(1-|\lambda|^2)^{1/2}}{1-\bar{\lambda} z} e_\lambda, \quad e_\lambda \in E, \quad \|e_\lambda\| = 1 \ . \tag{13.1}$$

We will call the vector $e_\lambda = \|\mathcal{D}_T x_\lambda\|^{-1} \mathcal{D}_T x_\lambda \in E$ a <u>spatial</u> (=<u>directional</u>) vector, or <u>angular director</u>, see [7].

The problem we are going to discuss looks as follows. Consider two statements on the system of the vectors of form (13.1): (i) the system is uniformly minimal; (ii) the system forms a Riesz (i.e. unconditional) basis. We want to find conditions on the geometry of angular directors e_λ necessary and sufficient for statements (i) and (ii) to be equivalent. In particular (and this is a precise formulation of the problem in [6]), does this equivalence hold provided all angular directors are contained in some finite-dimensional space (or, which is the same, $\dim E < \infty$).

In the scalar case $\dim E = 1$, when all directional vectors coincide, any uniformly minimal system of the vectors of form (13.1) forms a Riesz basis. This proposition is well known, see [7], and is in essence a reformultation of the Carleson interpolation theorem. It motivates the above formulation of the problem, which without this motivation might seem strange. The authors of the problem favored the opinion that the scalar case (see Theorem 14.3) is exceptional and the question posed has a negative answer. The main result of this chapter, Theorem

14.1, shows that it is not so, and in finite-dimensional case every uniformly minimal system of rational vector-valued functions of form (13.1) is a Riesz basis.

Before moving further, recall some concept used.

13.1. DEFINITION. A system $\{f_n\}$ of vectors in a Hilbert space is called a Riesz basis (or an unconditional basis) if an "approximate Parseval equality" holds, i.e.

$$c\,\Sigma\,|a_k|^2 \leqslant \|\Sigma\, a_k f_k\|^2 \leqslant C\cdot\Sigma\,|a_k|^2$$

for any finite collection of numbers a_k (Here and below the question of completeness is not discussed at all, so we always have in mind a basis in the closure of its own linear span).

It is easy to show (see for example [7], Lecture 4) that a family $\{f_n\}$ of vectors forms a Riesz basis if the so-called orthogonalizer $\mathcal{I}$, $\mathcal{I}(\Sigma\, c_k f_k) = \{c_k\} \in \ell^2$ is well-defined and maps isomorphically $\mathrm{span}(\{f_n\}_n)$ onto ℓ^2.

13.2. DEFINITION. A system $\{f_n\}$ of vectors is said to be uniformly minimal if

$$\inf_n \mathrm{dist}(\|f_n\|^{-1} f_n,\ \mathrm{span}(f_k : k \neq n)) \stackrel{def}{=} \delta > 0 \qquad (UM)$$

The number $\delta = \delta(\{f_n\})$ is called the constant of the uniform minimality of the system $\{f_n\}$.

It is easy to show that

$$\frac{1}{\delta} = \sup_n \|\mathcal{P}^n\|$$

where $\mathcal{P}^n$ denotes the skew projection onto $\mathcal{L}(f_n)$ with the kernel $\mathrm{span}(f_k : k \neq n)$ (compare with the definition of the uniform minimality in Section 9.1).

13.3. REMARK. Uniform minimality is generally speaking a considerable weaker property than that of being an unconditional basis. For example, the system $\{z^n\}_{n\in\mathbb{Z}}$ in the (scalar) weighted space $L^2(w)$ is uniformly minimal iff $w, \frac{1}{w} \in L^1$, but this system forms a Riesz basis iff $w, \frac{1}{w} \in L^\infty$.

13.4. DEFINITION. The system of vectors of the form (13.1) will be said <u>spatially compact</u> if the family of angular directors e_λ is relatively compact (in the norm if E).

This condition is certainly satisfied if the defect operator $\mathfrak{D}_T$ of our contraction T has finite rank ($\dim E<\infty$); in the general case it also expressed a certain degree of closeness of the operator to some isometric operator.

14. MAIN RESULT. DISCUSSION

The main result of this chapter is the following theorem proved in [20].

14.1. THEOREM. *Let*

$$\left\{ \frac{(1-|\lambda|^2)^{1/2}}{1-\bar{\lambda}z} e_\lambda \right\}_{\lambda\in\sigma}, \quad e_\lambda \in E, \quad \|e_\lambda\| = 1 \tag{14.1}$$

be a given spatially compact system of functions in the vector Hardy space $H^2(E)$; here σ is a subset of the disk $\mathbb{D}$. Then the system (14.1) forms a Riesz basis if and only if it is uniformly minimal.

14.2. DISCUSSION. The above theorem gives a positive answer to the question posed in [6] and discussed in Section 13.

Before proving the theorem we recall and discuss some known facts. Denote for the sake of briefity $\varphi_\lambda \stackrel{def}{=} (1-|\lambda|^2)^{1/2}/(1-\bar{\lambda}z)$. The system (14.1) in this notation will be rewritten as $\{\varphi_\lambda e_\lambda\}_{\lambda\in\sigma}$.

In the scalar case ($\dim E=1$, $H^2(E) = H^2$) the following description of uniformly minimal systems $\{\varphi_\lambda e_\lambda\}_{\lambda\in\sigma}$ is well-known (see [7]). Let $B = B^\sigma = \prod_{\lambda\in\sigma} b_\lambda$ where $b_\lambda \stackrel{def}{=} \frac{|\lambda|}{\lambda}(\lambda-z)/(1-\bar{\lambda}z)$, be the **Blaschke** product with zero set

σ (we suppose $B \equiv 1$ if $\sigma = \emptyset$ and $B \equiv 0$ if the product diverges), and let $B_\lambda \overset{def}{=} B / b_\lambda$. It is not hard to calculate (see [7]) that

$$\operatorname{dist}(\varphi_\lambda, \operatorname{span}(\varphi_\mu : \mu \in \sigma \setminus \{\lambda\})) = |B_\lambda(\lambda)| .$$

Thus the system $\{\varphi_\lambda\}_{\lambda \in \sigma}$ is uniformly minimal iff the so-called Carleson condition holds

$$\inf_{\lambda \in \sigma} |B_\lambda(\lambda)| \overset{def}{=} \delta(\sigma) > 0 \qquad (C)$$

and, moreover, $\delta(\sigma) = \delta(\{\varphi_\lambda\}_{\lambda \in \sigma})$ (see Definition 13.2).

The sets σ satisfying this condition are called Carleson sets. Such sets play an important role in analysis (see about this [2], [4], [7]). The following known result (see, for example, [7]) will be needed latter.

14.3. THEOREM. Let σ be a Carleson set. Then $\{\varphi_\lambda\}_{\lambda \in \sigma}$ is a Riesz basis, and, moreover

$$\|\mathcal{J}_\sigma\| \cdot \|\mathcal{J}_\sigma^{-1}\| \leqslant C(\delta(\sigma)) ;$$

here $\mathcal{J}_\sigma$ is the orthogonalizer of the family $\{\varphi_\lambda\}_{\lambda \in \sigma}$ (see Section 13.1) and the constant $C(\delta)$ is defined by the equality $C(\delta) = \delta^{-1} \cdot (1 - 64 \log \delta)$.

14.4. CARLESON MEASURES. It is well known (see [2], [4], [7]) that the Carleson condition (C) implies the imbedding theorem

$$H^2 \hookrightarrow L^2(\mu)$$

where $\mu = \sum_{\lambda \in \sigma} (1 - |\lambda|^2) \delta_\lambda$ (δ_λ is the Dirac measure at the point λ). The measures μ for which $H^2 \hookrightarrow L^2(\mu)$ (or, equivalently, $\int_{\mathbb{D}} |f|^2 d\mu \leqslant C \|f\|^2_{H^2}\ \forall f \in H^2$) are called Carleson measures. The best positive constant C in the above inequality is called the intensity of the Carleson measure μ.

We will rest upon the following description of Riesz bases of the form (14.1), obtained in [8] (see also [7]).

14.5. THEOREM. The family $\{\varphi_\lambda e_\lambda\}_{\lambda\in\sigma}$ is a Riesz basis iff it is uniformly minimal and the following two imbedding theorems are valid

$$1)\ \sum_{\lambda\in\sigma}(1-|\lambda|^2)\|P_\lambda f(\lambda)\|^2<\infty \qquad \forall f\in H^2(E)$$

$$2)\ \sum_{\lambda\in\sigma}(1-|\lambda|^2)\|Q_\lambda f(\bar\lambda)\|^2<\infty \qquad \forall f\in H^2(E)$$

where P_λ and Q_λ are some concrete (explicit formulae for P_λ and Q_λ are given in [7], [8] but they will not be needed here) orthogonal projections on E .

In [7], [8] a necessary and sufficient condition is given for the uniform minimality of the family $\{\varphi_\lambda e_\lambda\}_{\lambda\in\sigma}$ also. This is a vector analogue of the Carleson condition (C) . This condition is expressed in term of an inner function θ generating by the Lax - Beurling - Halmos Theorem (Theorem 2.1) the coinvariant (i.e. invariant under inverse shift S^*) subspace $\operatorname{span}(\varphi_\lambda e_\lambda:\lambda\in\sigma)=H^2(E)\ominus\theta H^2(E_*)$.

14.6. REMARK. It is obvious that for the imbedding theorems 1) and 2) to hold it suffices that the measure $\sum_{\lambda\in\sigma}(1-|\lambda|^2)\delta_\lambda$ be a Carleson measure, since in this case the following stronger imbedding theorems hold

$$1')\ \sum_{\lambda\in\sigma}(1-|\lambda|^2)\|f(\lambda)\|^2<\infty \qquad \forall f\in H^2(E)$$

$$2')\ \sum_{\lambda\in\sigma}(1-|\lambda|^2)\|f(\bar\lambda)\|^2<\infty \qquad \forall f\in H^2(E).$$

It is also known (V.I.Vasyunin, see [7]) that for a spatially compact Riesz basis of the form (14.1) the measure $\sum_{\lambda\in\sigma}(1-|\lambda|^2)\delta_\lambda$ is a Carleson measure.

14.7. EXAMPLE (N.K.Nikolskii, B.S.Pavlov, [8]). It is presented for the sake of completeness and it is not used in the proof of Theorem 14.1. Let σ be a countable subset of the disk $\mathbb{D}$ such that the measure $\mu=\sum_{\lambda\in\sigma}(1-|\lambda|^2)\delta_\lambda$ is not Carleson (such sets σ exist, see for example Lemma 20.1). Let also $\{e_\lambda\}_{\lambda\in\sigma}$ be an orthonormal system in E , $\dim E=\infty$ and

$e \perp e_\lambda$, $\forall \lambda \in \sigma$. Set $f_\lambda = \frac{1}{\sqrt{2}}(e + e_\lambda)$. It is evident that the system $\{\varphi_\lambda f_\lambda\}_{\lambda\in\sigma}$ is uniformly minimal and its constant of uniform minimality is at least $1/\sqrt{2}$, since $\delta(\{\varphi_\lambda f_\lambda\}_{\lambda\in\sigma}) \geq \delta(\{f_\lambda\}_{\lambda\in\sigma})$ and the latter constant is equal to $1/\sqrt{2}$.

Let $\mathcal{J}$ be the orthogonalizer of $\{\varphi_\lambda f_\lambda\}_{\lambda\in\sigma}$. Then

$$\|\mathcal{J}^{-1}\| = \sup_{c_\lambda : \Sigma|c_\lambda|^2=1} \Big\| \sum_{\lambda\in\sigma} c_\lambda \varphi_\lambda f_\lambda \Big\|_{H^2(E)} =$$

$$= \sup_{\substack{c_\lambda : \Sigma|c_\lambda|^2=1 \\ f\in H^2(E), \|f\|=1}} \Big| \sum_{\lambda\in\sigma} c_\lambda (\varphi_\lambda f_\lambda, f)_{H^2(E)} \Big| .$$

Taking the last supremum only on f's of the form $f = ge$, $g\in H^2$ and using that $(\varphi_\lambda f_\lambda, ge)_{H^2(E)} = (1-|\lambda|^2)^{1/2}\overline{g(\lambda)}\cdot(f_\lambda, e)_E = \frac{1}{\sqrt{2}}(1-|\lambda|^2)\overline{g(\lambda)}$, we obtain

$$\|\mathcal{J}^{-1}\| \geq \sup\Big\{\Big|\frac{1}{\sqrt{2}} \sum_{\lambda\in\sigma} (1-|\lambda|^2)^{1/2} c_\lambda \overline{g(\lambda)}\Big| : c_\lambda,\ \Sigma|c_\lambda|^2 = 1,\ g\in H^2,\ \|g\|=1\Big\} =$$

$$= \frac{1}{\sqrt{2}}\Big(\sup_{\substack{g\in H^2 \\ \|g\|=1}} \sum_{\lambda\in\sigma} (1-|\lambda|^2)\,|g(\lambda)|^2\Big)^{1/2} = \infty$$

because the corresponding measure $\mu = \sum_{\lambda\in\sigma}(1-|\lambda|^2)\delta_\lambda$ is not Carleson. Therefore the orthogonalizer $\mathcal{J}$ is not invertible and the system $\{\varphi_\lambda f_\lambda\}_{\lambda\in\sigma}$ is not a Riesz basis.

Choosing different sets σ one can obtain different interesting examples of uniformly minimal systems of the form (14.1) which are not Riesz bases. For example, in [8] the set σ is chosen to satisfy the Blaschke condition $\sum_{\lambda\in\sigma}(1-|\lambda|^2) < \infty$

and the system constructed is the system of eigenvectors of a contraction T (a model operator) whose defect operator $\mathcal{D}_T = (I - T^*T)^{1/2}$ belongs to the Hilbert-Schmidt class $\mathfrak{S}_2$.

The construction presented here will be used below, in Chapter V (see Sections 18.2, 20); the set σ in this case will be chosen to satisfy the sparseness condition (R) , see Section 19 condition (i), Section 20.1.

15. PROOF OF THEOREM 14.1

15.1. LEMMA. <u>Suppose that the system</u> $\{\varphi_\lambda e_\lambda\}_{\lambda\in\sigma}$ <u>is uniformly minimal</u>, $\delta(\{\varphi_\lambda e_\lambda\}_{\lambda\in\sigma}) \geq \delta > 0$ <u>and</u>

$$\| e_\lambda - e_\mu \| \leq (\delta/2)\cdot(C(\delta/2))^{-1}$$

<u>for some</u> $\delta > 0$, <u>where</u> $C(\delta)$ <u>is a constant from the theorem 14.3. Then</u> σ <u>is a Carleson set</u>, <u>and</u> $\delta(\sigma) \geq \delta/2$.

(Recall that $\delta(\{f_\lambda\})$ is the constant of the uniform minimality of the system $\{f_\lambda\}$ (see Definition 13.2) and $\delta(\sigma) = \delta(\{\varphi_\lambda\}_{\lambda\in\sigma})$; see Section 14.2).

Let us derive Theorem 14.1 from the lemma. Let $\delta \overset{def}{=\!=} \delta(\{\varphi_\lambda e_\lambda\}_{\lambda\in\sigma})$. We split the set σ into finitely many sets $\sigma_1, \dots, \sigma_n$ such that

$$\| e_\lambda - e_\mu \| \leq (\delta/2)\cdot(C(\delta/2))^{-1}, \quad \lambda,\mu\in\sigma_k .$$

By Lemma 15.1 the σ_k are Carleson sets, and hence $\sum_{\lambda\in\sigma}(1-|\lambda|^2)\delta_\lambda$ is a Carleson measure because it is a sum of finitely many Carleson measures $\mu_k = \sum_{\lambda\in\sigma_k}(1-|\lambda|^2)\delta_\lambda$. Therefore, the imbeddings 1), 2) from Theorem 14.5 hold and hence the system $\{\varphi_\lambda e_\lambda\}_{\lambda\in\sigma}$ is a Riesz basis. Theorem 14.1 is proved.

The idea of the proof of lemma 15.1 is very simple. Assume that $\delta(\sigma) < \delta/2$ and approximate a vector $\varphi_\lambda e_\lambda$ by the linear combination of the rest ones. This will contradict the assumption $\delta(\{\varphi_\lambda e_\lambda\}_{\lambda\in\sigma}) \geq \delta$.

As it will be showed below this programm can be carried out, because in the scalar case such approximation exists (see Section 14.2) and our case is "almost scalar", since the directional vectors e_λ are "close" to each other. The main trick is that to approximate $\varphi_\lambda e_\lambda$ we will take "not too many" vectors $\varphi_\mu e_\mu$, so that summarized error resulting from the dispersedness of the directional vectors e_μ will not spoil the scalar case estimate. The quantitative expression of this "not too many" is given by the following lemma

15.2. LEMMA. Let $0<\delta<1$ and let σ be a subset of the disk $\mathbb{D}$, such that $\delta(\sigma)<\delta$. Then there exists a finite subset $\tilde{\sigma}$ of σ such that

(i) $\delta(\tilde{\sigma})<\delta$

(ii) $\delta(\tilde{\sigma}\setminus\{\lambda\}) \geq \delta \quad \forall\lambda\in\tilde{\sigma}$.

PROOF. Let $\tilde{\sigma}$ be a finite subset of σ such that (i) holds (such a $\tilde{\sigma}$ exists, since $\delta(\sigma)<\delta$). If σ does not satisfy the condition (ii) we delete from it a point λ for which $\delta(\tilde{\sigma}\setminus\{\lambda\})<\delta$. Repeating this procedure finitely many times we obtain the desired $\tilde{\sigma}$, since for a singleton set ρ we have $\delta(\rho)=1$ •

15.3. PROOF OF LEMMA 15.1. Assume that $\delta(\sigma)<\delta/2$. Then by Lemma 15.2 there exists a finite subset $\tilde{\sigma}$ of σ such that

$$\delta(\tilde{\sigma})<\delta/2 \qquad \text{but } \delta(\tilde{\sigma}\setminus\{\lambda\})\geq\delta/2 \quad \forall\lambda\in\tilde{\sigma}.$$

By the first inequality,

$$\mathrm{dist}_{H^2}(\varphi_\lambda, \mathrm{span}(\varphi_\mu : \mu\in\tilde{\sigma}\setminus\{\lambda\})) < \delta/2 \tag{15.1}$$

for some $\lambda\in\tilde{\sigma}$. Let $P=P_{\tilde{\sigma}\setminus\{\lambda\}}$ be the orthogonal projection of H^2 onto $\mathrm{span}(\varphi_\mu : \mu\in\tilde{\sigma}\setminus\{\lambda\})$, and let

$$P\varphi_\lambda = \sum_{\mu\in\tilde{\sigma}\setminus\{\lambda\}} c_\mu\varphi_\mu \tag{15.2}$$

be the decomposition of $P\varphi_\lambda$ in the basis $\{\varphi_\mu\}_{\mu\in\tilde\sigma\setminus\{\lambda\}}$.

Set $f \overset{def}{=} \sum_{\mu\in\tilde\sigma\setminus\{\lambda\}} c_\mu \varphi_\mu e_\mu$ (the c_μ are from (15.2)) and estimate $\|\varphi_\lambda e_\lambda - f\|$:

$$\|\varphi_\lambda e_\lambda - f\| \leqslant \Big\|\Big(\sum_{\mu\in\tilde\sigma\setminus\{\lambda\}} c_\mu\varphi_\mu - \varphi_\lambda\Big)e_\lambda\Big\| + \Big\|\sum_{\mu\in\tilde\sigma\setminus\{\lambda\}} c_\mu\varphi_\mu(e_\lambda - e_\mu)\Big\| .$$

It follows from (15.1) that $\|\sum_{\mu\in\tilde\sigma\setminus\{\lambda\}} c_\mu\varphi_\mu - \varphi_\lambda\| < \delta/2$ and thus the first term is less than $\delta/2$. The second one can also be easily estimated. Let $\Delta_\mu = e_\lambda - e_\mu$, and let h_k be an orthonormal basis in E. Then

$$\Big\|\sum_\mu c_\mu\varphi_\mu\Delta_\mu\Big\|^2_{H^2(E)} = \sum_k \Big\|\sum_\mu c_\mu(\Delta_\mu, h_k)\varphi_\mu\Big\|^2_{H^2} \leqslant$$

$$\leqslant \sum_k \Big(\|\mathcal{J}^{-1}_{\tilde\sigma\setminus\{\lambda\}}\|^2 \cdot \sum_\mu |c_\mu(\Delta_\mu, h_k)|^2\Big) \leqslant$$

$$\leqslant \|\mathcal{J}^{-1}_{\tilde\sigma\setminus\{\lambda\}}\|^2 \cdot \Big(\sum_\mu |c_\mu|^2\Big)\cdot \max_\mu \|\Delta_\mu\|^2$$

(recall that $\mathcal{J}_\sigma$ denotes the orthogonalizer of the system $\{\varphi_\lambda\}_{\lambda\in\sigma}$). It follows from (15.2) and the definition of the orthogonalizer that $\{c_\mu\}_{\mu\in\tilde\sigma\setminus\{\lambda\}} = \mathcal{J}_{\tilde\sigma\setminus\{\lambda\}} P\varphi_\lambda$. Hence

$$\sum_\mu |c_\mu|^2 \leqslant \|\mathcal{J}_{\tilde\sigma\setminus\{\lambda\}}\|\cdot\|\varphi_\lambda\|^2 = \|\mathcal{J}_{\tilde\sigma\setminus\{\lambda\}}\|^2 .$$

Since $\delta(\tilde\sigma\setminus\{\lambda\}) \geqslant \delta/2$, it follows that (see Theorem 14.3)

$$\|\mathcal{J}_{\tilde\sigma\setminus\{\lambda\}}\|\cdot\|\mathcal{J}^{-1}_{\tilde\sigma\setminus\{\lambda\}}\| \leqslant C(\delta/2) .$$

Remembering that $\max_\mu \|\Delta_\mu\| \leqslant (\delta/2)\cdot(C(\delta/2))^{-1}$ by the hypothesis of Lemma 15.1 we easily see that

$$\Big\|\sum_{\mu\in\tilde\sigma\setminus\{\lambda\}} c_\mu\varphi_\mu(e_\lambda - e_\mu)\Big\| \leqslant \delta/2$$

and thus $\|\varphi_\lambda e_\lambda - f\| < \delta$. This contradicts the condition $\delta(\{\varphi_\lambda e_\lambda\}_{\lambda\in\sigma}) \geqslant \delta$ since $f \in \operatorname{span}(\varphi_\mu e_\mu : \mu\in\sigma\setminus\{\lambda\})$. The lemma and hence the theorem are proved. ● ●

CHAPTER V. ANGLES BETWEEN COINVARIANT SUBSPACES AND THE OPERATOR CORONA PROBLEM (THE SZ.-NAGY PROBLEM)

In this chapter the solution of a problem of B.Sz.-Nagy on left invertibility of an analytic operator-valued function posed in [6] is presented. This problem was solved by the author in [19]. The presentation here essentially follows [19].

The contents of this chapter is closely connected with that of the preceding one. The solution is reduced to the study of systems of vector-valued rational functions of the form (14.1).

16. STATEMENT OF THE PROBLEM

In the problem book [6] the following problem was posed by B. Sz.-Nagy. Let $\mathbb{F}$ be an operator-valued function in $H^\infty(E \to E_*)$ such that

$$\|\mathbb{F}(\lambda)e\| \geq \delta\|e\| \quad \forall e \in E \quad \forall \lambda \in \mathbb{D} \tag{16.1}$$

for some positive δ. Does there exist a bounded analytic function $G \in H^\infty(E_* \to E)$ such that

$$G\mathbb{F} \equiv I \quad ? \tag{16.2}$$

(Operators $\mathbb{F}(\lambda)$, $\lambda \in \mathbb{D}$ are not supposed to be onto, because otherwise $G = \mathbb{F}^{-1}$ and the problem is trivial).

The problem (16.2) (the "<u>Operator Corona Problem</u>" - this title will be explained below) is closely connected with many problems in operator theory treated in terms of the characteristic function θ_T of a contraction T. For instance, it was proved in [16] that the solvability of the problem (16.2) for $\mathbb{F} = \theta_T$ is equivalent to the similarity of T to some isometric operator. In other works (see for example [7], [1] and references there) it was established that the solvability of suitable "Operator Corona Problem" is equivalent to the convergence (summability) of spectral decompositions of contractions and also to solvability of free interpolation problems in vector and operator Hardy spaces. This was based on the connection of the Corona Problem with estimates of angles between coinvariant subspaces of the shift S discussed below in Section 17.

In [16] it was also proved that problem (16.2) has a solution $G \in H^\infty(E_* \to E)$ iff the vector Toeplitz operator $T_{F^\#}$ ($F^\#(\xi) = F(\bar{\xi})$, $T_F f \stackrel{def}{=} P_+ f$, $P_+ = P_{H^2}$ is the Riesz projection on vector Hardy space) is left invertible.

Note that the condition (16.1) is a simplest necessary one for the solvability of the Corona Problem (16.2).

If $\dim E = 1$, $\dim E_* < \infty$ the implication (16.1)$\Rightarrow$ (16.2) is nothing but the Carleson Corona Theorem proved originally in connection with a question on the maximal ideals space of H^∞ see [2], [4]. (This justifies the term "Operator Corona Problem"). Resting upon Wolf's proof of the Corona Theorem (see [2], [4]) V.A.Tolokonnikov [18] and M.Rosenblum [13] independently proved the implication (16.1)$\Rightarrow$(16.2) for $\dim E = 1$, $\dim E_* = \infty$. P.Fuhrmann (see for example [1]) proved this implication in the case $\dim E$, $\dim E_* < \infty$. Fuhrmann's construction and the result of V.A.Tolokonnikov mentioned above were used by V.I.Vasyunin [18] (see also [7]) to check the implication (16.1)$\Rightarrow$(16.2) in the case $\dim E < \infty$, $\dim E = \infty$, and to obtain upper estimates for the minimal norm of the solution of the Operator Corona Problem.

It will be shown below that in the general aase ($\dim E = \dim E_* = \infty$) the implication (16.1)$\Rightarrow$(16.2) is not valid. We also will present the lower bound for the minimal norm of solutions of Corona Problem (16.2) for finite-dimensional case ($\dim E < \infty$). The proof is based on geometrical reasons connected with estimates of angles between coinvariant spaces of the shift S.

17. CONNECTION OF THE CORONA PROBLEM WITH THE ANGLES BETWEEN COINVARIANT SUBSPACES

Any coinvariant (=invariant for the adjoint operator) subspace K of the shift operator S, $Sf=zf$, $f\in H^2(E)$ is generated, by the Lax - Beurling - Halmos Theorem 2.1, by an inner function $\theta\in H^\infty(E_*\to E)$ (E_* is an auxiliary Hilbert space), i.e.

$$K=K_\theta \overset{def}{=\!=} H^2(E)\ominus \theta H^2(E_*).$$

The following simple lemma is well-known (see [7], [17]).

17.1. LEMMA. <u>Let θ_1 and θ_2 be inner functions in $H^\infty(E_1\to E)$ and $H^\infty(E_2\to E)$ respectively. Then the angle between the spaces K_{θ_1} and K_{θ_2} is nonzero iff there exist two functions $\Psi_1\in H^\infty(E\to E_1)$ and $\Psi_2\in H^\infty(E\to E_2)$ such that</u>

$$\theta_1\Psi_1+\theta_2\Psi_2\equiv I .$$

<u>Moreover, the norm of the skew projection of $H^2(E)$ onto K_{θ_2} with the kernel K_{θ_1} is equal to the least possible norm of such Ψ_1 (or, indiferently Ψ_2).</u>

17.2. REMARK. Consider the function $F\in H^\infty(E\to E_1\oplus E_2)$ defined by

$$F(\lambda)e=\theta_1^*(\bar\lambda)e\oplus\theta_2^*(\bar\lambda)e,\quad e\in E\ ,\ \lambda\in\mathbb{D}. \tag{17.1}$$

Then by the above lemma the Corona Problem (16.2) for such a function has a bounded solution if and only if the angle between K_{θ_1} and K_{θ_2} is nonzero. Moreover, the minimal norm of such solution G is estimated by the norm of the skew projection $\mathcal{P}_{\theta_2\|\theta_1}$ onto K_{θ_2} with the kernel K_{θ_1} :

$$\|\mathcal{P}_{\theta_1\|\theta_2}\|\leqslant\|G\|_\infty\leqslant\|\mathcal{P}_{\theta_1\|\theta_2}\|+1$$

(recall that if the angle between two subspaces is α then the norm of the corresponding skew projection is $1/\sin\alpha$).

Therefore, to construct the counterexample and to obtain the lower estimate of the norm of a solution, we may, forgetting the original statement of the problem, estimate angles between coinvariant spaces.

Before formulating the main results we present for the sake of completeness

17.3. PROOF OF LEMMA 17.1. We will use the following well-known result (see [7], [15])

COMMUTANT LIFTING THEOREM. <u>Let</u> θ <u>be an inner function in</u> $H^\infty(E_*\to E)$, P <u>the orthogonal projection of</u> $L^2(E)$ <u>onto</u> K_θ <u>and</u> $T \overset{\text{def}}{=} PS^*\,|\,K_\theta$ <u>a model operator.</u>

<u>Then the commutant of</u> T (<u>i.e. the set</u> $\{A : AT=TA\}$) <u>is exactly the set of all operators</u> A <u>representable as</u>

$$Ax = PF^*x, \quad x\in K_\theta$$

<u>where</u> F <u>is a function in</u> $H^\infty(E\to E)$ <u>such that</u>

$$F\theta \in \theta H^\infty(E_*\to E_*) .$$

<u>Moreover</u>

$$\|A\| = \inf\{\|F+\theta h\|_\infty : h\in H^\infty(E\to E_*)\}$$

(note that adding θh, $h\in H^\infty(E\to E_*)$) to F we do not change the operator A), <u>and the infimum is attained.</u>

We apply this theorem to the model operator T acting in the space $K_\theta \overset{\text{def}}{=} \operatorname{span}(K_{\theta_1}, K_{\theta_2})$, $\theta_{1,2}\in H^\infty(E_{1,2}\to E)$, $\theta\in H^\infty(E_*\to E)$.

Let the angle between K_{θ_1} and K_{θ_2} be nonzero. Then the skew projection $\mathcal{P}$ onto K_{θ_1} with the kernel K_{θ_2} is bounded.

Since K_{θ_1} and K_{θ_2} are invariant subspaces of the inverse shift S^* (and hence for T) the projection $\mathcal{P}$ commutes with T and therefore there exists an operator-valued function $F\in H^\infty(E\to E)$, $F\theta\in\theta H^\infty(E_*\to E_*)$ such that

$$\mathcal{P} = P\mathbb{F}^*| K_\Theta .$$

Note that $\mathbb{F}^* K_{\Theta_1} \perp K_\Theta$ because $\mathcal{P}|K_{\Theta_1} = \mathbb{O}$. It follows from $\mathbb{F}\Theta \in \Theta H^\infty(E_* \to E_*)$ that

$$\mathbb{F}^* K_\Theta \perp \Theta H^2(E_*), \tag{17.2}$$

and therefore $\mathbb{F}^* K_{\Theta_1} \perp K_\Theta + \Theta H^2(E_*) = H^2(E)$. Hence

$$\mathbb{F} = \Theta_1 \Psi_1, \quad \Psi_1 \in H^\infty(E \to E_1). \tag{17.3}$$

Similarly, since $\mathcal{P}|K_{\Theta_2} \equiv I$, it follows that

$$(I - \mathbb{F}^*) K_{\Theta_2} \perp K_\Theta ,$$

and using (17.2) we obtain that

$$(I - \mathbb{F}^*) K_{\Theta_2} \perp H^2(E) .$$

Hence,

$$I - \mathbb{F} = \Theta_2 \Psi_2, \quad \Psi_2 \in H^\infty(E \to E_2) . \tag{17.4}$$

But (17.3) and (17.4) mean that

$$\Theta_1 \Psi_1 + \Theta_2 \Psi_2 \equiv I$$

Conversely, suppose now that there exist functions $\Psi_1, \Psi_2 \in H^\infty(E \to E_{1,2})$ such that

$$\Theta_1 \Psi_1 + \Theta_2 \Psi_2 \equiv I .$$

Set $\mathbb{F} \stackrel{def}{=\!=} \Theta_1 \Psi_1$ and define an operator A as follows:

$$Ax = P\mathbb{F}^* x, \quad x \in K_\Theta .$$

Since $\mathbb{F} = \Theta_1 \Psi_1$, $\Psi_1 \in H^\infty(E \to E_1)$, it follows that $A|K_{\Theta_1} = \mathbb{O}$, and since $I - \mathbb{F} = \Theta_2 \Psi_2$, $\Psi_2 \in H^\infty(E \to E_2)$, we have $A|K_{\Theta_2} = I$, i.e. an operator A is the projection $\mathcal{P}$ onto K_{Θ_2} with the kernel K_{Θ_1}.

It is obvious that

$$\|\mathcal{P}\| \leq \|\mathbb{F}\|_\infty = \|\Psi_1\|_\infty$$

and hence the angle between K_{θ_1} and K_{θ_2} is nonzero. ●

18. MAIN RESULTS

18.1. THEOREM. <u>Let</u> $0<\delta<1$, E <u>be an infinite-dimensional Hilbert space. There exist inner functions</u> $\theta_1, \theta_2 \in H^\infty(E_{1,2}\to E)$ <u>such that</u>

$$\|\theta_1^*(\lambda)e\|^2+\|\theta_2^*(\lambda)e\|^2\geqslant\delta^2\|e\|^2 \quad \forall\lambda\in\mathbb{D} \quad \forall e\in E \ ^{*)} \tag{18.1}$$

<u>but the angle between</u> K_{θ_1} <u>and</u> K_{θ_2} <u>is zero.</u>

The theorem obviously gives a counterexample to the implication (16.1)$\Longrightarrow$(16.2) in case $\dim E=\infty$ (see Remark 17.2).

18.2. SKETCH OF THE CONSTRUCTION OF SUBSPACES $K_{\theta_1}, K_{\theta_2}$. Set $E=E^1\oplus E^2$. Choose a set $\sigma\subset\mathbb{D}$ satisfying the sparseness condition (R) (see Section 19, (i)) which is not a Carleson set (see Section 14.4).

Let $\{e_\lambda\}_{\lambda\in\sigma}$ be an orthonormal basis in E^1 . We find a system $\{f_\lambda\}_{\lambda\in\sigma}$, $f_\lambda\in E^2$, $\|f_\lambda\|=C$, C being a constant depending on δ , such that the system of vector-valued rational functions $\{\varphi_\lambda f_\lambda\}_{\lambda\in\sigma}$ (recall that $\varphi_\lambda \overset{def}{=} (1-|\lambda|^2)^{1/2}/(1-\bar\lambda z)$ see Section 14.2) is uniformly minimal (with a suitable constant of uniform minimality), but its orthogonalizer is not bounded. (This system $\{\varphi_\lambda f_\lambda\}_{\lambda\in\sigma}$ will be dual in some sense to the one described above in Example 14.7). Set $K_{\theta_1}=\operatorname{span}(\varphi_\lambda e_\lambda : \lambda\in\sigma)$, $K_{\theta_2}=\operatorname{span}(\varphi_\lambda(e_\lambda\oplus f_\lambda) : \lambda\in\sigma)$. The inequality (18.1) will follow from the uniform minimality of $\{\varphi_\lambda f_\lambda\}$ and the sparseness of σ . The unboundedness of orthogonalizer will imply that the angle between K_{θ_1} and K_{θ_2} is zero.

Actually we will consider finite systems of vectors and deal with the corresponding estimates (for example a lower estimate of the norm of orthogonalizer instead of its unboundedness). We will do that by two reasons: firstly, to avoid the

*) This is nothing but the condition (16.1) for the function $\mathbb{F}$ defined by (17.1).

technical "infinite-dimensional" difficulties; secondly, which is more important, to obtain the lower estimates of the minimal norm of the solution of the Operator Corona Problem (16.2) in the case $\dim E < \infty$.

Let us give the precise statement.

18.3. THEOREM. Let a Hilbert space E, $\dim E = 2n$, $n > 0$ and δ, $0 < 1 < \delta$ be given. Then there exist inner functions θ_1, θ_2 in $H^\infty(E \to E)$ such that

$$\|\theta_1^*(\lambda)e\|^2 + \|\theta_2^*(\lambda)e\|^2 \geq \delta^2\|e\|^2 \quad \forall e \in E \;, \; \forall \lambda \in \mathbb{D} \tag{18.2}$$

but the norm of skew projection onto K_{θ_1} with the kernel K_{θ_2} is at least

$$C(n,\delta) = C \cdot \frac{\log^{1/2}(n\delta^2/(1-\delta)^2+1)}{\delta^2}(1-\delta)^{5/2}$$

where the constant C does not depend on n and δ.

This proposition obviously implies Theorem 18.1. Indeed, it is sufficient to set $K_{\theta_{1,2}} = \sum_{n=1}^{\infty} \oplus K_{\theta^{(n)}_{1,2}}$ where $\theta_1^{(n)}$, $\theta_2^{(n)} \in H^\infty(\mathbb{C}^{2n} \to \mathbb{C}^{2n})$ are inner functions from Theorem 18.3 with the same δ.

19. PROOF OF THEOREM 18.3

Fix δ, $0 < \delta < 1$. Choose a set $\sigma \subset \mathbb{D}$, $\operatorname{card} \sigma = n$ and a system of vectors $\{f_\lambda\}_{\lambda \in \sigma}$ in $\mathbb{C}^n$ such that

$$\text{(i)} \quad \left|\frac{\lambda - \mu}{1 - \bar{\mu}\lambda}\right| > \frac{2\delta}{1+\delta^2} \quad \forall \lambda, \mu \in \sigma, \; \lambda \neq \mu. \qquad *)$$

*) This condition means that the distances between the points of the set σ in the hyperbolic metric $\rho(\rho(\mu,\lambda) = \log[(1+\delta(\mu,\lambda))/(1-\delta(\mu,\lambda))]$, $\delta(\mu,\lambda) = \left|\frac{\mu-\lambda}{1-\bar{\mu}\lambda}\right|)$ are at least $2\rho_0$ ($\rho_0 = \log[(1+\delta)/(1-\delta)]$). Altogether, the condition that the points of a given set are uniformly separated in the hyperbolic metric $\inf_{\lambda \neq \mu} \rho(\lambda,\mu) > 0$ or, equivalently $\inf_{\lambda \neq \mu} \left|\frac{\mu-\lambda}{1-\bar{\mu}\lambda}\right| > 0$ which is known in the

(ii) $\| f_\lambda \| = 4\delta\cdot(1-\delta)^{-3/2}$ $\forall\lambda\in\sigma$.

(iii) The constant of uniform minimality of the system $\{\varphi_\lambda f_\lambda\}_{\lambda\in\sigma}$ is at least $1/\sqrt{2}$ (recall that $\varphi_\lambda=(1-|\lambda|^2)^{1/2}/(1-\bar\lambda z)$).

(iv) The norm of orthogonalizer of $\{\varphi_\lambda f_\lambda\}_{\lambda\in\sigma}$ is greater that the constant $C(n,\delta)$ from the theorem.

The existence of such a set σ and such a system $\{\varphi_\lambda f_\lambda\}_{\lambda\in\sigma}$ will be proved below in Section 20.

Let $\{e_\lambda\}_{\lambda\in\sigma}$ be an orthogonal basis in $E^1=\mathbb{C}^n$. Set $E=E^1\oplus E^2=\mathbb{C}^n\oplus\mathbb{C}^n$, and let θ_1, θ_1 be inner functions in $H^\infty(E\to E)$ which generate the following coinvariant subspaces

$$K_{\theta_1}\stackrel{def}{=\!=}\operatorname{span}(\varphi_\lambda e_\lambda:\lambda\in\sigma)$$

$$K_{\theta_2}\stackrel{def}{=\!=}\operatorname{span}(\varphi_\lambda(e_\lambda\oplus f_\lambda):\lambda\in\sigma).$$

Let us show that θ_1 and θ_2 are the functions we seek. Indeed, by (iv) there exist complex numbers c_λ, $\lambda\in\sigma$, $\Sigma|c_\lambda|^2=1$ such that

$$\Big\|\sum_{\lambda\in\sigma} c_\lambda\varphi_\lambda f_\lambda\Big\| < (C(n,\delta))^{-1}.$$

Put

$$f_1=\sum_{\lambda\in\sigma} c_\lambda\varphi_\lambda e_\lambda\,,\qquad f_2=\sum_{\lambda\in\sigma} c_\lambda\varphi_\lambda(e_\lambda\oplus f_\lambda).$$

It is obvious that $\|f_1\|^2=\sum_{\lambda\in\sigma}|c_\lambda|^2=1$, $\|f_2\|^2\ge\sum_{\lambda\in\sigma}|c_\lambda|^2=1$ and that

$$\|f_1-f_2\|<(C(n,\delta))^{-1}.$$

Hence the required estimate of the norm of the projection is true.

To prove estimate (18.2) we need the following lemma,

---

literature as the sparseness condition (R), plays an important role in many problems of analysis (free interpolation in H^∞, etc., see [2], [7]).

giving us an ability to reduce such an estimate to a simple geometrical problem.

19.1. LEMMA. <u>Let</u> θ <u>be an inner function in</u> $H^\infty(E_* \to E)$, <u>and</u> $e \in E$. <u>Then</u>

$$\|\theta^*(\lambda)e\| = dist(\varphi_\lambda e, K_\theta).$$

PROOF. Examine the vector Toeplitz operator T_{θ^*} with the symbol θ^*, $T_{\theta^*}: H^2(E) \to H^2(E_*)$,

$$T_{\theta^*} f \overset{def}{=\!=} P_+ \theta^* f \quad, \quad f \in H^2(E);$$

here P_+ is the orthogonal projection of $L^2(E_*)$ onto $H^2(E_*)$ (Riesz projection). It is obvious that $Ker\, T_{\theta^*} = K_\theta$. It is also easy to show that

$$T_{\theta^*} \varphi_\lambda e = \varphi_\lambda \theta^*(\lambda) e.$$

Thus, since T_{θ^*} acts isometrically on the subspace $(Ker\, T_{\theta^*})^\perp = \theta H^2(E_*)$ it follows that

$$dist(\varphi_\lambda e, K_\theta) = \|T_{\theta^*}\varphi_\lambda e\| = \|\varphi_\lambda \theta^*(\lambda) e\| = \|\theta^*(\lambda)e\|. \quad \bullet$$

19.2. VERIFICATION OF CONDITION (18.2).

To within a constant unitary factor the operator-valued function $\theta_1^* | E^2$ identically equals to I and matrix of the $\theta_1^* | E^1$ in the orthogonal basis $\{e_\lambda\}_{\lambda \in \sigma}$ has a diagonal form

$$diag\left\{\overline{\left(\frac{z-\lambda}{1-\bar\lambda z}\right)} : \lambda \in \sigma\right\}.$$

Thus, if a vector $e \in E$ satisfies $\|\theta_1^*(\lambda)e\| = \alpha\|e\|$, $\alpha < \delta$, then $|\lambda - \mu| / |1 - \bar\mu\lambda| \le \alpha$ for some $\mu \in \sigma$. Consequently, for the hyperbolic distance $\rho(\lambda, \mu)$ the following inequality holds: $\rho(\lambda,\mu) \le \log[(1+\alpha)/(1-\alpha)] \le \rho_0 \overset{def}{=\!=} \log[(1-\delta)/(1+\delta)]$. It follows from the sparseness condition (i) that $\rho(\lambda, \nu) \ge \rho_0$ for the other points $\nu \in \sigma$, $\nu \ne \mu$, or, equivalently $|\nu - \lambda| / |1 - \bar\nu\lambda| \ge \delta$, $\nu \in \sigma \setminus \{\mu\}$. Decomposing e as $e = e_1 + e_2$ where $e_1 = P_{\mathcal{L}(e_\mu)} e$ we get

$$\alpha^2\|e\|^2 = \|\theta_1^*(\lambda)e\|^2 = \|\theta_1^*(\lambda)e_1 + \theta_1^*(\lambda)e_2\|^2 = \|\theta_1^*(\lambda)e_2\|^2 \geq \delta\|e_2\|^2.$$

Hence $\sin \widehat{e\, e_\mu} = \frac{\|e_2\|}{\|e_1\|} \leq \frac{\alpha}{\delta}$, i.e. the vector e must be close to e_μ.

Let now $e = e_1 + e_2$, $e_{1,2} \in E^{1,2}$, $\|e\| = 1$, and

$$\|\theta_1^*(\lambda)e\|^2 + \|\theta_2^*(\lambda)e\|^2 < \delta^2,$$

for some point $\lambda \in \mathbb{D}$. Then it is obvious that

$$\|\theta_1^*(\lambda)e_1\| = \alpha\|e_1\|, \quad \alpha < \delta,$$

and

$$\|e_2\| < \delta.$$

Note that by lemma 19.1.

$$\|\theta_2^*(\lambda)e\| = \operatorname{dist}(\varphi_\lambda e, \operatorname{span}(\varphi_\mu(e_\mu \oplus f_\mu) : \mu \in \sigma)) \geq$$

$$\geq \operatorname{dist}(\varphi_\lambda e_1, \operatorname{span}(\varphi_\mu e_\mu : \mu \in \sigma)) = \|\theta_1^*(\lambda)e_1\| = \alpha\|e_1\|.$$

Thus

$$\delta^2 > \|\theta_1^*(\lambda)e\|^2 + \alpha^2\|e_1\|^2 = \alpha^2\|e_1\|^2 + \|e_2\|^2 + \alpha^2\|e_1\|^2$$

and therefore $\alpha < \delta/\sqrt{2}$. It is also easy to show that $\|\theta^*(\lambda)e\| = \alpha^2\|e_1\|^2 + \|e_2\|^2$, and hence $\|\theta_2^*(\lambda)e\| < \sqrt{\delta^2 - \alpha^2}\cdot\|e_1\|$. By Lemma 19.1 there exists a system of complex numbers c_ν such that

$$\Big\|\varphi_\lambda e - \sum_{\nu\in\sigma} c_\nu \varphi_\nu(e_\nu \oplus f_\nu)\Big\| < \sqrt{\delta^2 - \alpha^2}\,\|e_1\|. \tag{19.1}$$

As it was mentioned above the equality $\|\theta_1^*(\lambda)e_1\| = \alpha\|e_1\|$, $\alpha < \delta$ implies $\sin \widehat{e_1\, e_\mu} < \alpha/\delta$. Then

$$|c_\mu| = \|c_\mu\varphi_\mu\| \geq \|(e_1, e_\mu)\varphi_\lambda\| - \|(e_1, e_\mu)\varphi_\lambda - c_\mu\varphi_\mu\| >$$

$$> |(e_1, e_\mu)| - \sqrt{\delta^2 - \alpha^2}\cdot\|e_1\| > (\sqrt{1 - \alpha^2/\delta^2} - \sqrt{\delta^2 - \alpha^2})\|e_1\| =$$

$$= \frac{1-\delta}{\delta}\sqrt{\delta^2-\alpha^2}\cdot\sqrt{1-\|e_2\|^2} > \frac{1-\delta}{\delta}\sqrt{\delta^2-\delta^2/2}\cdot\sqrt{1-\delta^2} = \frac{1-\delta}{\sqrt{2}}\sqrt{1-\delta^2} > \frac{(1-\delta)^{3/2}}{\sqrt{2}} .$$

It follows from the uniform minimality of the system $\{\varphi_\nu f_\nu\}_{\nu\in\sigma}$ that

$$\Big\|\sum_{\nu\in\sigma} c_\nu\varphi_\nu f_\nu\Big\| \geq |c_\mu|\cdot\delta(\{\varphi_\nu f_\nu\}_{\nu\in\sigma})\|f_\mu\| >$$
$$> \frac{(1-\delta)^{3/2}}{\sqrt{2}}\cdot\frac{1}{\sqrt{2}}\cdot 4\delta\cdot(1-\delta)^{-3/2} = 2\delta ;$$

here we use also the properties (ii), (iii). But then

$$\Big\|\varphi_\lambda e - \sum_{\nu\in\sigma} c_\nu\varphi_\nu(e_\nu\oplus f_\nu)\Big\| \geq \Big\|\varphi_\lambda e_2 - \sum_{\nu\in\sigma} c_\nu\varphi_\nu f_\nu\Big\| \geq$$
$$\geq \Big\|\sum_{\nu\in\sigma} c_\nu\varphi_\nu f_\nu\Big\| - \|\varphi_\lambda e_2\| > 2\delta-\delta = \delta .$$

This contradicts to (19.1). Hence the inequality (18.2) holds and the theorem is proved to within the construction of the system $\{\varphi_\lambda f_\lambda\}_{\lambda\in\sigma}$ with the geometry described above in (i)-(iv).

20. CONSTRUCTION OF THE SYSTEM OF RATIONAL VECTOR-VALUED FUNCTIONS

Denote by $\mathrm{carl}\,\sigma$, where σ is a subset of $\mathbb{D}$ the Carleson intersity of the measure $\sum_{\lambda\in\sigma}(1-|\lambda|^2)\delta_\lambda$ (see Section 14.1)

$$\mathrm{carl}\,\sigma \overset{\mathrm{def}}{=\!=} \sup_{\substack{f\in H^2\\ \|f\|=1}} \sum_{\lambda\in\sigma}(1-|\lambda|^2)|f(\lambda)|^2 .$$

20.1. LEMMA. <u>For a given</u> δ, $0<\delta<1$, <u>and</u> $n>0$ <u>there exists a finite set</u> $\sigma\subset\mathbb{D}$, $\mathrm{card}\,\sigma = n$ <u>satisfying the sparseness condition (i) from Section 19 such that</u> $\mathrm{carl}\,\sigma \geq C\,\frac{\log(\delta^2/(1-\delta)^2+1)}{\delta^2}\cdot(1-\delta)^2$ <u>where</u> C <u>is an absolute constant (independent of</u> n <u>and</u> δ).

PROOF. It is well known (see [2], [4], [7]) that the Carleson intensity of a measure μ is equivalent in the sense of two-side estimates of $\sup_{\Delta}(\mu(\Delta)/\varepsilon)$, where the supremum is taken over all "squares" Δ with base on the circle $\mathbb{T}$,

$$\Delta = \Delta_{\varphi_0,\varepsilon} \stackrel{def}{=} \{ r\cdot e^{i\varphi}: 1-\varepsilon \leq r^2 \leq 1, |\varphi-\varphi_0| \leq \varepsilon\pi \}.$$

Consider such a "square" $\Delta = \Delta_{\varphi_0,\varepsilon}$. To prove the lemma it is sufficient to put into this "square" a set σ of n points for which the sparseness condition (i) from Section 19 holds and

$$\sum_{\lambda\in\sigma} (1-|\lambda|^2) \geq \varepsilon\cdot C \frac{\log(\delta^2/(1-\delta)^2+1)}{\delta^2}\cdot(1-\delta)^2 .$$

To do this we put the first point λ_1 on the upper bound of the "square" Δ (i.e. in such a way that $|\lambda_1| \leq |\lambda|$ for all $\lambda\in\Delta$). Remove from Δ the open hyperbolic disk d_1 of (hyperbolic) radius $2\rho_0$, $\rho_0 = \log\frac{1+\delta}{1-\delta}$ with the center at λ_1. Then we choose the second point $\lambda_2 \in \Delta\setminus d_1$ so that $|\lambda_2| \leq |\lambda|$ for all $\lambda\in\Delta\setminus d_1$. Remove from $\Delta\setminus d_1$ the hyperbolic disk d_2 of radius $2\rho_0$ with the center at λ_2. Choose $\lambda_3 \in \Delta\setminus d_1\setminus d_2$ so that $|\lambda_3| \leq |\lambda|$ for all $\lambda\in\Delta\setminus d_1\setminus d_2$ and so on. We obtain in such a way a set σ $\{\lambda_1,\lambda_2,\dots,\lambda_n\}$ satisfying the sparseness condition (i). It is evident that σ is contained in the "square" $\Delta_1 = \Delta\cap\{\lambda: |\lambda| \leq |\lambda_n|\}$ and the interior of Δ_1 is covered by the disks d_k, $k=1,\dots,n$.

The hyperbolic area of each (hyperbolic) disk d_k equals $4\pi\, sh^2\rho_0$ (see [11]; recall that the element of the hyperbolic area dH is defined by $dH = \frac{4r}{(1-r^2)^2}\, dr\, d\varphi$). Consequently, if Δ_2 is the "square" of the form $\Delta_2 = \Delta\cap\{\lambda: |\lambda| \leq r\}$ of hyperbolic area $n\cdot 4\pi\, sh^2\rho_0$ then $\Delta_1 \subset \Delta_2$. It is easy to show that

$$\Delta_2 = \{ r\cdot e^{i\varphi}: 1-\varepsilon \leq r^2 \leq 1-\frac{\varepsilon}{n\, sh^2\rho_0+1}, |\varphi-\varphi_0| \leq \varepsilon\pi \}.$$

Let $\mathbb{D}_1$ be a subset of Δ_2 of hyperbolic area

$4\pi\,sh^2\rho_0$ (=the area of d_k) such that $d_1 \cap \Delta_2 \subset \mathcal{D}_1$. Since $|\lambda| \geq |\lambda_1|$ for all $\lambda \in \mathcal{D}_1$,

$$\frac{1}{4\pi\,sh^2\rho_0} \int_{\mathcal{D}_1} (1-|\lambda|^2)\, dH(\lambda) \geq 1-|\lambda_1|^2$$

Let us choose a subset $\mathcal{D}_2 \subset E_1 \overset{def}{=\!=} \Delta_2 \setminus \mathcal{D}_1$ of the same hyperbolic area $4\pi\,sh^2\rho_0$ such that $d_2 \cap E_1 \subset \mathcal{D}_2$. By the construction of the points λ_k the inequality $|\lambda| \geq |\lambda_2|$ holds for all $\lambda \in \mathcal{D}_2$. Therefore

$$\frac{1}{4\pi\,sh^2\rho_0} \int_{\mathcal{D}_2} (1-|\lambda|^2)\, dH(\lambda) \geq 1-|\lambda_2|^2 .$$

Similarly, choosing a subset $\mathcal{D}_3 \subset E_2 \overset{def}{=\!=} E_1 \setminus d_1$ of the same area such that $d_3 \cap E_2 \subset \mathcal{D}_3$, we obtain

$$\frac{1}{4\pi\,sh^2\rho_0} \int_{\mathcal{D}_3} (1-|\lambda|^2)\, dH(\lambda) \geq 1-|\lambda_3|^2$$

and so on. We will obtain disjoint subsets $\mathcal{D}_1, \mathcal{D}_2, \dots, \mathcal{D}_n$ of Δ_2 that cover Δ_2 (perhaps except the set of zero hyperbolic area). Therefore

$$\int_{\Delta_2} (1-|\lambda|^2)\, dH(\lambda) = \sum_{k=1}^{n} \int_{\mathcal{D}_k} (1-|\lambda|^2)\, dH(\lambda)$$

and hence

$$\frac{1}{4\pi\,sh^2\rho_0} \int_{\Delta_2} (1-|\lambda|^2)\, dH(\lambda) \geq \sum_{k=1}^{n} (1-|\lambda_k|^2) = \sum_{\lambda \in \sigma} (1-|\lambda|^2) .$$

But it is easy to calculate that

$$\frac{1}{4\pi\,sh^2\rho_0} \int_{\Delta_2} (1-|\lambda|^2)\, dH(\lambda) = \varepsilon\, \frac{\log(n\, sh^2\rho_0+1)}{sh^2\rho_0} \geq$$

$$\geq \varepsilon C \cdot \frac{\log(n\delta^2/(1-\delta)^2+1)}{\delta^2}\,(1-\delta)^2$$

and the lemma is proved. ●

Take a set $\sigma\subset\mathbb{D}$, $\operatorname{card}\sigma=n$ from the above lemma. Let $E=\mathbb{C}^{n+1}$, and let vectors e, $\{e_\lambda\}_{\lambda\in\sigma}$ form an orthonormal basis in E. Set

$$h_\lambda=\frac{1}{\sqrt{2}}(e+e_\lambda).$$

The system $\{h_\lambda\}_{\lambda\in\sigma}$ is uniformly iminimal and its constant of uniform minimality $\delta(\{h_\lambda\}_{\lambda\in\sigma})$ equals $1/\sqrt{2}$. Therefore the system $\{\varphi_\lambda h_\lambda\}_{\lambda\in\sigma}$ is uniformly minimal and $\delta(\{\varphi_\lambda h_\lambda\}_{\lambda\in\sigma})\geqslant 1/\sqrt{2}$.

Let $\{\psi_\lambda\}_{\lambda\in\sigma}$ be the system of vectors in $K_\theta \overset{def}{=\!=} \operatorname{span}(\varphi_\lambda h_\lambda:\lambda\in\sigma)$ which is biorthogonal to the $\{\varphi_\lambda h_\lambda\}_{\lambda\in\sigma}$, i.e. such that

$$(\varphi_\lambda h_\lambda,\psi_\mu)=\begin{cases}0 & \lambda\neq\mu\\ 1 & \lambda=\mu\end{cases}.$$

Let $\mathcal{J}$ be the orthogonalizer of the system $\{\psi_\lambda\}_{\lambda\in\sigma}$. Then

$$\mathcal{J}f=\{(f,\varphi_\lambda h_\lambda)\}_{\lambda\in\sigma}\in\ell^2,\quad f\in K_\theta$$

and hence

$$\|\mathcal{J}\|^2=\sup_{\substack{f\in H^2(E)\\ \|f\|=1}}\sum_{\lambda\in\sigma}|(f,\varphi_\lambda h_\lambda)|^2\geqslant\sup_{\substack{f\in H^2\\ \|f\|=1}}\sum_{\lambda\in\sigma}|(fe,\varphi_\lambda h_\lambda)|^2=$$

$$=\sup_{\substack{f\in H^2\\ \|f\|=1}}\frac{1}{\sqrt{2}}\sum_{\lambda\in\sigma}(1-|\lambda|^2)|f(\lambda)|^2=\frac{1}{\sqrt{2}}\operatorname{card}\sigma\geqslant C\,\frac{\log(\delta^2n/(1-\delta)^2+1)}{\delta^2}(1-\delta)^2.$$

It is obvious that the geometry of the system (properties (ii)-(iv) from Section 19)

$$\{ 4\delta\cdot(1-\delta)^{-3/2}\cdot\|\psi_\lambda\|^{-1}\psi_\lambda \}_{\lambda\in\sigma}$$

coincides with the geometry of the system $\{\varphi_\lambda f_\lambda\}$ we are trying to construct. This is no a system of required form but the multiplication by the operator-valued function θ^* (recall that θ is an inner function in $H^\infty(E\to E)$) generating the subspace $K_\theta = span(\varphi_\lambda h_\lambda : \lambda\in\sigma)$ translates it into the system

$$\left\{ \frac{(1-|\lambda|^2)^{1/2}}{z-\lambda} f_\lambda \right\}_{\lambda\in\sigma}, \qquad f_\lambda\in E,\ \|f_\lambda\| = 4\delta\cdot(1-\delta)^{-3/2}$$

with the same geometry. The last system is translated by the unitary operator τ , $\tau f(\xi) = \bar{\xi} f(\bar{\xi})$, $\xi\in\mathbb{T}$, to the system we want to find:

$$\left\{ \frac{(1-|\lambda|^2)^{1/2}}{1-\bar{\lambda}z} f_\lambda \right\}_{\lambda\in\sigma} .$$

To finish the proof it remains to dispace the vectors f_λ into a n-dimensional space which is a trivial task. ●

21. REMARK

The lower estimate of the solution of Operator Corona Problem obtained in the above theorem may be rewritten, for a small δ , as $C\cdot(\log^{1/2}(\delta^2 n+1))/\delta^2$. Compare this estimate with the upper one obtained by V.I.Vasyunin [18] having a form $C\cdot n^2\cdot[\log^{3/2}(1/\delta)]\,/\delta^{2n}$ for small δ . Note that distance between these estimates is very large, as it should have been expected.

APPENDIX

22. SOME OPEN PROBLEMS

As it was said in the Introduction, the spectral theory of vector-valued functions is now too far from be completed. For the time being in this theory there are much more open problems than finished results. Here we present some problems arisen in the study of questions discussed in preceding chapters. Their solution would make much clearer many obscure aspects of the spectral theory.

22.1. PROBLEM. When the operator measure ($\dim E = \infty$) is an operator weight? Can the operator measure be decomposed into an "absolutely continuous" part (operator weight) and a "singular" one? (And what is it, a "singular" part?).

22.2. REMARK. If an operator measure W is such that the measures (We, e) are absolutely continuous for all $e \in E$, then W needs not to be an operator weight, see Example 2.1. Therefore the words "absolutely continuous" and "singular" in above section are in inverted commas.

22.3. PROBLEM. Find a constructive solution of the extremal Szegö problem. It would be interesting to express the distance $dist_{L^2(E,W)}(e, \mathcal{P}ol_A^{\circ})$ in terms of some analoque of the multiplicative integral, just us for the Kolmogorov extremal problem the distance $dist_{L^2(E,W)}(e, \mathcal{P}ol^{\circ})$ is expressed in terms of a usual integral of the weight W^{1}, see Theorem 6.1.

22.4. PROBLEM. Is the analoque of Lemma 4.7 true for Kolmogorov extremal problem, i.e. is it true that

$$dist_{L^2(E,W)}(e, \mathcal{P}ol^{\circ}) = \lim_{r \to 1^-} dist_{L^2(E,W_r)}(e, \mathcal{P}ol^{\circ})$$

Is this statement valid for operator measures?

If such a statement holds for operator measures we could solve Kolmogorov extremal problem for such measures.

Following problem is closely connected with Problem 22.1.

22.5. PROBLEM. Has the singular **component** (in sense of Problem 22.1) of the operator measure an influence on the solution of the Kolmogorov extremal problem? It is well-known that in scalar case the distance to find depends only on the absolutely continuous part of the measure (see [2], [4], [7]).

22.6. PROBLEM. Is the strong Szegö condition (SS) (the uniform Szegö condition $(\mathcal{U}S)$ equivalent to the star one $(SS)^*$ (respectively to $(\mathcal{U}S)^*$) in the case $\dim E = \infty$? In the case $\dim E < \infty$ this is so, see Section 5.

22.7. PROBLEM. Can an operator measure V satisfying the inequality (8.1) for some weight (measure) W fail to be a weight, i.e. can it have a singular part (in the sense of the Problem 22.1)?

22.8. PROBLEM. Is the condition

$$\sup_I \frac{1}{m(I)} \left\| \left[\int_I W\, dm \right]^{1/2} \cdot \left[\int_I W^{-1} dm \right]^{1/2} \right\| < \infty$$

where supremum is taken over all arcs $I \subset \mathbb{T}$ necessary and sufficient for the Riesz projection P_+ to be bounded on the weighted space $L^2(E, W)$?

This problem is of interest even in the matrix case.

22.9. MOTIVATION. It is easy to show that the above condition is a criterion of the uniform boundedness of all averaging operators

$$f \longmapsto f_I \overset{def}{=\!=} \frac{1}{m(I)} \int_I f\, dm$$

in the weighted space $L^2(E, W)$, where I denotes an arc of $\mathbb{T}$.

In the scalar case this condition is the well-known Muckenhoupt (A_2) condition (see [2], [4])

$$\sup_I \frac{1}{m(I)} \int_I w\,dm \cdot \int_I w^{-1}\,dm < \infty$$

which is equivalent to the boundedness of P_+ on $L^2(w)$.

22.10. REMARK. The author does not know whether the above "operator Muckenhoupt condition" from Problem 22.8 provides the uniform boundedness of Poisson or Feyér means on $L^2(E,W)$. In the scalar case the uniform boundedness of such means is proved easily than that of P_+ .

22.11. PROBLEM. V.I.Vasyunin showed(see [7]) that for any system of coinvariant subspaces K_θ in H^2 the uniform minimality is equivalent to the property of constituting a Riesz basis. Is the same true for coinvariant subspaces in $H^2(E)$, $\dim E < \infty$? *) (The case of one-dimensional spaces K_θ is studied in Chapter IV). Does there exist an analoque of the spatial compactness condition (see Definition 13.4) ensuring this equivalence in the case $\dim E = \infty$?

22.12. PROBLEM. Find necessary and sufficient condition (preferably of local nature) for solvability of Operator Corona Problem (16.2).

*) NOTE ADDED IN PROOF. A positive answer to this question has recently been obtained by the author in the case when all θ_n are inner functions in $H^\infty(E_n \to E)$, $\dim E < \infty$.

REFERENCES

1. Fuhrman P.: Linear systems and operators in Hilbert space, McGraw Hill, New York, 1981.
2. Garnett J.B.: Bounded analytic functions, Acad.Press, New York, 1981.
3. Helson H.: Lectures on invariant subspaces, Acad.Press, New York, 1964.
4. Koosis P.: Introduction to H^p spaces, Cambridge Univ.Press, Cambridge, 1980.
5. Koosis P.: Moyennes quadratiques ponderees de fonctions periodiques et de leurs conjuguees harminiques, C.R.Acad.Sci., Paris, Ser.A, 291 (1980), 255-257.
6. Linear and complex analysis problem book. 199 research problems, Lect.Notes Math., 1043 (1984).
7. Nikol'skii N.K.: Treatise on the shift operator, Springer Verlag, 1985.
8. Nikol'skii N.K. and Pavlov B.S.: Decomposition on eigenvectors of nonunitary operators and characteristic function, Zap.Nauchn.Sem.Leningrad.Otdel.Math.Inst.Steklov. (LOMI), 11 (1968), 90-143 (Russian).
9. Nikol'skii N.K. and Vasyunin V.I.: Notes on two functional models, Proc.conf. in honour of the Brahges proof of the Bieberbach conjecture, Colloq.Math.Publ., 1986, 113-141.
10. Nikol'skii N.K. and Vasyunin V.I.: A unified approach to function models and the transcription problem, LOMI preprint, E-5-86, Leningrad 1986.
11. Privalov I.I.: Introduction to the theory of complex variable, Moscow, Fizmatgiz, 1960 (Russian).
12. Rozanov Yu.A. Stationary stochastic processes, Gosudarstv. Izdat.Fiz.-Mat.Lit., Moscow, 1963.
13. Rosenblum M.: A corona theorem for countably many functions, Integr.Equat.Oper.Theory 3 (1980), 125-137.

14. Suciu I. and Valuşescu I.: Factorization theorem and prediction theory, Preprint INCREST No 2 / 1978, Bucuresti, 1978.
15. Sz.-Nagy B. and Foiaş C.: Analyse harmonique des opérateurs de l'espace de Hilbert, Akad.Kiodo, Budapest, 1967.
16. Sz.-Nagy B. and Foiaş C. On contractions similar to isometries and Toeplitz operators, Ann.Akad.Scient.Fennicae, Ser. AI, Matematica 2 (1976), 533-564.
17. Teodorescu R.: Sur les decompositions directes des contractions de l'espace de Hilbert, J.Funct.Anal., 18, No 4 (1975), 414-428.
18. Tolokonnikov V.A.: Estimates in Carleson's corona theorem. Ideals of the algebra H^∞, the problem of Szekefalvi-Nagy, Zap.Naučn.Sem.Leningrad.Otdel.Mat.Inst.Steklov (LOMI), 113 (1981), 178-198.
19. Treil' S.R.: Angles between coinvariant subspaces and operator corona problem (a Sz.-Nagy problem), Dokl.Akad.Nauk.SSSR (1988) (Russian, to appear).
20. Treil' S.R.: A spatially compact system of eigenvectors forms a Riesz basis if it is uniformly minimal, Dokl.Akad. Nauk.SSSR, 288, No 2 (1986), 308-312; English.transl.: Soviet.Math.Dokl., 33, No 3 (1986), 675-679.
21. Treil' S.R.: Extreme points of the unit ball of the operator Hardy space $H^\infty(E \to E_*)$, Zap.Naučn.Sem.Leningrad.Otdel.Mat. Inst.Steklov. (LOMI), 149 (1986), 160-164.
22. Treil' S.R.: Geometric aspects of the theory of Hankel and Toeplitz operators, Diss.LGU, Leningrad, 1985 (Russian).
23. Treil' S.R.: On error of the best prediction, Mathematical problems of control of stochastic and deterministic systems, Moscow, MAI, 1988 (Russian).
24. Treil' S.R.: Operator approach for weighted norm inequalities for singular integrals, Zap.Nauchn.Sem.Leningrad.Otdel.Mat. Inst.Steklov. (LOMI) 135 (1984), 150-174 (Russian).

Operator Theory:
Advances and Applications, Vol. 42

FORMULA FOR MULTIPLICITY OF CONTRACTIONS WITH FINITE DEFECT INDICES

V.I.Vasyunin

CONTENTS

1. Introduction. 282
2. Notation. 282
3. Main result. Statement and discussion. 284
4. Lower estimate for an operator without C_{10} part. . . . 286
5. Three auxiliary lemmas. 287
6. Upper estimate for an operator without C_{10} part. . . . 291
7. Complementation lemma. 294
8. Upper estimate in the general case. 296
9. Lower estimate in the general case. 299

1. INTRODUCTION

The multiplicity of a linear operator is invariant under various equivalence relations, such us unitary equivalence, linear similarity, quasisimilarity. Moreover, under some restrictions on a given operator (e.g. for normal, or Dunford scalar operators), one can define a local version of the multiplicity, so called multiplicity function, which uniquely determins the equivalent class of such an operator.

In the general case the search of a complete system of invariants under similarity or quasisimilarity presuppose the calculation of multiplicity as a first step. In this paper such a calculation is made for contractions with finite defect indices.

2. NOTATION.

Let T be a contraction on a Hilbert space H. We assume that T has finite defect indices, i.e.

$$\partial \overset{def}{=} rank(I-T^*T)<\infty; \quad \partial_* \overset{def}{=} rank(I-TT^*)<\infty .$$

Multiplicity of T, i.e. the number μ_T,

$$\mu_T \overset{def}{=} min\{dim\, E : span(T^n E : n \geq 0) = H\},$$

will be calculated in terms of the multiplicity function of the unitary part of T and the characteristic function θ of its completely nonunitary (c.n.u.) part. In this paper we shall use extensively the model theory of B.Sz.-Nagy and C.Foias, the main reference being their book [7]. Various descriptions of the similarity and quasisimilarity orbits for some special classes of operators can be found both in the monograph [7] and in the papers [1, 4, 8, 9, 12, 14, 16].

Describe now the parameters that determine the multiplicity of contractions with finite defect indices.

Let $T = \mathcal{U} \oplus T_0$ be the decomposition of a given contraction T into the unitary and c.n.u. parts, $\mathcal{U} = \mathcal{U}_s \oplus \mathcal{U}_a$ be the decomposition of $\mathcal{U}$ into the sum of the singular and absolutely continuous summands.

Let θ be the characteristic function of T_0, i.e. θ is a contractive $\partial_* \times \partial$ -matrix-valued function analytic in the unit disc. M_θ will denote a model operator with characteristic function θ. The model space $\mathcal{K}$, where M_θ acts, not always will be chosen in the Sz.-Nagy - Foiaş form. About various possibilities of choosing a model space see [5].

Put $r \overset{def}{=} \operatorname{Rank} \theta$ i.e. $\operatorname{rank} \theta(\zeta) = r$ for a.e. ζ, $|\zeta| = 1$. Let $\theta = \theta_{*0}\theta_{*1}$ be the canonical factorization, $\theta_{*0} = \theta_{10}\theta_{00}$ and $\theta_{*1} = \theta_{11}\theta_{01}$ be $*$-canonical factorizations (cf. [11], Thm.1.5, and [17], Lemma 3.2).

The operator with the characteristic function θ_{00} is a C_0 contraction and it is known (see e.g. [1]) how to calculate multiplicity of such an operator: $\mu_0 = r - r_0$. Here r is the order of matrix θ_{00} and r_0 denotes the maximal order of minors of θ_{00} that have no common inner divisor.

Let $\mu_a(\zeta)$ be the local multiplicity of $\mathcal{U}_a$, $\Delta_{11}(\zeta) \overset{def}{=} (I - \theta_{11}(\zeta)^* \theta_{11}(\zeta))^{1/2}$, $\mu_1(\zeta) \overset{def}{=} \operatorname{rank} \Delta_{11}(\zeta)$ and

$$\mu_{1a} \overset{def}{=} \sup_{|\zeta|=1} \{\mu_a(\zeta) + \mu_1(\zeta)\}.$$

The number μ_{1a} is the multiplicity of a direct sum of $\mathcal{U}_a$ and the contraction with the characteristic function θ_{11}.

Let us introduce another two auxiliary numbers ε and ε_*. First, we put

$$\varepsilon_* \overset{def}{=} \begin{cases} 1 & \text{if } r < \partial \\ 0 & \text{if } r = \partial \end{cases}.$$

Let $\mathcal{J}_\theta$ be the ideal in the Smirnoff class N^+ generated by the minors $M^{\mathbf{i}}$ of order r of the factor θ_{10} (here $\mathbf{i} = \{i_1, \dots, i_r\}$ is a multi-index of order r). Put

$$\varepsilon \overset{def}{=\!=} \begin{cases} 1 & \text{if } \mathcal{J}_\theta \neq N^+ \\ 0 & \text{if } \mathcal{J}_\theta = N^+ . \end{cases}$$

In other words $\varepsilon = 0$ if there exists a family f_i of bounded analytic functions in the unit disc such that $\sum_i f_i M^i(\theta_{10})$ is an outer function, and $\varepsilon = 1$ otherwise.

Some words about notation. First we discuss matrices and determinants. Columns of matrices will be marked by lower indices and rows by upper ones. So $(\theta_{10})^i_j$ is the entry of θ_{10} that stands in the j-th column and i-th row. The minor $M^i_j(\Phi)$ is the determinant of the matrix $\{\Phi^i_j\}^{i\in i}_{j\in j}$. We suppose that numbers in multi-indices are in the increasing order, i.e. $i_1 < i_2 < \ldots < i_k$ if $i = \{i_1, \ldots, i_k\}$. We omit one of the multi-indices if the minor has the maximal possible order, e.g. if Φ consists of m columns and $j = \{1, \ldots m\}$ we write M^i instead of M^i_j. By i' we denote the complementary index to i, i.e. $i' = \{1, \ldots, n\} \setminus i$ (we suppose that the size of a matrix is fixed). The number $\varepsilon_i = \pm 1$ denotes the sign of the permutation $\{i, i'\}$ of $\{1, \ldots, n\}$.

And finally some common notation. Let as usual $H^\infty(E \to E_*)$ denote the Hardy space of operator-valued functions, bounded and analytic in the unit disc, whose values are operators from E into E_*. Let $H^2(E)$ be the Hardy space of E-valued square summable analytic functions. For $E = \mathbb{C}^n$ we write H^2_n. The symbol P_H denotes an orthogonal projection onto H from any larger space. Sometimes for a c.n.u. contraction T with characteristic function θ we shall replace T by θ in notations containing T as an index, e.g. write μ_θ instead of μ_T, etc.

3. MAIN RESULT. STATEMENT AND DISCUSSION.

THEOREM

$$\mu_T = \max\{\mu_{u_s}, \ \partial_* - \tau + \max(\mu_0, \mu_{1a}, \varepsilon, \varepsilon_*)\}.$$

Before presenting a proof let us try to understand the structure of this formula. For our purposes a contraction can be thought of consisting of six components. The unitary part is naturally divided into two components, the singular one and the absolutely continuous one. The completely nonunitary part of a contraction can be regarded as the sum of four components corresponding to the four factors in the factorization $\Theta = \Theta_{10}\Theta_{00}\Theta_{11}\Theta_{01}$. The first component is a shift-like operator. The second one is a C_0 contraction. The third component is quasisimilar to an absolutely continuous unitary operator and it can be replaced by such a one. It is in this way that the common characteristic μ_{1a} appears. And the last component is and operator that behaves as the backward shift. The case when the operator under investigation really was such a direct sum was examined in [13]. That is, in [13] a partial case of the Theorem for $T = \mathcal{U} \oplus S_n \oplus T_0 \oplus S_m^*$ was obtained, where S_n is the unilateral shift of multiplicity n. For such an operator the multiplicity formula can be rewritten in the form

$$\mu_T = max\{\mu_{\mathcal{U}_s}, n + max(\mu_0, \mu_{1a}, \varepsilon_*)\}.$$

Here $\varepsilon_* = 1$ if $m > 0$ and $\varepsilon_* = 0$ if $m = 0$. In this simplified situation the parameter ε disappears. It is present only in the case when the C_{10} part of T is not quasi-similar to a unilateral shift.

Let $T = \mathcal{U}_s \oplus T_a$ be the decomposition in the sum of singular unitary operator and a contraction with absolutely continuous unitary dilation. Since in this case the lattice of the invariant subspaces of the sum splits up into the sum of lattices

$$\mathrm{Lat}\, T = \mathrm{Lat}\, \mathcal{U}_s \oplus \mathrm{Lat}\, T_a,$$

the multiplicity of the sum is equal to the maximum of the multiplicities:

$$\mu_T = max\{\mu_{\mathcal{U}_s}, \mu_{T_a}\}.$$

So from now on we shall suppose that T has absolutely contin-

uous unitary dilation.

Now let us try to explain (not leaving the intuitive level) when the multiplicity of a direct sum is the sum of multiplicities and when it is their maximum. Roughly speaking, if the spectra of operators overlap we have the first case; otherwise, the second possibility occurs. But "the overlapping" does not mean here the mere intersection of sets, it rather means a relative "density" of spectra. So the spectrum of $\mathcal{U}_S$ is very thin and chips off from everything else. The spectra of the C_0 part, of the C_{11} part (or the absolutely continuous unitary part) and of the C_{01} part (i.e. S^*-like operator) can be "separated" and therefore the multiplicity of their "sum" is equal to the maximum of multiplicities. The spectrum of the C_{01} part can be separated from the spectra of the other parts in the following sense. Let us take an arbtirary open subset of the unit disc which is separated (in the usual sense) from the spectra of C_0 and C_{11} parts. The eigenvectors of the operator corresponding to the eigenvalues from this subset span the whole subspace where the C_{01} part acts. The C_0 part can be "separated" from the C_{11} part by using its minimal function that kills the C_0 part and conserve the rest. The spectrum of the C_{10} part is the "densest" one and cannot be "separated" from other parts (except for the singular unitary operator of course) and the C_{10} part adds its multiplicity to the common multiplicity of the other parts.

As it was mensioned, it is enough to find the multiplicity of an operator with absolutely continuous unitary dilation. So we suppose that T is such an operator and prove that formula

$$\mu_T = \partial_* - r + max(\mu_0, \mu_{1a}, \varepsilon, \varepsilon_*) .$$

4. LOWER ESTIMATE FOR AN OPERATOR WITHOUT C_{10} PART

4.1. LEMMA. <u>Let $\Theta = \Theta_1\Theta_2$ be a regular factorization of the characteristic function of a c.n.u. contraction</u> T . <u>Let</u> $T = \begin{pmatrix} T_2 & * \\ 0 & T_1 \end{pmatrix}$ <u>be the corresponding triangulation on</u> $H_T = H_2 \oplus H_1$. <u>Then for an ar-</u>

bitrary operator A on H_A the following inequality holds:

$$\mu_{T\oplus A} \geqslant \mu_{T_1\oplus A} .$$

PROOF. The operator $T_1\oplus A$ is a compression of $T\oplus A$ to the coinvariant subspace $H_1\oplus H_A$. ▨

4.2. LEMMA. If $\tau=\partial_*$ and T is not the zero operator on a zero space then $\mu_T \geqslant max\{\mu_0, \mu_{1a}, \varepsilon_*\}$.

PROOF. In the case $\tau=\partial_*$ we have $\theta=\theta_{00}\theta_{11}\theta_{01}$. According to Lemma 4.1 $\mu_T\geqslant\mu_0$.

Let $\theta_{11}'\theta_{00}'$ be the $*$-canonical factorization of the function $\theta_{00}\theta_{11}$. Then $\theta_{11}'\cdot(\theta_{00}'\theta_{01})$ is the $*$-canonical factorization of θ , i.e. this is a regular factorization. Since $rank(I-\theta_{11}^*(\zeta)\theta_{11}(\zeta)) = rank(I-\theta_{11}^{*\prime}(\zeta)\theta_{11}'(\zeta))$ a.e., the multiplicity of $\mathcal{U}_a\oplus M_{\theta_{11}'}$, is equal to μ_{1a}. Therefore $\mu_T\geqslant\mu_{1a}$ according to Lemma 4.1.

The inequality $\mu_T\geqslant\varepsilon_*$ is obvious. ▨

5. THREE AUXILIARY LEMMAS.

5.1. LEMMA. <u>Let θ_1 and θ_2 be inner $n\times n$-matrix-functions with mutually prime determinants. Let $\theta=\theta_1\theta_2$ be a factorization of the characteristic function of an operator T on H . The corresponding triangulation is $T=\begin{pmatrix} T_2 & * \\ 0 & T_1\end{pmatrix}$ on $H=H_2\oplus H_1$ Then T is quasisimilar to $T_1\oplus T_2$</u> .

PROOF. Choose model spaces $\mathcal{K}_\theta = H_n^2 \ominus \theta H_n^2$ and $\mathcal{K}_\vartheta = H_{2n}^2 \ominus \vartheta H_{2n}^2$, where $\vartheta=\begin{pmatrix}\theta_1 & 0\\ 0 & \theta_2\end{pmatrix}$ is the characteristic function of operator $T_1\oplus T_2$. Then by direct calculation it is easy to verify that the following operators

$$X \overset{def}{=\!=} P_{\mathcal{K}_\vartheta}\begin{pmatrix} I \\ \theta_1^{ad}\end{pmatrix} \Big| \mathcal{K}_\theta : \mathcal{K}_\theta \longrightarrow \mathcal{K}_\vartheta ;$$

$$Y \overset{def}{=\!=} P_{\mathcal{K}_\theta}(\det\theta_2\cdot I, \theta_1) \mid \mathcal{K}_\vartheta : \mathcal{K}_\vartheta \to \mathcal{K}_\theta$$

are deformations (i.e. they have trivial kernel and co-kernel) and interwine the operators M_θ and M_ϑ . ◍

5.2. COROLLARY. $\mu_\theta = \max\{\mu_{\theta_1}, \mu_{\theta_2}\}$. ◍

5.3. LEMMA. <u>Let $\theta \in H^\infty(E \to E_*$ be an inner $*$-outer matrix-function and ω be a scalar inner function. Then almost all subspaces $\mathbb{F}$ of E_* with $\dim \mathbb{F} = \partial (\overset{def}{=} \dim E)$ have the following properties:</u>

1) <u>the minors of order $\partial - 1$ of the $\partial \times \partial$-matrix-function $P_{\mathbb{F}}\theta$ have no common inner divisor.</u>

2) <u>$\det P_{\mathbb{F}}\theta$ and ω have no common inner divisor.</u>

PROOF. Let $e_* \in E_*$ and let $e_*^\perp$ be the hyperplane orthogonal to e_* . Put $\vartheta \overset{def}{=} P_{e_*^\perp}\theta$. Let $\vartheta = \vartheta_i \vartheta_e$ be the canonical factorization and $\vartheta_i = \vartheta_{10}\vartheta_{00}$ be the $*$-canonical factorization. Note that ϑ_e is outer and $*$-outer for a.e. e_* because $\operatorname{Rank} \vartheta = \partial$ for a.e. e_* .

Since θ is $*$-outer, the minors of θ of order ∂ have no common inner divisor (see e.g. [6], Theorem on Outer Function). And therefore the minors of ϑ of order $\partial - 1$ have no common inner divisor, i.e. $\mu_{\vartheta_{00}} \le 1$. By the same reason $\det \vartheta_{00}$ and ω are mutually prime for a.e. e_* .

Hence we have proved the lemma for the case $\partial_* = \partial + 1$ ($\partial_* \overset{def}{=} \dim E_*$) . We complete the proof of the lemma by induction on $\partial_* - \partial$.

By the inductive assumption a.e. subspace $E \subset e_*^\perp$ of dimension ∂ has the following properties: 1) $\mu_{(P_{\mathbb{F}}\vartheta_{10})_i} \le 1$; 2) $\det(P_{\mathbb{F}}\vartheta_{10})$ and $\omega \cdot \det \vartheta_{00}$ are mutually prime. Let $\varphi \overset{def}{=} (P_{\mathbb{F}}\vartheta_{10})_e \vartheta_{00}$ and $\varphi = \varphi_i \varphi_e$ be the canonical factorization, then $\det \varphi_i = \det \vartheta_{00}$. Let us find the canonical factorization of $P_{\mathbb{F}}\theta$. Since $\mathbb{F} \subset e_*^\perp$, $P_{\mathbb{F}} P_{e_*^\perp} = P_{\mathbb{F}}$ and hence

$$P_{\mathbb{F}}\theta = P_{\mathbb{F}}\vartheta = (P_{\mathbb{F}}\vartheta_{10})\cdot\vartheta_{00}\vartheta_e = (P_{\mathbb{F}}\vartheta_{10})_i\,\varphi\,\vartheta_e = [(P_{\mathbb{F}}\vartheta_{10})_i\varphi_i][\varphi_e\vartheta_e].$$

Therefore $(P_{\mathbb{F}}\theta)_i = (P_{\mathbb{F}}\vartheta_{10})_i\,\varphi_i$ and according to the Corollary 5.2 we have

$$\mu_{(P_{\mathbb{F}}\theta)_i} = \max\{\mu_{(P_{\mathbb{F}}\vartheta_{10})_i}, \mu_{\varphi_i}\} \le 1 ,$$

Because $\mu_{\varphi_i} = \mu_{\vartheta_{00}} \leq 1$ and $\mu_{(P_F \vartheta_{10})_i} \leq 1$ by the inductive assumption. In other words the minors of $P_F \theta$ of the order $\partial - 1$ have no common inner divisor, i.e. the statement 1) of the lemma is verified. The second assertion follows from the same formula for $P_F \theta$: $\det(P_F\theta)_i = \det(P_F \vartheta_{10})_i \cdot \det \varphi_i = \det(P_F \vartheta_{10})_i \cdot \det \vartheta_{00}$ and therefore by the inductive assumption $\det(P_F \theta)$ and ω have no common inner factor. ▨

5.4. LEMMA. <u>For every outer</u> $*$-<u>inner function</u> Ω <u>in</u> $H^\infty(\mathbb{C}^n \to \mathbb{C}^m)$ <u>there exists a unique</u> (<u>up to a right constant unitary factor</u>) <u>inner</u> $*$-<u>outer function</u> θ <u>in</u> $H^\infty(\mathbb{C}^{n-m} \to \mathbb{C}^n)$ <u>such that</u>

$$\Omega\theta = 0 \tag{1}$$

and

$$\theta\theta^* + \Omega^*\Omega = I \;. \tag{2}$$

<u>And conversely, for every inner</u> $*$-<u>outer</u> θ <u>there exists a unique</u>(<u>up to a left constant unitary factor</u>) <u>outer</u> $*$-<u>inner</u> Ω <u>such that (1) and (2) hold</u>.

<u>For the minors of</u> θ <u>and</u> Ω <u>of maximal order the following relation holds</u>:

$$M_i(\Omega) = \alpha \varepsilon_i M^{i'}(\theta)$$

<u>where</u> α <u>is a unimodular constant</u>.

REMARK. Note that the first part of the lemma is in essence a reformulation of the corollary to Theorem [15] in Helson's book [3] . A statement very close to the needed one can also be found in [10] (Theorem 1). However we need a bit more, the uniqueness of θ and the formula for its minors. So we give here a complete proof.

PROOF. Let Ω be given and put $\Omega = (\Phi, \Psi)$ where Φ denotes the first m columns of Ω and Ψ is the rest. Then θ will be the inner part of the matrix

$$\begin{pmatrix} -\Phi^{ad} & \Psi \\ \det\Phi\cdot I \end{pmatrix}.$$

It is clear that (1) is valid because

$$(\Phi,\Psi)\begin{pmatrix} -\Phi^{ad} & \Psi \\ \det\Phi\cdot I \end{pmatrix} = -(\det\Phi)\cdot\Psi + \Psi(\det\Phi) = 0 \ .$$

Note that (1) together with condition on the sizes of Ω and θ is equivalent to (2). Indeed, if (1) is valid $\theta\theta^*$ and $\Omega^*\Omega$ are orthogonal projections with mutually ortogonal ranges. If $\mathrm{Rank}\,\Omega + \mathrm{Rank}\,\theta = n$ then (2) holds. And conversely, (2) implies

$$\Omega\theta = \Omega(\theta\theta^* + \Omega^*\Omega)\theta = (\Omega\theta)(\theta^*\theta) + (\Omega\Omega^*)(\Omega\theta) = 2\Omega\theta$$

hence $\Omega\theta = 0$.

Let us prove now the uniqueness of θ . Let $\Omega\theta_1 = 0$ and $\Omega\theta_2 = 0$. Then

$$\theta_1\theta_1^* = I - \Omega^*\Omega = \theta_2\theta_2^*$$

and since the θ_i are $*$-outer, we have

$$\theta_1(L^2_n \ominus H^2_n) = \mathrm{clos}\,\theta_1\theta_1^*(L^2_n \ominus H^2_n) = \mathrm{clos}\,\theta_2\theta_2^*(L^2_n \ominus H^2_n) = \theta_2(L^2_n \ominus H^2_n)$$

i.e. $\mathcal{U} \overset{def}{=\!=} \theta_1^*\theta_2$ is a unitary constant. Finally

$$\theta_2 = \theta_2\theta_2^*\theta_2 = \theta_1\theta_1^*\theta_2 = \theta_1\mathcal{U}.$$

This proves the uniqueness of θ .

To prove the equality for the minors note first that (1) and (2) mean that the matrix (θ, Ω^*) is unitary. Using the formula for minors of the inverse matrix (see e.g. [2])

$$M^{\mathbf{j}}_{\mathbf{i}}(A^{-1}) = \frac{\varepsilon_{\mathbf{i}}\,\varepsilon_{\mathbf{j}}\,M^{\mathbf{i}'}_{\mathbf{j}'}(A)}{\det A}$$

we get

$$M_{\mathbf{i}}(\Omega) = M_{\mathbf{i}}^{\{n-m+1,\, n-m+2,\dots,\, n\}}\left(\begin{pmatrix}\theta^*\\ \Omega\end{pmatrix}\right) =$$

$$= \varepsilon_{\mathbf{i}}\,\alpha\, M^{\mathbf{i}'}_{\{1,2,\dots,\,n-m\}}((\theta,\Omega^*)) = \varepsilon_{\mathbf{i}}\,\alpha\, M^{\mathbf{i}'}(\theta)\ ,$$

where $\alpha = \varepsilon_{\mathbf{j}}(\det(\theta,\Omega^*))^{-1}$ is a unimodular function. Since

$$\alpha = \frac{\varepsilon_{\mathbf{i}} M_{\mathbf{i}}(\Omega)}{M^{\mathbf{i}}(\theta)}$$

and the $M^{\mathbf{i}'}(\theta)$ have no common inner factor, α belongs to the Smirnoff class and therefore α is a an inner function. However the minors $M_{\mathbf{i}}(\Omega)$ also have no common inner factor and therefore α is a constant. ▨

The relation between minors of θ and Ω allows us for a given Ω to choose such a θ (which will be denoted by $\hat{\Omega}$) that the equality

$$\det\begin{pmatrix}\Omega\\ A\end{pmatrix} = \det(A\hat{\Omega})$$

holds for any $(n-m)\times m$-matrix A . Indeed,

$$\det\begin{pmatrix}\Omega\\ A\end{pmatrix} = \sum_{\mathbf{i}} \varepsilon_{\mathbf{i}'} M_{\mathbf{i}'}(\Omega) M_{\mathbf{i}}(A) = \alpha \sum_{\mathbf{i}} M^{\mathbf{i}}(\theta) M_{\mathbf{i}}(A) = \alpha \det(A\theta)\,,$$

and it is enough to put $\hat{\Omega} = \theta\mathcal{U}$, where $\mathcal{U}$ is a unitary constant with $\det \mathcal{U} = \bar{\alpha}$.

Analogously, for a given θ we choose $\hat{\theta}$ such that

$$\det(\theta, A) = \det(\hat{\theta} A)$$

for any $n\times m$-matrix A .

6. UPPER ESTIMATE FOR AN OPERATOR WITHOUT C_{10} PART.

6.1. LEMMA. <u>If</u> $\tau = \partial_*$ <u>then</u> $\mu_T \leqslant \max\{\mu_0, \mu_{1a}, \varepsilon_*\}$.

PROOF. If T is the zero operator on the zero space then both sides of the inequality are zeros. So we exclude this case and prove the inequality $\mu_T \leq \max\{\mu_0, \mu_{1a}, 1\}$.

Under the condition $\tau = \partial_*$ we have the following factorization of θ: $\theta = \theta_{00}\theta_{11}\theta_{01}$. According to Lemma 5.3 there exists a decomposition of E into the orthogonal sum $E = H^2_\tau \oplus H^2_{\partial-\tau}$ such that for the corresponding decomposition $\theta = (\vartheta, \vartheta')$ we have:
1) $\mu_{\vartheta_i} \leq 1$; 2) $\det\vartheta$ and $\det\theta_{00}$ have no common inner divisor. Let $\varphi \overset{def}{=\!=} \theta_{00}\theta_{11}\vartheta$ and let $\varphi = \varphi_{*e}\cdot\varphi_{*i}$ be the $*$-canonical decomposition. Corollary 5.2 implies

$$\mu_{\varphi_{*i}} = \max\{\mu_{\theta_{00}}, \mu_{\theta_i}\} \leq \max\{\mu_0, 1\}.$$

Let $\Gamma = \hat{\theta}_{01}$. The matrix $\Gamma = \begin{pmatrix}\gamma \\ \gamma'\end{pmatrix}$ maps $H^2_{\partial-\tau}$ into $E = H^2_\tau \oplus H^2_{\partial-\tau}$. Put $\Delta_{11} = (I - \theta_{11}^*\theta_{11})^{1/2}$ and

$$Y = \begin{pmatrix} \varphi_{*e} & 0 & 0 \\ \Delta_{11}\vartheta\varphi_{*i}^* & I & 0 \\ \gamma^*\varphi_{*i}^* & 0 & I \end{pmatrix},$$

Y acts from $L^2_\tau \oplus \operatorname{clos}\Delta_{11}L^2_\tau \oplus L^2_{\partial-\tau}$ to itself.

For a c.n.u. contraction T with characteristic function θ we choose the following functional model:

$$\mathcal{H} = \begin{pmatrix} L^2_\tau \\ \operatorname{clos}\Delta_{11}L^2_\tau \\ L^2_{\partial-\tau} \end{pmatrix}, \qquad \pi = \begin{pmatrix} \theta \\ \Delta_{11}\theta_{01} \\ \Gamma^* \end{pmatrix}, \qquad \pi^* = \begin{pmatrix} I \\ 0 \\ 0 \end{pmatrix}$$

i.e. $M = P_{\mathcal{K}}\, z\,|\,\mathcal{K}$ and $\mathcal{K} = \mathcal{H} \ominus [\pi H^2_\partial \oplus \pi^*(L^2_\tau \ominus H^2_\tau)]$ (for details see [5]).

Let us take the operator $M_1 = P_{\mathcal{K}_1}\, z\,|\,\mathcal{K}_1$ on the space $\mathcal{K}_1 = (H^2_\tau \ominus \varphi_{*i}H^2_\tau) \oplus \operatorname{clos}\Delta_{11}L^2_\tau \oplus (L^2_{\partial-\tau} \ominus H^2_{\partial-\tau})$ and verify that the operator X, $X \overset{def}{=\!=} P_{\mathcal{K}}Y\,|\,\mathcal{K}_1$ has dense range and intertwines M with M_1; i.e. $MX = XM_1$.

The following two inclusions

$$Y\begin{pmatrix} \varphi_{*i}H^2_\tau \\ 0 \\ H^2_{\partial-\tau} \end{pmatrix} = \begin{pmatrix} \varphi \\ \Delta_{11}\vartheta \\ \gamma^* \end{pmatrix} H^2_\tau + \begin{pmatrix} 0 \\ 0 \\ H^2_{\partial-\tau} \end{pmatrix} =$$

$$=\pi(H^2_{\tau}\oplus 0)+\pi\Gamma H^2_{\partial-\tau}\subset \pi H^2_{\partial}$$

and

$$Y\left(\mathcal{K}_1\oplus\begin{pmatrix}\varphi_{*i}H^2_{\tau}\\ 0\\ H^2_{\partial-\tau}\end{pmatrix}\right)=Y\begin{pmatrix}H^2_{\tau}\\ \operatorname{clos}\Delta_{11}L^2_{\tau}\\ L^2_{\partial-\tau}\end{pmatrix}\subset\begin{pmatrix}H^2_{\tau}\\ \operatorname{clos}\Delta_{11}L^2_{\tau}\\ L^2_{\partial-\tau}\end{pmatrix}=\mathcal{K}\oplus\pi H^2_{\partial}$$

imply the intertwining property: $MX=XM_1$. Further

$$\operatorname{clos}Y\begin{pmatrix}H^2_{\tau}\\ \operatorname{clos}\Delta_{11}L^2_{\tau}\\ L^2_{\partial-\tau}\end{pmatrix}=\begin{pmatrix}\operatorname{clos}\varphi_{*e}H^2_{\tau}\\ \operatorname{clos}\Delta_{11}L^2_{\tau}\\ L^2_{\partial-\tau}\end{pmatrix}=\mathcal{K}\oplus\pi H^2_{\partial}$$

and hence

$$\operatorname{clos}X\mathcal{K}_1=\operatorname{clos}P_{\mathcal{K}}Y\left(\mathcal{K}_1\oplus\begin{pmatrix}\varphi_{*i}H^2_{\tau}\\ 0\\ H^2_{\partial-\tau}\end{pmatrix}\right)=\operatorname{clos}P_{\mathcal{K}}(\mathcal{K}\oplus\pi H^2_{\partial})=\mathcal{K}\ .$$

Therefore the operator $I\oplus X$ intertwines $\mathcal{U}_a\oplus M$ with $\mathcal{U}_a\oplus M_1$ and has dense range. This implies the following estimate for the multiplicity: $\mu_{\mathcal{U}_a\oplus T}\leq\mu_{\mathcal{U}_a\oplus M_1}$. However M_1 is the direct sum of a C_0 contraction with characteristic function φ_{*i} , a unitary operator with local spectral multiplicity $\mu_1(\zeta)=\operatorname{rank}\Delta_{11}(\zeta)$ and a backward shift $S^*_{\partial-\tau}$. The main theorem of [13] supplies us with the multiplicity of such a direct sum:

$$\mu_{\mathcal{U}_a\oplus M_1}=\max\{\mu_{\varphi_{*i}},\mu_{\mathcal{U}_a\oplus\mathcal{U}_1},\varepsilon_*\},$$

i.e.

$$\mu_{\mathcal{U}_a\oplus T}\leq\max\{\mu_0,\mu_{1a},1\}\ .$$

This finishes the proof of the lemma.

Lemma 6.1. together with Lemma 4.2 give us the multiplicity of operators, whose characteristic function θ satisfies the condition $\partial_*=\tau$. This condition means that T , in a sense, has no part of the unilateral shift type. Now we show that if T has such a part then its multiplicity should be added to the resulting formula.

7. COMPLEMENTATION LEMMA.

7.1. LEMMA. <u>For a given matrix-function</u> θ <u>in</u> $H^\infty(\mathbb{C}^m \to \mathbb{C}^n)$ $(n \geq m)$ <u>the following are equivalent.</u>

1) <u>there exist</u> $\Omega \in H^\infty(\mathbb{C}^n \to \mathbb{C}^m)$ <u>and a scalar outer function</u> δ <u>such that</u> $\Omega\theta = \delta I$;

2) <u>there exists</u> $\theta' \in H^\infty(\mathbb{C}^{n-m} \to \mathbb{C}^n)$ <u>such that the</u> $n \times n$-<u>matrix-function</u> (θ, θ') <u>is outer</u>;

3) <u>there exists a family</u> $\{f_i\}$ <u>of scalar</u> H^∞-<u>functions such that</u> $\sum_i f_i M^i(\theta)$ <u>is an outer function.</u>

PROOF. Note that Ω in 1) has to be outer, becouse

$$\operatorname{clos} \Omega H^2_n \supset \operatorname{clos} \Omega \theta H^2_m = \operatorname{clos} \delta H^2_m = H^2_m \ .$$

Let $\Omega = \Omega_{*e} \Omega_{*i}$ be $*$-canonical factorization and put $\theta' = \hat{\Omega}_{*i}$. Then

$$\det(\theta, \theta') = \det(\Omega_{*i}\theta) = \det(\Omega\theta) / \det\Omega_{*e} = \delta^n / \det\Omega_{*e} \ ,$$

i.e. $\det(\theta, \theta')$ is outer. So 1) implies 2).

It is clear that 2) implies 3), it is enough to choose $f_i = \varepsilon_i M_i(\theta')$. So we have to prove 3)$\Rightarrow$1) only. We shall prove this by induction on m.

For $m=1$ statements 1) and 3) are identical in this case $\Omega_i = f_i$. Fix m and suppose 3)$\Rightarrow$1) is true for an arbitrary $\theta \in H^\infty(\mathbb{C}^m \to \mathbb{C}^n)$ and prove the implication for $\theta \in H^\infty(\mathbb{C}^{m+1} \to \mathbb{C}^n)$. Let θ_k denote the k-th column of θ and ϑ be the matrix θ from which the first column is deleted, i.e. $\theta = (\theta_1, \vartheta)$. Clearly assertion 3) is valid for ϑ and therefore by inductive assumption there exist a matrix-function ω and a scalar outer function δ such that $\omega\vartheta = \delta I$.

Let $\{e_k\}$ be the standard basis vectors in $\mathbb{C}^n$. Put $\varphi_k \overset{def}{=} \sum_i f_i M^i(e_k), \vartheta)$ and let φ be the row with entries φ_k. Then

$$\varphi\theta_1 = \sum_k \varphi_k \theta_1^k = \sum_i f_i \sum_k \theta_1^k M^i(e_k, \vartheta) =$$

$$= \sum_i f_i M^i(\theta_1, \vartheta) = \sum_i f_i M^i(\theta) \overset{def}{=\!=} \delta_1 \quad ,$$

where δ_1 is an outer function. Further $\varphi\vartheta = 0$ because for $k>1$ $M^i(\theta_k, \vartheta) = 0$, and therefore

$$\varphi\theta_k = \sum_i f_i M^i(\theta_k, \vartheta) = 0 .$$

Finally put

$$\Omega = \begin{pmatrix} \delta\varphi \\ \omega(\delta_1 I - \theta_1\varphi) \end{pmatrix} .$$

Verifying the equality $\Omega\theta = \delta\delta_1 I$ we complete the proof:

$$\Omega\theta = \begin{pmatrix} \delta\varphi \\ \omega(\delta_1 I - \theta_1\varphi) \end{pmatrix} (\theta_1, \vartheta) =$$

$$= \begin{pmatrix} \delta\delta_1 & 0 \\ \omega\theta_1(\delta_1 I - \varphi\theta_1) & \omega\delta_1\vartheta \end{pmatrix} = \delta\delta_1 I . \quad ⊘$$

7.2. COROLLARY. Let $\theta \in H^\infty(\mathbb{C}^m \to \mathbb{C}^n)$ be an inner *-outer function and $T = M_\theta$. Then the assertions 1)-3) of Lemma 7 are equivalent to each of the following two assertions:

4) T is quasisimilar to the unilateral shift of multiplicity $n-m$;

5) $\mu_T = n - m$.

PROOF. The equivalence 4)$\Longleftrightarrow$1) is just the result of Lemma 1 in [15]. The equivalence 5)$\Longleftrightarrow$2) is almost obvious, namely, projections of columns of the matrix θ' are a cyclic set. ⊘

8. UPPER ESTIMATE IN THE GENERAL CASE.

8.1. LEMMA. $\mu_T \leqslant \partial_* - r + \max\{\mu_0, \mu_{1a}, \varepsilon, \varepsilon_*\}$.

PROOF. If $\mu_0 = \mu_{1a} = \varepsilon = \varepsilon_* = 0$ then T is a c.n.u. contraction with inner $*$-outer characteristic function, which satisfies condition 3) of Lemma 7.1. The Corollary 7.2 gives us $\mu_T = \partial_* - r$ So we may assume that $\max\{\mu_0, \mu_{1a}, \varepsilon, \varepsilon_*\} \geqslant 1$.

Let us choose, by using Lemma 5.3, a subspace $\mathbb{F} \subset E_*$ of dimension r such that in the decomposition

$$\Theta_{10} = \begin{pmatrix} \vartheta^\perp \\ \vartheta \end{pmatrix} \qquad \text{in} \quad E_* = \begin{pmatrix} \mathbb{F}^\perp \\ \mathbb{F} \end{pmatrix}$$

minors of ϑ of the order $r-1$ have no common inner divisor (i.e. $\mu_{\vartheta_i} = 1$), as well as $\det \vartheta$ and $\det \Theta_{00}$.

As in the proof of Lemma 6.1 we work in the model representation

$$\mathcal{H} = \begin{pmatrix} L^2(E_*) \\ \operatorname{clos} \Delta_{11} L^2_r \\ L^2_{\partial - r} \end{pmatrix}, \quad \pi = \begin{pmatrix} \Theta \\ \Delta_{11}\Theta_{01} \\ \Gamma^* \end{pmatrix}, \quad \pi_* = \begin{pmatrix} I \\ 0 \\ 0 \end{pmatrix}$$

where $\Gamma = \hat{\Theta}_{01}$.

Let us take a subspace $\operatorname{clos} P_{\mathcal{K}} \begin{pmatrix} \mathbb{F}^\perp \\ 0 \\ 0 \end{pmatrix}$ of dimension not greater than $\partial_* - r$ and verify that the compression of M_θ to the subspace

$$\mathcal{K}' \overset{\text{def}}{=\!=} \left\{ M_\theta^n P_{\mathcal{K}} \begin{pmatrix} \mathbb{F}^\perp \\ 0 \\ 0 \end{pmatrix} : n \geqslant 0 \right\}^\perp$$

has multiplicity not greater than $\max\{\mu_0, \mu_{1a}, 1\}$. Then the lemma will be proved.

Denote $T' = P_{\mathcal{K}'} M_\theta | \mathcal{K}'$ and calculate the characteristic function θ' of c.n.u. part of T' .

Since

$$\operatorname{clos}\left\{ M_\Theta^n P_{\mathcal{K}} \begin{pmatrix} \mathbb{F}^\perp \\ 0 \\ 0 \end{pmatrix} : n \geqslant 0 \right\} \oplus \pi H^2(E) =$$

$$= \operatorname{span}\left\{ z^n \begin{pmatrix} F^\perp \\ 0 \\ 0 \end{pmatrix}, \begin{pmatrix} \Theta \\ \Delta_{11}\Theta_{01} \\ \Gamma^* \end{pmatrix} H^2(E) \right\} = \begin{pmatrix} H^2(F^\perp) \\ \operatorname{clos} \begin{pmatrix} \vartheta\Theta_{00}\Theta_{11}\Theta_{01} \\ \Delta_{11}\Theta_{01} \\ \Gamma^* \end{pmatrix} H^2(E) \end{pmatrix} ,$$

we have

$$\mathcal{K}' = H_{\mathcal{U}_a} \oplus \begin{pmatrix} H^2(F) \\ \operatorname{clos} \Delta_{11} L^2_r \\ L^2_{\partial - r} \end{pmatrix} \ominus \operatorname{clos} \begin{pmatrix} \vartheta\Theta_{00}\Theta_{11}\Theta_{01} \\ \Delta_{11}\Theta_{01} \\ \Gamma^* \end{pmatrix} H^2(E) .$$

Let us construct contractive outer functions $A \in H^\infty(\mathbb{C}^r \to \mathbb{C}^r)$ and $B \in H^\infty(E \to E)$ that satisfy the following equalities

$$A^*A = \Delta_{11}^2 + \Theta_{11}^* \Theta_{00}^* \vartheta^* \vartheta \Theta_{00} \Theta_{11}$$

$$B^*B = \Theta_{01}^* A^* A \Theta_{01} + \Gamma\Gamma^* .$$

Wiener theorem (see e.g. [3] or [7]) allows to do this because

$$\log\det(\Delta_{11}^2 + \Theta_{11}^*\Theta_{00}^*\vartheta^*\vartheta\Theta_{00}\Theta_{11}) \geqslant \log\det\Theta_{11}^*\Theta_{11} + \log\det\vartheta^*\vartheta \in L^1$$

and

$$\log\det(\Theta_{01}^* A^* A\Theta_{01} + \Gamma\Gamma^*) = \log\det A^*A \in L^1 .$$

Let us denote further

$$\Phi_1 \overset{\text{def}}{=\!=} \vartheta\Theta_{00}\Theta_{11}A^{-1} \in H^\infty(\mathbb{C}^r \to F), \qquad \Phi_2 \overset{\text{def}}{=\!=} \Delta_{11}A^{-1} \in L^\infty(\mathbb{C}^r \to \mathbb{C}^r)$$

$$\Psi_1 \overset{\text{def}}{=\!=} A\Theta_{01}B^{-1} \in H^\infty(E \to \mathbb{C}^r), \qquad \Psi_2 \overset{\text{def}}{=\!=} \Gamma^* B^{-1} \in L^\infty(E \to \mathbb{C}^{\partial - r}) .$$

Then we have

$$\begin{pmatrix} \vartheta\Theta_{00}\Theta_{11} \\ \Delta_{11} \end{pmatrix} = \begin{pmatrix} \Phi_1 \\ \Phi_2 \end{pmatrix} A , \qquad \Phi_1^*\Phi_1 + \Phi_2^*\Phi_2 = I ,$$

$$\begin{pmatrix} A\theta_{01} \\ \Gamma^* \end{pmatrix} = \begin{pmatrix} \Psi_1 \\ \Psi_2 \end{pmatrix} B, \qquad \Psi_1^*\Psi_1 + \Psi_2^*\Psi_2 = I$$

and hence

$$\operatorname{clos}\begin{pmatrix} \vartheta\theta_{00}\ \theta_{11}\theta_{01} \\ \Delta_{11}\theta_{01} \\ \Gamma^* \end{pmatrix} H^2(E) = \operatorname{clos}\begin{pmatrix} \Phi_1 & 0 \\ \Phi_2 & 0 \\ 0 & I \end{pmatrix}\begin{pmatrix} A\theta_{01} \\ \Gamma^* \end{pmatrix} H^2(E) =$$

$$= \begin{pmatrix} \Phi_1 & 0 \\ \Phi_2 & 0 \\ 0 & I \end{pmatrix} \operatorname{clos}\begin{pmatrix} \Psi_1 \\ \Psi_2 \end{pmatrix} BH^2(E) = \begin{pmatrix} \Phi_1\Psi_1 \\ \Phi_2\Psi_1 \\ \Psi_2 \end{pmatrix} H^2(E) .$$

Therefore we have a model representation for the c.n.u. part of T':

$$\mathcal{H}' = \begin{pmatrix} H^2(F) \\ \operatorname{clos}\Delta_{11}L^2_r \\ L^2_{\partial - r} \end{pmatrix}, \qquad \pi' = \begin{pmatrix} \Phi_1\Psi_1 \\ \Phi_2\Psi_1 \\ \Psi_2 \end{pmatrix}, \qquad \pi'_* = \begin{pmatrix} I \\ 0 \\ 0 \end{pmatrix}$$

and the characteristic function θ' of T' is equal to the pure part of $(\pi'_*)^*\pi' = \Phi_1\Psi_1$.

Let k be the rank of constant unitary part of $\Phi_1\Psi_1$. Since $\Phi_1\Psi_1 \in H^\infty(E \to F)$, $\dim E = \partial$, $\dim F = r$ and $\operatorname{Rank}\Phi_1\Psi_1 = r$, we have

$$\theta' \in H^\infty(\mathbb{C}^{\partial-k} \to \mathbb{C}^{r-k}) \qquad \text{and} \quad r' \overset{\text{def}}{=\!=} \operatorname{Rank}\theta' = r-k$$

i.e.

$$\partial' = \partial - k \ , \quad \partial'_* = r-k \ , \quad r' = r-k .$$

Applying Lemma 6.1 to T' we get

$$\mu_{T'} \leqslant \max\{\mu'_0, \mu'_{1a}, \varepsilon'_*\} .$$

Let us verify that $\mu'_{1a} = \mu_{1a}$ and $\mu'_0 \leqslant \max\{\mu_0, 1\}$.

Note first that parameters μ_0 and $\mu_1(\zeta)$ are the same for a function and its pure part, so we can regard $\Phi_1 \Psi_1$ instead of θ'.

Let us check that Ψ_1 is outer $*$-inner. Indeed, since $\Psi_1 B = A\theta_{01}$ and $A\theta_{01}$ is outer, Ψ_1 is outer. On the other hand

$$0 \leq I - \Psi_2 \Psi_2^* = \Gamma^*\Gamma - \Psi_2 \Psi_2^* = \Psi_2 (BB^* - I)\Psi_2^* \leq 0$$

i.e. $\Psi_2 \Psi_2^* = I$ and since $\Psi_1^* \Psi_1$ is equal to $I - \Psi_2^* \Psi_2$, it is an orthoprojection of rank $\partial - \operatorname{rank} \Psi_2 = \partial - (\partial - r) = r$. Hence Ψ_1 is a coisometry, i.e. Ψ_1 is $*$-inner. So μ_0' and μ_1' depend on Φ_1 only.

$$\mu_1'(\zeta) = \operatorname{rank}(I - \Phi_1^*(\zeta)\Phi_1(\zeta)) = \operatorname{rank} \Phi_2^*(\zeta)\Phi_2(\zeta) =$$

$$= \operatorname{rank} A^*(\zeta)\Phi_2^*(\zeta)\Phi_2(\zeta)A(\zeta) = \operatorname{rank} \Delta_{11}^2(\zeta) = \mu_1(\zeta) .$$

Now turn to μ_0'. It is determined by the inner part of Φ_1.

$$(\Phi_1)_i = (\Phi_1 A)_i = (\vartheta\,\theta_{00}\,\theta_{11})_i = (\vartheta\,\theta_{00})_i = \vartheta_i\,\theta_{00}' ,$$

where θ_{00}' is the inner factor of $\vartheta_e\,\theta_{00}$. For θ_{00}' we have $\mu_{\theta_{00}'} = \mu_{\theta_{00}} = \mu_0$ and $\det\theta_{00}' = \det\theta_{00}$, and since ϑ is chosen so that $\mu_{\vartheta_i} \leq 1$ and $\det\vartheta$ and $\det\theta_{00}$ have no common inner factor, Corollary 5.2 implies that equality

$$\mu_0' = \mu_{\vartheta_i \theta_{00}'} = \max\{\mu_{\vartheta_i}, \mu_{\theta_{00}'}\} \leq \max\{1, \mu_0\} .$$

Therefore $\mu_{T'} \leq \max\{\mu_0, \mu_{1a}, 1\}$. This finishes the proof of the lemma. ▨

9. LOWER ESTIMATE IN THE GENERAL CASE.

9.1. LEMMA $\mu_{\theta_{10}\,\theta_{00}} \geq \partial_* - r + \mu_0$.

PROOF. The columns $\{X_1, \ldots, X_\mu\}$ of a matrix-function X form a cyclic set for the model operator M with an inner characteristic function θ iff the matrix (θ, X) is outer. Indeed

$$\operatorname{span}\{M^n X_i : n \geq 0,\ 1 \leq i \leq \mu\} = H^2(E_*) \ominus \theta H^2(E) \Longleftrightarrow \operatorname{span}\{M^n X_i\} \oplus$$

$$\oplus\, \theta H^2(E) = H^2(E_*) \Longleftrightarrow \operatorname{span}\{z^n X_i, z^n \theta_j\} = H^2(E_*) \Longleftrightarrow (\theta, X) \quad \text{is outer.}$$

In our case $\theta = \theta_{10}\theta_{00}$ and hence

$$(\theta, X) = (\theta_{10}, X)\begin{pmatrix} \theta_{00} & 0 \\ 0 & I_\mu \end{pmatrix}.$$

Here the index μ marks the size of the identity matrix.

Let the columns of X form a cyclic set. Since $\partial_* \times (r+\mu)$-matrix (θ, X) is outer, its minors of order ∂_* have no common inner divisor (see [6], Theorem on Outer Function). Therefore the minors of $\begin{pmatrix} \theta_{00} & 0 \\ 0 & I_\mu \end{pmatrix}$ of order ∂_*, hence the minors of θ_{00} of order $\partial_* - \mu$, have no common inner divisor. This implies (cf. e.g. [1], [6]) the an estimate $\mu_0 \leq r - (\partial_* - \mu)$. Hence $\mu \geq \partial_* - r + \mu_0$. ▨

9.2. LEMMA $\mu_T \geq \partial_* - r + \max\{\mu_0, \mu_{1a}, \varepsilon, \varepsilon_*\}$.

PROOF. Corollary 7.2. asserts that $\mu_{\theta_{10}} = \partial_* - r$ iff $\varepsilon = 0$. Therefore $\mu_T \geq \partial_* - r + \varepsilon$. Lemma 9.1 together with Lemma 4.1 imply the estimate

$$\mu_T \geq \mu_{\theta_{10}\theta_{00}} \geq \partial_* - r + \mu_0 .$$

Let $\Omega = \hat{\theta}_{10}$. Define an operator X,

$$X : \begin{pmatrix} H^2_{\partial_*} \\ \operatorname{clos} \Delta_{11} L^2_r \\ L^2_{\partial - r} \end{pmatrix} \longrightarrow \begin{pmatrix} H^2_{\partial_* - r} \\ \operatorname{clos} \Delta_{11} L^2_r \end{pmatrix}$$

by the following matrix

$$X = \begin{pmatrix} \Omega & 0 & 0 \\ -\Delta_{11}\theta_{11}^{ad}\theta_{00}^*\theta_{10}^* & \det \theta_{11} & 0 \end{pmatrix},$$

where θ_{11}^{ad} is the matrix adjoint to θ_{11}, i.e. $\theta_{11}^{ad}\theta_{11} = \theta_{11}\theta_{11}^{ad} = \det\theta_{11}\cdot I$.

As earlier we regard M as the compression of multiplication by z to the subspace

$$\mathcal{K} = \begin{pmatrix} H^2_{\partial_*} \\ \operatorname{clos}\Delta_{11}L^2_{r} \\ L^2_{\partial-r} \end{pmatrix} \ominus \pi H^2_{\partial}, \quad \pi = \begin{pmatrix} \theta \\ \Delta_{11}\theta_{01} \\ \Gamma^* \end{pmatrix},$$

where $\Gamma = \hat{\theta}_{01}$. It is clear that X has dense range, $X\pi = 0$ and $X \mid \mathcal{K}$ intertwines the operator M with multiplication by z on the space $H^2_{\partial_*-r} \oplus \operatorname{clos}\Delta_{11}L^2_{r}$, i.e. with the operator $S_{(\partial_*-r)} \oplus \mathcal{U}_{\mu_1(\zeta)}$, where $\mathcal{U}_{\mu_1(\zeta)}$ is the unitary operator with absolutely continuous spectrum of local multiplicity $\mu_1(\zeta)$. Therefore the multiplicity of $M \oplus \mathcal{U}_a$ majorizes the multiplicity of $S_{(\partial_*-r)} \oplus \mathcal{U}_{\mu_1(\zeta)} \oplus \mathcal{U}_a$, which is equal to $\partial_* - r + \mu_{1a}$ (cf. [13]).

And finally we have to prove that $\mu_T \geq \partial_* - r + \varepsilon_*$. So we can suppose that $\mu_0 = \mu_{1a} = \varepsilon = 0$. Since $\varepsilon = 0$. Lemmas 5.4 and 7.1 imply the existence of a $*$-inner function Ω' such that the matrix-function $\begin{pmatrix}\Omega \\ \Omega'\end{pmatrix}$ is outer. Put

$$Y = \begin{pmatrix} \Omega & 0 \\ \Omega' & 0 \\ 0 & I \end{pmatrix} : \begin{pmatrix} H^2_{\partial_*} \\ L^2_{\partial-r} \end{pmatrix} \longrightarrow \begin{pmatrix} H^2_{\partial_*-r} \\ H^2_{r} \\ L^2_{\partial-r} \end{pmatrix}.$$

In our case, since $\mu_0 = \mu_1 = 0$, we have $\theta = \theta_{10}\theta_{01}$ and we can suppose that $\varepsilon_* = 1$, i.e. $\partial > r$ and the factor θ_{01} is nontrivial. We shall use the model space

$$\mathcal{K} = \begin{pmatrix} H^2_{\partial_*} \\ L^2_{\partial-r} \end{pmatrix} \ominus \begin{pmatrix} \theta \\ \Gamma^* \end{pmatrix} H^2_{\partial}.$$

Let A be an outer $\partial\times\partial$ -matrix-function defined by the equality

$$A^*A = \Gamma\Gamma^* + \theta^*\Omega'^*\Omega'\theta$$

(such a function exists because $\log\det(\Gamma\Gamma^* + \theta^*\Omega'^*\Omega'\theta) =$
$= \log\det \theta_{10}^*\Omega'^*\Omega'\theta_{10} = \log\det\theta_{10}^*(\Omega^*, \Omega'^*)\begin{pmatrix}\Omega\\ \Omega'\end{pmatrix}\theta_{10} \in L^1$)
and put

$$\Phi = \Omega'\theta A^{-1}, \quad \Psi = \Gamma^* A^{-1}$$

Then

$$\operatorname{clos}\begin{pmatrix}\Omega'\theta\\ \Gamma^*\end{pmatrix} H^2_\partial = \begin{pmatrix}\Phi\\ \Psi\end{pmatrix} H^2_\partial$$

and therefore

$$Y\begin{pmatrix}\theta\\ \Gamma^*\end{pmatrix} H^2_\partial \subset \begin{pmatrix} 0 \\ \begin{pmatrix}\Phi\\ \Psi\end{pmatrix} H^2_\partial \end{pmatrix}.$$

Let

$$\mathcal{K}_1 \overset{def}{=\!=} \begin{pmatrix} H^2_{\partial_*-r} \\ \begin{pmatrix} H^2_r \\ L^2_{\partial-r}\end{pmatrix} \ominus \begin{pmatrix}\Phi\\ \Psi\end{pmatrix} H^2_\partial \end{pmatrix},$$

$$M_1 \overset{def}{=\!=} P_{\mathcal{K}_1} z \,|\, \mathcal{K}_1 = S_{(\partial_*-r)} \oplus M_\Phi ,$$

$$Z \overset{def}{=\!=} P_{\mathcal{K}_1} Y \,|\, \mathcal{K} .$$

Since Y has dense range, $\operatorname{clos} Z\mathcal{K} = \mathcal{K}_1$. Therefore the intertwining relation $ZM = M_1 Z$ implies the estimate $\mu_T \geqslant \mu_{M_1}$. But the theorem in [13] gives us $\mu_{M_1} = \partial_* - r + \mu_\Phi \geqslant \partial_* - r + 1$ because Φ is not a unitary constant and M_Φ is not the zero operator on the zero space.

So we get $\mu_T \geqslant \partial_* - r + \varepsilon_*$ and finish the proof of the Lemma. ▨

Lemma 9.2. together with Lemma 8.1 and the note on a singular unitary part (see Section 3) imply the equality we wanted to prove

$$\mu_T = \max\{\mu_{u_0}, \partial_* - r + \max(\mu_0, \mu_{1a}, \varepsilon, \varepsilon_*)\}.$$

ACKNOWLEDGEMENTS. I wish to thank to S.V.Kislyakov and B.M.Solomyak for reading the manuscript.

REFERENCES

1. Bercovici, H. and Voiculescu, D.: Tensor operations on characteristic functions of C_0 contractions, Acta Sci.Math.(Szeged), 39, fasc.3/4 (1977), 205-231.
2. Gantmacher, F.P. Theory of Matrices, Nauka, Moscow, 1967 (Russian).
3. Helson, H.: Lectures on Invariant Subspaces, Academic Press, N.Y. - London, 1964.
4. Kriete III, T.L.: Similarity of canonical models, Bull.AMS, 76, N 2 (1970), 326-330.
5. Nikolskii, N.K. and Vasyunin, V.I.: A unified approach to function models, and the transcription problem, preprint E-5-86, LOMI, Leningrad, 1986.
6. Nikol'skii, N.K.: Treatise on the Shift Operator, Springer - Verlag, 1986.
7. Sz.-Nagy, B. and Foias, C.: Harmonic analysis of operators on Hilbert space, North Holland - Akadémiai Kiadó, Amsterdam - Budapest, 1970.
8. Sz.-Nagy, B. and Foias, C.: Modèle de Jordan pour une classe d'opérateurs de l'espace de Hilbert, Acta Sci.Math. (Szeged), 31, fasc. 1/2 (1970), 91-115.
9. Sz.-Nagy, B. and Foias, C.: Jordan model for contractions of class C.o, ibid., 36, fasc.3/4 (1974), 305-322.
10. Uchiyama, M.: Contractions and unilateral shifts, ibid., 46, fasc. 1/4 (1983), 345-356.
11. Uchiyama, M.: Contractions with (σ, c) defect operators, J. Operator Theory, 12, N 2 (1984), 221-233.
12. Vasyunin, V.I. and Makarov, N.G.: On quasi-similarity of model contractions with non-equal defects. Zap.naučn.Sem.Le-

ningrad.Otdel.Mat.Inst.Steklov.(LOMI), 149 (1986), 24-37 (Russian).

13. Vasyunin, V.I. and Karayev, M.T.: The multiplicity of some contractions, Zap.naučn.Sem.Leningrad.Otdel.Mat.Inst.Steklov. (LOMI), 157 (1987), 23-29 (Russian).

14. Wu, P.Y.: Jordan model for weak contractions, Acta Sci. Math. (Szeged), 40, fasc. 1/2 (1978), 189-196.

15. Wu, P.Y.: When is a contraction quasi-similar to an isometry?, ibid., 44, fasc. 1/2 (1982), 151-155.

16. Wu, P.Y.: Which $C_{\cdot 0}$ contraction is quasi-similar to its Jordan model?, ibid,, 47, fasc. 3/4 (1984), 449-455.

17. Wu, P.Y.: Toward a characterization of reflexive contractions, J.Operator Theory, 13, N 1 (1985), 73-86.

Operator Theory:
Advances and Applications, Vol. 42

RIEMANN SURFACE MODELS OF TOEPLITZ OPERATORS

D.V.Yakubovich

CONTENTS

0. Introduction. 307
Chapter 1. The Similarity theorem. 313
1. The Similarity theorem in H^2 313
2. Toeplitz operators on other spaces. 321
3. Proofs of Theorems 1.5 and 2.4. 324
4. Some properties of E^p and of the related function spaces. 338
Chapter 2. Commutant. 343
5. The definition of the reduced ultraspectrum and commutativity of the commutant. 343
6. A complete description of commutant. 356
Chapter 3. Classification theorems for Toeplitz operators
7. 369
Chapter 4. Invariant subspaces of multiplication operators on Riemann surfaces. 375
8. Test atlases, standard subspaces and the function $\varphi_{\alpha,w}$. 375
9. The two-sheeted surfaces. 392

10. A description of invariant subspaces of M_{π} 396
<u>Chapter 5</u>. Invariant and hyperinvariant subspaces of Toeplitz operators. 406
11. 406
References 413

We consider a Toeplitz operator $T_{\mathbb{F}}$ whose symbol $\mathbb{F}$ is continuous on the unit circle, analytic in the unit disc except for a finite number of poles and has non-negative winding number with respect to any point in $\mathbb{C}$. It is shown that $T_{\mathbb{F}}$ is similar to the multiplication operator by the projection Π on a certain function space $H^2_{\mathbb{F}}(\sigma_*)$ on the so-called ultraspectrum σ_* of $T_{\mathbb{F}}$ which is a Riemann surface projecting into the spectrum $\sigma(T_{\mathbb{F}})$. The ultraspectrum σ_* and the space $H^2_{\mathbb{F}}(\sigma_*)$ are evaluated explicitly. As a consequence of this theorem, we describe the **commutant**, invariant and hyperinvariant subspaces of $T_{\mathbb{F}}$ for a wide class of symbols. In particular, suppose that $\mathbb{F}$ has a finite number of selfintersections and some additional conditions hold. We show that in this case the commutant of $T_{\mathbb{F}}$ is isomorphic to $H^\infty(\sigma_c)$ for a Riemann surface σ_c and hence is abelian.

Some criteria for similarity of $T_{\mathbb{F}_1}$ and $T_{\mathbb{F}_2}$ are obtained if $\mathbb{F}_1$ and $\mathbb{F}_2$ are symbols of the above type. We prove also an analogue of Fon Neumann's inequality for $T_{\mathbb{F}}$.

In Chapter 4 a characterization of the invariant subspaces of multiplication by the projection Π on a Riemann surface (R, Π) subject to some geometric condition is given.

0. INTRODUCTION.

As a rule, the classical spectral theory is based on the possibility of "splitting the spectrum" of an operator which takes place for some classes of operators. This is the case if the operator is spectral in the sense of Dunford or decomposable in the sense of Colojoara - Foias. It is known that many operators whose spectrum is contained in a curve satisfy these properties. However many important classes of operators (weighted shifts, singular integral operators, Wiener - Hopf operators, etc.) have in general "heavy" spectra that can not split in the classic sense. A characteristic feature of these operators is that they (or their adjoints) have eigenspaces analytically depending on spectral parametre if it ranges over some subdomains of $\mathbb{C}$. M.J.Cowen and R.G.Douglas [11, 12] (as well as other authors) have studied some vast classes of such operators from the point of view of differential geometry. They obtained complete unitary invariants for operators of these classes. Here a much more concrete class of Toeplitz operators which has been described briefly is studied. We use this class to demonstrate some topological effects which can occur in studying spectral properties of operators with "heavy" spectra.

Instead of unitary equivalence, we deal with similarity. Recall that (bounded) operators T_1, T_2 acting on Banach spaces are called similar if there exists a linear isomorphism U such that $UT_1 = T_2 U$. We say that U implements similarity between T_1 and T_2 .

To describe our results, some notions are needed. Let $\mathbb{T}$ denote the unit circle and $\mathbb{D}$ the unit disc. For $F \in L^\infty(\mathbb{T})$, the Toeplitz operator T_F is defined on H^2 by

$$T_F x = P_+(F \cdot x), \quad x \in H^2,$$

where P_+ is the orthogonal projection from $L^2(\mathbb{T})$ onto H^2 . We call F the symbol of T_F . It will be assumed that basic properties of Toeplitz operators with continuous symbols are familiar to the reader, see for instance [14, 23]. We say that the symbol F has positive winding if $F \in C(\mathbb{T})$ and $wind_F(\lambda) \geq 0$ for every $\lambda \in \mathbb{C} \setminus F(\mathbb{T})$. Here $wind_F(\lambda)$ is the winding number of the curve $\zeta \to F(\zeta)$, $\zeta \in \mathbb{T}$, with respect to λ .

In Chapter 1 we define the ultraspectrum $\sigma_* = \sigma_*(T_F)$ of T_F which is a Riemann surface with possible branching. It must be noticed here that the term "Riemann surface" will always mean a pair (R, Π) where R is an abstract Riemann surface (not necessarily connected) and $\Pi : R \to \mathbb{C}$ is a branched covering map called projection. In Chapter 1, $\mathbb{F}$ is supposed to satisfy the above conditions as well as some other requirements that are less restrictive (see 1.1 below). Put $M_\Pi f = \Pi \cdot f$ for any function f on σ_*. We prove the following Similarity theorem: $T_{\mathbb{F}}$ is similar to the multiplication operator M_Π acting on a certain Hilbert space $H^2_{\mathbb{F}}(\sigma_*)$ of analytic functions on σ_*. We give also a modification of this theorem for Toeplitz operators on Banach spaces of analytic functions in simply connected domains with piecewise smooth boundary.

The author's work was inspired by the papers of D.Clark, J.Morrel and D.Wang [5, 6, 8,37] where some similarity models were cnnstructed for symbols "with loops" (not necessarily of positive winding).The main geometric condition on $\mathbb{F}$ in these papers was that the closures of every two bounded components of $\mathbb{C} \setminus \mathbb{F}(\mathbb{T})$ intersect in a finite number of points (for example, $\mathbb{F}(\mathbb{T})$ can be the curve in Fig.1.) A model for $\mathbb{F}(z) = az + bz^{-1}$ passing an ellipse had been suggested by P.Duren [15] .

Figure 1

We use many ideas of D.Clark. Note however that the statement of D.Clark concerning a similarity theorem for $T_{\mathbb{F}}$ in the case $\mathbb{F}(\mathbb{T}) = \tau(\mathbb{T})$ for some analytic on $\overline{\mathbb{D}}$ function τ [6] is shown to be incorrect [32].

We want to outline the idea that leads to our result on similarity. Let us agree first that the pairing on H^2 is bilinear and is given by

$$\langle x, y \rangle = \sum_{n \geq 0} \hat{x}(n) \hat{y}(n) = \int_{\mathbb{T}} x(\zeta)\, y(\zeta^{-1})\, \zeta^{-1} d\zeta .$$

We define the adjoint operator according to this pairing. Then $\sigma(T^*) = \sigma(T)$ for every T. For Toeplitz operators, $T^*_{\mathbb{F}} = T_{\mathbb{F}^T}$, where $\mathbb{F}^T(z) = \mathbb{F}(z^{-1})$.

Recall that for every symbol $\mathbb{F}$ with positive winding and every $\lambda \notin \mathbb{F}(\mathbb{T})$ we have

$$\dim \operatorname{Ker}(T_F^* - \lambda I) = \operatorname{wind}_F(\lambda) \geq 0 .$$

Hence one can find open subsets $\mathcal{O}_n$ of $\sigma(T_F)$ and analytic families $\{h_\lambda\}_{\lambda \in \mathcal{O}_n}$ of eigenvectors of T_F^* satisfying $T_F^* h_\lambda \equiv \equiv \lambda h_\lambda$. D.Clark's and J.Morrel's formulae in [8] provide a concrete example of such families. Moreover, looking at these formulas we arrive at a "complete" analytic family $\{h_\delta\}_{\delta \in \sigma_*}$ of eigenvectors of T_F^* where σ_* is a set naturally endowed with the Riemann surface structure. The projection Π of σ_* to $\mathbb{C}$ is defined by

$$T_F^* h_\delta = \Pi(\delta) h_\delta , \quad \delta \in \sigma_* . \tag{0.1}$$

By the "completeness" of $\{h_\delta\}$ we mean here the fact that the vectors $\{h_\delta : \delta \in \Pi^{-1}\{\lambda\}\}$ for every $\lambda \in \operatorname{int} \sigma(T_F)$ (except for a finite set of λ 's) span $\operatorname{Ker}(T_F^* - \lambda I)$.

Define an operator U acting on H^2 by the rule

$$(Ux)(\delta) = \langle x, h_\delta \rangle, \quad \delta \in \sigma_* . \tag{0.2}$$

The image UH^2 consists of analytic functions on σ_* . These functions must satisfy some linear relations on the fibres $\Pi^{-1}\{\lambda\}$, $\lambda \in \sigma(T_F)$, since the eigenvectors $\{h_\delta : \delta \in \Pi^{-1}\{\lambda\}\}$ are in general linearly dependent. The space $H_F^2(\sigma_*)$ is exactly the space of all functions u in the Hardy class $H^2(\sigma_*)$ such that $u | \Pi^{-1}\{\lambda\}$ satisfies the above relations for every $\lambda \in \sigma(T_F)$. To prove the Similarity theorem, we show that U is an isomorphism of H^2 onto $H_F^2(\sigma_*)$ such that

$$UT_F = M_\Pi U . \tag{0.3}$$

The last equality is an immediate consequence of (0.1) and (0.2). To prove that U is an isomorphism, we reduce the equation $Ux = v$ to a sequence of Cousin problems that have unique solutions.

The choice of the ultraspectrum σ_* seems to be rather arbitrary. However σ_* contains an essential part σ_{**} which is defined in Chapter 2 and is called the "reduced ultraspectrum". The reduced ultraspectrum is a closed and open subset of the ultraspectrum and so is a Riemann surface too. In Chapter 3, we prove the following theorem: Suppose that $\mathbb{F}, \mathbb{F}_1$ satisfy the assumptions of Chapter 1 and each curve $\zeta \to \mathbb{F}(\zeta)$, $\zeta \to \mathbb{F}_1(\zeta)$ $(\zeta \in \mathbb{T})$ has a finite number of self-intersections. Then $T_{\mathbb{F}}$ and $T_{\mathbb{F}_1}$ are similar iff 1) the reduced ultraspectra $\sigma_{**}(T_{\mathbb{F}})$ and $\sigma_{**}(T_{\mathbb{F}_1})$ are isomorphic; 2) there exists β such that $\beta, 1/\beta \in H^{\infty}(\sigma_{**}(T_{\mathbb{F}}))$ and the mapping $u \to \beta \cdot u$ transfers the fiber linear relations in $H^2_{\mathbb{F}}(\sigma_{**}(T_{\mathbb{F}}))$ into the fiber linear relations in $H^2_{\mathbb{F}_1}(\sigma_{**}(T_{\mathbb{F}_1}))$.

In Chapter 2, we describe a relation between σ_{**} and the commutant $\{T_{\mathbb{F}}\}' = \{A : T_{\mathbb{F}}A = AT_{\mathbb{F}}\}$ of $T_{\mathbb{F}}$. We introduce the "commutant surface" σ_c of $T_{\mathbb{F}}$ and find a full description of $\{T_{\mathbb{F}}\}'$ in terms of U, σ_{**} and σ_c. In particular, $\{T_{\mathbb{F}}\}'$ is shown to be isomorphic to $H^{\infty}(\sigma_c)$ as a Banach algebra and hence is abelian. Note that this fact depends crucially on the assumption that the curve $\zeta \to \mathbb{F}(\zeta)$, $\zeta \in \mathbb{T}$, has a finite number of self-intersections which is made in Chapter 2. It is well-known that for the symbol $\mathbb{F}(z) = z^n$, $n > 1$, which does not satisfy this assumption the commutant $\{T_{\mathbb{F}}\}'$ is not abelian.

In Chapter 5, we give a description of the lattice $\mathrm{Lat}(T_{\mathbb{F}})$ of invariant subspaces of $T_{\mathbb{F}}$ (i.e., such closed subspaces $\mathcal{E}$ of H^2 that $T_{\mathbb{F}}\mathcal{E} \subset \mathcal{E}$). Here it is supposed additionally that $\mathbb{F} \in C^2(\mathbb{T})$, $1/\mathbb{F}'$ is bounded near $\mathbb{T}$ and that

(H) every bounded component Ω_i of $\mathbb{C} \setminus \mathbb{F}(\mathbb{T})$ has a common boundary arc with a component Ω_j of $\mathbb{C} \setminus \mathbb{F}(\mathbb{T})$ such that $\mathrm{wind}_{\mathbb{F}}(\Omega_j) < \mathrm{wind}_{\mathbb{F}}(\Omega_i)$ where $\mathrm{wind}_{\mathbb{F}}(\Omega_i) \stackrel{def}{=} \mathrm{wind}_{\mathbb{F}}(\lambda)$ for every $\lambda \in \Omega_i$ ("absence of holes").

Due to the Similarity theorem, the problem reduces to the description of the invariant subspaces of M_{Π} in $H^2_{\mathbb{F}}(\sigma_*)$. In Chapter 4, a more general problem of the characterization of the invariant subspaces of multiplication operators on Riemann surfaces is discussed. For Toeplitz operators, we show that every $\mathcal{E} \in \mathrm{Lat}(T_{\mathbb{F}})$ can be characterized in terms of "traces" of the functions $U x$, $x \in \mathcal{E}$, on "test" sets having a simple structure, and of the decay of the functions $U x$ "above" the self-intersection points of $F(\mathbb{T})$. We give also a description of hyperinvariant subspaces of T_F (i.e. invariant under every operator in $\{T_F\}'$) in terms of σ_{**} , σ_* and U .

Note that for a wide class of Riemann surfaces R , subspaces of $H^2(R)$ invariant under multiplication by every function in $H^\infty(R)$ have been described by Sarason [43], Hasumi [45] and Voichick [46]. The problem discussed in Chapter 4 has however a different nature.

Note also that existence of invariant subspaces and decomposability properties of Toeplitz operators (for much wider classes of symbols than here) were studied for instance in works of Dyn'kin [17] and Peller [24], [25].

The techniques of Chapters 4 and 5 come back to works by A.L.Vol'berg and B.M.Solomyak [29-31] and to Wermer's paper [38]. However, we arrive at some new difficulties and leave some questions unanswered.

For rational symbols F , many results of this paper are stated also in [39] without proofs.

I wish to express my deep gratitude to my supervisor prof. Nikol'skiĭ N.K. and to A.A.Borichev, B.M.Solom'yak, V.V.Peller and S.V.Kisliakov who had made many suggestions and improvements.

SOME NOTATIONS:

$\hat{\mathbb{C}} = \mathbb{C} \cup \{\infty\}$ - extended complex plane;

$\mathbb{D} = \{z : |z| < 1\}$, $\mathbb{T} = \{z : |z| = 1\}$,

$\mathcal{H}(\Omega)$ - the space of analytic functions on a domain Ω in $\hat{\mathbb{C}}$ or on a Riemann surface;

$\{T\}' = \{A : AT = TA\}$ - commutant of an operator T ;

$\operatorname{Lat}(T)$ - the lattice of invariant subspaces of T ;

$C_A = C(\mathbb{T}) \cap H^\infty$ - the disk-algebra;

$\mathcal{L}(\mathcal{H})$ - the algebra of bounded operators on a Banach space $\mathcal{H}$;

χ_A - the characteristic function of a set A .

CHAPTER 1. THE SIMILARITY THEOREM

1. THE SIMILARITY THEOREM IN H^2.

1.1. THE CONDITIONS ON SYMBOL. It will be always assumed that:

(i) $F = P/Q$, $P \in C_A$, Q is a polynomial with roots in $\mathbb{D}$, and F is locally univalent on $\mathbb{D}$ near $\mathbb{T}$. The latter means that every point ζ of the unit circle $\mathbb{T}$ has a neighbourhood V such that the function $F|\mathbb{D} \cap V$ is univalent.

(ii) The curve $t \to F(e^{it})$, $0 \le t \le 2\pi$, has non-negative winding number $\operatorname{wind}_F(\lambda)$ about each point λ in $\mathbb{C} \setminus F(\mathbb{T})$.

(iii) $F(\Xi) \cap F(\mathbb{T}) = \emptyset$, where $\Xi = \{z \in \mathbb{D} : F'(z) = 0\}$.

(iv) Each point $\lambda \in \mathbb{C}$ has at least $\deg Q$ preimages in the unit disc $\mathbb{D}$ under F (counted with their multiplicities).

It follows from the argument principle that the number of preimages of λ equals $\operatorname{wind}_F(\lambda) + \deg Q$ if $\lambda \in \mathbb{C} \setminus F(\mathbb{T})$. Hence the condition (iv) must be verified only for $\lambda \in F(\mathbb{T})$. The condition (ii) implies (iv) if $F(\mathbb{T})$ is a curve "in a generic position" (that is, if $F(\mathbb{T})$ has only a finite number of simple transversal self-intersections and has no overlapping arcs).

It will be more convenient first to define the ultraspectrum σ_* and then to give an expression for the function $\sigma \mapsto h_\sigma$ on it.

1.2. THE DEFINITION OF THE ULTRASPECTRUM $\sigma_*(T_F)$. Denote

$$k = \deg Q + 1 .$$

Consider first an auxiliary metric space Φ_0, whose points are subsets

$$\delta = \{d_1, \dots, d_k\}$$

of $\overline{\mathbb{D}} \setminus F^{-1}(F(\Xi))$ having k elements such that

$$F(d_1) = F(d_2) = \dots = F(d_k) \stackrel{def}{=\!=} \Pi(\delta).$$

The metric in Φ_0 is determined by

$$\rho(\delta, \delta') = \min_{\tau} \sum_i | d_i - d'_{\tau(i)} |,$$

where τ ranges over all rearrangements of $\{1, \dots, k\}$. The function Π will be referred to as the projection. Put

$$\sigma_*^0 = \{ \delta \in \Phi_0 : \delta \subset \mathbb{D} \}.$$

One can easily see that each point δ of σ_*^0 has a neighbourhood $\mathcal{V}(\delta)$ such that $\Pi | \mathcal{V}(\delta)$ is a homeomorphism onto an open set in $\mathbb{C}$. The maps $\Pi | \mathcal{V}(\delta)$ define a Riemann surface structure on σ_*^0 .

It is well known that $\sigma(T_F) = F(\mathbb{T}) \cup \{\lambda : \operatorname{wind}_F(\lambda) \neq 0\}$ for every $F \in C(\mathbb{T})$. Hence (ii) yields

$$\sigma(T_F) \setminus F(\Xi \cup \mathbb{T}) \subset \Pi(\sigma_*^0) \subset \sigma(T_F) \setminus F(\Xi). \tag{1.1}$$

The ultraspectrum $\sigma_* = \sigma_*(T_F)$ of T_F is obtained from σ_*^0 by adding a finite number of points projecting into $F(\Xi) \cap \sigma(T_F)$. Suppose that ζ_i are points of $F(\Xi) \cap \sigma(T_F)$. Choose discs V_i with centres at ζ_i such that $V_i \cap F(\Xi \cup \mathbb{T}) = \emptyset$. Let W_{ij} be the components of $\Pi^{-1}(V_i)$; they are n_{ij}- sheeted coverings of $V_i \setminus \{\zeta_i\}$ for some $n_{ij} \in \mathbb{N}$. We form the surface σ_* by adding some new point δ_{ij} to every W_{ij} . Set

$$\Pi(\delta_{ij}) = \zeta_i$$

and

$$\eta_{ij}(\delta) = (\Pi(\delta) - \zeta_i)^{1/n_{ij}}$$

Then the functions η_{ij} on W_{ij} are by definition local parameter near δ_{ij} (every continuous branch of η_{ij} on W_{ij} can be taken). Together with the functions $\Pi \mid U(\delta)$ these functions define a Riemann surface structure on σ_*. The projection Π is a branched covering map and the points δ_{ij} are the only possible branched points of Π.

To every δ_{ij} there corresponds an unordered collection $\alpha(\delta_{ij}) = (d_1, \dots, d_k)$ of points in $\mathbb{D}$ which can be obtained from sets $\delta \in \sigma_*^o$ by passing to the limit along an arbitrary curve in σ_* ending at δ_{ij}. For instance, suppose that $k=2$, $z_1 \in \mathbb{D}$, $\zeta_1 = F(z_1)$, and $F = G^4$ in a neighbourhood of z_1 for some univalent G with $G(z_1) = 0$. Then the sets

$$W_{11} = \{\{d_1, d_2\} : G(d_1) = -G(d_2) \neq 0\},$$

$$W_{12} = \{\{d_1, d_2\} : G(d_1) = \pm i\, G(d_2) \neq 0\}$$

are examples of W_{1j}. One can see that $n_{11} = 2$, $n_{12} = 4$ and $\alpha(\delta_{11}) = \alpha(\delta_{12}) = (0, 0)$.

The set $\Phi \overset{def}{=\!=} \Phi_0 \cup \sigma_*$ is endowed with a topology in a natural way (the base of the topology is formed by open sets in σ_* and open sets in Φ_0). It can easily be seen that the closure $\overline{\sigma_*}$ of σ_* in Φ is compact. If $F(\mathbb{T})$ is a curve in a generic position then $\overline{\sigma_*} = \Phi$ but in other cases Φ may have some isolated points. The symbol $\partial \sigma_*$ will denote the boundary of σ_* in Φ. Every $\delta \in \partial \sigma_*$ has a neighbourhood in $\partial \sigma_*$ which projects homeomorphically onto a Jordan arc in $\mathbb{C}$. Hence $\overline{\sigma_*}$ is a compact Riemann surface with boundary $\partial \sigma_*$. Note that the positive orientation on $\partial \sigma_*$ after projection to $\mathbb{C}$ agrees with the orientation of the curve $t \to F(e^{it})$, $0 \leq t \leq 2\pi$.

It follows from (1.1) and (iv) that

$$\sigma(T_F) \setminus F(\mathbb{T}) \subset \Pi(\sigma_*) \subset \sigma(T_F) \tag{1.2}$$

and

$$\Pi(\overline{\sigma_*}) = \overline{\Pi(\sigma_*)} = \sigma(T_F). \tag{1.3}$$

1.3. EIGENVECTORS OF $T_{\mathbb{F}}^*$. Consider the polynomial

$$\omega(z) = z^{k-1} Q(z^{-1}), \tag{1.4}$$

where Q is the polynomial from (i). Set

$$h_\delta(z) = \omega(z) \prod_{s=1}^{k} (1 - d_s z)^{-1} \in H^2 \tag{1.5}$$

if $\delta = \{d_1, \dots, d_k\} \in \mathfrak{S}_*^\circ$. Let $\lambda = \Pi(\delta)$. Then

$$T_{\mathbb{F}}^* h_\delta - \lambda h_\delta = P_+((F^T(z) - \lambda)\, Q(z^{-1}) z^{k-1} \prod_{s=1}^{k} (1 - d_s z)^{-1}) =$$

$$= P_+ z^{-1} \left[(P^T - \lambda Q^T) \prod_{s=1}^{k} (z^{-1} - d_s)^{-1} \right] = 0$$

since the function in square brackets is analytic on $\hat{\mathbb{C}} \setminus \mathbb{D}$. This yields (0.1).

The formula (1.5) was used in [8] in the case when $\mathbb{F}(\mathbb{T})$ is a simple curve. It is straightforward that the eigenvector h_δ is an analytic vector-function of δ, $\delta \in \mathfrak{S}_*^\circ$. If it is assumed that $(d_1, \dots, d_k) = \alpha(\delta)$ in (1.5) for $\delta \in \mathfrak{S}_* \setminus \mathfrak{S}_*^\circ$, then the vector-function $\delta \to h_\delta$ becomes analytic on the whole ultra-spectrum $\mathfrak{S}_*$ (since it is continuous in δ_{ij}).

1.4. THE SPACES $H^2(\mathfrak{S}_*)$ AND $H^2_{\mathbb{F}}(\mathfrak{S}_*)$. If A is a connected Riemann surface, then the Hardy class $H^2(A)$ is defined as the space of functions u, $u \in \mathcal{H}(A)$, such that $|u|^2$ has a harmonic majorant on A. The norm

$$\|u\|_{H^2(A)} = \inf_{v} |v(\delta)|^{1/2},$$

where δ is a fixed point of A and the infimum is taken over all harmonic majorants v of $|u|^2$, equips $H^2(A)$ with a Hilbert space structure (see [36]). The papers [19, 22] contain a systematic study of classes H^p on Riemann surfaces.

We define the linear space $H^2(\mathfrak{S}_*)$ as the set of functions $u \in \mathcal{H}(\mathfrak{S}_*)$ such that $u | A_i \in H^2(A_i)$ for every component A_i of the surface σ_* . Let $H^2_F(\sigma_*)$ denote the subspace of $H^2(\sigma_*)$ consisting of the functions u such that the equation

$$\begin{aligned}(d_3-d_2)\,u(d_2,d_3,d_4,\ldots,d_{k+1})+(d_2-d_1)\,u(d_1,d_2,d_4,\ldots,d_{k+1})+\\ +(d_1-d_3)\,u(d_1,d_3,d_4,\ldots,d_{k+1})=0\end{aligned} \tag{1.6}$$

holds for every (distinct) $d_1,\ldots,d_{k+1}$ satisfying

$$F(d_1) = F(d_2) = \ldots = F(d_{k+1}) \notin F(\Xi)\,. \tag{1.7}$$

This equation links the values of u on fibres $\Pi^{-1}(\lambda)$ for $\lambda \in$ $\in \operatorname{int}\sigma(T_F)$. If $k=1$, then $H^2_F(\sigma_*) = H^2(\sigma_*)$ by definition. Note that σ_* is isomorphic the unit disc $\mathbb{D}$ in this case.

If $k>1$, then the surface σ_* can have an infinite number of connected components (for example, if $F(\mathbb{T})$ is the curve with infinite number of intersections represented by Figure 2, and $k=2$). However as shown in i. 3.6 below, a finite set of components, say $A_1,\ldots,A_n$, exists with the following properties: if $u\in\mathcal{H}(\sigma_*)$, u satisfies (1.6), and $u\,|\,A_i \in H^2(A_i)$ for $i\le n$, then $u\in H^2_F(\sigma_*)$; moreover, if $u\,|\,A_i \equiv 0$ for $i\le n$, then $u\equiv 0$. Hence one may consider $H^2_F(\sigma_*)$ as a closed subspace of $\bigoplus_{i=1}^{n} H^2(A_i)$, and therefore as a Hilbert space itself.

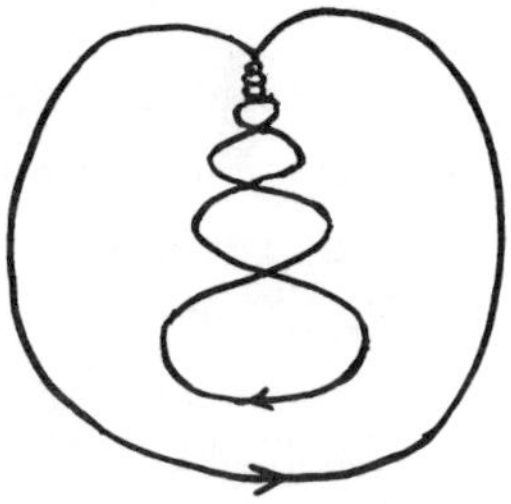

Figure 2

The multiplication operator M_Π is obviously a bounded operator on $H^2_F(\sigma_*)$.

1.5. SIMILARITY THEOREM FOR H^2 . <u>Formula (0.2) defines a linear isomorphism</u> U <u>between</u> H^2 <u>and</u> $H^2_F(\sigma_*)$ <u>such that</u>

$$U T_F U^{-1} = M_\Pi \ .$$

This theorem is proved in Section 3.

1.6. COROLLARY. The eigenvectors of the operator $T_F^* = T_{F^T}$ (here $F^T(z) \overset{def}{=} F(z^{-1})$) are complete in H^2.

To prove this, suppose that $x \in H^2$ and $\langle x, h \rangle = 0$ for every eigenvector h of T_F^*. Then $Ux = 0$, and $x = 0$ due to the theorem.

1.7. COROLLARY. Let $G = \Pi(\sigma_*(T_F))$. Then T_F admits the $H^\infty(G)$-calculus. That is a bounded multiplicative map from $H^\infty(G)$ to $\mathcal{L}(H^2)$ exists which coincides with the natural one on rational functions with poles off $\sigma(T_F)$. In particular,

$$\|\varphi(T_F)\| \leqslant C \max_{\lambda \in \sigma(T_F)} |\varphi(\lambda)| \tag{1.8}$$

for such rational φ.

Indeed, $H^\infty(G)$ - calculus for M_Π can be constructed using the formula

$$\varphi(M_\Pi)(u) \overset{def}{=} (\varphi \circ \Pi) \circ u, \quad u \in H^2_F(\sigma_*).$$

For some other classes of symbols, the inequality (1.8) follows from the papers [5, 7, 17, 25]. In [4], the best constant C is calculated in some cases. It follows from [35] that for some $F \in L^\infty(\mathbb{T})$, the inequality (1.8) does not hold for any C.

1.8. EXAMPLES OF CALCULATING σ_*. If the curve $F(\mathbb{T})$ has a finite number of self-intersections then ultraspectrum σ_* can be computed as follows. First a triangulation τ of $\sigma(T_F)$ must be constructed such that $F(\mathbb{T})$ is contained in the union of the arcs of τ and $F(\Xi) \cap \sigma(T_F)$ is contained in the vertices of τ. Then the pre-image $F^{-1}(\tau)$ of τ is a partition of $\overline{\mathbb{D}}$ which defines some triangulation $\Pi^{-1}(\tau)$ of $\overline{\sigma}_*$. Indeed, there is a one-to-one correspondence between triangles of $\Pi^{-1}(\tau)$ and sets of K triangles of $F^{-1}(\tau)$ which are sticked together by F. So the order of sticking the triangles in $\Pi^{-1}(\tau)$ follows from the combinatorial scheme of the partition $F^{-1}(\tau)$. Let us consider some simple cases.

1) If F has no poles, then $k=1$ and (σ_*, Π) is a Riemann surface of $(F|\mathbb{D})^{-1}$ (or equivalently, the surface $(\mathbb{D}, F)$). The Similarity theorem is obvious in this case.

2) Let $F_0(z)=(z+\frac{1}{2})^2$, let $a\in\mathbb{D}$, $a\neq-\frac{1}{2}$ and $F_0(a)\notin F(\mathbb{T})$. We define

$$F(z) = F_0(z) + \varepsilon(z-a)^{-1},$$

where $\varepsilon\neq 0$ and ε is small enough (see Figure 3a). Then the conditions (i)-(iv) are satisfied and $k=2$. There is a point z_ε near a such that $F'(z_\varepsilon)=0$. Hence $wind_F(F(z_\varepsilon))= =wind_{F_0}(F_0(a))$ for small ε . If $wind_{F_0}(F_0(a))=1$, then $\sigma_*(T_F)$ has two components A and B (Fig.3b). The component A has unique branch point (whose projection ζ_0 is near to $F_0(-\frac{1}{2})=0$). The projection of ∂A is $F(\mathbb{T})$. The component B has one sheet and projects homeomorphically onto the set $\{\lambda: wind_F(\lambda)= =2\}$. It can be proved that the mapping $u\mapsto u|A$ is a linear isomorphism of $H^2_F(\sigma_*)$ onto $H^2(A)$. Therefore T_F is similar to a multiplication operator $M_{\Pi|A}$ on $H^2(A)$. Let φ be the conformal mapping of $\mathbb{D}$ onto A . Then T_F is similar to the analytic Toeplitz operator $T_{\Pi\circ\varphi}$.

If $wind_{F_0}(F_0(a))=2$, then the ultraspectrum $\sigma_*(T_F)$ is connected and has two branch points (Fig.3c). It will be shown in Chapter 3 that T_F is not similar to any analytic Toeplitz operator in this case.

3) Suppose that $F(z)=z^2+az^{-1}$ and $a>0$. If $a<1$, then F satisfies conditions (i), (ii), (iv) (see Fig.4a). The equation $F'(z)=0$ has three solutions z_1, z_2, z_3 which are the cubic roots of $a/2$. Thus (iii) is true iff $1-a\neq 3(a/2)^{2/3}$. If $1-a> >3(a/2)^{2/3}$, then $wind_F(F(z_i))=2$, the ultraspectrum is connected and has three branch points projecting to $F(z_i)$ (Fig.4b). If $(1-a)<3(a/2)^{2/3}$, then $wind_F(F(z_i))=1$. The ultraspectrum splits into three "sheets" A_1, A_2, A_3 (Fig.4c).

There is a distinction between the cases 4b and 4c from the operator theory point of view. That is, put

$$\mathcal{E} = \{ x\in H^2: Ux|A_1=0\}.$$

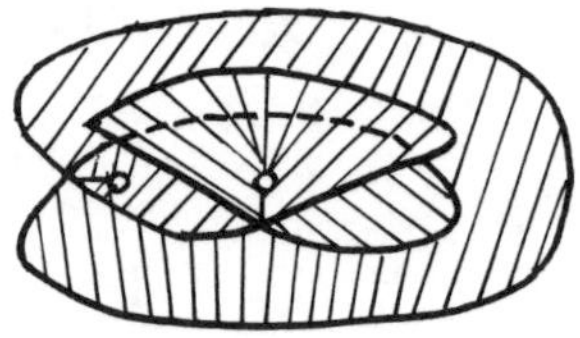

Fig.3c

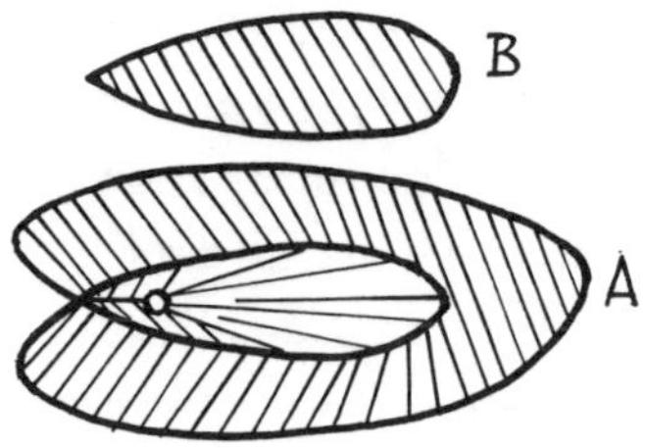

Fig.3b

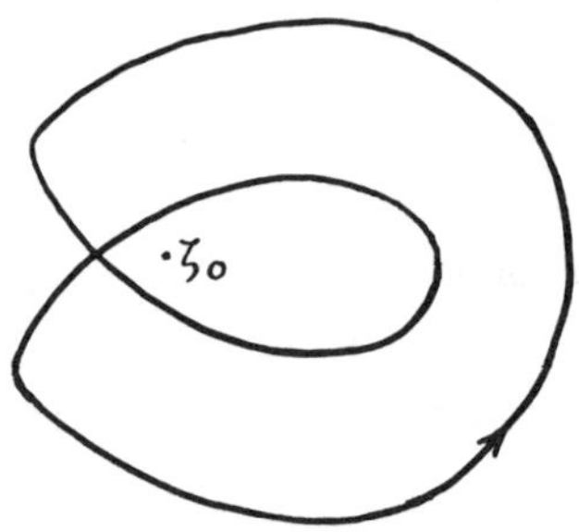

Fig.3a

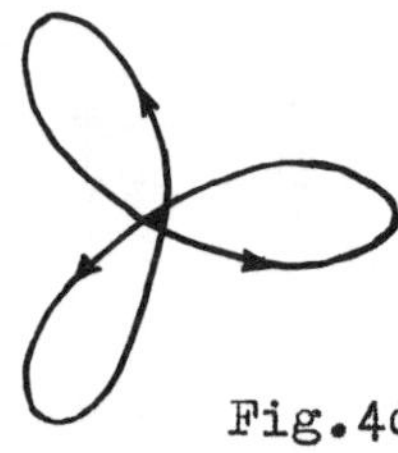

Fig.4d

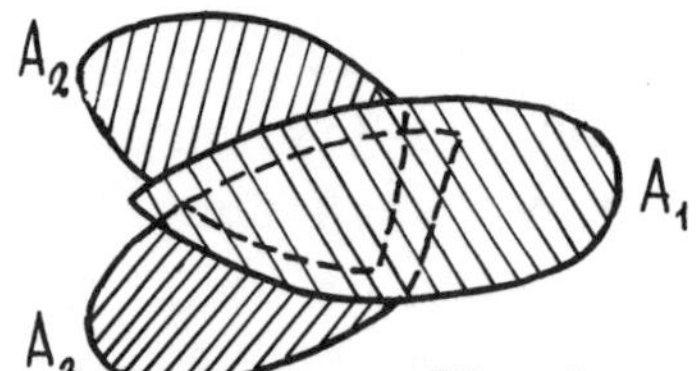

Fig.4c

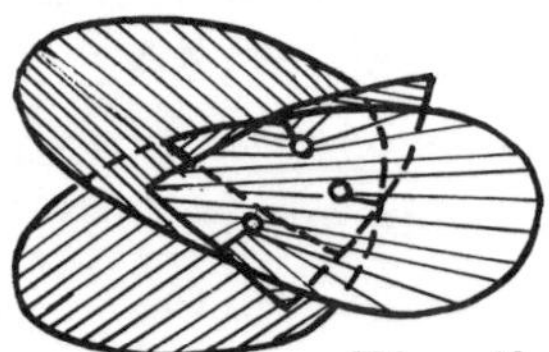

Fig.4b

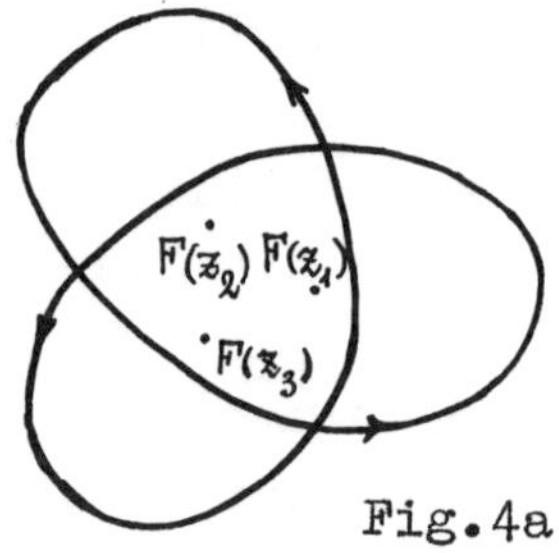

Fig.4a

in the case 3c). Then $\xi \in \mathrm{Lat}(T_F)$, $\xi \neq 0$, and

$$\sigma(T_F \mid \xi) = \mathrm{clos}\,\Pi(A_2 \cup A_3) \neq \sigma(T_F) .$$

It follows easily from the Similarity Theorem 1.5 that $\sigma(T_F \mid \xi) = \sigma(T_F)$ for every $\xi \in \mathrm{Lat}(T_F)$, $\xi \neq 0$ if the ultraspectrum of T_F is connected (as in the case 4b).

If $a = 1$, then the conditions (i)-(iii) are satisfied and (iv) is not (see Fig.4d). This is the case of "symbol with loops" considered by D.Clark [5, 7]. The ultraspectrum σ_* splits into three connected components corresponding to three loops of $F(\mathbb{T})$. It can be seen from the considerations of D.Clark that U implements a similarity between T_F and multiplication by Π on the Smirnov class $E^2(\sigma_*)$. Since $E^2(\sigma_*)$ is smaller than $H^2(\sigma_*)$ in this case, the condition (iv) is essential in the Similarity theorem.

4) Let F satisfy (i)-(iii) and map $\mathbb{T}$ onto a curve γ. Suppose that the correspondence between $\mathbb{T}$ and γ is one-to-one excepting the **self-intersection** points of γ and some closed subarc α of γ which F runs three times (twise in on direction and once in the other). If γ is a "curve with loops" (in particular, γ may be a simple closed curve), then T_F is similar to a direct sum of analytic Toeplitz operators and some normal operator whose spectrum coincides with α (see [7]). The condition (iv) is violated for $\lambda\in\alpha$ in this case. However if γ is a general curve with positive winding and $\alpha\subset \mathrm{int}\,\sigma(T_F)$, then (iv) is valid (one can easily see that correspondent F does exist). A similarity model for T_F is provided by Theorem 1.5 now. Observe that for any non-zero invariant subspace K of T_F and any $f\in K$, $f\not\equiv 0$, the spectrum of $T_F\,|\,K$ contains $\Pi(\{\delta:(Uf)(\delta)\neq 0\})$ and hence can not be equal to an arc.

2. TOEPLITZ OPERATORS ON OTHER SPACES.

We include a variant of the similarity theorem for Toeplitz operators on Banach analytic spaces which was stated in the author's paper [40].

2.1. A CLASS OF SPACES. Let $\mathcal{N}$ be the collection of all bounded domains in $\mathbb{C}$ with piecewise-smooth boundary (not necessarily simply-connected). For $G\in\mathcal{N}$, let $\mathcal{A}(G)$ denote the space of those functions in $\mathcal{H}(G)$ that can be continued analytically into a neighbourhood of $\overline{G}$. A collection $B=\{B(G)\}_{G\in\mathcal{N}}$ of Banach spaces will be called a bundle of Banach spaces if the following is satisfied for every $G, G_1, G_2 \subset \mathfrak{N}$:

1) $B(G)$ is a Banach space of analytic functions on G and $\mathcal{A}(G) \subset B(G) \subsetneq \mathcal{H}(G)$;

2) $x \in B(G)$, $a \in \mathcal{A}(G) \Longrightarrow a \cdot x \in B(G)$;

3) $G_1 \subset G$, $x \in B(G) \Longrightarrow x|G_1 \in B(G_1)$;

4) $G = G_1 \cup G_2$, $x \in \mathcal{H}(G)$, $x|G_i \in B(G_i) \Longrightarrow x \in B(G)$;

5) the invariance under conformal changes of variables: if $G \in \mathfrak{N}$, φ is a conformal mapping defined in a neighbourhood of $\overline{G}$ and $x \in B(G)$, then $x \circ \varphi^{-1} \in B(\varphi(G))$.

As an example one can consider Smirnov classes $E^p(G)$, $1 \leq p \leq \infty$, analytic Lipschitz classes $Lip_\alpha(G)$, $0 < \alpha < \infty$, Bergman spaces of functions summable in the power p with respect to area measure and others.

Let Ω be a connected set in $\mathfrak{N}$ such that $0 \in \Omega$ and $\partial\Omega$ is a simple closed curve.

2.2. THE CONDITIONS ON SYMBOL F AND THE DEFINITION OF T_F. Assume that:

1) F is a meromorphic function on $\overline{\Omega}$ and has no poles on $\partial\Omega$;

2) The curve $\gamma = F|\partial\Omega$ has non-negative winding number with respect to every $\lambda \in \mathbb{C} \setminus F(\partial\Omega)$;

3) $F(\Xi) \cap F(\partial\Omega) = \emptyset$ and $F(0) \notin F(\Xi \cup \partial\Omega)$; here $\Xi = \{ z \in \overline{\Omega} : F'(z) = 0 \}$;

4) γ has a finite number of self-intersections and has no "overlapping arcs";

5) All self-intersections of γ are double and transversal.

Suppose that F is analytic on some neighbourhood V of $\partial\Omega$. Define Toeplitz operator T_F on $B(\Omega)$ by

$$(T_F x)(z) = \frac{1}{2\pi i} \int_{\Gamma_z} \frac{F(\zeta) x(\zeta)}{\zeta - z} d\zeta, \quad x \in B(\Omega), \ z \in \Omega;$$

here $\Gamma_z \subset \Omega$ denotes a positive-oriented contour which is homotopic to $\partial\Omega$ in $V \setminus \{z\}$. The definition is correct and T_F proves to be a bounded operator on $B(\Omega)$ (see [40]).

We note that Toeplitz operators in Banach spaces corresponding to subdomains of $\mathbb{C}$ (including multiply connected domains) have been studied in [1, 20, 27] and in other papers.

2.3. THE ULTRASPECTRUM σ_* AND THE MODEL SPACE $B_F(\sigma_*)$. Let $k-1$ denote the number of poles of F in Ω. The ultraspectrum $\sigma_*(T_F)$ and the projection Π are defined as in i.1.2, with $\mathbb{D}$ replaced by Ω. Choose any open subsets $G_0, \dots, G_k \subset \sigma_*$ with piecewise-smooth boundaries posessing the properties: $\bigcup_0^n G_i = \sigma_*$; $\operatorname{clos} \Pi(G_j) \cap F(\Xi) = \emptyset$ and $\Pi \mid G_j$ is injective for $j \geq 1$; the closure of G_0 in $\overline{\sigma_*}$ is contained in σ_*. The space

$$B(\sigma_*) = \{ u \in \mathcal{H}(\sigma_*) : \|u\|_{B(\sigma_*)} \stackrel{def}{=} \|u \mid G_0\|_{H^\infty(G_0)} +$$

$$+ \sum_{i=1}^{n} \|u \circ (\Pi \mid G_i)^{-1}\|_{B(\Pi(G_i))} < \infty \}$$

is a Banach space since it is easily identified with a closed subspace of $H^\infty(G_0) \oplus \bigoplus_{i=1}^{n} B(\Pi(G_i))$. The norm in $B(\Omega)$ depends on the choice of G_i but it can be deduced from the closed graph theorem that all these norms are equivalent. Let $B_F(\sigma_*)$ denote the subspace of $B(\sigma_*)$ of functions u which satisfy (1.6) for every distinct $d_1, \dots, d_{K+1} \in \Omega \setminus F^{-1}(F(\Xi))$ such that (1.7) holds. The multiplication operator M_Π acts on $B_F(\sigma_*)$ and is bounded there.

The following statement is a variant of Theorem 1.5.

2.4. SIMILARITY THEOREM FOR BANACH SPACES. <u>Suppose that B is a bundle of Banach spaces and Ω is a domain containing the origin such that $\partial\Omega$ is a simple piecewise smooth curve. If F satisfies the conditions 1)-5), then the operator T_F on $B(\Omega)$ is similar to M_Π, acting on $B_F(\sigma_*)$.</u>

Put $\Omega^T = \mathbb{C} \setminus \{ z : z^{-1} \in \overline{\Omega} \}$. Note that the isomorphism U between $B(\Omega)$ and $B_{\mathbb{F}}(\sigma_*)$ which implements the similarity can be defined by formulae (0.2), (1.5) if we agree that

$$\langle x, \varphi \rangle \overset{def}{=\!=} \int_\Gamma \varphi(\zeta) x(\zeta^{-1}) \frac{d\zeta}{\zeta}$$

for $x \in B(\Omega)$ and $\varphi \in \mathcal{A}(\overline{\Omega^T})$, where Γ is a positive oriented contour homotopic to $\partial\Omega^T$ on the domain of φ .

3. PROOFS OF THEOREMS 1.5 AND 2.4.

3.1. THE PROOF OF THEOREM 1.5 will be divided into several steps. Note first that

$$U = VT_{\omega^T} \tag{3.1}$$

where ω is defined by (1.4) and V is an operator from H^2 to $H^2_{\mathbb{F}}(\sigma_*)$ given by

$$(Vx)(\delta) \overset{def}{=\!=} \langle x, \prod_{s=1}^{k} (1-d_1 z)^{-1} \rangle, \quad \delta = \{d_1, \ldots, d_k\} \in \sigma_*^\circ . \tag{3.2}$$

We have

$$(Vx)(\delta) = \langle x, \sum_{s=1}^{k} c_s(\delta)(1-d_s z)^{-1} \rangle = \sum_{s=1}^{k} c_s(\delta)\, x(d_s) , \tag{3.3}$$

where

$$c_s(\delta) = d_s^{k-1} \prod_{j \neq s} (d_s - d_j)^{-1} . \tag{3.4}$$

The formulae (3.2) and (3.3) allow us to define $(Vx)(\delta)$ for every $x \in \mathcal{H}(\mathbb{D})$. If $x \in H^2$, then the function Vx is analytic on σ_*° and bounded in a neighbourhood of $\sigma_* \setminus \sigma_*^\circ$ because of (3.2). Hence $Vx \in \mathcal{H}(\sigma_*)$ for every $x \in H^2$. It follows that $Vx \in \mathcal{H}(\sigma_*)$ for every $x \in \mathcal{H}(\mathbb{D})$ since such x can be approximated by functions in H^2 uniformly on compact subsets of $\mathbb{D}$.

Define the space $\mathcal{H}_{\mathbb{F}}(\sigma_*)$ as the set of those $u \in \mathcal{H}(\sigma_*)$

which satisfy (1.6) for all $d_1,\dots,d_{K+1}$ such that (1.7) holds. The following two lemmas will be useful.

3.2. LEMMA. <u>For</u> $x\in\mathcal{H}(\mathbb{D})$ <u>the inclusion</u> $x\in H^2$ <u>is equivalent to</u> $Vx\in H^2(\sigma_*)$.

3.3. LEMMA. V <u>is a bijection of</u> $\mathcal{H}(\mathbb{D})$ <u>onto</u> $\mathcal{H}_F(\sigma_*)$.

Theorem 1.5 is an easy consequence of these two lemmas. Indeed, for every $d_1,\dots,d_{K+1}\in\mathbb{D}$ such that (1.7) holds we have the following relation between the eigenvectors h_δ :

$$(d_1-d_2)h_{\{d_1,d_2,d_4,\dots,d_{K+1}\}}+(d_2-d_3)h_{\{d_2,d_3,d_4,\dots,d_{K+1}\}}+ \tag{3.5}$$
$$+(d_3-d_1)h_{\{d_1,d_3,d_4,\dots,d_{K+1}\}}=0$$

since it is a multiple of the obvious identity

$$(d_1-d_2)(1-d_3z)+(d_2-d_3)(1-d_1z)+(d_3-d_1)(1-d_2z)=0\ .$$

Hence for every $x\in H^2$ the function $u=Ux$ satisfies (1.6), and the inclusion $UH^2\subset H^2_F(\sigma_*)$ follows. Lemmas 3.2 and 3.3 imply that V is a bijection between H^2 and $H^2_F(\sigma_*)$. Since $\mathrm{wind}_{\omega T}(0)=0$, $T_{\omega T}$ is invertible (see [14]). It follows from (3.1) and the closed graph theorem that U is a linear isomorphism between H^2 and $H^2_F(\sigma_*)$. The interwining property $UT_F=M_\Pi U$ is a consequence of (0.1):

$$(UT_F x)(\delta)=\langle T_F x,h_\delta\rangle=\langle x,T_F^* h_\delta\rangle=$$
$$=\Pi(\delta)\langle x,h_\delta\rangle=(M_\Pi Ux)(\delta)\ .$$

The following lemma will be used several times later.

3.4. LEMMA. <u>Let</u> $\lambda_0\in\mathbb{C}\setminus F(\Xi)$ <u>and let</u> $n=\mathrm{card}\,F^{-1}\{\lambda_0\}$. <u>There exist an open disc</u> W <u>with centre at</u> λ_0 , <u>closed sets</u> $\mathcal{D}_j$ <u>and continuous injective functions</u> $d_j:\mathcal{D}_j\to\overline{\mathbb{D}}\ (1\le j\le n)$ <u>such that</u>:

a) $(\mathbb{F}\circ \mathfrak{d}_j)(\lambda) \equiv \lambda$, $\lambda \in \mathcal{D}_j$, $1 \leq j \leq n$;
b) $\mathbb{F}^{-1}\{\lambda\} = \{\mathfrak{d}_j(\lambda) : \mathcal{D}_j \ni \lambda\}$ for all $\lambda \in W$;
c) $\mathcal{D}_j$ contains W and $\mathcal{D}_j \cap \mathbb{F}(\mathbb{T}) = \emptyset$ if $\mathfrak{d}_j(\lambda_0) \notin \mathbb{T}$;
d) $\mathfrak{d}_j(\partial \mathcal{D}_j \cap W) \subset \mathbb{T}$ if $\mathfrak{d}_j(\lambda_0) \in \mathbb{T}$.

PROOF. Let $\mathbb{F}^{-1}\{\lambda_0\} = \{d_1, \dots, d_n\} \subset \overline{\mathbb{D}}$. The conditions (i), (iv) on $\mathbb{F}$ imply that there are open G_j and closed disjoint domains $\mathcal{Y}_j$, $\mathcal{Y}_j \subset \overline{\mathbb{D}} \setminus \mathbb{F}^{-1}(\mathbb{F}(\mathbb{T}))$, $(1 \leq j \leq n)$ such that $d_j \in G_j \cap \overline{\mathbb{D}} \subset \mathcal{Y}_j$ and $\mathbb{F} | \mathcal{Y}_j$ is injective for all j . Take a disc W with centre at λ_0 such that $W \cap \mathbb{F}(\overline{\mathbb{D}} \setminus \bigcup_j G_j) = \emptyset$. Put $\mathfrak{d}_j = (\mathbb{F} | \mathcal{Y}_j)^{-1}$ and $\mathcal{D}_j = \mathbb{F}(\mathcal{Y}_j)$. If $\lambda \in \partial \mathcal{D}_j \cap W$ and $z = \mathfrak{d}_j(\lambda)$, then $\mathbb{F}(z) = \lambda$, which yields

$$z \in \partial \mathcal{Y}_j \cap (\textstyle\bigcup_i G_i) = \partial \mathcal{Y}_j \cap G_j \subset \mathbb{T} .$$

This proves d). The statements a), b) are obvious and c) is true if W is taken to be sufficiently small. ▨

Let $\{\mathfrak{A}(\Omega)\}_{\Omega \in \mathcal{N}}$ be classes of analytic functions, such that $\mathfrak{A}(\Omega) \subset \mathcal{H}(\Omega)$ for $\Omega \in \mathcal{N}$. Suppose that $\Omega_0 \in \mathcal{N}$ and $f \in \mathcal{H}(\Omega_0)$. **We say that f belongs to $\mathfrak{A}(\Omega)$ locally** ($f \in \mathfrak{A}_{loc}(\Omega)$) if each $\zeta \in \partial\Omega$ has a neighbourhood W such that

$$f | W \cap \Omega \subset \mathfrak{A}(W \cap \Omega) .$$

3.5. PROOF OF LEMMA 3.2. 1) Let $x \in H^2$; we must prove that $Vx \in H^2(\sigma_*)$. Let w be a harmonic majorant of $|x|^2$. Put

$$\ell(\delta) = \sum_{i=1}^{k} w(d_i)$$

if $\delta = (d_1, \dots, d_k) \in \sigma_*^0$. The function ℓ is harmonic on σ_*^0 and can be continued continuously to σ_* . Hence ℓ is harmonic on σ_* . Choose a neighbourhood G of $\sigma_* \setminus \sigma_*^0$ such that $\overline{G} \subset \sigma_*$. Then

$$|(Vx)(\delta)| < A_1 \ell(\delta) \qquad \text{on } \sigma_* \setminus G$$

for some constant A_1 , since the functions c_s are bounded on

$\sigma_* \setminus G$. Thus,

$$|(Vx)(\sigma)| < A_1 \ell(\sigma) + A \qquad \text{on } \sigma_*$$

for some constant A_2 , since $Vx|G$ is bounded.

2) Conversely, let $x \in \mathcal{H}(\mathbb{D})$ and $Vx \in H^2(\sigma_*)$. It suffices to show that x belongs to H^2 locally (the equality $H^2 = H^2_{loc}(\mathbb{D})$ is almost trivial; some hints for its proof are given in 4.9 below). Let $\zeta \in \mathbb{T}$. The conditions (iii) and (iv) imply that $F(\zeta)$ has $k-1$ different pre-images $\zeta_2, \dots, \zeta_k$ in $\mathbb{D}$ under the mapping F . There are branches $d_2, \dots, d_k$ of F^{-1} which are defined in neighbourhood W of $F(\zeta)$ and satisfy the condition $d_i(F(\zeta)) = \zeta_i$ for $i=2,\dots,k$. Choose a neighbourhood V_ζ of ζ such that F is univalent on $V_\zeta \cap \mathbb{D}$ and $F(V_\zeta \cap \mathbb{D}) \subset W$. Define an analytic function $\tau_\zeta : V_\zeta \cap \mathbb{D} \to \sigma_*$ by the formula

$$\tau_\zeta(z) \stackrel{def}{=} \{z, d_2(z), \dots, d_k(z)\} .$$

It follows from (3.3) that

$$x(d_1) = c_1(\tau_\zeta(d_1))^{-1}\Big[(Vx)(\tau_\zeta(d_1)) - \sum_{s=2}^{k} c_s(\tau_\zeta(d_1))\, x(d_s(d_1))\Big]$$

if $d_1 \in V_\zeta \cap \mathbb{D}$. We may assume that W and V_ζ are so small that the sets $d_i(W)$ are contained in $r\mathbb{D}$ for some $r<1$ and the mutual distancies between the sets V_ζ and $d_i(W)$, $i=1,\dots,k-1$, are positive. Then $|c_1 \circ \tau_\zeta|$ is bounded away from zero on V_ζ . Since $c_s \circ \tau_\zeta$ and $x \circ d_s$ belong to $H^\infty(V_\zeta)$ and $(Vx)\circ\tau_\zeta$ belongs to $H^2(V_\zeta)$ because of the assumption, it follows that $x \in H^2(V_\zeta)$. Thus $x \in H^2$ since ζ is arbitrary.∎

3.5. REMARKS ON THE DEFINITION OF $H^2_F(\sigma_*)$. The above proof and Lemma 3.3 imply two assertions which will be used in the subsequent chapters. It is convenient to place them here.

1) Let $u \in \mathcal{H}(\sigma_*)$. It can be deduced from i. 2) of the above proof that $x \in \mathcal{H}(\mathbb{D})$ and $Vx \in H^2_{loc}(\sigma_*)$ imply $x \in H^2$.

Hence it follows from Lemmas 3.2 and 3.3 that

$$\mathcal{H}_F(\sigma_*)\cap H^2_{loc}(\sigma_*)=H^2(\sigma_*).$$

It can be proved directly that $H^2_{loc}(\sigma_*) = H^2(\sigma_*)$ but we do not use this fact.

2) Choose a finite subcovering $\{V_{\zeta_i}\}_{1\le i\le n}$ of the covering $\{V_\zeta\}_{\zeta\in\mathbb{T}}$ of the unit circle $\mathbb{T}$. Every set $\tau_{\zeta_i}(V_{\zeta_i})$ is connected, and so it is contained in some component A_i of σ_*. The proof of Lemma 3.2 and Lemma 3.3 imply that the norm $\|u\|_{H^2(\cup_1^n A_i)}$ on $H^2_F(\sigma_*)$ is equivalent to $\|V^{-1}u\|_{H^2}$. Hence the components $A_1,\dots,A_n$ can be used to define a Hilbert space structure on $H^2_F(\sigma_*)$ (see 1.4).

3.7. OUTLINE OF THE PROOF OF LEMMA 3.3. Let us introduce Riemann surfaces σ_j for $1\le j\le k$ and their projections Π_j onto $\hat{\mathbb{C}}$; they are defined analogously to (σ_*,Π), with k replaced by j. Note that $\sigma_1=\mathbb{D}$ and $\sigma_k=\sigma_*$. If $j\le k-1$, the surface (σ_j,Π_j) is a covering of the whole Riemann sphere $\hat{\mathbb{C}}$ in contrast with σ_* (see (iv)). Henceforth we add the multiple poles of F into Ξ. Then every branch point of every σ_j projects into $F(\Xi)$.

Let $j\ge 2$. We say that $u\in\mathcal{H}_F(\sigma_j)$ if $u\in\mathcal{H}(\sigma_j)$ and the identity

$$(d_1-d_2)u(d_1,d_2,d_4,\dots,d_{j+1})+(d_2-d_3)u(d_2,d_3,d_4,\dots,d_{j+1})+$$
$$+(d_3-d_1)u(d_1,d_3,d_4,\dots,d_{j+1})=0,\quad d_1,d_2,\dots,d_{j+1}\in\mathbb{D}\setminus F^{-1}(F(\Xi)) \tag{3.6}$$

is satisfied which is quite analogous to (1.6).

Suppose that $x\in H^2$ and $u=Vx\in H^2_F(\sigma_*)$. Set

$$u_j(\delta)=\langle x,\prod_1^j(1-d_s z)^{-1}\rangle \tag{3.7}$$

if $\delta=\{d_1,\dots,d_j\}\in\sigma_j^0$ and $1\le j\le k$. It is immediate that

$$\begin{aligned} &d_1 u_j(d_1, d_3, \ldots, d_{j+1}) - d_2 u_j(d_2, d_3, \ldots, d_{j+1}) = \\ &= (d_1 - d_2) u_{j+1}(d_1, d_2, d_3, \ldots, d_{j+1}) \end{aligned} \tag{3.8}$$

if $d_1, \ldots, d_{j+1} \in \mathbb{D}$ and $\mathbb{F}(d_1) = \ldots = \mathbb{F}(d_{j+1})$. To prove Lemma 3.3 it suffices to check the following two assertions:

(A) Given $x \in \mathcal{H}(\mathbb{D})$, form the functions $u_1 = x$, $u_2, \ldots, u_k$ following formula (3.8). Then $u_j \in \mathcal{H}_{\mathbb{F}}(\sigma_j)$ for $j \geq 2$ and $u_k = \nabla x$.

(B) Suppose that $1 \leq j < k$ and $u_{j+1} \in \mathcal{H}_{\mathbb{F}}(\sigma_{j+1})$. Then the equation (3.8) in which $u_j \in \mathcal{H}(\sigma_j)$ is an unknown function has exactly one solution. This solution belongs to $\mathcal{H}_{\mathbb{F}}(\sigma_*)$ if $j \geq 2$.

3.8. PROOF OF (A). The formula (3.8) which expresses u_{j+1} in terms of u_j has singularities on $\Pi_{j+1}^{-1}(\mathbb{F}(\Xi))$ only. (A) is obvious if $x \in H^2$ since (3.7) follows from (3.8) by induction on j. Hence u_j satisfies (3.6), is analytic on σ_j° and bounded near $\Pi_j^{-1}(\mathbb{F}(\Xi))$, and the formula $u_k = \nabla x$ holds.

Let $x \in \mathcal{H}(\mathbb{D})$, $x^{(n)} \in H^2$ and $x^{(n)}$ tend to x uniformly on the discs $r\overline{\mathbb{D}}$, $r < 1$. Let $u_1^{(n)} = x^{(n)}$, $u_2^{(n)}, \ldots, u_k^{(n)}$ be the functions corresponding to $x^{(n)}$. Then $u_j^{(n)}(d_1, \ldots, d_j)$ does not depend on the order of $d_1, \ldots, d_j$ by (3.7). The limit argument shows that $u_j(d_1, \ldots, d_j)$ does not depend on the order of $d_1, \ldots, d_j$ too, and that $u_j \in \mathcal{H}(\sigma_j^\circ)$. The formulae $u_j \in \mathcal{H}_{\mathbb{F}}(\sigma_j^\circ)$ and $u_k = \nabla x$ can be deduced from the limit argument too.

Let $\delta \in \sigma_j \setminus \sigma_j^\circ$ and C be a small circle with centre at $\Pi_j(\delta)$. Since $u_j^{(n)}$ tend to u_j uniformly on $\Pi_j^{-1}(C)$, it follows that u_j has no singularity in δ. ▨

The statement (B) will be proved by reducing (3.8) to a certain Cousin problem on $\hat{\mathbb{C}}$.

3.9. THE SOLUTION ON FIBRES $\Pi_j^{-1}(\lambda)$. Fix some j, $1 \leq j < k$, and some $u_{j+1} \in \mathcal{H}_{\mathbb{F}}(\sigma_{j+1})$. Let us multiply each side of (3.8) by $d_3 d_4 \ldots d_{j+1}$ and introduce an unknown function v_j:

$$v_j(d_1, \ldots, d_j) = d_1 \ldots d_j \, u_j(d_1, \ldots, d_j). \tag{3.9}$$

It follows that

$$v_j(d_1, d_3, \dots, d_{j+1}) - v_j(d_2, d_3, \dots, d_{j+1}) = \\ = (d_1 - d_2) d_3 d_4 \dots d_{j+1} u_{j+1}(d_1, d_2, \dots, d_{j+1}) \tag{3.10}$$

if $d_1, \dots, d_{j+1} \in \mathbb{D}$ and $F(d_1) = \dots = F(d_{j+1}) \notin F(\Xi)$.

These equalities define some differences between values of v_j on fibres $\Pi_j^{-1}\{\lambda\}$, $\lambda \in \hat{\mathbb{C}}$, since in our situation $\Pi_j(d_1, d_3, \dots, d_{j+1}) = \Pi_j(d_2, d_3, \dots, d_{j+1})$. We show first that (3.10) can be solved on each fibre $\Pi_j^{-1}\{\lambda\}$ individually.

Let $\lambda \in \hat{\mathbb{C}} \setminus F(\Xi)$. Consider two-dimensional cell complex R_λ whose vertices are points of $\Pi_j^{-1}\{\lambda\}$, that is j-element subsets of $F^{-1}\{\lambda\} \cap \mathbb{D}$. The pairs (σ_1, σ_2) of vertices of R_λ that correspond to some equation (3.10) will be called edges of R_λ. Finally each three vertices linked with three edges will form a side of R_λ.

We introduce a one-dimensional cochain φ on R_λ by means of defining its values on edges:

$$\varphi(\{d_1, d_3, \dots, d_{j+1}\}, \{d_2, d_3, \dots, d_{j+1}\}) = \\ = (d_2 - d_1) d_3 d_4 \dots d_{j+1} u_{j+1}(d_1, d_2, \dots, d_{j+1}). \tag{3.11}$$

Suppose that $\sigma_0, \dots, \sigma_n$ are vertices of R_λ such that (σ_i, σ_{i+1}) is an edge for $i = 0, \dots, n-1$. We observe that every two points of R_λ can be linked with some "broken line" $\sigma_0 \dots \sigma_n$ of this kind. Consider the chain

$$\psi = (\sigma_0, \sigma_1) + (\sigma_1, \sigma_2) + \dots + (\sigma_{n-1}, \sigma_n). \tag{3.12}$$

If $v_j \mid \Pi_j^{-1}\{\lambda\}$ satisfies (3.10), then

$$\varphi(\psi) = v_j(\sigma_n) - v_j(\sigma_0).$$

It follows easily that (3.10) has a solution on $\Pi_j^{-1}\{\lambda\}$ iff $\varphi(\psi) = 0$ for every closed chain ψ (i.e. for every ψ such that

$\sigma_0 = \sigma_n$). The following lemma will be proved later.

3.10. LEMMA ON R_λ . The one-dimensional homology group of is trivial.

In view of this lemma, every closed chain ψ has a decomposition into a finite sum

$$\psi = \sum \partial K_i$$

where K_i are some sides of R_λ . Hence it suffices to show that $\varphi(\partial K) = 0$ for any side $K = (\sigma_1, \sigma_2, \sigma_3)$ of R_λ . Two different cases are possible:

1) $\sigma_i = \{d_1, d_2, \ldots, d_{j+1}\} \setminus \{d_i\}, \quad i = 1, 2, 3$. Then

$$\varphi(\partial K) = \varphi((\sigma_1, \sigma_2)) + \varphi((\sigma_2, \sigma_3)) + \varphi((\sigma_3, \sigma_1)) =$$
$$= d_4 \ldots d_{j+1} u_{j+1}(d_1, d_2, \ldots, d_{j+1})[(d_1 - d_2)d_3 + (d_2 - d_3)d_1 + (d_3 - d_1)d_2] = 0 .$$

2) $\sigma_i = \{d_i, d_4, \ldots, d_{j+2}\}, \quad i = 1, 2, 3$. Then

$$\varphi(\partial K) = d_4 \ldots d_{j+2}[(d_2 - d_1) u_{j+1}(d_1, d_2, d_4, \ldots, d_{j+2}) +$$
$$+ (d_3 - d_2) u_{j+1}(d_2, d_3, d_4, \ldots, d_{j+2}) + (d_3 - d_1) u_{j+1}(d_1, d_3, \ldots, d_{j+1})] = 0 ,$$

since $u_{j+1} \in \mathcal{H}_F(\sigma_{j+1})$.

We conclude that (3.10) has a solution on $\Pi_j^{-1}\{\lambda\}$ for every $\lambda \in \hat{\mathbb{C}} \setminus F(\Xi)$.

3.11. THE FUNCTION $w(\sigma, \sigma')$. Let $\sigma, \sigma' \in R_\lambda$ and let ψ be a chain like (3.12) with endpoints σ and σ' (that is, with $\sigma_0 = \sigma$ and $\sigma_n = \sigma'$). The preceding arguments show that

$$w(\sigma, \sigma') \overset{def}{=\!=} \varphi(\psi)$$

does not depend on the choice of ψ .

Suppose now that $\sigma, \sigma', \sigma'' \in R_\lambda$. Choose chains ψ_1, ψ_2, ψ_3 which connect σ and σ', σ' and σ'', σ'' and σ correspondingly. Then $\psi_1 + \psi_2 + \psi_3$ is a closed chain, and we get the cocycle equation

$$w(\sigma, \sigma') + w(\sigma', \sigma'') + w(\sigma'', \sigma) = 0. \tag{3.13}$$

The function $w(\sigma, \sigma')$ is defined for all σ, σ' such that

$\Pi_j(\sigma) = \Pi_j(\sigma')$. The equations (3.10) are equivalent to

$$v_j(\sigma) - v_j(\sigma') = w(\sigma, \sigma'), \quad \forall \sigma, \sigma' \in \sigma_j^\circ : \Pi_j(\sigma) = \Pi_j(\sigma'). \tag{3.14}$$

For every branches $\tilde{\sigma}$, $\tilde{\sigma}'$ of Π_j^{-1} defined on some set W , $W \subset \hat{\mathbb{C}} \setminus \mathbb{F}(\Xi)$, the function $\lambda \mapsto w(\tilde{\sigma}(\lambda), \tilde{\sigma}'(\lambda))$ is analytic. (the term "branch" will always mean a continuous branch). To prove this, suppose that $\lambda_0 \in W$. Since $\lambda_0 \notin \mathbb{F}(\Xi)$, it is easy to find some branches

$$\tilde{\sigma} = \tilde{\sigma}_0, \tilde{\sigma}_1, \ldots, \tilde{\sigma}_{n-1}, \tilde{\sigma}_n = \tilde{\sigma}'$$

of Π_j^{-1} which are defined in a neighbourhood W' of λ_0 with the following property: ($\sigma_i(\lambda)$, $\sigma_{i+1}(\lambda)$) is an edge of R_λ for $i = 0, \ldots, n-1$ and for all $\lambda \in W'$. The chain $\Psi(\lambda)$ corresponding to these edges satisfies

$$w(\tilde{\sigma}(\lambda), \tilde{\sigma}'(\lambda)) \equiv \varphi(\Psi(\lambda)), \quad \lambda \in W'$$

and we are done. Note also that $w(\sigma, \sigma')$ is bounded if $\Pi_j(\sigma)$ is in a small neighbourhood of $\mathbb{F}(\Xi)$.

3.12. LEMMA ON THE SOLUTION OF THE PROBLEM (3.14) NEAR BRANCH POINTS. Let $\lambda_0 \in \mathbb{F}(\Xi)$ and let L be an open disc with centre at λ_0 such that $\bar{L} \cap \mathbb{F}(\Xi \cup \mathbb{T}) = \{\lambda_0\}$. Then the problem (3.14) (or equivalently (3.10)) has a "partial" solution v_L which is defined and analytic on $\Pi_j^{-1}(L)$.

If $\lambda_0 = \infty$, then we take $L = \{\lambda : |\lambda| > R\}$, where R is sufficiently large.

PROOF. Let G be one of the components of $\Pi_j^{-1}(L)$. It is an m-sheeted branched covering of L for some m . We first solve (3.14) on G .

The unique branch point of $(G, \Pi_j | G)$ projects into λ_0 . Let η be any branch of $(\Pi_j - \lambda_0)^{1/m}$ (or $(\Pi_j - \lambda_0)^{-1/m}$ if $\lambda_0 = \infty$) on G . Put $\varepsilon = exp(2\pi i/m)$. The function η maps G homeomorphically onto some disc B with centre at the origin. We introduce an unknown function y on B defined by

$$y \circ \eta = v_j | G ,$$

and a function g on B ,

$$g(\eta(\delta)) \stackrel{def}{=\!=} w(\delta, \delta') ,$$

where $\eta(\delta') = \varepsilon \eta(\delta)$. The equations (3.14) considered as equations on $v_j | G$ are equivalent to

$$y(\varepsilon z) - y(z) = g(z), \qquad z \in B \setminus \{0\} \tag{3.15}$$

because of (3.13). The properties of the function $w(\cdot,\cdot)$ imply that g is analytic on B . Hence g can be written in the form

$$g(z) = \sum_{n \geq 0} g_n z^n .$$

It follows from (3.13) and from the solvability of (3.15) for every fixed z that

$$\sum_{s=0}^{m-1} g(\varepsilon^s z) \equiv 0 \qquad \text{on } B .$$

Hence $g_n = 0$ if $n = \ell \cdot m$ and $\ell = 0, 1, \ldots$. So y can be taken as

$$y(z) = \sum_{n \geq 0,\, n \neq \ell \cdot m} g_n z^n / (\varepsilon^n - 1) .$$

This series obviously converges on B and yields a partial solution $v_j | G$ of (3.14). Since $\Pi_j(G) = L$ and (3.13) holds, a solution $v_j | \Pi_j^{-1}(L)$ of (3.14) can be constructed uniquely which coincides on G with the solution obtained. It follows trivially that $v_j \in \mathcal{H}(\Pi_j^{-1}(L))$. ▨

If $v_j^\circ | \Pi_j^{-1}(L)$ is a partial solution of (3.14), the general solution of (3.14) on $\Pi_j^{-1}(L)$ is given by

$$v_j = v_j^\circ + f \circ \Pi_j ,$$

where $f \in \mathcal{H}(L)$ is arbitrary.

3.13. THE SOLUTION OF (3.14). Denote by M the union of some discs with centres at the points of $F(\Xi)$ which are disjoint and do not intersect with $F(\mathbb{T})$. Take some sets $W_1, \dots, W_n$ forming some open covering of σ_j together with $\Pi_j^{-1}(M)$. Assume that the functions $\Pi_j | W_i$, $1 \leq i \leq n$, have inverse functions $\varkappa_i$ which are analytic.

To prove that such covering exists, choose some open disks $B_1, \dots, B_s$ with centres in Ξ so that $F(\cup B_i) \subset M$. The set $\mathbb{D} \setminus \cup B_i$ can be covered by a finite number of open sets $K_1, \dots, K_\ell$ such that $F | K_i$ is univalent for every i. The sets $\{\{d_1, \dots, d_j\} : d_i \in K_{m_i}, 1 \leq i \leq j\}$, where m_i are arbitrary indices with $m_1 < m_2 < \dots < m_j$, can be taken for $W_1, \dots, W_n$.

Let v_j° be a partial solution of (3.14) on $\Pi_j^{-1}(M)$ which exists by the above lemma. Any solution of (3.14) satisfies

$$v_j(\varkappa_i(\lambda)) = f_i(\lambda), \quad 1 \leq i \leq n,$$

$$v_j(\sigma) = v_j^\circ(\sigma) + f_{n+1}(\Pi(\sigma)), \quad \Pi(\sigma) \in M$$

for some $f_1, \dots, f_{n+1}$. Put

$$M_i = \Pi_j(W_i), \quad 1 \leq i \leq n$$

and $M_{n+1} = M$. If v_j is a solution of (3.14), then f_i are defined and analytic on M_i. Since $\Pi_j(\sigma_j) = \hat{\mathbb{C}}$, the domains M_i cover the Riemann sphere $\hat{\mathbb{C}}$.

The equations (3.14) are equivalent to the following conditions on f_i:

$$f_s(\lambda) - f_t(\lambda) = w_{st}(\lambda) \quad \text{if} \quad \lambda \in M_s \cap M_t$$

for $1 \leq s < t \leq n+1$, where

$$w_{st}(\lambda) = \begin{cases} w(\varkappa_s(\lambda), \varkappa_t(\lambda)), & t < n+1, \\ v_j^\circ(\varkappa_s(\lambda)), & t = n+1. \end{cases}$$

Thus the problem (3.14) reduces to a Cousin problem on $\hat{\mathbb{C}}$ (see [18]). This problem has a solution if and only if the w_{ij} satisfy the cocycle equation

$$w_{rs} + w_{st} = w_{rt} \quad \text{in} \quad W_r \cap W_s \cap W_t \ , \ r<s<t \ .$$

The last condition follows from (3.13) and from the fact that v_j^0 is a partial solution of (3.14) in $\Pi_j^{-1}(M_{n+1})$. Any two solutions of a Cousin problem in $\hat{\mathbb{C}}$ differ by a constant. Hence the problem (3.10) is solvable and any two of its **solutions** differ by a constant.

3.14. THE SOLUTION OF THE PROBLEM (3.8). We start with the following remark. Let $\delta_1, \delta_2 \in \sigma_j$ be two sets which contain the origin and differ by only one element. Then (3.10) implies that $v_j(\delta_1) = v_j(\delta_2)$. Hence $v_j(\delta_1) = v_j(\delta_2)$ for every $\delta_1, \delta_2 \in \sigma_j$ containing zero, if v_j is a solution of (3.10). Since any such δ_1, δ_2 project to $F(0)$, there is a solution v_j of (3.10) such that $v_j(\delta) = 0$ for every $\delta \in \sigma_j$ containing the origin. It follows from the assumption $F(0) \notin F(\Xi)$ that the function u_j defined by

$$u_j(d_1, \ldots, d_j) = (d_1 d_2 \ldots d_j)^{-1} \cdot v_j(d_1, \ldots, d_j)$$

(see (3.9)) is analytic on σ_j. The equation (3.8) is a consequence of (3.10) on $\sigma_j^0 \setminus \Pi_j^{-1}\{F(0)\}$; for the points of $\Pi_j^{-1}\{F(0)\}$ it follows by continuity. Therefore (3.8) has a solution u_j. This solution is obviously unique.

To prove that $u_j \in \mathcal{H}_{F}(\sigma_j)$ if $j>1$, take some $\delta = \{d_1, \ldots, d_{j+1}\} \in \sigma_{j+1}^0$ and let $\delta_j = \delta \setminus \{d_j\}$. Then (3.8) implies that

$$(d_1 - d_2)\, u_j(\delta_3) + (d_2 - d_3)\, u_j(\delta_1) + (d_3 - d_1)\, u_j(\delta_2) =$$

$$= \left[d_1 u_j(\delta_3) - d_3 u_j(\delta_1) \right] + \left[d_2 u_j(\delta_1) - d_1 u_j(\delta_2) \right] +$$

$$+ \left[d_3 u_j(\delta_2) - d_2 u_j(\delta_3) \right] =$$

$$= u_{j+1}(\delta) \left[(d_1 - d_3) + (d_2 - d_1) + (d_3 - d_2) \right] = 0 \ ,$$

which proves (3.6). This completes the proof of the theorem (modulo Lemma 3.8) if $\mathbb{F}(0)\notin\mathbb{F}(\Xi)$.

3.15. THE CASE IF $\mathbb{F}(0)\in\mathbb{F}(\Xi)$ reduces to the known case by means of a conformal mapping

$$\tau(z)=\frac{z-\mu}{1-\bar{\mu}z}$$

of the unit disc $\mathbb{D}$ onto itself. Choose $\mu\in\mathbb{D}$ so that $\mathbb{F}(\tau(0))\notin\mathbb{F}(\Xi)$. Let

$$w(z)=(1-\bar{\mu}z)^{-1}.$$

Define an operator G on H^2 by

$$Gx\overset{def}{=\!=}w\cdot(x\circ\tau).$$

Then $GT_{\mathbb{F}}G^{-1}=T_{\mathbb{F}\circ\tau}$ and $(1+|\mu|^2)^{-1/2}G$ is a unitary operator; see more general statements in [9]. We show that $T_{\mathbb{F}}$ and M_{Π} are similar using the analogous property of $T_{\mathbb{F}\circ\tau}$ which is already known. The surfaces $\sigma_*=\sigma_*(T_{\mathbb{F}})$ and $\sigma_*^{\tau}=\sigma_*(T_{\mathbb{F}\circ\tau})$ can be identified by means of the mapping

$$\delta\to\delta^{\tau}=\tau^{-1}(\delta),\qquad \delta\in\sigma_*^{\circ}(T_{\mathbb{F}}).$$

The formula

$$h_{\delta\tau}^{\tau}=w\cdot\frac{1}{h_{\delta}(\tau(0))}\cdot(h_{\delta}\circ\tau)$$

connects eigenvectors $h_{\delta\tau}^{\tau}$ of $T_{\mathbb{F}\circ\tau}^{*}$ with eigenvectors h_{δ} of $T_{\mathbb{F}}$. Define an operator $L:\mathcal{H}(\sigma_*)\longrightarrow\mathcal{H}(\sigma_*^{\tau})$ by

$$(Lu)(\delta^{\tau})\overset{def}{=\!=}h_{\delta}(\tau(0))^{-1}u(\delta).$$

Since $|h_{\delta}(\tau(0))|^{-1}$ is bounded and bounded away from zero, L is an isomorphism between $H^2(\sigma_*)$ and $H^2(\sigma_*^{\tau})$. The operator L converts the relations (1.6) into the corresponding relations for $H^2_{\mathbb{F}\circ\tau}(\sigma_*^{\tau})$ (to see this it is convenient to use that τ^{-1}

does not change the cross ratio of $d_1, d_2, d_3, -\mu^{-1}$). Hence L is an isomorphism between $H^2_F(\sigma_*)$ and $H^2_{F\circ\tau}(\sigma_*^\tau)$.

Let $U_F : H^2 \to H^2_F(\sigma_*)$ and $U_{F\circ\tau} : H^2 \to H^2_{F\circ\tau}(\sigma_*^\tau)$ be the operators corresponding to F and $F\circ\tau$. Then

$$(1-|\mu|^2)\, U_{F\circ\tau} G = L U_F .$$

Indeed, we have for $x \in H^2$:

$$(1-|\mu|^2)(U_{F\circ\tau} Gx)(\delta^\tau) = (1-|\mu|^2) \langle w \cdot x\circ\tau,\ w\cdot(h_\delta\circ\tau)/h_\delta(\tau(0)) \rangle =$$

$$= (h_\delta(\tau(0)))^{-1} \langle x, h_\delta \rangle = (L U_F x)(\delta^\tau)$$

since $\tau' = (1-|\mu|^2)\, w^2$. Thus $U_F = (1-|\mu|^2)^{2-1} L^{-1} U_{F\circ\tau} G$ is an isomorphism between H^2 and $H^2_F(\sigma_*)$. ▨

3.16. PROOF OF LEMMA 3.10. We prove only that each closed broken line $\alpha = A_0 \dots A_\ell$, where $A_\ell = A_0$ (or the corresponding 1-cycle)is a boundary. This is the only assertion which has been used. In fact, this assertion is equivalent to the statement of the lemma since each 1-cycle can be decomposed easily into a sum of 1-cycles of the above type.

It can be assumed that an arbitrary finite set M instead of $F^{-1}\{\lambda\}$ is taken in the definition of the complex (so that the vertices of the complex are subsets of M having j elements). We use induction on $n = \mathrm{card}(M)$.

The assertion is trivial if $n = j$. Let $n > j$. Take some $b \in M \setminus A_0$. Suppose that A_i is the first point of α such that $b \in A_i$. If $b \notin A_{i+1}$, replace α by the broken line $\alpha' = A_0 A_1 \dots A_{i-1} A_{i+1} \dots A_\ell$ ($A_{i-1} A_{i+1}$ is an edge in this case). In the other case let $\alpha' = A_0 A_1 \dots A_{i-1} B A_{i+1} \dots A_\ell$, where $B = (A_i \cup A_{i+1}) \setminus \{b\}$. The 1-cycles corresponding to α and α' differ in a boundary of a triangle in each case. Note that α' has less points containing b than α. The continuation of this process yields a broken line $\alpha^{(m)}$ homologous to α whoes points do not contain b. Since $\alpha^{(m)}$ is a closed broken line in the complex corresponding to $M \setminus \{b\}$, it is a boundary by the induction hypothesis. Thus so is α . ▨▨

3.17. PROOF OF THEOREM 2.4 is held following the same scheme, with replacement of $\mathbb{D}$ by Ω . Lemma 3.2 can be replaced by the following Lemma 5 from [40]: in the conditions of Theorem 2.4, for every $x \in \mathcal{H}(\Omega)$ we have $x \in B(\Omega) \Longleftrightarrow Vx \in B(\sigma_*)$. ▨▨

We remark that in [40], the fact that U is injective (which implies that eigenvectors of $T_{\mathbb{F}}^*$ are dense in $B(\Omega)$) is proved in a way which differs from the above one. However, it was used there that $\mathbb{F}$ is analytic on a neighbourhood of $\partial\Omega$.

It follows from Theorem 1.5 and 2.4 that if the hypotheses of Theorem 2.4 are fulfilled, then the model spaces $H_{\mathbb{F}}^2(\sigma_*)$ and $E_{\mathbb{F}}^2(\sigma_*)$ coincide (and have equivalent norms). Here $E_{\mathbb{F}}^2(\sigma_*)$ is the space corresponding to the Hardy-Smirnov spaces $E^2(G)$, $G \in \mathcal{N}$. In the next section, we give some weaker conditions on $\mathbb{F}$ ensuring that $H_{\mathbb{F}}^2(\sigma_*) = E_{\mathbb{F}}^2(\sigma_*)$. We include also some technical lemmas needed in Chapter 4.

4. SOME PROPERTIES OF E^p AND OF THE RELATED FUNCTION CLASSES.

4.1. THE DEFINITION OF E^p (see [26]). Suppose Ω is a domain bounded by a rectifiable Jordan curve and let $p>0$. A function $f \in \mathcal{H}(\Omega)$ belongs to the Hardy - Smirnov class $E^p(\Omega)$ iff there is a sequence of curves γ_n converging to $\partial\Omega$ and satisfying

$$\|f\|_{L^p(\gamma_n)} = \int_{\gamma_n} |f(z)|^p |dz| < C$$

for all n . The functions in $E^p(\Omega)$ have nontangential limit values a.e. on $\partial\Omega$. The space $E^p(\Omega)$ is a Banach space with the norm $\|f\|_p = \|f\|_{L^p(\partial\Omega)}$ if $p \geq 1$; for $0<p<1$ it is a linear topological space with the metric $f \longrightarrow \|f\|_p^p$.

If $p \geq 1$, then a function in $E^p(\Omega)$ can be recovered by its boundary values by means of the Cauchy integral:

$$f(z) = \frac{1}{2\pi i} \int_{\partial\Omega} f(\zeta)(\zeta - z)^{-1} d\zeta .$$

In the sequel φ denotes a fixed conformal mapping of $\mathbb{D}$ onto Ω.

4.2. A RELATION WITH H^p. Let $p>0$ and $f \in \mathcal{H}(\Omega)$. Then $f \in E^p(\Omega)$ iff $(f\circ\varphi)\cdot(\varphi')^{1/p} \in H^p$. In particular, $E^p(\mathbb{D}) = H^p$.

4.3. THE SMIRNOV CLASS DOMAINS. FACTORIZATION. We say that Ω belongs to the Smirnov class (S) if the polynomials are dense in $E^p(\Omega)$ for $p \geqslant 1$. This is equivalent to the fact that φ' is a outer function [26]. The class (S) contains all domains with piecewise-smooth boundaries [26]. We shall assume here $\Omega \in (S)$.

We say that a function f on Ω is outer (inner) if $f\circ\varphi$ is outer (inner) in $\mathbb{D}$. The inner-outer factorization in $\mathbb{D}$ is defined in [16, 21, 26]. It follows easily from 4.2 that every $f \in E^p(\Omega)$ admits a unique factorization $f = f_i f_e$, where f_i is inner and f_e is outer in Ω. The inner functions in Ω can be characterized by the conditions $f \in H^\infty(\Omega)$, $|f| = 1$ a.e. on $\partial\Omega$.

We define the Smirnov class $\mathcal{D}(\Omega)$ and the Nevanlinna class $N(\Omega)$ by

$$\mathcal{D}(\Omega) = \{ g_1/g_2 : g_1 \in H^\infty(\Omega), \quad g_2 \quad \text{is outer in } \Omega \}$$

and

$$N(\Omega) = \{ g_1/g_2 : g_1, g_2 \in H^\infty(\Omega) \};$$

then $E^p(\Omega) \subset \mathcal{D}(\Omega) \subset N(\Omega)$ for all $p>0$.

4.4. THE PHRAGMEN - LINDELÖF PRINCIPLE. <u>If</u> $f \in \mathcal{D}(\Omega)$, $f|\partial\Omega \in L^p(\partial\Omega)$ (<u>where</u> $0<p<\infty$), <u>then</u> $f \in E^p(\Omega)$.

This follows from the well-known case $\Omega = \mathbb{D}$ and from 4.2.

4.5. CARLESON CONTOURS. <u>If</u> γ <u>is a rectifiable curve, then the Carleson constant of</u> γ <u>is</u>

$$c(\gamma) \overset{def}{=\!=} \sup\{ r^{-1} m\{ z \in \gamma : |z-\zeta| < r\} : \zeta \in \mathbb{C},\ r>0 \};$$

here m stands for the arc length measure. For every positive p and A there exists B such that $\|f|\gamma\|_{L^p(\gamma)} \leq B\|f\|_{E^p(\Omega)}$ for all $f\in E^p(\Omega)$ and all $\gamma\subset\bar{\Omega}$ with $c(\gamma)<A$.

For $p>1$ this follows from the David techniques [13]. The general case reduces easily to the case $p>1$ by means of the factorization of f and considering a power of f_e.

4.6. AN IMBEDDING LEMMA. Let Ω, Ω' be simply connected domains with piecewise-smooth boundaries and let $p>0$. Then $f|\Omega'\in E^p(\Omega')$ for all $f\in E^p(\Omega)$.

The proof can be obtained by using 4.5.

4.7. AN "ANALYTIC STICKING" LEMMA. Suppose $\Omega, \Omega_1, \Omega_2$ are simply connected domains with piecewise smooth boundaries such that $\bar{\Omega} = \bar{\Omega}_1 \cup \bar{\Omega}_2$. If $p>0$, $f\in\mathcal{H}(\Omega)$, $f|\Omega_i \in E^p(\Omega_i)$ for $i=1,2$, then $f\in E^p(\Omega)$. This can be derived from 4.5 and from definition of $E^p(\Omega)$.

4.8. A CHARACTERIZATION OF $\mathcal{D}(\Omega)$. $f\in\mathcal{D}(\Omega)$ iff $\exists\{f_n\}_{n\geq 1}$, $f_n\in H^\infty(\Omega)$: $|f_1|<|f_2|<|f_3|<\dots$ on Ω and $f_n\longrightarrow f$ pointwise in Ω. This follows from standard facts on factorization in $\mathbb{D}$ [26].

Let $\mathfrak{A}(G)$ be classes of analytic functions defined for some reasonable domains G in $\mathbb{C}$ (or on a Riemann surface), and let Ω be one of these domains with boundary $\partial\Omega$. We say that f belongs to $\mathfrak{A}(\Omega)$ locally ($f\in\mathfrak{A}_{loc}(\Omega)$) if $f\in\mathcal{H}(\Omega)$ and every $\zeta\in\partial\Omega$ has a (reasonable) neighbourhood $V(\zeta)$ such that

$$f|V(\zeta)\cap\Omega\in\mathfrak{A}(V(\zeta)\cap\Omega).$$

4.9. A LOCAL DESCRIPTION OF H^2. $f\in H^2 \Longleftrightarrow f\in H^2_{loc}(\mathbb{D})$.

The proof can be obtained using 4.7 and 4.2.

4.10. A LOCAL DESCRIPTION OF $\mathcal{D}(\Omega)$: $f\in\mathcal{D}(\Omega) \Longleftrightarrow f\in\mathcal{D}_{loc}(\Omega)$.

The implication $f\in\mathcal{D} \Longrightarrow f\in\mathcal{D}_{loc}$ follows from 4.8. The inverse implication is obtained by multiplying of f by a proper outer function and application of 4.9.

4.11. DEFINITION OF E^2 CLASSES ON RIEMANN SURFACES. Let R be a Riemann surface with boundary ∂R and $\Pi: R \to \mathbb{C}$ be its projection to $\mathbb{C}$. Suppose that (R, Π) has a finite number of branch points, they do not lie on ∂R, and $\Pi(\partial R)$ is a piecewise-continuous curve. Choose open domains $G_1, \dots, G_n$ in R with piecewise-smooth boundaries such that $\bigcup_0^n G_i = R$, $\bar{G}_0 \cap \partial R = \emptyset$, and such that $\bar{G}_j$ does not contain branch points and projects to $\mathbb{C}$ injectively for $j \geqslant 1$. We define the Hardy - Smirnov class $E^2(G)$ as the set of $u \in \mathcal{H}(R)$ such that $u | G_j \in E^2(G_j)$ for $j \geqslant 1$. Supplied with the norm $\| u | G_0 \|_{H^\infty} + \sum_{j=1}^n \| u | G_j \|_{E^2(G_j)}$ the class $E^2(R)$ becomes a Banach space. An easy application of the closed graph theorem shows that up to an equivalence of norms $E^2(R)$ does not depend on the choice of G_j. Moreover, one of the equivalent norms on $E^2(R)$ is

$$\| u \| = \Big(\int_{\partial R} | u(\sigma) |^2 \, | d\Pi(\sigma) | \Big)^{1/2}$$

(see Lemma 4.6 in [31]). This norm is used in the sequel. For $R = \sigma_*$, the above definition coincides with the general definition of $B_F(\sigma_*)$ for the bundle $B(G) \overset{def}{=} E^2(G)$, $G \in \mathcal{N}$. Set

$$E_F^2(\sigma_*) = E^2(\sigma_*) \cap \mathcal{H}_F(\sigma_*).$$

4.12. LEMMA. <u>Suppose that</u> $a < | F'(z) | < b$ <u>for all</u> z <u>in the annulus</u> $r < |z| < 1$, <u>where</u> $a > 0$, $b < \infty$ <u>and</u> $r < 1$ <u>are constants</u>. Then $E_F^2(\sigma_*) = H_F^2(\sigma_*)$.

PROOF. Let V denote the operator defined in 3.1. By Lemmas 3.2 and 3.3, it suffices to check that

$$\forall x \in \mathcal{H}(\mathbb{D}) : x \in H^2 \iff Vx \in E^2(\sigma_*).$$

Fix domains $G_0, \dots, G_n$ and suppose that $x \in H^2$, $j \geqslant 1$. Choose branches $\alpha_1, \dots, \alpha_k$ of $F^{-1} \circ \Pi$ so that $\sigma = \{ \alpha_1(\sigma), \dots, \alpha_k(\sigma) \}$ for all $\sigma \in G_j$. We have $x \circ \alpha_1 \in E^2(G_j)$ for $s = 1, \dots, k$ due to the condition on F' (see 4.2). Note that the functions c_s from (3.4) are bounded near $\partial \sigma_*$. Hence $Vx \in E^2(G_j)$ due to

(3.3), and thus $Vx \in E^2(\sigma_*)$. The proof of the inverse implication is completely analogous to the proof of Lemma 3.2, i.2. Using the condition on $\mathbb{F}'$ we get $x \in E^2_{loc}(\mathbb{D})$, $x \in H^2$. ▨

CHAPTER 2. COMMUTANT.

The following new conditions on $\mathbb{F}$ will be needed in this chapter:

(v) The curve $\zeta \to \mathbb{F}(\zeta)$, $\zeta \in \mathbb{T}$, has a finite number of self-intersections. This means that each point of $\mathbb{F}(\mathbb{T})$ except for a finite set has only one pre-image on $\mathbb{T}$ under $\mathbb{F}$;

(vi) All self-intersections of $\mathbb{F}(\mathbb{T})$ are simple and transversal;

(vii) The zeroes of $\mathbb{F}'$ in $\mathbb{D}$ have order at most 2, **and the function $\mathbb{F}|\Xi$ is injective.**

The fact that the commutant $\{T_{\mathbb{F}}\}'$ is abelian will be proved under the conditions (i)-(v). A complete description of $\{T_{\mathbb{F}}\}'$ will require the conditions (i)-(vii).

5. THE DEFINITION OF THE REDUCED ULTRASPECTRUM AND COMMUTATIVITY OF THE COMMUTANT.

5.1. THE METHOD OF CONSTRUCTING THE REDUCED ULTRASPECTRUM σ_{**}.

Let E be a function on a domain G in $\mathbb{C}$ (or on some Riemann surface) whose values are subspaces of H^2 of some fixed finite dimension n. We say that the space-valued function E is analytic if every $\mu_0 \in G$ has a neighbourhood $\mathcal{U}$ such that

$$E(\mu) = \operatorname{span}\{a_j(\mu) : 1 \leqslant j \leqslant n\}$$

for $\mu \in \mathcal{U}$, where the a_j are some analytic vector-functions on $\mathcal{U}$. Consider the family $\mathfrak{A}$ of analytic space-valued functions defined on subregions of $\mathbb{C}$ which is formed according the rules: (1) The functions $E(\lambda) = \operatorname{Ker}(T_{\mathbb{F}}^* - \lambda I)$ defined on the components of $\mathbb{C} \setminus \mathbb{F}(\mathbb{T})$, are in $\mathfrak{A}$; (2) If a function E is in

$\mathfrak{A}$ then every single-valued analytic continuation of E is also in $\mathfrak{A}$; (3) If $E_1, E_2 \in \mathfrak{A}$ and W is an open set contained in the domains of E_1, E_2, then the function $\lambda \to E_1(\lambda) \cap E_2(\lambda)$, $\lambda \in W$, is in $\mathfrak{A}$ whenever it is analytic; (4) Every function in $\mathfrak{A}$ is obtained using (1)-(3) a finite number of times.

Every $E(\lambda)$ for every $E \in \mathfrak{A}$ and every λ is obviously invariant under the commutant $\{T_F^*\}'$. It will be shown that if $E \in \mathfrak{A}$ and $E(\lambda)$ is one-dimensional then $E(\lambda)$ is spanned by h_σ for some $\sigma \in \sigma_*$. The reduced ultraspectrum except for a finite number of points is the union of such points σ . The most important step of the proof of the commutativity of $\{T_F^*\}'$ is to show that the vectors $\{h_\sigma : \sigma \in \sigma_{**} \cap \Pi^{-1}\{\lambda\}\}$ span $\operatorname{Ker}(T_F^* - \lambda I)$ for every $\lambda \in \mathbb{C} \setminus \mathbb{F}(\Xi \cup \mathbb{T})$.

5.2. LEMMA. <u>Let</u> $a_1, \dots, a_{n+1}$ <u>be single-valued analytic vector functions on a connected Riemann surface</u> G .

a) <u>Suppose that the vectors</u> $a_1(\mu), \dots, a_{n+1}(\mu)$ <u>are linearly dependent for every</u> $\mu \in V$, <u>and the set</u> V <u>has a limit point in</u> G . <u>Then these vectors are linearly dependent for every</u> $\mu \in G$.

b) <u>If moreover the vectors</u> $a_1(\mu), \dots, a_n(\mu)$ <u>are linearly independent for all</u> $\mu \in G$, <u>then</u>

$$a_{n+1}(\mu) = \sum_{i=1}^{n} c_i(\mu)\, a_i(\mu) , \tag{5.1}$$

<u>where the</u> c_i <u>are analytic functions on</u> G .

PROOF. a) The vectors $a_i(\mu)$ are linearly dependent iff $\det\left[\langle a_i(\mu), e_j\rangle\right]_{i,j=1}^{n+1} = 0$ for all $e_1, \dots, e_{n+1} \in H^2$. This determinant is an analytic function of μ .

b) The assumptions imply that the equation (5.1) can be solved uniquely for each μ . Take the inner product of both sides of (5.1) with some $e_1, \dots, e_n \in H^2$ and solve the system obtained. To conclude that c_i are analytic in an arbitrary point μ_0 of G , it suffices to choose $e_1, \dots, e_n$ in a proper way. ▨

It follows from Lemma 5.2, a) that an analytic continuation of a space-valued function along some path is unique whenever it exists.

5.3. THE SURFACES σ_ℓ^o AND THE FUNCTIONS $\mu \to E_\mu$ ON THEM. It will be easier from the technical point of view to consider the functions $E(\lambda)$ from i.5.1 as defined on some Riemann surfaces. The precise definition of the surface σ_{**} will differ formally from the definition of i.5.1.

Define $\sigma_{**} = \sigma_*$ if $k=1$. Assume that $k>1$. Consider the Riemann surfaces $(\sigma_\ell^o, \Pi_\ell)$ for $k \leq \ell \leq k-1+\max\{wind_{\mathbb{F}}(\lambda) : \lambda \in \mathbb{C} \setminus \mathbb{F}(\mathbb{T})\}$; their definition can be obtained from the definition of the surface (σ_*^o, Π) by replacing k with ℓ. Set

$$E_\mu = \vee\{h_\delta : \delta \subset \mu\}$$

if $\mu \in \sigma_\ell^o$. If μ is a set $\{d_1, \ldots, d_\ell\}$ and $\ell < k$, then put formally $E_\mu = 0$. It is obvious that the space-valued function $\mu \to E_\mu$ is analytic on σ_ℓ^o and the inclusion $E_\mu \subset Ker(T_{\mathbb{F}}^* - \Pi_\ell(\mu) I)$ holds.

Let $\lambda \in \mathbb{C} \setminus \mathbb{F}(\Xi)$. We associate with every subspace N of $Ker(T_{\mathbb{F}}^* - \lambda I)$ the set

$$\beta(N) = \{\delta \in \sigma_* \cap \Pi^{-1}\{\lambda\} : h_\delta \in N\}.$$

It can be seen easily that $\beta(E_\mu) = \{\delta \in \sigma_* : \delta \subset \mu\}$ (see the next lemma).

5.4. LEMMA ON E_μ. 1) <u>If</u> $\mu \in \sigma_\ell^o$ <u>and</u> $\ell \geq k$, <u>then</u> $\dim(E_\mu) = \ell - k + 1$;

2) <u>If</u> $\mu_1 \in \sigma_{\ell_1}^o$, $\mu_2 \in \sigma_{\ell_2}^o$ <u>and</u> $\Pi_{\ell_1}\mu_1 = \Pi_{\ell_2}\mu_2$, <u>then</u> $E_{\mu_1} \cap E_{\mu_2} = E_{\mu_1 \cap \mu_2}$;

3) $\beta(E_\mu) = \{\delta \in \sigma_* : \delta \subset \mu\}$;

4) <u>Suppose that</u> $\mu, \mu_1, \mu_2 \subset \mathbb{F}^{-1}\{\lambda\} \cap \mathbb{D}$ <u>and</u> $card(\mu_1) \geq k$, $card(\mu_2) \geq k$; <u>here</u> λ <u>is a point of</u> $\mathbb{C} \setminus \mathbb{F}(\Xi)$. <u>Then</u> $E_\mu = E_{\mu_1} + E_{\mu_2}$ <u>iff</u> $\mu = \mu_1 \cup \mu_2$ <u>and</u> $card(\mu_1 \cap \mu_2) \geq k-1$.

PROOF. Let $\lambda \notin \mathbb{F}(\Xi)$ and let $d_1, \ldots, d_s$ be the roots of the equation $\mathbb{F}(z) = \lambda$ in $\mathbb{D}$.

1) One can assume that $\mu = \{d_1, \ldots, d_\ell\}$. We put $\delta_i = \{d_1, d_2, \ldots, d_{k-1}, d_i\}$ for $k \leq i \leq \ell$; then $h_{\delta_i} \in E_\mu$. The vectors h_{δ_i} are linearly independent since so are the rational functions

$\omega^{-1} h_{\delta_i}$ (see (1.5)). Therefore $dim(E_\mu) \geq \ell - k + 1$. On the other hand,

$$E_\mu \subset E'_\mu \overset{def}{=} \{ p \cdot \omega \cdot \prod_1^s (1 - d_i z)^{-1} : \quad p \text{ is a polynomial, } deg(p) \leq s - k + 1, \; \forall d \in \{d_1, \dots, d_s\} \setminus \mu \quad (1 - d \cdot z) | p \}.$$

If $F(0) = d_i, i \geq K$, then the condition $deg\, p \leq s - k$ must be added. We have $dim(E'_\mu) = \ell - k + 1$. Thus $E_\mu = E'_\mu$ and $dim(E_\mu) = \ell - k + 1$.

The statements 2), 3) follows from the equation $E_\mu = E'_\mu$ which holds for $0 \leq card(\mu) < k$ as well.

4) Let $E_\mu = E_{\mu_1} + E_{\mu_2}$. Then $E_{\mu_1} \cup E_{\mu_2} \subset E_\mu \subset E_{\mu_1 \cup \mu_2}$ and thus $\mu = \mu_1 \cup \mu_2$. By subtracting two obvious equalities one can see that

$$[card\, \mu - dim(E_{\mu_1} + E_{\mu_2})] + [card(\mu_1 \cap \mu_2) - dim(E_{\mu_1 \cap \mu_2})] = = (card_{\mu_1} - dim(E_{\mu_1})) + (card\, \mu_2 - dim(E_{\mu_2})). \tag{5.2}$$

It follows from i.1) that if $\eta \subset F^{-1}\{\lambda\} \cap \mathbb{D}$, then the equation $dim(E_\eta) - card(\eta) = k - 1$ is equivalent to $card(\eta) \geq \geq k - 1$. Thus (5.2) shows that $card(\mu_1 \cap \mu_2) \geq k - 1$.

Conversely, if $\mu = \mu_1 \cup \mu_2$ and $card(\mu_1 \cap \mu_2) \geq k - 1$, then $E_{\mu_1} + E_{\mu_2} \subset E_\mu$ and

$$dim(E_{\mu_1} + E_{\mu_2}) = card(\mu) + k - 1 = dim(E_\mu)$$

due to (5.2). Hence $E_{\mu_1} + E_{\mu_2} = E_\mu$. ▨

Item 2 of this lemma implies that an intersection of the spaces E_η, $\eta \in \cup_\ell \sigma^\circ_\ell$, is a space E_η too.

5.5. DEFINITION OF $\mathcal{C}$-SURFACES AND OF THE REDUCED ULTRASPECTRUM. Some of the connected components of the surfaces σ°_ℓ, $\ell \geq k$, will be called $\mathcal{C}$-surfaces according to the following two rules:

1) A component A of some surface σ_ℓ^0 is a $\mathcal{C}$-surface if the set

$$V \overset{def}{=\!=} \{\eta : E_\eta = Ker(T_{\mathbb{F}}^* - (\Pi_\ell \eta) I\}$$

has non-empty interior in A ;

2) Suppose that A_1 and A_2 are some $\mathcal{C}$-surfaces, $A_1 \subset \sigma_{p_1}^0$, $A_2 \subset \sigma_{p_2}^0$, and $\Pi_{p_1}(\eta_1) = \Pi_{p_2}(\eta_2)$, $\ell \overset{def}{=\!=} card(\eta_1 \cap \eta_2) \geq k$ for some $\eta_1 \in A_1$, $\eta_2 \in A_2$. Then the connected component B of σ_ℓ^0 which contains the point $\eta_1 \cap \eta_2$ is also a $\mathcal{C}$-surface.

Every $\mathcal{C}$-surface must be obtained by using these rules finitely many times.

In the notation of the rule 2), let $\eta_j = \tilde{\eta}_j(\lambda_0)$ for some branches $\tilde{\eta}_j$ of $\Pi_{p_j}^{-1}$ defined in a neighbourhood $\mathcal{U}$ of $\lambda_0 \overset{def}{=\!=} \Pi_{p_1}(\eta_1)$ (here $j=1,2$). Such branches always exist and may be constructed with the help of the branches of $\mathbb{F}^{-1}$ near λ_0. According to Lemma 5.2, 2), we have $\tilde{\eta}_1(\lambda) \cap \tilde{\eta}_2(\lambda) \in B$ and

$$E_{\tilde{\eta}_1(\lambda)} \cap E_{\tilde{\eta}_2(\lambda)} = E_{\tilde{\eta}_1(\lambda) \cap \tilde{\eta}_2(\lambda)}$$

for all $\lambda \in \mathcal{U}$.

Let σ_{**}^0 equal the union of $\mathcal{C}$-surfaces which are contained in σ_*^0. <u>The reduced ultraspectrum</u> σ_{**} is defined as the closure of σ_{**}^0 in σ_*. It follows that $\sigma_{**} \setminus \sigma_{**}^0$ is contained in $\Pi^{-1}(\mathbb{F}(\Xi))$.

Note that for every $\ell \geq k$, the function $\eta \longrightarrow E_\eta$ cannot be continued analytically accross the boundary $\partial \sigma_\ell$ of the surface σ_ℓ^0 (to a larger Riemann surface). To prove this fact, suppose that this function can be continued analytically into a neighbourhood of some η_0, $\eta_0 \in \partial \sigma_\ell$, in a larger Riemann surface R (one can easily construct such R that $R \supset \bar{\sigma}_\ell^0$). It can be assumed that $\eta_0 = \{d_1, \ldots, d_\ell\}$, where $d_1 \in \mathbb{T}$ and $d_2, \ldots, d_\ell \in \mathbb{D}$. Then it is easy to check that for any vector-function e, $e(\eta) \in E_\eta$, defined in a neighbourhood of η_0 in σ_ℓ^0, we have

$$w\text{-}\lim_{\eta\to\eta_0} e(\eta) = 0 \iff \lim_{\eta\to\eta_0} e(\eta) = 0. \tag{5.3}$$

Put $\delta(\lambda) = \{\mathfrak{d}_1(\lambda), \ldots, \mathfrak{d}_k(\lambda)\}$ where $\mathfrak{d}_j$ are branches of Π^{-1} such that $\mathfrak{d}_j(\Pi(\eta_0)) = d_j$; the $\mathfrak{d}_j$ exist by Lemma 3.4. If $e(\eta) = \| h_{\delta(\Pi(\eta))}\|^{-1} h_{\delta(\Pi(\eta))}$, then the left-hand side of (5.3) is satisfied and the strong limit on the right-hand side does not exist. This contradiction proves our statement.

Comparing the rules 1) and 2) of i.5.5 with the rules (1)-(3) of i.5.1 and bearing in mind Lemma 5.4, 2), we see that the spaces E_η , where η ranges over C-surfaces, actually coincide with the spaces $E(\lambda)$ from 5.1. This means that the collection of analytic space-valued functions $E(\lambda)$ from 5.1 defined on open subsets of $\mathbb{C}\setminus \mathbb{F}(\Xi)$ coinsides with the collection of functions $\lambda \to E_{\rho_\ell(\lambda)}$ for all $\ell \geqslant K$ and all branches ρ_ℓ of Π_ℓ^{-1} whose values are contained in C-surfaces. This justifies the unformal definition of σ_{**} made in i.5.1.

5.6. PROPOSITION. Let M be a component of σ_ℓ° for some $\ell \geqslant k$ and let $\{T_{\mathbb{F}}^*\}' E_\eta \subset E_\eta$ for all η from some open non-empty subset of M . Then $\{T_{\mathbb{F}}^*\}' E_\eta \subset E_\eta$ for all $\eta \in M$.

PROOF. Put

$$\mathcal{V} = \operatorname{int}\{\eta\in M : \{T_{\mathbb{F}}^*\}' E_\eta \subset E_\eta\}.$$

Let $\eta_0 \in \overline{\mathcal{V}}$. Take some analytic vector-functions $e_1(\eta),\ldots,e_{\ell-k+1}(\eta)$ which form a basis for E_η for all η in a neighbourhood $\mathcal{U}$ of η_0 . The inclusion $\{T_{\mathbb{F}}^*\}' E_\eta \subset E_\eta$ takes place if and only if $e_1(\eta),\ldots,e_{\ell-k+1}(\eta)$, $Ke_j(\eta)$ are linearly dependent for all j and all $K\in \{T_{\mathbb{F}}^*\}'$. Hence it takes place for all $\eta\in\mathcal{U}$ by Lemma 5.2. Thus $\mathcal{U}\subset\mathcal{V}$, which implies that $\mathcal{V}$ is closed. Since $\mathcal{V}$ is open and non-empty, $\mathcal{V} = M$. ▨

5.7. COROLLARY. If $\eta\in M$ and M is a C-surface, then E_η is invariant for $\{T_{\mathbb{F}}^*\}'$.

This follows immediately from the above preposition, the definition of C-surfaces and the remark following this definition. ▨

5.8. DECOMPOSABLE SURFACES. Let M, N, P be C-surfaces containing in $\sigma_m^\circ, \sigma_n^\circ, \sigma_p^\circ$ respectively. Suppose that $\eta \in M$, $\eta_1 \in N$, $\eta_2 \in P$ and $\Pi_m \eta = \Pi_\ell \eta_1 = \Pi_p \eta_2$. If $E_\eta = E_{\eta_1} + E_{\eta_2}$ and

$$dim(E_{\eta_1}) = dim(E_{\eta_2}) = dim(E_\eta) - 1$$

(i.e., $n = p = m-1$) then we say that the surface M is decomposable in η. In this case $m \geq k+1$.

5.9. LEMMA. <u>If a C-surface M is decomposable in some point then it is decomposable in each point</u>.

PROOF. Let M be decomposable in η and let N, P, η_1, η_2 be as above. Then $card(\eta_1) = card(\eta_2) = m-1$ and $\eta_1 \cup \eta_2 = \eta$ by Lemma 5.4, 4). Take an arbitrary point $\mu \in M$ and some open connected simply connected set W so that $\eta, \mu \in W \subset M$. There are branches $\alpha_1, \ldots, \alpha_m$ of $F^{-1} \circ \Pi_m$ defined on W such that $\nu = \{\alpha_1(\nu), \ldots, \alpha_m(\nu)\}$ for all $\nu \in W$. Hence

$$\eta_s = \{\alpha_i(\eta) : i \neq j_s\},$$

$s = 1, 2$, for some j_1, j_2. Set

$$\nu_s(\nu) = \{\alpha_i(\nu) : i \neq j_s\}, \quad s = 1, 2, \quad \nu \in W.$$

Lemma 5.4, i.4 shows that $E_\nu = E_{\nu_1(\nu)} + E_{\nu_2(\nu)}$ for all $\nu \in W$. Since W, N, P are connected, it follows that $\nu_1(\nu) \in N$, $\nu_2(\nu) \in P$ for $\nu \in W$. Thus M is decomposable in all points of W and, in particular, in μ. ▨

So the C-surfaces can be divided into decomposable (in every point) and undecomposable (in each point).

5.10. LEMMA. <u>Suppose that M is a C-surface, $M \subset \sigma_m^\circ$ and $m > k$. Then M is decomposable</u>.

PROOF. Let Ψ be the set of self-intersections of $F(\mathbb{T})$. We consider two cases.

1) Suppose that for every $\eta_0 \in \partial M$ a continuous function d_{η_0} with values in $\mathbb{T}$ exists which is defined on a neighbourhood V_{η_0} of η_0 in ∂M and satisfies $d_{\eta_0}(\eta) \in \eta$ for all $\eta \in V_{\eta_0}$. This will be shown to contradict the assumption $m > k$. Indeed, for all $\eta \in \partial M \setminus \Pi_m^{-1}(\Psi)$ a unique $d \in \mathbb{T}$ exists such that $d \in \eta$. Hence the functions d_{η_0} with the above properties can be found uniquely. Therefore they "stick" into a continuous function $d : \partial M \to \mathbb{T}$.

For every $\eta_0 \in \partial M$ the set $d_{\eta_0}(V_{\eta_0})$ contains a neighbourhood of $d_{\eta_0}(\eta_0)$. Besides this, ∂M is compact. Thus the set $d(\partial M)$ is closed and open in $\mathbb{T}$ and so $d(\partial M) = \mathbb{T}$.

Choose any λ_0 in $(\partial\sigma(T_F)) \setminus \Psi$. Let $\lambda_0 = F(\zeta_0)$, $\zeta_0 \in \mathbb{T}$, and $\zeta_0 = d(\eta_0)$, $\eta_0 \in \partial M$ (so that $\Pi_m(\eta_0) = \lambda_0$). Choose also some η in M near η_0. Putting $\lambda = \Pi_m(\eta)$, we see that $\mathrm{wind}_F(\lambda) = 1$, $\mathrm{card}\, F^{-1}\{\lambda\} = k$, and $\eta \subset F^{-1}\{\lambda\}$. This gives the desired contradiction.

2) Suppose to the contrary that the function d_{η_0} with the above **properties** does not exist for some $\eta_0 \in \partial M$. Put $\lambda_0 = \Pi_m(\eta_0)$, then $\lambda_0 \in \Psi$. Functions $d_1, \dots, d_n$ and a neighbourhood V of λ_0 correspond to λ_0 as in Lemma 3.4. It can be supposed that the diameter of V is small enough and $V \cap F(\mathbb{T})$ is the union of some arcs α_i ending at λ_0. Each α_i is contained in some $\partial\mathcal{D}_j$. Suppose that some small neighbourhood of η_0 in ∂M projects into arcs α_{i_1} and α_{i_2}. Let $\alpha_{i_1} \subset \partial\mathcal{D}_{j_1}$ and $\alpha_{i_2} \subset \partial\mathcal{D}_{j_2}$. Then $j_1 \neq j_2$ due to the assumption. We change the enumeration so that $i_1 = j_1 = 1$, $i_2 = j_2 = 2$ (see Fig.5).

Put

$$\tilde{\eta}_J(\lambda) = \{ d_i(\lambda) : i \in J \}$$

for every index set J if $\lambda \in \bigcap_{j \in J} \mathcal{D}_j$. If J has ℓ elements, then $\tilde{\eta}_J$ is a continuous branch of Π_ℓ^{-1}.

We define the index set $\mathcal{I}$ by

$$\eta_0 = \tilde{\eta}_{\mathcal{I}}(\lambda_0).$$

Note that if $\mathcal{J}$ has m elements, $\mathcal{J} \neq \mathcal{I}$ and $\lambda \in \mathbb{C}$, then the distance between η_0 and $\tilde{\eta}_{\mathcal{J}}(\lambda)$ is bounded away from zero. It follows that the set

$$W \overset{def}{=\!=} \tilde{\eta}_{\mathcal{I}}(\cap_{j\in\mathcal{I}} \mathcal{D}_j)$$

is a neighbourhood of η_0 in $\overline{M}$. Some neighbourhood of λ_0 in $\Pi_m(\partial W)$ coincides with $\alpha_1 \cup \alpha_2$.

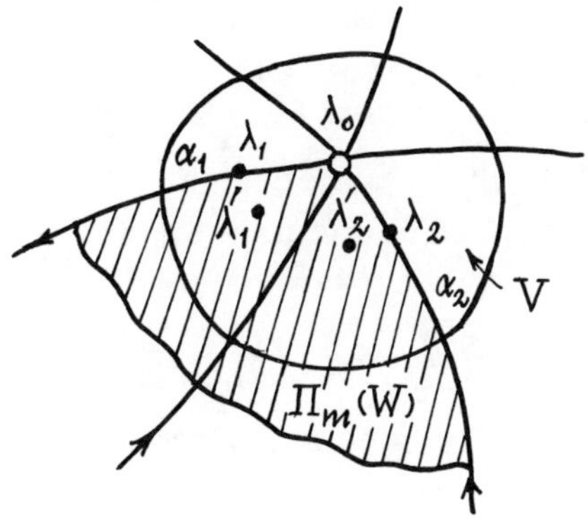

Figure 5

Choose points $\lambda_1 \in \alpha_1 \setminus \{\lambda_0\}$, $\lambda_2 \in \alpha_2 \setminus \{\lambda_0\}$ and then there are $\lambda_1', \lambda_2' \in \Pi_m(W)$ such that the distances $|\lambda_i - \lambda_i'|$ are small enough. It is assumed that for $i = 1,2$, λ_i' belongs to the connected component of $\mathbb{C} \setminus \mathbb{F}(\mathbb{T})$ whose boundary contains λ_i. Put

$$\mathcal{J}_i = \{ j : \lambda_i \in \mathcal{D}_j \}$$

for $i=1,2$. Then $\mathbb{F}^{-1}\{\lambda_i\} = \{\tilde{d}_j(\lambda_i) : j \in \mathcal{J}_i\}$ and $\mathcal{J}_i \supset \mathcal{I}$. For λ near λ_i, we have

$$\mathbb{F}^{-1}\{\lambda\} \cap \mathbb{D} = \tilde{\eta}_{\mathcal{J}_1}(\lambda) \qquad \text{if } \lambda \in \mathit{int}\, \Pi_m(W)$$

and

$$\mathbb{F}^{-1}\{\lambda\} \cap \mathbb{D} = \tilde{\eta}_{\mathcal{J}_1 \setminus \{1\}}(\lambda) \qquad \text{if } \lambda \notin \Pi_m(W)$$

(since $\alpha_1 \subset \partial \mathcal{D}_1$).

Put

$$\mu_i(\lambda) = \tilde{\eta}_{\mathcal{J}_i \setminus \{i\}}(\lambda)$$

and

for $i=1,2$. Let N_1, N_2 be the connected components of the surface σ^o_{m-1} containing $\mu_1(\lambda'_1)$ and $\mu_2(\lambda'_2)$ respectively. Then $\mu_i(\lambda) \in N_i$ for all λ near λ_i. For such λ we have $E_{\mu_i(\lambda)} = \operatorname{Ker}(T^*_F - \lambda I)$ if it is additionally assumed that $\lambda \notin \Pi_m(W)$. Hence N_1, N_2 are C-surfaces.

Next we make use of the relation

$$\mu_i(\lambda'_i) \cap \tilde{\eta}_{\mathfrak{X}}(\lambda'_i) = \tilde{\eta}_{\mathfrak{X}\setminus\{i\}}(\lambda'_i) .$$

Since $\tilde{\eta}_{\mathfrak{X}}(\lambda'_i) \in M$ and M is a C-surface, the points $\tilde{\eta}_{\mathfrak{X}\setminus\{1\}}(\lambda'_1)$ and $\tilde{\eta}_{\mathfrak{X}\setminus\{2\}}(\lambda'_2)$ belong to some C-surfaces too. The point $\tilde{\eta}_{\mathfrak{X}\setminus\{2\}}(\lambda'_1)$ belongs to the same C-surface as $\tilde{\eta}_{\mathfrak{X}\setminus\{2\}}(\lambda'_2)$ since the set $\tilde{\eta}_{\mathfrak{X}\setminus\{2\}}(\Pi_m(W))$ is connected. We have

$$\operatorname{card}(\tilde{\eta}_{\mathfrak{X}\setminus\{1\}}(\lambda'_1) \cap \tilde{\eta}_{\mathfrak{X}\setminus\{2\}}(\lambda'_1)) = m-2 \geq k-1 ,$$

which gives $E_{\tilde{\eta}_{\mathfrak{X}}(\lambda'_1)} = E_{\tilde{\eta}_{\mathfrak{X}\setminus\{1\}}(\lambda'_1)} + E_{\tilde{\eta}_{\mathfrak{X}\setminus\{2\}}(\lambda'_1)}$ by Lemma 5.4, i.4. This shows that M is decomposable. ▨

5.11. COMBINATORIAL TRIPLES AND COMBINATORIAL BASES. To deal with the linear relations (1.6), the following notions will be useful.

Points $\delta_1, \delta_2, \delta_3 \in \Pi^{-1}\{\lambda\}$ will be called a combinatorial triple if they have the form $\delta_i = \{d_1, d_2, \dots, d_{k+1}\} \setminus \{d_i\}$ here $d_1, \dots, d_{K+1}$ are points from the unit disc $\mathbb{D}$ with the property $F(d_1) = \dots = F(d_{k+1})$. Every single relation of the form (1.6) is connected with a combinatorial triple, and vice versa.

The following property will often be used. If $\tilde{\delta}_1, \tilde{\delta}_2, \tilde{\delta}_3$ are branches of Π^{-1} defined on a domain $\mathcal{U}$ and $f_1(\lambda)$, $f_2(\lambda)$, $f_3(\lambda)$ form a combinatorial triple for some $\lambda \in \mathcal{U}$, then the same takes place for all $\lambda \in \mathcal{U}$.

Let A be a subset of $\Pi^{-1}\{\lambda\}$ for some $\lambda \in \mathbb{C} \setminus F(\Xi)$. We shall say that A is combinatorially closed if for any combinatorial triple $\delta_1, \delta_2, \delta_3$ we have $\delta_1, \delta_2 \in A \Rightarrow \delta_3 \in A$. Suppose that $A, B \subset \Pi^{-1}\{\lambda\}$. We say that A generates B if B is contained in the smallest combinatorially closed set which contains A. In this case $span\{h_\delta : \delta \in B\} \subset span\{h_\delta : \delta \in A\}$ (see (3.5)). If A generates $\beta(E_\mu)$ and has $card(\mu) - k+1$ elements, then A is called a combinatorial basis in $\beta(E_\mu)$. Lemma 5.4, i.1, implies that $\{h_\delta : \delta \in A\}$ is a basis in E_μ in the last case.

It follows easily from Lemma 5.10 that the sum of spaces spanned by h_δ, $\delta \in \sigma_{**} \cap \Pi^{-1}\{\lambda\}$, coincides with $Ker(T_{\mathbb{F}}^* - \lambda I)$ for every $\lambda \notin \mathbb{F}(\Xi)$. We need a stronger statement.

5.12. LEMMA. Let $\lambda \notin \mathbb{F}(\Xi)$. Then the set $\Pi^{-1}\{\lambda\} \cap \sigma_{**}$ contains a combinatorial basis for $\beta(Ker(T_{\mathbb{F}}^* - \lambda I))$.

PROOF. The following sharper statement is true: the set $\beta(E_\eta)$ has a combinatorial basis contained in σ_{**} for every C-surface M and every $\eta \in M$. We prove the latter by induction on $card(\eta)$. The statement is obvious if $card(\eta) = k$. Suppose that $card(\eta) = \ell > k$. Then the construction of i.2 of the last proof can be applied to M. Since M is connected, it follows easily that the statement depends on M and not on η. So we assume that $\eta = \tilde{\eta}_\varkappa(\lambda_1')$. The induction hypothesis enables us to choose combinatorial bases τ_1 in $\beta(E_{\tilde{\eta}_{\varkappa \setminus \{1\}}(\lambda_1')})$ and τ_2 in $\beta(E_{\tilde{\eta}_{\varkappa \setminus \{2\}}(\lambda_1')})$ contained in σ_{**}. There must be a point δ in τ_2 such that $\mathfrak{a}_1(\lambda_1') \in \delta$. Then $\tau_1 \cup \{\delta\}$ is a combinatorial basis in $\beta(E_\eta)$. This concludes the induction step.

It remains to apply the above statement to $\eta = \Pi^{-1}\{\lambda\}$, since this point lies on some C-surface. ▨

5.13. THEOREM ON COMMUTATIVITY OF THE COMMUTANT. Suppose that conditions (i)-(v) are fulfilled. Then

1) To every $K \in \{T_{\mathbb{F}}\}'$ there corresponds a function φ_K which is analytic on σ_{**} and such that

$$K^* h_\delta = \varphi_K(\delta) h_\delta, \quad \forall \delta \in \sigma_{**}; \tag{5.4}$$

2) Each function $u \in H^2_{\mathbb{F}}(\sigma_*)$ is determined uniquely by its values on σ_{**};

3) For every $K \in \{T_{\mathbb{F}}\}'$ the equation

$$(UKx)(\delta) = \varphi_K(\delta)(Ux)(\delta), \quad \delta \in \sigma_{**}, \tag{5.5}$$

holds for $x \in H^2$; here U is the isomorphism from the Similarity theorem 1.5;

4) The commutant $\{T_F\}'$ is abelian.

PROOF. 1) Let $K \in \{T_F\}'$. By Corollary 5.7, a function φ_K defined on σ°_{**} exists with the property $K^* h_\delta = \varphi_K(\delta) h_\delta$ for all $\delta \in \sigma^{\circ}_{**}$. It follows that $|\varphi_K(\delta)| \leq \|K\|$ for $\delta \in \sigma^{\circ}_{**}$. Hence the function φ_K can be continued analytically to σ_{**}.

The statement 2) follows easily from Lemma 5.12 and relations (1.6).

The statement 3) is a reformulation of 1).

4) For every K, $L \in \{T_F\}'$ and $x \in H^2$ we have

$$(UKLx)(\delta) = \varphi_K(\delta)\varphi_L(\delta)(Ux)(\delta) = (ULKx)(\delta),$$

$\delta \in \sigma_{**}$. It follows from 2) that $UKLx = ULKx$. Hence $KL = LK$. ▨

5.14. COROLLARY. The commutant $\{T_F\}'$ does not contain non-zero compact operators.

To show this, note that for every non-empty open V, $V \subset \sigma_{**}$ the space $span\{h_\delta : \delta \in V\}$ is infinite-dimensional, as can easily be seen by considering rational functions $\omega^{-1} h_\delta$ (see 1.5). Therefore if $K \in \{T_F\}'$ and $\varphi_K \neq 0$, then the space $span\{Ker(K^* - \lambda I) : |\lambda| > \varepsilon\}$ is infinite-dimensional for some $\varepsilon > 0$. Hence K is not a compact operator. ▨

Other cases when this corollary is true for analytic Toeplitz operators can be found in [10, 28, 34] and in other papers.

Sharpening Lemma 5.10 one can show that this theorem is true even if it is only assumed instead of (v) that $F(\alpha) \neq F(\beta)$ for every distinct arcs α and β of $\mathbb{T}$.

The proof that $\{T_F\}'$ is abelian would be much shorter if we could show that the eigenvectors h_δ such that $wind_F(\Pi(\delta)) = 1$

are **complete in** H^2[28]. Indeed, if $wind_{F}(\Pi(\delta))=1$ and $K\in\{T_F^*\}'$, then

$$T_F^* K^* h_\delta = K^* T_F^* h_\delta = \Pi(\delta) K^* h_\delta$$

and thus $K^* h_\delta = \mu_K(\delta) h_\delta$ for some function μ_K. Hence for δ such that $wind_F(\Pi(\delta))=1$ we obtain

$$K^* L^* h_\delta = L^* K^* h_\delta .$$

The **completeness of corresponding** h_δ implies that $K^* L^* = L^* K^*$. For curves $F(\mathbb{T})$ with simple geometry, this property is true. The following example shows that it is not true in general.

5.15. EXAMPLE OF A SYMBOL F SATISFYING (i)-(vii) SUCH THAT

$$\mathcal{L} \stackrel{def}{=} span\{Ker(T_F^* - \lambda I) : dim\, Ker(T_F^* - \lambda I) = 1\} \neq H^2 .$$

This example is given by a function F with one pole; the curve $F(\mathbb{T})$ is presented on Fig.6b. The function F can be described in the following way. Consider a Riemann surface R which can be imagined as the exterior $\hat{\mathbb{C}} \setminus \bar{\mathbb{D}}$ of the disc $\mathbb{D}$ with four shoots added. The beginnings of shoots are marked

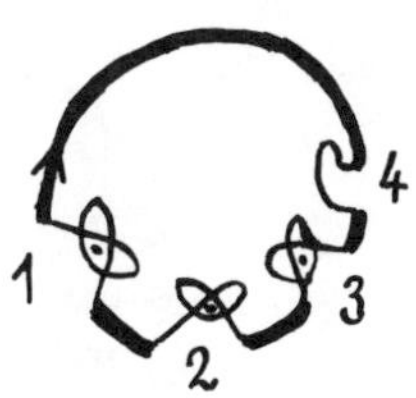

Fig. 6a

Fig.6b

with figures 1-4; the Fig.6a shows these shoots diminished. Each of the shoots 1-3 has one branch point and the shoot 4 has no branch points. Since R is homeomorphic to $\mathbb{D}$, a conformal mapping $\tilde{F}: \mathbb{D} \to R$ exists. The function F is a composition of $\tilde{\tilde{F}}$ with the projection of $\tilde{R}$ onto $\hat{\mathbb{C}}$.

We have $K=2$ and $\omega(z) \equiv 1$ in our case (see i.1.3). Formulae (3.1), (3.3), (3.4) imply that

$$x \in \mathcal{X}^{\perp} \Longleftrightarrow d_1 x(d_1) = d_2 x(d_2), \qquad \forall \delta = \{d_1, d_2\} : \operatorname{wind}_F(\Pi(\delta)) = 1 .$$

Form a surface $\tilde{R}$ by sticking the sheet 4 to the sheet containing ∞. Let $\nu: R \to \tilde{R}$ be the corresponding mapping. Then every function of the form $z^{-1}(g \circ \nu \circ \tilde{F})(z)$, where $g \in H^{\infty}(\tilde{R})$, $g(\infty)=0$, belongs to $\mathcal{X}^{\perp}$. A non-zero function g on $\tilde{R}$ can be found since the surface $\tilde{R}$ is not compact. ▨

6. A COMPLETE DESCRIPTION OF COMMUTANT.

Hereafter the symbol F will be supposed to satisfy (i)-(vii).

The formula (5.5) shows that to describe $\{T_F\}'$ it is useful to obtain a description of traces $u|\sigma_{**}$ of functions $u \in H^2_F(\sigma_*)$. The following three lemmas are aimed to do this. We shall need some new notions here.

6.1. CLASSES $\mathcal{H}_F(\sigma_{**})$ AND $H^2_F(\sigma_{**})$. Class $\mathcal{H}_F(\sigma_{**})$ is the set of those functions $v \in \mathcal{H}(\sigma_{**})$ which can be continued to a function u on $\sigma_{**} \cup \sigma_*^{\circ}$ satisfying (1.6) on σ_*°. This condition is obviously equivalent to some linear relations on v on each fiber $\Pi^{-1}\{\lambda\} \cap \sigma_{**}$. The continuation of v in the above sense is unique by Lemma 5.12.

Put $H^2_F(\sigma_{**}) = H^2(\sigma_{**}) \cap \mathcal{H}_F(\sigma_{**})$.

6.2. COMBINATORIAL SEQUENCES. A sequence $\delta_1, \dots, \delta_n$ of points of some fiber $\Pi^{-1}\{\lambda\}$ ($\lambda \in \mathbb{C}$) will be called a "combinatorial sequence" if every point δ_j belongs to σ_{**} or forms a combinatorial triple (i.5.11) with some two preceding points. Every point $\delta \in \sigma_*^{\circ}$ can be included into some combinatorial sequence because of Lemma 5.12.

6.3. LEMMA. If $v \in \mathcal{H}_F(\sigma_{**})$ and u is a continuation of v in the sense of 6.1, then $u \in \mathcal{H}_F(\sigma_*)$.

PROOF. Suppose that $\delta_0 \in \sigma_* \setminus \Pi^{-1}(F(\Xi))$ and put $\lambda_0 = \Pi(\delta_0)$. Consider the branches α_j , $1 \le j \le n$ of F^{-1} and the sets V, $\mathcal{D}_1, \dots, \mathcal{D}_n$ that correspond to λ_0 by Lemma 3.4. We use the notation $\delta_J(\lambda) = \{\alpha_j(\lambda) : j \in J\}$ for $\lambda \in \mathbb{C}$ and $J \subset \{1, \dots, n\}$.

Put $A = \{j : \alpha_j(\lambda_0) \in \mathbb{T}\}$. The condition (vi) shows that card $(A) \le 2$ and that $\lambda_1 \in V$ exist such that $\lambda_1 \notin \mathcal{D}_j$ for $j \in A$. Suppose that $\delta_0 = \delta_{J_0}(\lambda_0)$ and consider the point $\delta_{J_0}(\lambda_1)$. One can find a combinatorial sequence $\delta_{J_1}(\lambda_1), \dots, \delta_{J_r}(\lambda_1)$ with $J_r = J_0$. Since $A \cap J_i = \emptyset$ for $1 \le i \le r$, it follows that $\delta_{J_1}(\lambda), \dots, \delta_{J_r}(\lambda)$ are defined and form a combinatorial sequence for all $\lambda \in V$. Using (1.6) and the assumption $u \in \mathcal{H}(\sigma_{**})$, we conclude successively that the functions $u \circ \delta_{J_i}$ are analytic for $i = 1, 2, \dots, r$. Hence u is analytic in a neighbourhood of δ_0 .

It remains to prove that u is analytic on $\Pi^{-1}(\zeta)$ for every $\zeta \in F(\Xi)$. Fix some ζ and some disc V centered at ζ such that $V \cap F(\Xi \cup \mathbb{T}) = \{\zeta\}$. We show that u is analytic on $\Pi^{-1}(V)$. Considering $F - \zeta$ instead of F , we arrive at the case $\zeta = 0$.

Due to (vi), F has a unique multiple root (say, d°). The multiplicity p of d equals 2 or 3. Let n be the number of the pre-images of 0 under F counted with their multiplicities. There are single-valued analytic functions $\alpha_1, \dots, \alpha_n$ defined on some disc W with centre at 0 such that

$$F \circ \alpha_j(\mu) \equiv \mu^p$$

and $F^{-1}\{\mu^p\} = \{\alpha_1(\mu), \dots, \alpha_n(\mu)\}$ for $\mu \in W$. Put

$$\eta_J(\mu) = \{\alpha_j(\mu) : j \in J\}$$

if J is an index set. If $\Pi(\delta) = \mu^p$ and $\mu \in W \setminus \{0\}$, then $\delta = \eta_J(\mu)$ for some $J \subset \{1, 2, \dots, n\}$. We show that all functions of the form $u\circ\eta_J$, $J\subset\{1,\dots,n\}$, are bounded.

For any J, $J\subset\{1,\dots,n\}, card(J)=k$, we can construct a sequence $\eta_{J_1}(\mu),\dots,\eta_{J_m}(\mu)$ which is combinatorial for each $\mu\in W\setminus\{0\}$ and satisfies $J_m=J$. Note that if some points $\eta_{J^{(1)}}(\mu)$, $\eta_{J^{(2)}}(\mu)$, $\eta_{J^{(3)}}(\mu)$ form a combinatorial triple for $\mu\in W\setminus\{0\}$ and the functions $u\circ\eta_{J^{(1)}}$, $u\circ\eta_{J^{(2)}}$ are bounded, then so is $u\circ\eta_{J^{(3)}}$. To prove this we may assume that $J^{(i)}=$ $=\{1,2,\dots,k+1\}\setminus\{i\}$, $i=1,2,3$. We have

$$u\circ\eta_{J^{(3)}}=\frac{(d_2-d_3)(u\circ\eta_{J^{(1)}})+(d_3-d_1)(u\circ\eta_{J^{(2)}})}{d_2-d_1} .$$

If $d_2(0)\neq d_1(0)$, then the right-hand side is bounded on W. Suppose that $d_2(0)=d_1(0)$; then $d_2(0)=d_1(0)=d^\circ$. Since $p=2$ or $p=3$, it follows that an ε exists such that $\sigma_{J^{(1)}}(\mu)\equiv\sigma_{J^{(2)}}(\varepsilon\mu)$ and $\varepsilon^p=1$. Hence the numerator of the right-hand side fraction is zero for $\mu=0$. Since d_2-d_1 has first-order zero in 0 , the boundedness of $u\circ\eta_{J^{(3)}}$ on W follows.

We conclude successively that the functions $u\circ\tilde\eta_{J_1},\dots,u\circ\tilde\eta_{J_m}$ are bounded on $W\setminus\{0\}$. Therefore the function u is bounded on $\Pi^{-1}(V\setminus\{0\})$ and consequently is analytic on $\Pi^{-1}(V)$. ▨

6.4. LEMMA. <u>The mapping</u> $u\to u|\sigma_{**}$ <u>is an isomorphism of</u> $H^2_{\mathbb{F}}(\sigma_*)$ <u>onto</u> $H^2_{\mathbb{F}}(\sigma_{**})$.

PROOF. This mapping is one-to-one by the Theorem 5.13, i.2). If $u\in H^2_{\mathbb{F}}(\sigma_*)$, then obviously $u|\sigma_{**}\in H^2_{\mathbb{F}}(\sigma_{**})$. Conversely, if $v\in H^2_{\mathbb{F}}(\sigma_{**})$, then $v=u|\sigma_{**}$ for some $u\in\mathcal{H}_{\mathbb{F}}(\sigma_*)$ because of the previous lemma. It remains to verify that u belongs to the class H^2 locally (see Remark 3.6, 1)).

Suppose that $\sigma_0\in\partial\sigma_*$ and $\lambda_0=\Pi(\sigma_0)$. Let d_j be branches of $\mathbb{F}^{-1}$ defined on some sets $\mathcal{D}_j$ containing λ_0 (see Lemma 3.4) and let the symbols f_J, A, V have the same sense as described in the beginning of the previous proof. Assume that $\sigma_0=f_{J_0}(\lambda_0)$ and put $A'=A\setminus J_0$. Take a point $\lambda_1\in\bigcap_{j\in J_0}\mathcal{D}_j\setminus\mathbb{F}(\mathbb{T})$ such that $\lambda_1\notin\mathcal{D}_j$ for $j\in A'$. This is possible since λ_0 is at the worst a double transversal intersection point.

Choose a combinatorial sequence $f_{J_1}(\lambda_1), \ldots, f_{J_p}(\lambda_1)$ such that $J_p = J_0$. The points $f_{J_p}(\lambda)$ for $\lambda \in \bigcap_{j \in J_\ell} \mathcal{D}_j$ form a neighbourhood of σ_0 in σ_*. (An analogous fact has been used already in the proof of Lemma 5.10). The functions $f_{J_1}, \ldots, f_{J_{p-1}}$ are defined for these λ because of the choice of λ_1. We derive succesively from (1.6) and the condition $u|\sigma_{**} \in H^2(\sigma_{**})$ that $u \circ f_{J_\ell} \in H^2$, $\ell = 1, \ldots, p$. Hence u belongs to H^2 locally in σ_0. ▨

6.5. COROLLARY. <u>The formula</u>

$$(U_{**}x)(\sigma) = \langle x, h_\sigma \rangle, \quad \sigma \in \sigma_{**},$$

<u>defines an isomorphism of</u> H^2 <u>onto</u> $H^2_{\mathbb{F}}(\sigma_{**})$ <u>such that</u>

$$U_{**} T_{\mathbb{F}} U_{**}^{-1} = M_{\Pi | \sigma_{**}}.$$

<u>Here</u> $M_{\Pi|\sigma_{**}}$ <u>is multiplication by</u> Π <u>on</u> $H^2_{\mathbb{F}}(\sigma_{**})$.

This follows immediately from Theorem 5.13 and the previous lemma. ▨

In view of Theorem 5.13, 3), to describe $\{T_{\mathbb{F}}\}'$ it remains now to describe the functions $\varphi \in H^\infty(\sigma_{**})$ such that the multiplication operator M_φ acts from $H^2_{\mathbb{F}}(\sigma_{**})$ onto itself. The following lemma is the first step in this direction. It will be applied to vectors h_σ, $\sigma \in \sigma_{**} \cap \Pi^{-1}\{\lambda\}$, for a fixed $\lambda \in \mathbb{C} \setminus \mathbb{F}(\Xi)$.

6.6. LEMMA ON VECTORS IN $\mathbb{C}^n$. <u>Suppose that</u> $x_1, \ldots, x_\ell$ <u>are vectors in</u> $\mathbb{C}^n$. <u>Then there exists a finite collection</u> $\{N_1, \ldots, N_r\}$ <u>of subspaces in</u> $\mathbb{C}^n$ <u>such that</u>:

1) $\forall i \; \exists j : x_i \in N_j$;

2) $\forall j \; N_j \cap \operatorname{span}\{N_i : i \neq j\} = 0$;

3) <u>For any subspaces</u> $\{N'_j\}_1$ <u>satisfying 1) and 2), every</u> N_j <u>is contained in some</u> N'_i;

4) <u>Every operator</u> K <u>on</u> $\mathbb{C}^n$ <u>with eigenvectors</u> $x_1, \ldots, x_\ell$ <u>has</u> $N_1, \ldots, N_r$ <u>as eigenspaces</u>.

<u>The spaces</u> N_j <u>are linear spans of some subsets of</u> $\{x_1, \ldots, x_\ell\}$

<u>which can be calculated using only the numbers</u> $dim(span_{i\in J}\, x_i)$ <u>for all</u> $J\subset\{1,\dots,\ell\}$.

The proof will be carried out by induction on ℓ . The induction base is trivial. Let $N_1^{(\ell)},\dots,N_r^{(\ell)}$ satisfy 1)-4) and let $x_{\ell+1}$ be a new vector in $\mathbb{C}^n$. We take the spaces $N_1^{(\ell)},\dots,N_r^{(\ell)},(x_{\ell+1})$ as $N_j^{(\ell+1)}$ if $x_{\ell+1}\notin span\{N_j^{(\ell)}: 1\le j\le r\}$. Otherwise represent $x_{\ell+1}$ in the form

$$x_{\ell+1} = y_{i_1} + \dots + y_{i_p}$$

where $1\le i_1<i_2<\dots<i_p\le s$, $y_{i_j}\in N_{i_j}^{(\ell)}$, $y_{i_j}\ne 0$. Such a representation exists and is unique by the property 2) of $N_i^{(\ell)}$. We take the subspaces $N_i^{(\ell)}$ for $i\ne i_j$ $(1\le j\le p)$ **together** with the space $span\{N_{i_j}^{(\ell)}: 1\le j\le p\}$ as $N_i^{(\ell+1)}$ in this case. It is routine to derive the properties 1)-3) of the spaces $N_i^{(\ell+1)}$ (with respect to $x_1,\dots,x_{\ell+1}$) from the corresponding properties of $N_i^{(\ell)}$. To prove 4), suppose that K is an operator on $\mathbb{C}^n$ satisfying $K|N_i^{(\ell)}=\mu_i I$ (the induction hypothesis) and $Kx_{\ell+1}=\mu x_{\ell+1}$. We have

$$\mu(y_{i_1}+\dots+y_{i_p}) = \mu x_{\ell+1} = \mu_{i_1} y_{i_1} + \dots + \mu_{i_p} y_{i_p} ,$$

which yields $\mu=\mu_{i_1}=\dots=\mu_{i_p}$. **Hence** $span\{N_{i_j}^{(\ell)}\}$ is an eigenspace for K , which proves 4).

To pass from $N_i^{(\ell)}$ to $N_i^{(\ell+1)}$ it suffices to have information on the linear dependence of $x_{\ell+1}$ on $\{x_i : i\in J\}$ for all subsets J of $\{1,\dots,\ell\}$. This implies the last assertion of the lemma. ▨

Note that the property 3) shows in particular that the subspaces N_i can be found uniquely.

6.7. AN EQUIVALENCE RELATION ON σ_{**} . Suppose that $\lambda\in\sigma(T_F)\setminus F(\Xi)$. The vectors $\{h_\delta : \delta\in\sigma_{**}\cap\Pi^{-1}\{\lambda\}\}$ span the space $Ker(T_F^*-\lambda I)$. Last lemma shows that this space decomposes into a direct sum of spaces $N_i(\lambda)$ whose union contains all described vectors. We write $\delta_1\sim\delta_2$ if $\delta_1,\delta_2\in\sigma_{**}$, $\Pi(\delta_1)=\Pi(\delta_2)=\lambda$ and $h_{\delta_1}, h_{\delta_2}$ belong to the same space $N_i(\lambda)$. After performing this construction for all $\lambda \notin F(\Xi)$ we obtain an equivalence relation on σ_{**}. The following assertion explains why this definition is useful.

6.8. LEMMA. <u>The formula</u>

$$(UKU^{-1}w)(\delta) \overset{def}{=\!=} \varphi(\delta)w(\delta), \quad w\in H^2_F(\sigma_*), \quad \delta\in\sigma_{**}, \tag{6.1}$$

<u>defines a one-to-one correspondence between the operators</u> K <u>in</u> $\{T_F\}'$ <u>and the functions</u> $\varphi\in H^\infty(\sigma_{**})$ <u>that take equal values at equivalent points.</u>

PROOF. If $K\in\{T_F\}'$, then a function $\varphi_K\in H^\infty(\sigma_{**})$ exists that satisfies (5.5) which is equivalent to (5.4). Lemma 6.6 applied to vectors h_δ, $\delta\in\sigma_{**}\cap\Pi^{-1}\{\lambda\}$, and to $K^*|Ker(T_F^*-\lambda I)$ gives the implication $\delta_1\sim\delta_2 \Rightarrow \varphi(\delta_1)=\varphi(\delta_2)$.

Conversely, let φ be a function from $H^\infty(\sigma_{**})$ such that $\delta_1\sim\delta_2\Rightarrow\varphi(\delta_1)=\varphi(\delta_2)$. Let $v\in H^2_F(\sigma_{**})$; we must show that $\varphi v\in H^2_F(\sigma_{**})$. By Lemma 6.4, $v=u|\sigma_{**}$ for some $u\in H^2_F(\sigma_*)$. Take an arbitrary $\lambda\in\sigma(T_F)\setminus F(\Xi)$. Define an operator Φ_λ on $Ker(T_F^*-\lambda I)$ by the rule: for every $N_i(\lambda)$ and every $h_\delta\in N_i(\lambda)$ set

$$\Phi_\lambda|N_i(\lambda)\equiv\varphi(\delta)I.$$

The condition imposed on φ ensures that Φ_λ is well-defined. We have

$$\Phi_\lambda h_\delta=\varphi(\delta)h_\delta,\quad \delta\in\sigma_{**}\cap\Pi^{-1}\{\lambda\}.$$

Let $u\in Ux$, where $x\in H^2$. Set

$$y(\delta)=\langle x,\Phi_\lambda h_\delta\rangle$$

for $\delta\in\sigma_*\cap\Pi^{-1}\{\lambda\}$. The function y is a continuation of the function $M_\varphi v$ to the fiber $\Pi^{-1}\{\lambda\}$ satisfying (1.6) due to (3.5). This implies that $M_\varphi v\in H^2_F(\sigma_{**})$.

Hence M_φ is a bounded operator on $H^2_F(\sigma_{**})$. It obviously belongs to $\{M_{\Pi|\sigma_{**}}\}'$. Put

$$K = U_{**}^{-1} M_\varphi U_{**}$$

(see Corollary 6.5). Then K belongs to $\{T_F\}'$, and K and φ are connected by (6.1). ▨

We need some topological properties of the equivalence $\sim$ on σ_{**}.

6.9. LEMMA. Suppose that W is a simply connected open subset of $\mathbb{C}$, with $W \cap F(\Xi \cup \mathbb{T}) = \emptyset$. There are single-valued continuous branches $\mathfrak{d}_i$, $i \in L$ (L being a finite index set) of the function $F^{-1}|W$ such that $F^{-1}\{\lambda\} = \{\mathfrak{d}_i(\lambda), \lambda \in L\}$ for all $\lambda \in W$. Put $f_J(\lambda) = \{\mathfrak{d}_i(\lambda) : i \in J\}$ if $J \subset L$.

1) **There exists a set** Λ which has no accumulation points in W such that for every J_1, J_2 with $card(J_1) = card(J_2) = k$ the equivalence of the points $f_{J_1}(\lambda), f_{J_2}(\lambda)$ does not depend on λ if $\lambda \in W \setminus \Lambda$.

2) If $\lambda_0 \in \Lambda$ and $f_{J_1}(\lambda_0) \sim f_{J_2}(\lambda_0)$, then $f_{J_1}(\lambda) \sim f_{J_2}(\lambda)$ for every $\lambda \notin \Lambda$.

PROOF. The existence of the $\mathfrak{d}_i$ follows from the monodromy theorem. Suppose that $J_1, \dots, J_n$ are index sets. Lemma 5.2, a) implies that the vectors $h_{f_{J_1}(\lambda)}, \dots, h_{f_{J_n}(\lambda)}$ are either linearly dependent for all λ or are linearly independent for all λ outside some set $\Lambda(J_1, \dots, J_n)$ which is discrete in W. Take Λ to be the union of the sets $\Lambda(J_1, \dots, J_n)$ for all choices of $J_1, \dots, J_n$ and for all n which do not exceed the number of sheets of σ_* over W. Then for every n and every $J_1, \dots, J_n$ the quantity $\dim span\{h_{J_i(\lambda)} : 1 \le i \le n\}$ is constant for $\lambda \in W \setminus \Lambda$. Therefore the statement 1) is a consequence of the last assertion of Lemma 6.6.

It follows that there is a partition of the set of all k-element subsets of L into sets $\mathfrak{A}_1, \dots, \mathfrak{A}_t$ such that the spaces $N_i(\lambda)$ coincide with the spaces

$$M_i(\lambda) \overset{def}{=} span\{h_{f_J(\lambda)} : J \in \mathfrak{A}_i\}$$

if $\lambda \in W \setminus \Lambda$. Suppose now that $\lambda_0 \in \Lambda$. The union of the spaces $M_i(\lambda_0)$, $1 \leq i \leq t$, contains every h_σ with $\sigma \in \sigma_{**} \cap \Pi^{-1}\{\lambda_0\}$. Lemma 5.2, a) implies that $\dim M_i(\lambda_0) \leq \dim M_i(\lambda)$ if $\lambda \notin \Lambda$. Hence

$$\sum_{i=1}^{t} \dim M_i(\lambda_0) \leq \sum_{i=1}^{t} \dim M_i(\lambda) = \dim \operatorname{Ker}(T_{\mathbb{F}}^* - \lambda_0 I)$$

(here $\lambda \in W \setminus \Lambda$). Since $\operatorname{span}_i\{M_i(\lambda_0)\} = \operatorname{Ker}(T_{\mathbb{F}}^* - \lambda_0 I)$, we conclude that the vectors h_σ, $\sigma \in \Pi^{-1}\{\lambda_0\}$, and the spaces $M_i(\lambda_0)$ satisfy properties 1)-2) from Lemma 6.6, with $\mathbb{C}^n$ replaced by $\operatorname{Ker}(T_{\mathbb{F}}^* - \lambda_0 I)$. Therefore every $N_j(\lambda_0)$ is contained in some $M_i(\lambda_0)$, $M_i(\lambda_0)$, and 2) follows. ▨

6.10. THE DEFINITION OF THE COMMUTANT SURFACE σ_c . The set $\sigma(T_{\mathbb{F}}) \setminus \mathbb{F}(\Xi \cup \mathbb{T})$ can be covered by a finite number of simply connected domains W_j . Let Λ be the union of the correspond "exceptional" sets Λ_j . All accumulation points of Λ lie in $\mathbb{F}(\Xi \cup \mathbb{T})$. Pick a triangulation τ of $\sigma(T_{\mathbb{F}})$ consisting of (open) curvilinear triangles R_j . We assume that the points of $\mathbb{F}(\Xi) \cap \sigma(T_{\mathbb{F}})$ and the self-intersection points of $\mathbb{F}(\mathbb{T})$ are the vertices of τ , $\mathbb{F}(\mathbb{T}) \subset \cup_j \partial R_j$, and $\operatorname{int}(R_\ell) \cap \cap \mathbb{F}(\mathbb{T}) = \emptyset$ for all ℓ . The components $\widetilde{R}_j$ of the sets $\Pi^{-1}(R) \cap \sigma_{**}$ where R ranges over triangles of τ form a triangulation of the reduced ultraspectrum σ_{**} . It may be assumed that $\partial \widetilde{R}_j \cap \cap \Lambda = \emptyset$ for all j . The commutant surface is obtained from the surface σ_{**} by means of sticking some of its sheets which is held according to the following rules:

1) Triangles $\widetilde{R}_i, \widetilde{R}_j$ are sticked together (by means of $(\Pi|\widetilde{R}_j)^{-1}(\Pi|\widetilde{R}_i)$ if they lie above some triangle R_ℓ , $R_\ell \in \tau$, and contain equivalent points. Moreover, the correspondent sides of $\widetilde{R}_i, \widetilde{R}_j$ are sticked too whenever they belong to σ_{**} (and not to $\partial \sigma_{**}$).

2) Let b_i be a common side of $\widetilde{R}_{i_1}, \widetilde{R}_{i_2}$ and b_j be a common side of $\widetilde{R}_{j_1}, \widetilde{R}_{j_2}$. If the triangles $\widetilde{R}_{i_1}, \widetilde{R}_{j_1}$ have been sticked (with b_i sticked to b_j), then we also stick $\widetilde{R}_{i_2}$ and $\widetilde{R}_{j_2}$.

This process terminates after a finite number of steps and

yields a set σ_c (which is a quotient set of σ_{**}). Let $\Pi_c : \sigma_* \to \sigma_c$ be the corresponding mapping. The equation

$$\rho_c \circ \Pi_c = \Pi \mid \sigma_{**}$$

obvisouly defines a mapping $\rho_c : \sigma_c \to \mathbb{C}$. The topology induced by Π_c endows the set σ_c with a structure of a two-dimensional manifold. Indeed, every point of $\sigma_c \setminus \rho_c^{-1}(\mathbb{F}(\Xi))$ has a neighbourhood which is homeomorphic to a disc because of the rule 1). As for points of $\rho_c^{-1}(\mathbb{F}(\Xi))$, this assertion follows from a simple combinatorial argument which is omitted. There is a unique Riemann structure on σ_c such that the function ρ_c is analytic, σ_c endowed with this Riemann structure will be called the commutant surface. The mapping $\Pi_c : \sigma_{**} \to \sigma_c$ is certainly holomorphic.

6.11. THE COMMUTANT DESCRIPTION THEOREM. Suppose that the symbol $\mathbb{F}$ has properties (i)-(vii). Then the algebras $H^\infty(\sigma_c)$ and $\{T_{\mathbb{F}}\}'$ are isomorphic. The isomorphism $\mathcal{K} : H^\infty(\sigma_c) \to \{T_{\mathbb{F}}\}'$ can be defined by

$$(U\mathcal{K}(\psi)U^{-1}v)(\delta) = \psi(\Pi_c(\delta))\cdot v(\delta), \quad v \in H^2_{\mathbb{F}}(\sigma_*), \delta \in \sigma_{**} . \tag{6.2}$$

Introducing some new norm in H^2 (which is equivalent to the usual one) and the corresponding operator norm in $\{T_{\mathbb{F}}\}'$ we can convert $\mathcal{K}$ into an isometric isomorphism.

PROOF. $\mathcal{K}$ is obviously linear and multiplicative and the relation $\mathcal{K}(1) = I$ holds. In view of Lemma 6.8, to prove that $\mathcal{K}$ is an isomorphism, it suffices to check that the class of functions $\{\psi \circ \Pi_c : \psi \in H^\infty(\sigma_c)\}$ equals the class of those functions $\varphi \in H^\infty(\sigma_{**})$ which take equal values at equivalent points. The rule 1) of the definition of σ_c yields

$$\delta_1 \sim \delta_2 \Longrightarrow \Pi_c(\delta_1) = \Pi_c(\delta_2) \Rightarrow (\psi \circ \Pi_c)(\delta_1) = (\psi \circ \Pi_c)(\delta_2) .$$

Conversely, let φ be a function from $H^\infty(\sigma_c)$ which takes equal values at equivalent points. We have to show that $\varphi = \psi \circ \Pi_c$

for some $\psi \in H^{\infty}(\sigma_c)$.

Let us first prove that $\Pi_c(\delta_1) = \Pi_c(\delta_2)$ implies $\varphi(\delta_1) = \varphi(\delta_2)$. It suffices to verify that $\varphi \circ (\Pi | \tilde{R}_i)^{-1} = \varphi \circ (\Pi | \tilde{R}_j)^{-1}$ for every triangles $\tilde{R}_i, \tilde{R}_j$ which have been sticked together. Recall that $\Pi(\tilde{R}_i) = \Pi(\tilde{R}_j)$ in this case. If $\tilde{R}_i, \tilde{R}_j$ have been sticked according to the first rule, then $(\Pi|\tilde{R}_i)^{-1}(\lambda) \sim (\Pi|\tilde{R}_j)^{-1}(\lambda)$ for all $\lambda \in (\operatorname{int} \Pi(R_i)) \setminus \Lambda$ in view of Lemma 6.9. Hence

$$\varphi \circ (\Pi | \tilde{R}_i)^{-1} \equiv \varphi \circ (\Pi | \tilde{R}_j)^{-1}$$

on $\operatorname{int} \Pi(\tilde{R}_i)$. Now suppose that the triangles $\tilde{R}_{i_1}$, $\tilde{R}_{i_2}$, $\tilde{R}_{j_1}$, $\tilde{R}_{j_2}$ are as in the second rule. Then $\varphi \circ (\Pi|\tilde{R}_{i_1})^{-1} \equiv \varphi \circ (\Pi|\tilde{R}_{j_1})^{-1}$ implies $\varphi \circ (\Pi|\tilde{R}_{i_2})^{-1} \equiv \varphi \circ (\Pi|\tilde{R}_{j_2})^{-1}$ by the uniqueness theorem for analytic functions. Formally a third case exists: the triangles $\tilde{R}_i$, $\tilde{R}_j$ can be sticked together since they are sticked to some third triangle $\tilde{R}_\ell$. We have

$$\varphi \circ (\Pi| \tilde{R}_i)^{-1} \equiv \varphi \circ (\Pi | \tilde{R}_\ell)^{-1} \equiv \varphi \circ (\Pi| \tilde{R}_j)^{-1}$$

in this case. Thus an obvious induction argument concludes the proof.

As a consequence, we get some bounded function ψ on σ_c such that $\varphi = \psi \circ \Pi_c$. It is easy to see that $\psi \in \mathcal{H}(\sigma_c \setminus \rho_c(\Xi))$. Therefore $\psi \in H^{\infty}(\sigma_c)$. Hence $\mathcal{K}$ is an isomorphism of algebras $H^{\infty}(\sigma_c)$ and $\{T_F\}'$. The equality $\rho_c \circ \Pi_c = \Pi$ implies that $\mathcal{K}(\rho_c) = T_F$.

Let us introduce a new norm in H^2 by

$$\|x\|_F \overset{def}{=\!=} \|U_{**} x\|_{H^2(\sigma_{**})} .$$

Then

$$\|\mathcal{K}(\psi) x\|_F = \|(\psi \circ \Pi_c)\cdot(U_{**}x)\|_{H^2(\sigma_{**})} \leq \|\psi\|_\infty \|x\|_F ,$$

which yields the inequality $\|\mathcal{K}(\psi)\| \leq \|\psi\|_\infty$. The equality follows from the fact that $U_{**}x$ is a constant function if $x = T^{-1}_{\omega T}(const)$

(see formulae (3.1), (3.2)). ▨

There are many papers concerning commutant of Toeplitz operators for analytic symbols; the cases when (v) is violated are considered there too. C.Cowen [10] and J.Thomson [34] give conditions on $\mathbb{F}$ implying $\{T_{\mathbb{F}}\}' = \{T_B\}'$ for a finite Blaschke product B such that $\mathbb{F} = G \circ B$ for G analytic on $\mathbb{D}$. It is sufficient for example that $\mathbb{F}$ be analytic on $\overline{\mathbb{D}}$ [34] or that the inner factor of $\mathbb{F} - \mathbb{F}(\alpha)$ be a finite Balschke product for some $\alpha \in \mathbb{D}$ [10].

C.Cowen finds in [10] a correspondence between $\{T_B\}'$ and bounded analytic functions on the Riemann surface of $B^{-1} \circ B$ (note that $\{T_B\}'$ is not abelian if B is not a Blaschke factor). Abrahamse and Ball give in [2] an example of infinite Blaschke products φ and ψ such that the double commutant $\{T_\varphi, T_\psi\}''$ is isomorphic to $H^\infty(\{z : \frac{1}{2} < |z| < 1\})$ as a Banach algebra.

6.12. REMARKS. 1) The theorem remains true if the condition (vi) is replaced by

(vi') $a < |\mathbb{F}'(z)| < b$ if $r < |z| < 1$ for some constant $a > 0$, $b < \infty$, $r < 1$.

To prove this, it suffices to show that under the conditions (i)-(v), (vi'), (vii) the mapping $u \mapsto u|\sigma_{**}$ is an isomorphism of $H^2_{\mathbb{F}}(\sigma_*)$ onto $E^2_{\mathbb{F}}(\sigma_{**})$ (We can use this fact instead of Lemma 6.4 in the proof of the theorem). Consider the triangulation τ and the triangulation $\Pi^{-1}(\tau)$ of $\sigma_*(T_{\mathbb{F}})$ which consists of the components of the sets $\Pi^{-1}(R)$, $R \in \tau$. Recall that $H^2_{\mathbb{F}}(\sigma_*) = E^2_{\mathbb{F}}(\sigma_*)$ by Lemma 4.12. By 6.3, each $v \in E^2_{\mathbb{F}}(\sigma_{**})$ has a form $v = u|\sigma_{**}$ for some $u \in \mathcal{H}_{\mathbb{F}}(\sigma_*)$. Hence the following implications hold

$$v \in E^2_{\mathbb{F}}(\sigma_{**}) \Longrightarrow v|\widetilde{R} \in E^2(\widetilde{R}) \quad \forall \widetilde{R} \in \Pi^{-1}(\tau),\ \widetilde{R} \subset \sigma_{**} \Longrightarrow$$

$$\Longrightarrow v|\widetilde{R} \in E^2(\widetilde{R}) \quad \forall \widetilde{R} \in \Pi^{-1}\tau \Longrightarrow u \in E^2_{\mathbb{F}}(\sigma_*) .$$

The first and the third implications are consequences of properties 4.6, 4.7 of E^2- classes. To prove the second implication, we observe that every triangle $\widetilde{R}$, $\widetilde{R}\in\Pi^{-1}(\tau)$, can be included into a "combinatorial sequence" of triangles $\{\widetilde{R}_i\}_{i=1}^{n}$ such that $\Pi(\widetilde{R}_i) = \Pi(R)$ for all i and $\widetilde{R}_n = R$. The relation (1.6) yields successively $u|\widetilde{R}_i\in E^2(\widetilde{R}_i)$, $i=1,\dots,n$. ▨

2) It is easy to verify that if (vii) is violated in Theorem 6.11 then (6.2) provides an algebra isomorphism between $\{T_F\}'$ and some subalgebra in $H^\infty(\sigma_c)$ of finite codimension. However the author does not know an example where this subalgebra is proper.

6.13. EXAMPLES. 1) The surface σ_{**} contains the set $\{\delta\in\sigma_{**}: wind_F(\Pi(\delta))=1\}$. Hence $\sigma_* = \sigma_{**}$ for symbols considered in 1.8 corresponding to Figures 3c, 4b, 4c. This implies easily that $\sigma_c(T_F)$ is isomorphic to $int\,\sigma(T_F)$, and

$$\{T_F\}' = \{\varphi(T_F):\ \varphi\in H^\infty(int\,\sigma(T_F))\} \tag{6.3}$$

in these cases.

2) Suppose that F is analytic in $\mathbb{D}$ (that is, $k=1$, see 1.8, example 1). We have $\sigma_{**} = \sigma_*$ too. However now σ_c is isomorphic to σ_* (so that there is no sticking at all). Thus $\{T_F\}'$ equals $\{T_\varphi:\varphi\in H^\infty\}$ which is in general a wider class than the set of $H^\infty(int\,\sigma(T_F))$ - functions of T_F .

3) If F is the symbol corresponding to Fig.3b, then $\sigma_{**} = \sigma_* = A$. The commutant $\{T_F\}'$ is wider than the set of $H^\infty(int\,\sigma(T_F))$-functions of T_F . In particular, $T_F-\lambda I$ has a square root iff $\lambda\notin int\,\sigma(T_F)$ or $\lambda = \zeta_0$ (see 1.8, example 2).

4) Let F_0 be an analytic function satisfying (i)-(vii) such that $F_0(\mathbb{T})$ is the curve drawn on Fig.7. Pick some distinct points $a_1,\dots,a_n$ of $\mathbb{D}$ such that $F_0(a_j)\notin F_0(\mathbb{T})$. Put $F(z)=F_0(z)+\sum_{j=1}^{n} c_j(z-a_j)^{-1}$, where c_j are small non-zero numbers. Let Ω_1,Ω_2 be the components of $\{\lambda\in\mathbb{C}\setminus F(\mathbb{T}): wind_F(\lambda)=2\}$. Let τ be a function from C_A satisfying (i)-(vii) such that $F(\mathbb{T})=\tau(\mathbb{T})$ and $(R,\widetilde{\Pi})$ be the Riemann surface of τ^{-1} (so that $\widetilde{\Pi}(\partial R) = F(\mathbb{T})$. Note that the ultraspectrum $\sigma_*(T_F)$ has a

branch point projecting into Ω_ℓ iff $F_0(a_j) \in \Omega_\ell$ for some j. It follows easily that the commutant surface $\sigma_c(T_F)$ can be obtained from R by sticking sheets of $(R, \tilde{\Pi})$ projecting into such components Ω_ℓ that $F_0(a_j) \in \Omega_\ell$ for some j. We conclude that(6.3) holds iff $F_0(a_{i_1}) \in \Omega_1$ and $F_0(a_{i_2}) \in \Omega_2$ for some i_1, i_2.

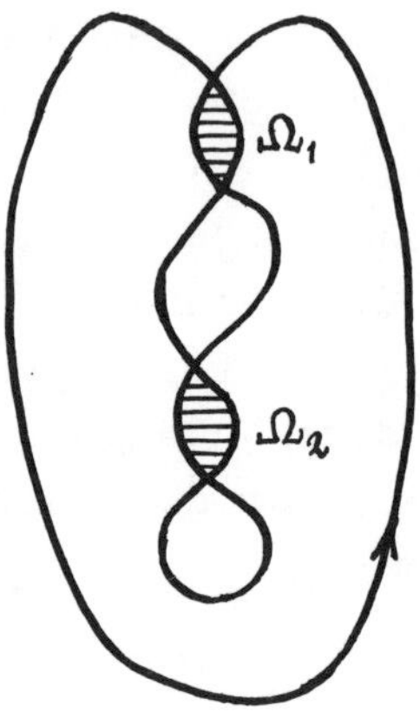

Figure 7

CHAPTER 3. CLASSIFICATION THEOREMS FOR TOEPLITZ OPERATORS.

Let $\mathbb{F}, \mathbb{F}_1$ satisfy the conditions (i)-(iv). Some criteria for similarity of $T_{\mathbb{F}}$ and $T_{\mathbb{F}_1}$ will be obtained here.

7.1. THE LOCAL OPERATORS A_λ. Similarity preserves the essential spectrum and the Fredholm index of an operator. Hence for $T_{\mathbb{F}}$ and $T_{\mathbb{F}_1}$ to be similar the following condition is necessary:

$$(M)\quad \mathbb{F}(\mathbb{T}) = \mathbb{F}_1(\mathbb{T}) \quad \text{and } wind_{\mathbb{F}}(\lambda) = wind_{\mathbb{F}_1}(\lambda) \quad \text{for all } \lambda \notin \mathbb{F}(\mathbb{T}).$$

We shall suppose in the sequel that (M) is satisfied.

Let (σ_*, Π) and (σ_*', Π_1) denote the ultraspectra of $T_{\mathbb{F}}$ and $T_{\mathbb{F}_1}$ respectively. Theorem 1.5 yields that $T_{\mathbb{F}}$ and $T_{\mathbb{F}_1}$ are similar iff M_Π and M_{Π_1}, acting on $H^2_{\mathbb{F}}(\sigma_*)$ and $H^2_{\mathbb{F}_1}(\sigma_*')$ respectively are similar. Suppose that

$$S: H^2_{\mathbb{F}}(\sigma_*) \longrightarrow H^2_{\mathbb{F}_1}(\sigma_*')$$

is an isomorphism such that $S M_\Pi = M_{\Pi_1} S$. We have

$$S^* Ker(M^*_{\Pi_1} - \lambda I) = Ker(M^*_\Pi - \lambda I)$$

for all $\lambda \in \mathbb{C}$ in this case. Put

$$\Lambda = \mathbb{F}(\Xi) \cup \mathbb{F}_1(\Xi).$$

Take some $\lambda \in \sigma(T_{\mathbb{F}}) \setminus \Lambda$. Let ℓ_λ denote the (finite-dimensional) space of functions on $\Pi^{-1}\{\lambda\}$ satisfying (1.6), and ℓ_λ' denote the analogous space of functions on $\Pi_1^{-1}\{\lambda\}$. Then for $\lambda \notin \mathbb{F}(\mathbb{T})$ and

$$u | \Pi^{-1}\{\lambda\} \in \ell_\lambda , \qquad v | \Pi_1^{-1}\{\lambda\} \in \ell'_\lambda$$

for $u \in H^2_{\mathbb{F}}(\sigma_*)$, $v \in H^2_{\mathbb{F}_1}(\sigma'_*)$.

Each functional $\varphi \in Ker(M^*_\Pi - \lambda I)$ has a form

$$\langle u, \varphi \rangle = \langle u | \Pi^{-1}\{\lambda\}, \psi \rangle \quad , \quad u \in H^2_{\mathbb{F}}(\sigma_*)$$

for some $\psi \in \ell^*_\lambda$ which is determined uniquely. This allows us to identify the spaces $Ker(M^*_\Pi - \lambda I)$ and ℓ^*_λ, as well as $Ker(M^*_{\Pi_1} - \lambda I)$ and ℓ'^*_λ. Then the operator

$$A_\lambda = (S^* | \ell'^*_\lambda)^* \tag{7.1}$$

is an isomorphism of ℓ_λ onto ℓ'_λ. Let $\delta' \in \Pi_1^{-1}\{\lambda\}$ and $\psi_{\delta'}$, $\psi_{\delta'} \in \ell'^*_\lambda$, denote the point evaluation at δ': $\langle v, \psi_{\delta'} \rangle = v(\delta')$ for $v \in \ell'_\lambda$. For every $u \in H^2_{\mathbb{F}}(\sigma_*)$ we have

$$(Su)(\delta') = \langle Su, \psi_{\delta'} \rangle = \langle u, S^*\psi_{\delta'} \rangle =$$

$$= \langle u | \Pi^{-1}\{\lambda\}, S^*\psi_{\delta'} \rangle = \langle A_\lambda(u|\Pi^{-1}\{\lambda\}), \psi_{\delta'} \rangle .$$

Hence S can be defined by

$$(Su)(\delta') = (A_\lambda(u|\Pi^{-1}\{\lambda\}))(\delta') , \tag{7.2}$$

where $\delta' \in \sigma'_*$ and $\lambda = \Pi_1(\delta') \notin \Lambda$. We come to the following statement.

7.2. THEOREM. <u>Suppose</u> $\mathbb{F}, \mathbb{F}_1$ <u>satisfy the conditions (i)-(iv),</u> (M). <u>Let</u> $U: H^2 \to H^2_{\mathbb{F}}(\sigma_*)$, $U_1: H^2 \to H^2_{\mathbb{F}_1}(\sigma'_*)$ <u>denote the isomorphisms corresponding to</u> $T_{\mathbb{F}}$ <u>and</u> $T_{\mathbb{F}_1}$. <u>Then an operator</u> $W \in \mathcal{L}(H^2)$ <u>implements a similarity between</u> $T_{\mathbb{F}}$ <u>and</u> $T_{\mathbb{F}_1}$ <u>iff</u> $W = U_1^{-1} S U$, <u>where</u> $S: H^2_{\mathbb{F}}(\sigma_*) \to H^2_{\mathbb{F}_1}(\sigma'_*)$ <u>is an isomorphism defined by formulae having the form (7.2)</u>. <u>In particular</u>, $T_{\mathbb{F}}$ <u>is similar to</u> $T_{\mathbb{F}_1}$ <u>if and only if there exist operators</u>

$\{A_\lambda\}_{\lambda\in\sigma(T_F)\setminus\Lambda}$ <u>that define an isomorphism</u> $S: H^2_{\mathbb{F}}(\sigma_*) \to H^2_{\mathbb{F}_1}(\sigma'_*)$.

PROOF. Suppose that (7.2) defines an isomorphism S **of** $H^2_{\mathbb{F}}(\sigma_*)$ **onto** $H^2_{\mathbb{F}}(\sigma'_*)$ **for some family** $\{A_\lambda\}_{\lambda\in\sigma(T_{\mathbb{F}})\setminus\Lambda}$ **. Put** $W = U_1^{-1} S U$ **. It is straightforward that** $M_{\Pi_1} S = S M_\Pi$ **. Hence**

$$W T_{\mathbb{F}} W^{-1} = U_1^{-1} S M_\Pi S^{-1} U_1 = U_1^{-1} M_{\Pi_1} U_1 = T_{\mathbb{F}_1} .$$

The converse has already been proved. ▨

This theorem reduces the problem of similarity between $T_{\mathbb{F}}$ and $T_{\mathbb{F}_1}$ to finding a family of finite-dimensional operators $\{A_\lambda\}$ with certain analytic properties and certain growth conditions. In many cases, the following more explicit criterion can be applied.

7.3. THEOREM. <u>Suppose that</u> $\mathbb{F}$ <u>and</u> $\mathbb{F}_1$ <u>satisfy (i)-(vii). Put</u> $\sigma_{**} = \sigma_{**}(T_{\mathbb{F}})$ <u>and</u> $\sigma'_{**} = \sigma_{**}(T_{\mathbb{F}_1})$. <u>Then</u> $T_{\mathbb{F}}$ <u>and</u> $T_{\mathbb{F}_1}$ <u>are similar iff</u>

1) (σ_{**}, Π) <u>and</u> (σ'_{**}, Π_1) <u>are isomorphic, that is there exists a Riemann surface isomorphism</u> τ <u>of</u> σ_{**} <u>onto</u> σ'_{**} <u>such that</u> $\Pi = \Pi_1 \circ \tau$;

2) <u>There exists a function</u> β <u>on</u> σ_{**} , **with** $\beta, \beta^{-1} \in H^\infty(\sigma_{**})$, <u>such that the mapping</u>

$$u \mapsto (\beta\circ\tau^{-1})(u\circ\tau^{-1}) \tag{7.3}$$

<u>is an isomorphism of</u> ℓ_λ <u>onto</u> ℓ'_λ <u>for all</u> $\lambda\notin\Lambda$.

PROOF. Under the conditions 1) and 2), (7.3) defines an isomorphism of $H^2_{\mathbb{F}}(\sigma_{**})$ onto $H^2_{\mathbb{F}_1}(\sigma'_{**})$. This isomorphism splits $M_{\Pi|\sigma_{**}}$ and $M_{\Pi_1|\sigma'_{**}}$. It defines an isomorphism of $H^2_{\mathbb{F}}(\sigma_*)$ onto $H^2_{\mathbb{F}_1}(\sigma'_*)$ by Lemma 6.4. Hence $T_{\mathbb{F}}$ and $T_{\mathbb{F}_1}$ are similar.

To prove the converse, suppose that

$$T_{\mathbb{F}_1} = W T_{\mathbb{F}} W^{-1}$$

for some invertible operator W on H^2 . The remark made at the end i.5.5 implies that to every η from a C-surface for $T_{\mathbb{F}}$

there corresponds some η_1 from a C-surface for T_{F_1} such that $E_\eta = W^* E_{\eta_1}$. It is assumed here that η does not project to Λ. In particular, we can assign to every $\delta \in \sigma_{**} \setminus \Pi^{-1}(\Lambda)$ a point

$$\delta' \stackrel{def}{=} \tau(\delta) \in \sigma'_{**}$$

so that

$$span(h_\delta) = W^* span(h'_{\delta'}), \tag{7.4}$$

where $h'_{\delta'}$ is the eigenvector of the type (1.5) corresponding to T_{F_1}. The mapping τ is obsiously well-defined and the relation

$$\Pi = \Pi_1 \circ \tau$$

holds on $\sigma_{**} \setminus \Pi^{-1}(\Lambda)$. Moreover, the symmetry between T_F and T_{F_1} implies that τ is a bijection of $\sigma_{**} \setminus \Pi(\Lambda)$ onto $\sigma'_{**} \setminus \Pi_1(\Lambda)$. It is easy to verify that τ, τ^{-1} **are continuous. Hence** τ can be continued to a homeomorphism of (σ_{**}, Π) onto (σ'_{**}, Π). This yields 1).

From (7.4), we have

$$W^* h'_{\tau(\delta)} = \beta(\delta) h_\delta \qquad \text{on } \sigma_{**} \setminus \Pi^{-1}(\Lambda)$$

for some $\beta \in \mathcal{H}(\sigma_{**} \setminus \Pi^{-1}(\Lambda))$. Recall that $(Ux)(\delta) = \langle x, h_\delta \rangle$ and $(U_1 x)(\delta') = \langle x, h'_{\delta'} \rangle$ for $x \in H^2$. Hence for the linear isomorphism S,

$$S \stackrel{def}{=} U_1 W U^{-1}$$

we have the formula

$$Su | \sigma_{**} = (\beta \circ \tau^{-1})(u \circ \tau^{-1}), \quad u \in H^2_F(\sigma_*).$$

It remains to show that β, $1/\beta \in H^\infty(\sigma_{**})$. Since the constant functions belong to $H^2_F(\sigma_*)$, it follows that

$$\beta \circ \tau^{-1} \in H^2_F(\sigma'_{**}). \tag{7.5}$$

Let $\zeta \in \mathbb{T}$ be arbitrary and let $\mathcal{Y}$ be a neighbourhood of ζ. Take any analytic function $\rho : \mathcal{Y} \cap \mathbb{D} \to \sigma_{**}$ of the form

$$\rho(z) = \{z, \mathfrak{d}_2(z), \ldots, \mathfrak{d}_k(z)\},$$

where $\mathfrak{d}_j$ are branches of F^{-1} and $\operatorname{clos} \mathfrak{d}_j(\mathcal{Y} \cap \mathbb{D}) \subset \mathbb{D}$ for $j = 2, \ldots, k$. It follows from (7.5), (1.6) and from Lemma 5.12 that to get $\beta \in H^\infty(\sigma_{**})$ it suffices to show that $\beta \circ \rho$ is bounded for all such ρ and all sufficiently small $\mathcal{Y}$.

We use the formula $U = VT_{\omega^T}$ (see (1.4), (3.1)-(3.4)). Take any $x \in H^2$ and put

$$v = SUT_{\omega^T}^{-1} x = SVx .$$

Since $v \in H^2_F(\sigma'_{**})$, we have

$$v \circ \tau \circ \rho = (\beta \circ \rho)[(Vx) \circ \rho] \in H^2(\mathcal{Y} \cap \mathbb{D}) .$$

Substitute expression (3.3) for V in this relation. Note that $\beta \circ \rho \in H^2(\mathcal{Y} \cap \mathbb{D})$ and that

$$(c_s \circ \rho)(x \circ \mathfrak{d}_s) \in H^\infty(\mathcal{Y} \cap \mathbb{D})$$

for $s = 2, \ldots, K$ if $\mathcal{Y}$ is small enough. Hence $(\beta \circ \rho)(c_1 \circ \rho) x \in H^2(\mathcal{Y} \cap \mathbb{D})$. As $|c_1 \circ \rho| > \varepsilon$ on $\mathcal{Y}$ for some $\varepsilon > 0$, we conclude that $(\beta \circ \rho) x \in H^2(\mathcal{Y} \cap \mathbb{D})$ for all $x \in H^2$. Thus $\beta \circ \rho \in H^\infty(\mathcal{Y}_1 \cap \mathbb{D})$ for a smaller neighbourhood $\mathcal{Y}_1$ of ζ. This implies that $\beta \in H^\infty(\sigma_{**})$.

The operator S^{-1} is defined by the formula

$$S^{-1} v | \sigma_{**} = (1/\beta)(v \circ \tau), \quad v \in H^2_F(\sigma'_*) .$$

Hence an analogous argument yields $1/\beta \in H^\infty(\sigma_{**})$. ▨

7.4. EXAMPLES. 1) Put $F_0(z) = (z + \frac{1}{2})^2$. Let

$$F(z) = F_0(z) + \sum_{j=1}^{n} \varepsilon_j (z - a_j)^{-1}$$

be a small perturbation of F_0. Here $a_j \in \mathbb{D}$, the points $F_0(a_j)$ are pairwise distinct, $F_0(a_j) \notin F_0(\mathbb{T})$, and ε_j are non-zero and small enough. The curves $F(\mathbb{T})$ and $F_0(\mathbb{T})$ have similar form, like the one shown on Fig.3a. Put $G_\ell = \{\lambda : \mathrm{wind}_F(\lambda) = \ell\}$ for $\ell = 1,2$. It is easy to see that $\sigma_{**}(T_F)$ coincides with the connected component of $\sigma_*(T_F)$ containing $\Pi^{-1}(G_1)$ in our case. If $\mathrm{wind}_{F_0}(F_0(a_j)) = 2$ for some j, then the surface $\sigma_{**}(T_F)$ has more than two sheets over G_2. Hence T_F is not similar to any analytic Toeplitz operator whose symbol satisfies (i)-(vii). Indeed, there are non-trivial linear fiber relations on $\sigma_{**}(T_F)$ whereas no such relations correspond to analytic Toeplitz operators.

Consider analytic vector-functions that can be obtained by an analytic continuation of eigenvectors $\{h_\sigma : \sigma \in \Pi^{-1}(G_1)\}$ of T_F^*. One can check that some three values of these vector-functions are linearly dependent. This implies easily that T_F is not similar to any analytic Toeplitz operator at all.

Suppose now that $\mathrm{wind}_{F_0}(F_0(a_j)) = 1$ for all j. Then the reduced ultraspectrum $\sigma_{**}(T_F)$ has a unique branch point, say, σ_0. The point $\Pi(\sigma_0)$ is in G_2, near to zero. There exists a conformal mapping ψ of $\mathbb{D}$ onto $\sigma_{**}(T_F)$. Since $H^2_F(\sigma_{**}) = H^2(\sigma_{**})$ in this case, it follows that T_F is similar to $T_{\Pi \circ \psi}$.

Note that Theorem 7.3 implies that if T_F is similar to an analytic Toeplitz operator T_φ with φ satisfying (i)-(vii), then $\varphi'(z) = 0$ and $\varphi(z) = \Pi(\sigma_0)$ for some $z \in \mathbb{D}$. In particular, analytic Toeplitz operators T_{φ_1}, T_{φ_2} satisfying (i)-(vii), with $\varphi_1(\mathbb{T}) = \varphi_2(\mathbb{T}) = F(\mathbb{T})$ and

$$\dim \operatorname{Ker}(T_{\varphi_i}^* - \lambda I) \equiv \dim \operatorname{Ker}(T_F^* - \lambda I)$$

for all $\lambda \in \mathbb{C} \setminus F(\mathbb{T})$ are not similar in general. This gives one more counterexample to a theorem in [6] (see also [32]).

2) For $F(z) = z^2 + az^{-1}$, we have $\sigma_* = \sigma_{**}$ both in cases 3b) and 3c) (see 1.8, Example 3). Since there are non-trivial fiber linear relations in $H^2_F(\sigma_*)$ in this case, we conclude that T_F is not similar to any analytic Toeplitz operator.

CHAPTER 4. INVARIANT SUBSPACES OF MULTIPLICATION OPERATORS ON RIEMANN SURFACES.

In Chapter 5 we shall be concerned with invariant and hyperinvariant subspaces of T_F if F satisfies (i)-(vi), (H) (see the introduction), and

(viii) $F \in C^2(\mathbb{T})$, $1/F'$ is bounded on the annulus $\{z : r \leq |z| < 1\}$ for some $r < 1$.

Bearing in mind Similarity Theorem 1.5, we study first the lattice of invariant subspaces of a multiplication operator on a Riemann surface.

8. TEST ATLASES, STANDARD SUBSPACES AND THE FUNCTION $\varphi_{\alpha,w}$.

8.1. CONDITIONS ON A RIEMANN SURFACE. Suppose that R is a compact Riemann surface (not necessarily connected) with boundary ∂R and Π is the projection of $\overline{R} = R \cup \partial R$ to $\mathbb{C}$. Let γ denote the curve $\Pi(\partial R)$; here the usage of the term "curve" is not canonic since γ can be a union of intersecting curves. If $\lambda \in \mathbb{C} \setminus \gamma$, the valence $\varkappa(\lambda)$ is defined as the number of preimages of λ on R counted with their multiplicities. The function $\varkappa$ is locally constant on $\mathbb{C} \setminus \gamma$. We call a point λ of γ an intersection point of γ if there is no neighbourhood W of γ such that $W \cap \gamma$ is a Jordan arc.

The following conditions will be imposed on R:

(a) R is a Riemann surface with a finite number of components and the valence function $\varkappa$ is bounded on $\mathbb{C}$.

(b) γ has a finite number of intersection points and can be decomposed into a finite union of C^2-smooth arcs.

(c) The natural positive orientation on ∂R after projection to $\mathbb{C}$ agrees with some orientation on γ.

(d) Every point ζ which is an intersection point of γ has a neighbourhood W divided by γ into four domains (say, $G_1, \dots, G_4$; see Fig.8). No one of the components of $\Pi^{-1}(W)$ projects onto three of these domains.

(e) Let Δ be the projection of the branch point set of (R, Π). Then Δ is finite and $\gamma \cap \Delta = \emptyset$.

(H') Each bounded component Ω_i of $\mathbb{C} \setminus \gamma$ has a common boundary arc with some component Ω_j of $\mathbb{C} \setminus \gamma$ such that $\varkappa(\Omega_j) < \varkappa(\Omega_i)$.

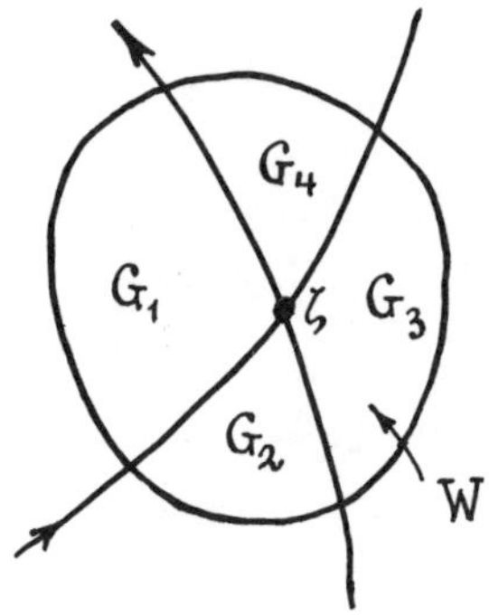

Figure 8

Note that (b) does not require that different arcs of ∂R project onto different arcs of γ. Moreover, the ultraspectrum (σ_*, Π) satisfies (a)-(e), (H') if F is a symbol satisfying (i)-(vi), (viii), (H). For example, (H') is a consequence of (H) and of the fact that $\varkappa(\lambda) = \binom{K}{\mathrm{wind}_F(\lambda)+K-1}$ if $R = \sigma_*$.

Denote by $\mathcal{O}$ the set of all intersection points of γ.

8.2. TEST SETS AND TEST ATLASES. An open set in R (or in $\mathbb{C}$) will be called admissible if its boundary is a union of piecewise C^2-smooth simple curves. Let V be an admissible subset of R and let $\mathcal{E} \in \mathrm{Lat}(M_\Pi)$. The closed subspace

$$tr_V \mathcal{E} \overset{def}{=\!=} clos_{E^2(V)} \{u|V : u \in \mathcal{E}\}$$

of $E^2(V)$ will be referred to as <u>the trace of</u> $\mathcal{E}$ <u>on</u> V. Obviously, $tr_V \mathcal{E} \in \mathrm{Lat}(M_{\Pi|V})$.

Our way of describing $\mathrm{Lat}(M_\Pi)$ will be concerned with certain collections of admissible subsets of R. We associate with such a collection μ and every $\mathcal{E} \in \mathrm{Lat}(M_\Pi)$ the subspace

$$S(\mathcal{E}, \mu) \overset{def}{=\!=} \{u \in E^2(R) : u|V \in tr_V \mathcal{E} \quad \forall V \in \mu\}.$$

We have $S(\mathcal{E}, \mu) \in \mathrm{Lat}(M_\Pi)$ and $S(\mathcal{E}, \mu) \supset \mathcal{E}$.

If V is an admissible subset of R, $\partial\Pi(V) = \Pi(\partial V)$, and $\partial\Pi(V)$ is a simple closed curve, then V will be called a

<u>test set</u>. For a test set V, the number ℓ of pre-images of a point $\lambda \in \Pi(V)$ on V (counted with multiplicities) does not depend on λ. We say that V is an ℓ-fold test set in this case and write $\ell = \ell(V)$.

A collection $\mu = \{V_i\}$ of test sets will be referred to as a <u>test atlas</u> if there exists $\varepsilon > 0$ such that for every test set V with $diam(\Pi(V)) < \varepsilon$ one can find $V' \in \mu$ such that $V' \supset V$.

A test atlas always exists and can be constructed for instance in the following way. Take some discs $\mathcal{D}_i$ covering $\Pi(\overline{R})$ so that for each j at most one of the sets $\mathcal{D}_j \cap \Delta$, $\mathcal{D}_j \cap \gamma$, $\mathcal{D}_j \cap \mathcal{O}$ is not empty, and $card(\mathcal{D}_j \cap \mathcal{O}) \leq 1$ holds. Consider all domains $\mathcal{F}_m$ in $\mathbb{C}$ satisfying $\partial \mathcal{F}_\ell \subset \partial \mathcal{D}_j \cup (\gamma \cap \mathcal{D}_j)$ for some j. The collection of test sets V such that $\Pi(V) = \mathcal{F}_m$ for some m is an example of a finite test atlas on R.

If V is a test set, then the lattice of $M_{\Pi|V}$ in $E^2(V)$ has a description analogous to the Lax-Halmos theorem on the lattice of the multiple shift (Lemma 8.3 below). Hence $S(\mathcal{E},\mu)$ in a sense has a familiar structure if μ is a test atlas. We shall show that in fact $S(\mathcal{E},\mu)$ does not depend on the test atlas μ. Our aim is to find necessary and sufficient conditions on a function $u \in S(\mathcal{E},\mu)$ that guarantee $u \in \mathcal{E}$. This will be done in Theorem 10.3 which is the main result of this chapter.

An important tool is the function $\varphi_{\alpha,w}$ defined in 8.12 for $\alpha \in L^2(\partial R)$ and $w \in E^2(R)$. Theorem 8.13 below shows that under certain assumptions on α, w this function is an analytic continuation of the function $\lambda \mapsto \langle (M_\Pi - \lambda)^{-1} w, \alpha \rangle$, $\lambda \notin \Pi(R)$, to $\hat{\mathbb{C}} \setminus \mathcal{O}$.

8.3. LEMMA. <u>Suppose that</u> V <u>is an</u> n<u>-fold test set and</u> $\mathcal{X}$ <u>is a subspace of</u> $E^2(V)$. <u>Then</u> $\mathcal{X} \in Lat(M_\Pi)$ <u>iff there exist</u> r, $0 \leq r \leq n$ <u>and</u> $y_1, \ldots, y_r \in H^\infty(V)$ <u>satisfying</u>

$$\sum_{\Pi(\delta)=\lambda} y_i(\delta)\overline{y_j(\delta)} = \delta_{ij} \tag{8.1}$$

<u>for a.e.</u> $\lambda \in \partial\Pi(V)$ <u>such that</u> $\mathcal{X}$ <u>admits a representation</u>

$$\mathcal{X} = \{ \sum_{i=1}^{r} (x_i \circ \Pi) y_i : x_1, \ldots, x_r \in E^2(\Pi(V)) \} . \tag{8.2}$$

Here δ_{ij} is the Kronecker delta.

PROOF. If (8.1) is fulfilled, then (8.2) defines a linear isometric operator from $\overset{r}{\underset{1}{\oplus}} E^2(\Pi V)$, with image $\mathcal{L}$. Hence $\mathcal{L}$ is closed. Obviously, $\mathcal{L} \in \operatorname{Lat}(M_{\Pi/V})$.

To prove the converse, we introduce the set $C = \mathbb{T} \times \{1, 2, \ldots, n\}$, which is the union of n disjoint copies of $\mathbb{T}$. Let p be the canonic projection of C onto $\mathbb{T}$. Consider a conformal mapping τ of $\overline{\mathbb{D}}$ onto $\overline{\Pi(V)}$ and let $\tilde{\tau} : C \to \partial V$ be any piecewise continuous bijection with the property $\Pi \circ \tilde{\tau} = \tau \circ p$. Define a unitary operator $G : L^2(\partial V) \to L^2(C)$ by

$$Gu \overset{def}{=\!=} (\tau' \circ p)^{1/2} (u \circ \tilde{\tau}) . \tag{8.3}$$

Since V is a test set, it follows that $\Pi(V)$ is a domain of the Smirnov class, so that τ' is an outer function from H^1. Hence every real power of τ' is defined and boundary values of τ' are defined a.e. on $\mathbb{T}$.

The subspace $G\mathcal{L}$ is invariant for $M_{\tau \circ p}$ and hence for M_p. It follows that it has the form $G\mathcal{L} = \mathcal{L}_1 \oplus \mathcal{L}_2$, where $\mathcal{L}_1$ is a reducing subspace for M_p and $\mathcal{L}_2 = \Theta \overset{r}{\underset{1}{\oplus}} H^2$ [23]. Here Θ is an analytic $n \times r$-matrix-function which defines an isometric operator for almost all $\zeta \in \mathbb{T}$. Observe that for $x = Gu \in \mathcal{L}_1$, we have $(\tau \circ p - \lambda)^{-1} x \in \mathcal{L}_1$ for all $\lambda \notin \partial \Pi(V)$. Thus $(\Pi - \lambda)^{-1} u \in E^2(V)$ for all $u \in G^{-1} \mathcal{L}_1$ and all $\lambda \in \Pi(V)$. This implies $\mathcal{L}_1 = 0$.

Let θ_i, $1 \le i \le r$, be the columns of Θ. Set

$$w(\lambda) = (\tau'(\tau^{-1}(\lambda)))^{-1/2} \tag{8.4}$$

for $\lambda \in \overline{\Pi(V)}$, and

$$y_i = (w \circ \Pi) G^{-1} \theta_i . \tag{8.5}$$

Then $\theta_i \in G\mathcal{L}$, $w \circ \Pi$ is analytic on V, and $w \cdot E^2(\Pi(V)) = H^2(\Pi(V))$, $(w \circ \Pi) \cdot E^2(V) = H^2(V)$. So $y_i \in H^2(V)$. The formulae (8.3)-(8.5) and $\Pi \circ \tilde{\tau} = \tau \circ p$ imply that $y_i = \theta_i \circ \tilde{\tau}^{-1}$ a.e. on ∂V. Hence $y_i \in H^\infty(V)$ and (8.1) is valid. Finally,

$$\mathcal{X} = G^{-1}\mathcal{X}_{\mathcal{V}} = \{ G^{-1} \sum_1^{r} (x_i \circ p)\theta_i : x_1, \ldots, x_r \in H^2 \} =$$

$$= \{ \sum_1^{r} (x_i \circ \tau^{-1} \circ \Pi)(\theta_i \circ \tilde{\tau}^{-1})/(w \circ \Pi) : x_1, \ldots, x_r \in H^2 \} =$$

$$= \{ \sum_1^{r} (x_i' \circ \Pi) y_i : x_1', \ldots, x_r' \in E^2(\Pi(V)) \} . \quad ▨$$

8.4. PREPOSITION . Suppose that V is a Riemann surface with a projection Π which satisfies (a)-(e). Let W be the polynomially-convex hull of $\Pi(V)$. If $\mathcal{E} \in \operatorname{Lat} M_{\Pi|V}$, $x \in \mathcal{E}$, $a \in \mathcal{D}(W)$ and $(a \circ \Pi) x \in E^2(V)$, then $(a \circ \Pi) x \in \mathcal{E}$.

PROOF. If $a \in H^\infty$, then the result follows by weak approximation of a by polynomials. In the general case it follows from the parametric representation of functions of class $\mathcal{D}$ [26] that $|a_n| \uparrow |a|$ and $a_n \to a$ on W and $a_n \to a$ a.e. on ∂W for some $a_n \in H^\infty(W)$. The dominated convergence theorem yields $(a_n \circ \Pi) x \to (a \circ \Pi) x$ in $E^2(V)$. ▨

8.5. CAUCHY INTEGRALS AND SPACES OF WEAK TYPE. Let Ω be an admissible domain and m the arc length measure on $\partial\Omega$. Recall the definitions of the weak type spaces:

$$L^{1,\infty}(m) = \{ f : \|f\|_{1,\infty} \overset{def}{=\!=} \sup_{t>0} t\, m\{|f|>t\} < \infty \},$$

$$L_0^{1,\infty}(m) = \{ f \in L^{1,\infty}(m) ; m\{|f|>t\} = o(t) \quad \text{for } t \to \infty \}.$$

We shall need the corresponding analytic spaces

$$E^{1,\infty}(\Omega) = \{ f \in \mathcal{D}(\Omega) : f | \partial\Omega \in L^{1,\infty}(m) \},$$

$$E_0^{1,\infty}(\Omega) = \{ f \in \mathcal{D}(\Omega) : f | \partial\Omega \in L_0^{1,\infty}(m) \}.$$

The topology in all these spaces is induced by the complete invariant metric $d(f,g)=\|f-g\|_{1,\infty}^{1/2}$. The triangle inequality follows from the relation

$$m(\{|f_1+f_2|>y\}\leq \|f_1\|_{1,\infty}x_1^{-1}+\|f_2\|_{1,\infty}x_2^{-1},$$

where $y>0$ and $x_j=\|f_j\|_{1,\infty}^{1/2}y(\|f_1\|_{1,\infty}^{1/2}+\|f_2\|_{1,\infty}^{1/2})$.

For a piecewise C^2-smooth curve Γ and a function $f\in L^1(\Gamma)$, the **Cauchy** integral

$$\mathcal{K}_{f,\Gamma}(z)=\frac{1}{2\pi i}\int_\Gamma \frac{f(\zeta)d\zeta}{\zeta-z}$$

defines an analytic function on $\hat{\mathbb{C}}\setminus\Gamma$.

If Ω is a component of $\hat{\mathbb{C}}\setminus\Gamma$, then the function $\mathcal{K}_{f,\Gamma}$ has non-tangential boundary values for a.e. $\zeta\in\partial\Omega$ which will be denoted by $\mathcal{K}_{f,\Gamma,\Omega}(\zeta)$. **For** $z\in\Omega$, $\mathcal{K}_{f,\Gamma,\Omega}(z)\overset{def}{=\!=}\mathcal{K}_{f,\Gamma}(z)$.

8.6. THE IMBEDDING LEMMA FOR $E_0^{1,\infty}$. If G_1, G_2 are admissible **domains and** $G_1\subset G_2$, then $f|G_2\in E_0^{1,\infty}(G_2)$ for every $f\in E_0^{1,\infty}(G_1)$, and the inequality $\|f|G_2\|_{1,\infty}\leq C\|f\|_{1,\infty}$ holds for some $C>0$.

PROOF. Consider linear operator L which acts on $L^1(\partial G_1)$ by the formula

$$(Lf)(z)=\mathcal{K}_{f,\Gamma,\Omega}(z),\quad z\in\partial G_2 .$$

From the standard results on Cauchy integrals and from the Carleson imbedding lemma 4.6 it follows that L is a continuous operator from $L^p(\partial G_1)$ into $L^p(\partial G_2)$ for $1<p<\infty$. By Interpolation **Theorem** 3.15 from [32], we have

$$\|Lg\|_{2,\infty}\leq C'\|g\|_{2,\infty} \tag{8.6}$$

for some $C'>0$ and all $g\in L^{2,\infty}$. Suppose that $f\in E^{1,\infty}(G_1)$ and let $(f)_e$ be the outer part of f. Applying (8.6) to $(f)_e^{1/2}$, we get the imbedding lemma for the class $E^{1,\infty}$. The imbedding lemma for $E_0^{1,\infty}$ follows using approximation of f by H^∞-functions. ▨

8.7. COROLLARY. Let Γ be a curve in $\mathbb{C}$ and let G be an admissible domain such that $\Gamma \cap G = \emptyset$. Then $\mathcal{K}_{f,\Gamma} | G \in E_0^{1,\infty}(G)$ and

$$\| \mathcal{K}_{f,\Gamma} | G \|_{1,\infty} \leq C \| f \|_1$$

for all $f \in L^1(\Gamma)$.

This fact is well-known if $G = \mathbb{D}$ and $\Gamma = \mathbb{T}$. We obtain by the method of a conformal mapping that the same is valid if $\Gamma = \partial G$ is a C^2-smooth curve. Now the general case follows easily with the help of 8.6. ▨

8.8. A.B.ALEXANDROV'S STICKING THEOREM. Suppose that $\Omega, \Omega_1, \Omega_2$ are admissible domains, $\Omega_1 \cap \Omega_2 = \emptyset$, $\bar{\Omega} = \Omega_1 \cup \Omega_2$. If $f_j \in E_0^{1,\infty}(\Omega_j)$, $j=1,2$ and $f_1 = f_2$ a.e. on $\partial\Omega_1 \cap \partial\Omega_2$, then $f_j = f | \Omega_j$ for some $f \in E_0^{1,\infty}(\Omega)$.

Recall that a function g on $\mathbb{T}$ is said to satisfy the Muckenhoupt condition (A_2) if

$$\sup_{I \subset \mathbb{T}} |I|^{-2} \Big(\int_I |g| \, dm \Big) \Big(\int_I |g^{-1}| \, dm \Big) < \infty ,$$

here I means a subarc of $\mathbb{T}$. Theorem 8.8 is implied by the following statement.

8.9. LEMMA. Let Ω be an admissible domain and φ a conformal mapping of $\mathbb{D}$ onto Ω. If $\varphi' \in (A_2)$, then every $f \in E_0^{1,\infty}(\Omega)$ can be represented as the Cauchy (A)-integral

$$f(z) = \frac{1}{2\pi i} (A) \int_{\partial\Omega} \frac{f(\zeta)\, d\zeta}{\zeta - z} \overset{def}{=\!=} \lim_{A \to +\infty} \int_{\partial\Omega} \frac{f(\zeta) \chi_{\{|f|<A\}}}{\zeta - z} d\zeta .$$

For $\Omega = \mathbb{D}$, this is Theorem 6, b) from [3]. In the general case, the lemma reduces to the equality

$$0 = \frac{1}{2\pi i} (A) \int_{\partial\Omega} F(\zeta)\, d\zeta ,$$

where $F(\zeta) = (f(\zeta) - f(z))(\zeta - z)^{-1}$, which can be proved in the same way as Lemma 4 of [3]. This lemma implies Theorem 8.8 if the conformal mappings $\psi_j : \mathbb{D} \to \Omega_j$ satisfy $\psi_j' \in (A_2)$. Indeed, it follows that $f \in \mathcal{H}(\Omega)$ in this case. In the general case we conclude from the previous argument that f is analytic in every interiour point of $\alpha = \overline{\Omega}_1 \cap \overline{\Omega}_2$ where this arc is C^2-smooth. Let $\zeta \in \alpha$ and α not C^2-smooth in ζ. Take small domains Ω_1', Ω_2' with C^2-smooth boundaries such that $\partial\Omega_1' \cap \partial\Omega_2'$ is an arc and ζ is its interiour point. It follows that $f \in E_0^{1,\infty}(\Omega_i')$ for $i=1,2$ and thus f is analytic in ζ.

8.10. CAUCHY INTEGRALS OVER ∂R are defined for $\alpha \in L^2(\partial R)$ and $v \in E^2(R)$ by

$$K_{\alpha v}(\lambda) \overset{def}{=\!=} \frac{1}{2\pi i} \int_{\partial R} \frac{\alpha(\delta)\, v(\delta)}{\Pi(\delta) - \lambda}\, d\Pi(\delta), \qquad \lambda \in \mathbb{C} \setminus \gamma$$

(it is assumed that ∂R is positively oriented). Each linear functional φ on $E^2(R)$ has the form $\varphi(v) = \langle v, \alpha \rangle$ for some $\alpha \in L^2(\partial R)$, where

$$\langle v, \alpha \rangle \overset{def}{=\!=} \int_{\partial R} \alpha(\delta)\, v(\delta)\, d\Pi(\delta).$$

We observe that

$$K_{\alpha v}(\lambda) = \frac{i}{2\pi} \sum_{n=1}^{\infty} \lambda^{-n-1} \langle \alpha, \Pi^n \cdot v \rangle \tag{8.7}$$

if $|\lambda| > max\{|\zeta| : \zeta \in \gamma\}$.

Let $(\alpha v)_\Sigma$ be the "projection" of the function $\alpha v | \partial R$ to the complex plane, that is

$$(\alpha v)_\Sigma(\lambda) \overset{def}{=\!=} \sum_{\delta \in \partial R,\, \Pi(\delta) = \lambda} \alpha(\delta)\, v(\delta).$$

Then $K_{\alpha v} = \mathcal{K}_{(\alpha v)_\Sigma, \gamma}$. Hence $K_{\alpha v} | \Omega \in E_0^{1,\infty}(\Omega)$ for every component Ω of $\widehat{\mathbb{C}} \setminus \gamma$. The function $K_{\alpha v} | \Omega$ as well as its boundary values on $\partial\Omega$ will be denoted as $K_{\alpha v, \Omega}$. If Ω_i, Ω_j are adjacent components of $\mathbb{C} \setminus \gamma$ and $\varkappa(\Omega_i) < \varkappa(\Omega_j)$, then

$$K_{\alpha v,\Omega_j} = K_{\alpha v,\Omega_i} + (\alpha v)_\Sigma \qquad \text{a.e. on } \partial\Omega_i \cap \partial\Omega_j \tag{8.8}$$

by the Plemjel-Privalov formula [26].

The use of the Cauchy integrals is based on the following simple fact. Suppose that $\mathcal{E} \in \mathrm{Lat}(M_\Pi)$ and $w \in E^2(R)$. Let Ω_0 denote the unbounded component of $\mathbb{C} \setminus \gamma$. Then

$$w \in \mathcal{E} \iff K_{\alpha w} \equiv 0 \qquad \text{in } \Omega_0 \quad \forall \alpha \in \mathcal{E}^\perp, \tag{8.9}$$

where

$$\mathcal{E}^\perp = \{\alpha \in E^2(\partial R) : \langle \alpha, v \rangle = 0 \quad \forall v \in \mathcal{E}\}.$$

This follows from the Hahn-Banach theorem and from (8.7).

The next lemma can be found in Solomyak's paper [30] or in Vol'berg and Solomyak's paper [31] in slightly different forms.

8.11. LEMMA. <u>Suppose that</u> $\mathcal{E} \in \mathrm{Lat}(M_\Pi)$, $\alpha \in L^2(\partial R)$, $\alpha \perp \mathcal{E}$. <u>Then a discrete set</u> Λ <u>in</u> $\mathbb{C} \setminus \gamma$ <u>exists such that</u>

$$\mathrm{rank}\left[K_{\alpha v}(\lambda), v(\delta_j)\right]_{v \in \mathcal{E}}^{\Pi(\delta_j)=\lambda} \le \mathrm{rank}\left[v(\delta_j)\right]_{v \in \mathcal{E}}^{\Pi(\delta_j)=\lambda} \tag{8.10}$$

<u>for all</u> $\lambda \notin \gamma \cup \Delta \cup \Lambda$; <u>here</u> Δ <u>is the projection of the set of branch points of</u> R <u>defined in</u> (d) .

The matrices in (8.10) have $\varkappa(\lambda)+1$ and $\varkappa(\lambda)$ columns respectively. Both of them have infinitely many rows; the rank of such matrices can be defined as the size of the maximal non-zero minor.

THE PROOF is quite analogous to the proof of Lemma 2 in [30]. It is a consequence of (H') and of the following statements which can easily be verified. If Ω_j are the components of $\mathbb{C} \setminus \gamma$, then: (1) the inequality (8.10) holds for all $\lambda \in \Omega_0$; (2) both parts of (8.10) are constant on each Ω_j except for a set Λ_j which is dicrete in Ω_j ; (3) the relation (8.10) is valid on $\Omega_j \setminus \Lambda_j$ for some Λ_j discrete in Ω_j iff it is valid on

$\partial\Omega_j$ (in the sense that if the matrix on the left-hand side has a minor of n-th order which is non-degenerate a.e. on $\partial\Omega_j$ for some n, then so does the matrix on the right-hand side; (4) suppose that Ω_i, Ω_j have a common boundary arc α and satisfy $\varkappa(\Omega_i) < \varkappa(\Omega_j)$; then (8.10) for boundary values on α of functions in Ω_i implies the same property for Ω_j. The statement (4) is a consequence of (c) and (8.8). Condition (H') ensures that (1)-(4) imply the statement of the lemma for every Ω_j, if $\Lambda = \cup\Lambda_j$. ▨

8.12. DEFINITION OF $\varphi_{\alpha,w}$. Let $\mathcal{E} \in \mathrm{Lat}(M_\Pi)$ and Ω be a component of $\mathbb{C}\setminus\gamma$. It is easy to see that

$$\mathrm{rank}\,[v(\delta_j)]_{v\in\mathcal{E}}^{\Pi(\delta_j)=\lambda} = r$$

for all $\lambda\in\Omega\setminus\Lambda$, where Λ is discrete in Ω and r is a non-negative integer. We shall use the notation

$$r(\mathcal{E}, G) \overset{def}{=\!=} r(\mathcal{E}, \Omega) \overset{def}{=\!=} r,$$

if $\emptyset \neq G\subset\Omega$.

Choose a connected simply-connected admissible domain W, $W\subset\Omega\setminus\Lambda$. There exist functions $v_1,\dots,v_r\in\mathcal{E}$ and branches f_j of $\Pi^{-1}|W$ satisfying

$$\det\,[v_i\circ f_j]_{i,j=1}^{r} \not\equiv 0 \qquad \text{in } W \tag{8.11}$$

Let $\alpha\in\mathcal{E}^\perp$. It follows from (8.10) and (8.11) that there exist unique $\beta_1,\dots,\beta_r\in\mathcal{N}(W)$ such that

$$K_{\alpha v} = \beta_1\cdot(v\circ f_1)+\beta_2\cdot(v\circ f_2)+\dots+\beta_r\cdot(v\circ f_r) \tag{8.12}$$

in W, if $v = v_1, v_2,\dots, v_r$. Moreover, (8.10) shows that (8.12) is true for all $v\in\mathcal{E}$.

Suppose that $w\in S(\mathcal{E},\mu)$, where μ is a test atlas. Put

$$\varphi_{\alpha,w}(\lambda) = K_{\alpha w}(\lambda) - \sum_{j=1}^{r} \beta_j(\lambda)(w \circ \tilde{\delta}_j)(\lambda) . \tag{8.13}$$

This definition makes sense for all $\lambda \notin \gamma \cup \Delta$. If λ is in the unbounded component of $\mathbb{C} \setminus \gamma$, put $\varphi_{\alpha,w}(\lambda) = K_{\alpha w}(\lambda)$.

After the substitution of $v = v_i$ in (8.12) and elimination of $\beta_1, \dots, \beta_r$ from the system obtained formula (8.13) yields

$$\varphi_{\alpha,w} = \frac{\det\left[K_{\alpha v_i}, v_i \circ \tilde{\delta}_j\right]_{0 \le i \le r}^{1 \le j \le r}}{\det\left[v_i \circ \tilde{\delta}_j\right]_{1 \le i \le r}^{1 \le j \le r}} ; \tag{8.14}$$

here $v_0 \overset{def}{=\!=} w$.

8.13. THEOREM ON $\varphi_{\alpha,w}$. <u>Let</u> $\alpha \in \xi^{\perp}$ <u>and</u> $w \in S(\xi, \mu)$ <u>where</u> $\xi \in \in \mathrm{Lat}(M_{\Pi})$ <u>and</u> μ <u>is a test atlas on</u> R. **<u>Then</u>** $\varphi_{\alpha,w}$ **<u>is well-defined</u>** <u>by (8.13) and analytic on</u> $\mathbb{C} \setminus \mathcal{O}$. <u>Moreover</u>, $\varphi_{\alpha,w} | \Omega \in \mathcal{D}(\Omega))$ <u>for every component</u> Ω <u>of</u> $\mathbb{C} \setminus \gamma$.

PROOF. We first verify that $\varphi_{\alpha,w}$ is well-defined. Let W, $W' \subset \Omega$, the functions $\beta_i \in \mathcal{N}(W)$ and branches $\tilde{\delta}_i$ of $(\Pi | W)^{-1}$ correspond to W and $\beta_i' \in \mathcal{N}(W')$, $\tilde{\delta}_i'$ correspond to W'. Here $r = r(\xi, \Omega)$. Suppose that $w \in S(\xi, \mu)$ and $\lambda_0 \in W \cap W'$. We need to check the equality

$$\sum_{j=1}^{r} \beta_j(\lambda_0)(w \circ \tilde{\delta}_j)(\lambda_0) = \sum_{j=1}^{r} \beta_j'(\lambda_0)(w \circ \tilde{\delta}_j')(\lambda_0) . \tag{8.15}$$

By the definition of the test atlas, a test set V, $V \in \mu$, exists such that $\lambda_0 \in \Pi(V)$ and $\varkappa(\lambda_0) = \ell(V)$. Choose an admissible domain A satisfying $\lambda_0 \in A \subset W \cap W' \cap \Pi(W)$. There are $v_n \in \xi$ such that $v_n | V \to W | V$ in $E^2(V)$ if $n \to \infty$. We have $v_n \circ \tilde{\delta}_j \to \to w \circ \tilde{\delta}_j$ and $v_n \circ \tilde{\delta}_j' \longrightarrow w \circ \tilde{\delta}_j'$ in $E^2(A)$. On the other hand,

(8.12) implies that

$$\sum_{j=1}^{r} \beta_j(\lambda)(v_n \circ f_j)(\lambda) = \sum_{j=1}^{r} \beta_j'(\lambda)\,(v_n \circ f_j')(\lambda) \tag{8.16}$$

for $\lambda \in A$ and for all n. Take some $\psi \in H^\infty(A)$, $\psi \not\equiv 0$, with the properties $\psi \cdot \beta_j$, $\psi \cdot \beta_j' \in H^\infty(A)$, $j = 1, \dots, r$. Multiply both parts of (8.16) by ψ, pass to the limit in $E^2(A)$ and then divide by ψ. We come to (8.15).

Suppose that Ω is a component of $\mathbb{C} \setminus \gamma$ and $V \in \mu$, with $\varkappa(\Omega) = \ell(V)$. Take an arbitrary admissible simply connected set W containing in $(\Omega \cap \Pi(V)) \setminus \Delta$. We show that $\varphi_{\alpha,w} \in \mathcal{D}(W)$ Since $K_{\alpha w} \in E_0^{1,\infty}(\Omega)$, we are to prove that $\sum \beta_j (w \circ f_j) \in \mathcal{D}(W)$. Choose $\psi \in H^\infty(W)$, $\psi \not\equiv 0$, such that $\psi \cdot \beta_j \in H^\infty(W)$ for $j = 1, \dots, r$. The last relation is equivalent to

$$\sum \psi \beta_j \cdot (w \circ f_j) \in (\psi)_i\, E^2(W), \tag{8.17}$$

where $(\psi)_i$ is the inner part of ψ. Take some $v_n \in \mathcal{E}$ satisfying $v_n \to w$ in $E^2(V)$. Then

$$\sum \psi \beta_j \cdot (v_n \circ f_j) \in (\psi)_i\, E^2(W), \tag{8.18}$$

since $\varphi_{\alpha v_n} \equiv 0$ for all n. It remains to pass to the limit in (8.18) as $n \to \infty$ bearing in mind that $(\psi)_i\, E^2(W)$ is a closed subspace of $E^2(W)$.

For every point λ of Ω a test set $V \in \mu$ exists such that $\ell(V) = \varkappa(\lambda)$ and $\lambda \in \Pi(V)$. Hence $\varphi_{\alpha,w}$ is analytic on $\Omega \setminus \Delta$. The definition of the test atlas implies as well that every $\lambda \in \partial\Omega$ has a neighbourhood G such that $G \cap \Omega \subset \Omega \cap \Pi(V)$ for some $V \in \mu$ with $\varkappa(\Omega) = \Pi(V)$. If we check that $\varphi_{\alpha,w}$ is analytic on $\Delta \cap \Omega$, it would follow that $\varphi_{\alpha,w} | \Omega \in \mathcal{D}_{loc}(\Omega)$ and $\varphi_{\alpha,w} | \Omega \in \mathcal{D}(\Omega)$.

Let $\zeta \in \Delta \cap \Omega$. Take $V \in \mu$ such that $\zeta \in \Pi(V)$ and $\varkappa(\zeta) = \ell(V)$. Choose a positively oriented contour Γ which surrounds ζ and is contained in $\Pi(V)$. Choose then simply-connected admissible W_1, W_2 such that $\Gamma \subset W_1 \cup W_2$ and $\overline{W_1} \cup \overline{W_2} \subset \Omega \setminus \Delta$. Some functions $\beta_j^{(i)} \in \mathcal{N}(W_i)$ and branches $\tilde{f}_j^{(i)}$ of Π^{-1}/W_i correspond to W_i

for $i=1,2$. Each $\beta_j^{(i)}$ has a finite number of poles in W_i. Hence we may slightly change Γ so that it contain no such **poles.**

The expression for

$$(\Phi w)(\lambda) \overset{def}{=\!=} \varphi_{\alpha,w}(\lambda) - K_{\alpha w}(\lambda)$$

(see (8.12)) depends on $w|V$ only. Moreover, Φ is continuous as an operator from $E^2(V)$ **to** $C(\Gamma)$. If $n \geqslant 0$, then

$$\int_\Gamma (\Phi w)(\lambda)\,\lambda^n\,d\lambda = 0$$

for all $w \in \xi$ and thus for all $w \in S(\xi, \mu)$. Hence both functions Φw and $\varphi_{\alpha,w}$ are analytic in ζ.

Now take arbitrary $\lambda_0 \in \gamma \setminus \mathcal{O}$. To **complete the proof we have to prove** that $\varphi_{\alpha,w}$ can be extended analytically to a neighbourhood of λ_0. Let Ω_1, Ω_2 be the components of $\mathbb{C} \setminus \gamma$ whose boundaries contain λ_0. To be definite suppose that $\varkappa(\Omega_1) < \varkappa(\Omega_2)$. Choose sets $V_1, V_2 \in \mu$ and an arc Γ with the properties: $\lambda_0 \in \Pi(V_1)$, $\varkappa(\Omega_1) = \ell(V_1)$, $\varkappa(\Omega_2) = \ell(V_2)$, $\lambda_0 \in \Gamma \subset \subset (\partial \Pi(V_2)) \cap \gamma$, $\Gamma \subset \Pi(V_1)$. They exist by the definition of the test atlas. We show first that the boundary values of $\varphi_{\alpha,w}$ coincide a.e. on Γ.

Take simply connected admissible sets W_i, $W_i \subset (\Omega_i \cap \Pi(V_i)) \setminus \Delta$, such that $\Gamma \subset \partial W_i$ (here $i=1,2$). Let $\beta_j^{(i)} \in \mathcal{N}(W_i)$ be the corresponding functions and $f_j^{(i)}$ be the corresponding branches of $\Pi^{-1}|W_i$ involved in the definition of $\varphi_{\alpha,w}|W_i$. Here $1 \leqslant j \leqslant r(\xi, \Omega_i) \overset{def}{=\!=} r_i$. Let $f_1, \dots, f_t$ be all branches of $\Pi^{-1}|W_2$ **and** $f_1, \dots, f_s$ those of them which satisfy $f_j(\Gamma) \subset \partial R$. Put

$$(\Psi w)(\lambda) = \sum_{j=1}^{r_1} \beta_j^{(1)}(w \circ f_j^{(1)}) - \sum_{j=1}^{r_2} \beta_j^{(2)}(w \circ f_j^{(2)}) + \sum_{j=1}^{s} (\alpha \circ f_j)(w \circ f_j)$$

for $\lambda \in \Gamma$. By (8.8) the boundary values of $\varphi_{\alpha,w}$ coincide a.e. on Γ iff $\Psi w = 0$ a.e. on Γ. Set

$$\rho(\lambda) = max\{|\beta_j^{(1)}(\lambda)|,\ |\beta_j^{(2)}(\lambda)|,\ |\alpha(f_j(\lambda))|\}$$

for $\lambda\in\Gamma$. The expression for Ψw depends on $w|U_2$ only and Ψ is continuous from $E^2(V_2)$ into $L^2(\Gamma,\rho^{-1}|d\lambda|)$. We have $\Psi w=0$ for all $w\in\mathcal{E}$ since $\varphi_{\alpha,w}\equiv 0$ for these w. Hence $\Psi w=0$ a.e. on Γ for all $w\in\mathcal{S}(\mathcal{E},\mu)$. By A.B.Alexandrov's Sticking Theorem 8.8 it remains to verify that $\varphi_{\alpha,w}|\Gamma'\in L_0^{1,\infty}$ for some open arc Γ' such that $\lambda_0\in\Gamma'\subset\Gamma$. Let n denote the maximal integer with the following property:

$$det\left[v_i\circ f_j^{(1)}\right]_{i,j=1}^{r_1}\ \vdots\ (\lambda-\lambda_0)^n \tag{8.19}$$

for all $v_1,\dots,v_{r_1}\in\mathcal{E}$ and for every choice of branches $f_1,\dots,f_{r_1}^{(1)}$ of the function $\Pi^{-1}|V_1$. Passing to the $E^2(V_1)$ -limit in (8.19) we conclude that (8.19) is true for every $v_1,\dots,v_{r_1}\in\mathcal{S}(\mathcal{E},\mu)$ as well. Choose $v_1,\dots,v_{r_1}$ and branches $f_1^{(1)},\dots,f_{r_1}^{(1)}$ of $\Pi^{-1}|V_1$ so that the corresponding determinant is not divisible by $(\lambda-\lambda_0)^{n+1}$. Put $v_0=w$. By (8.14), we have

$$\varphi_{\alpha,w}=\sum_{s=0}^{r_1}\frac{det\left[v_i\circ f_j^{(1)}\right]_{0\le i\le r_1,\, i\ne s}^{1\le j\le r_1}}{det\left[v_i\circ f_j^{(1)}\right]_{1\le i\le r_1}^{1\le j\le r_1}}K_{\alpha v_s,\Omega_1} \tag{8.20}$$

Note that $K_{\alpha v_s,\Omega_1}\in L_0^{1,\infty}(\Gamma)$ for all s and the coefficients at $K_{\alpha v_1,\Omega_1}$ are analytic in λ_0. Hence $\varphi_{\alpha,w}|\Gamma_1\in L_0^{1,\infty}(\Gamma_1)$ for an open arc Γ_1 such that $\lambda_0\in\Gamma_1\subset\Gamma$. ▨

8.14. NOTE. Let ζ be a point of $\mathcal{O}$ and W a neighbourhood of ζ which is divided by γ into four domains $G_1,\dots,G_4$ (see (d) and Fig.8). Suppose that $\varkappa$ attains its minimal value in G_3. Then $\varphi_{\alpha,w}\in E_0^{1,\infty}(G_3)$.

Indeed, all branches f_j of $(\Pi|G_3)^{-1}$ can be continued analytically to a neighbourhood of ζ. Therefore the functions v_i in

(8.20) can be chosen so that the coefficients at the Cauchy integrals be analytic in ζ . This proves that $\varphi_{\alpha,w}|G_3$ belongs to $E_\bullet^{1,\infty}$ locally near ζ . As for the other points of ∂G_3 , $\varphi_{\alpha,w}|G_3$ belongs to $E_o^{1,\infty}$ locally in their neighbourhoods due to the proof of the theorem. ▨

8.15. COROLLARY. Let μ be a test atlas and let $\mathcal{E}\in \mathrm{Lat}(M_\Pi)$, $w\in S(\mathcal{E},\mu)$, $\alpha\in L^2(\partial R)$. Then $\varphi_{\alpha,w}\equiv 0$ iff $\alpha\perp \Pi^n\cdot w$ for all $n\geq 0$.

This follows from the fact that $\varphi_{\alpha,w}\equiv K_{\alpha w}$ on the unbounded component of $\mathbb{C}\setminus\gamma$ and from (8.7). ▨

8.16. COROLLARY. Let $\mathcal{E}\in \mathrm{Lat}(M_\Pi)$ and $w\in S(\mathcal{E},\mu)$, where μ is a test atlas on R . Then $w\in\mathcal{E}$ provided $\varphi_{\alpha,w}$ is analytic on $\mathcal{O}$ for all $\alpha\in\mathcal{E}^\perp$.

Indeed, under the last condition $\varphi_{\alpha,w}$ is analytic on $\hat{\mathbb{C}}$ and $\varphi_{\alpha,w}(\infty) = K_{\alpha w}(\infty) = 0$. Hence $\varphi_{\alpha,w}\equiv 0$ for all $\alpha\perp\mathcal{E}$ and thus $\alpha\perp w$ for all $\alpha\perp\mathcal{E}$ (see (8.7)). ▨

So to find a description of invariant subspaces it is useful to find a condition on w which is necessary and sufficient for $\varphi_{\alpha,w}$ to be analytic on $\mathcal{O}$.

8.17. COROLLARY. If R is such that $\mathcal{O}=\emptyset$, then $\mathcal{E}=S(\mathcal{E},\mu)$ for every test atlas μ .

This follows immediately from 8.16 and 8.13. ▨

8.18. COROLLARY. Let $\mathcal{E}\in \mathrm{Lat}(M_\Pi)$. Then $S(\mathcal{E},\mu_1)=S(\mathcal{E},\mu_2)$ for every test atlases μ_1,μ_2 .

To prove this, suppose that $w\in S(\mathcal{E},\mu_1)$ and $V\in\mu_2$. Then

$$w|X\cap V\in tr_{X\cap V}\mathcal{E} = tr_{X\cap V}tr_V\mathcal{E}$$

for every $X\in\mu$. Note that the sets $X\cap V$ for $X\in\mu_1$ form a test atlas on V . Since Riemann surface V satisfies the hypotheses of Corollary 8.17, we obtain $w|V\in tr_V\mathcal{E}$. Hence $w\in S(\mathcal{E},\mu_2)$, and thus $S(\mathcal{E},\mu_1)\subset S(\mathcal{E},\mu_2)$, $S(\mathcal{E},\mu_1)=S(\mathcal{E},\mu_2)$. ▨

Henceforth the space $S(\mathcal{E},\mu)$ which is independent of the choice of the test atlas μ will be called the <u>standard subspace corresponding to</u> $\mathcal{E}$ and will be denoted as $St(\mathcal{E})$. We have $St(\mathcal{E}) \supset \mathcal{E}$ and $St(St(\mathcal{E})) = St(\mathcal{E})$ for every $\mathcal{E} \in Lat(M_\Pi)$.

8.19. THEOREM. <u>Suppose that</u> ν <u>is a collection of admissible subsets of</u> R <u>such that</u> $\nu \supset \mu$ <u>for some test atlas</u> μ. <u>Besides, suppose that to every</u> $\zeta \in \mathcal{O}$ <u>its neighbourhood</u> W_ζ <u>and a set</u> V_ζ, $V_\zeta \in \nu$ <u>correspond with the property</u> $\Pi^{-1}(W_\zeta) \subset V_\zeta$. <u>Then</u>

$$\mathcal{E} = S(\mathcal{E}, \nu)$$

<u>for every</u> $\mathcal{E} \in Lat(M_\Pi)$.

REMARK. This is a kind of localization theorem since maximum of the diametres of the sets $\Pi(V)$, $V \in \nu$, can be aribtrary small. Hence if (H') is fulfilled, each invariant subspace of M_Π on $E^2(R)$ is determined by its traces on subsets of R which may be arbitrary small (in a proper sense).

PROOF. By 8.16, it suffices to check that $\varphi_{\alpha,w}$ is analytic on $\mathcal{O}$ whenever $\alpha \in \mathcal{E}^\perp$ and $w \in S(\mathcal{E}, \mu)$. Let $\zeta \in \mathcal{O}$. It can be assumed that is an admissible set which is divided by γ into some domains G_s $(1 \le s \le 4)$ and satisfies $W_\zeta \cap \Delta = \emptyset$. Choose for every G_s functions $v_i^{(s)} \in \mathcal{E}$ and branches $f_j^{(s)}$ of $\Pi^{-1}|G_s$ involved in the definition of $\varphi_{\alpha,w}$ (see 8.12 and (8.14)); here $r_s = r(\mathcal{E}, C_s)$ and $1 \le i, j \le r_s$. Put

$$det_s = det\,[v_i^{(s)} \circ f_j^{(s)}]$$

for $1 \le s \le 4$. Then $det_s \not\equiv 0$ in G_s by the assumptions of 8.12.

The functions det_s have nontangential boundary values a.e. on γ. The circle $C_a = \{\lambda : |\zeta - \lambda| = a\}$ intersects γ non-tangentially in four points for small a since γ is piecewise C^2-smooth. Hence we may fix two circles C_{a_1} and C_{a_2} contained in W_ζ so that

$$\inf\{|det_s(\lambda)| : \lambda \in C_{a_i} \cap G_s\} > 0 \qquad (8.21)$$

for $i=1,2$ and $s=1,\dots,4$. Choose a closed contour Γ surrounding ζ which lies between C_{a_1} and C_{a_2} . Assume that $a_1<a_2$.

The relation $w\in S(\xi,\nu)$ implies that there are $h_n\in\xi$ such that $h_n|V_\zeta\to w|V_\zeta$ in $E^2(V_\zeta)$ if $n\to\infty$. Put

$$\alpha^{(1)}=\alpha\cdot(\chi_{W_\zeta}\circ\Pi)$$

and

$$\alpha^{(2)}=\alpha\cdot(\chi_{\mathbb{C}\setminus W_\zeta}\circ\Pi).$$

Then $\alpha^{(1)},\alpha^{(2)}\in L^2(\partial R)$ and $\alpha^{(1)}+\alpha^{(2)}=\alpha$. It follows from (8.14) that

$$\varphi_{\alpha,h_n}(\lambda)=K_{\alpha^{(1)}h_n}(\lambda)+K_{\alpha^{(2)}h_n}(\lambda)+\frac{\psi_{h_n,s}(\lambda)}{det_s(\lambda)}$$

for $\lambda\in G_s$, where

$$\psi_{h,s}\overset{def}{=\!=}\sum_{t=1}^{r_s}(-1)^t(h\circ\tilde{\delta}_t^{(s)})\,det\left[K_{\alpha v_i},\,v_i\circ\tilde{\delta}_j^{(s)}\right]_{1\le i\le r_s}^{1\le j\le r_s,\ j\ne t}.$$

Here $n\geq 0$ and $s=1,\dots,4$. For some $p<1$, the function $\psi_{h,s}$ belongs to $E^p(G_s)$ for every $h\in E^2(V_\zeta)$, and $\psi_{h_n,s}\to\psi_{w,s}$ in $E^p(G_s)$ if $n\to\infty$. Note also that $K_{\alpha^{(1)}h_n}\to K_{\alpha^{(1)}w}$ in $E^q(G_s)$ for every $q<1$. Put

$$\tilde{\varphi}_{\alpha,h}(\lambda)=K_{\alpha^{(1)}h}(\lambda)+\frac{\psi_{h,s}(\lambda)}{det_s(\lambda)}\qquad(8.22)$$

for $\lambda\in G_s$ and $h\in E^2(V_\zeta)$. The functions $K_{\alpha^{(2)}h}$ are analytic on W_ζ for all $h\in E^2(R)$. Since $\varphi_{\alpha,h_n}\equiv 0$ for all n , it follows that $\tilde{\varphi}_{\alpha,h_n}$ are analytic on W_ζ . The function $\tilde{\varphi}_{\alpha,w}$ is analytic on $W_\zeta\setminus\{\zeta\}$ since $\varphi_{\alpha,w}$ is analytic there. Formulae (8.21), (8.22) imply that $\tilde{\varphi}_{\alpha,h_n}\to\tilde{\varphi}_{\alpha,w}$ in $L^p(C_{a_i})$ for some $p>0$. Hence $\|\tilde{\varphi}_{\alpha,h_n}-\tilde{\varphi}_{\alpha,w}\|_{E^p(A)}\to 0$, where A is the annulus $\{\lambda: a_1<|\zeta-\lambda|<a_2\}$. In particular, $\|\tilde{\varphi}_{\alpha,h_n}-\tilde{\varphi}_{\alpha,w}\|_{C(\Gamma)}\to 0$ for $n\to\infty$.

Therefore

$$\int_{\Gamma} \lambda^t \varphi_{\alpha,w}(\lambda)\,d\lambda = \int_{\Gamma} \lambda^t K_{\alpha^{(2)}w}(\lambda)\,d\lambda +$$

$$+\int_{\Gamma} \lambda^t \tilde{\varphi}_{\alpha,w}(\lambda)\,d\lambda = \lim_{n\to\infty} \int_{\Gamma} \lambda^t \tilde{\varphi}_{\alpha,h_n}(\lambda)\,d\lambda = 0$$

for all integer $t \geq 0$. Thus $\varphi_{\alpha,w}$ is analytic in ζ. ▨

Note that the condition (H') is essential in the above theorem (as well as in Theorem 10.3 below). This can be seen from the example $R=\{z : r<|z|<1\}$, $\Pi(z) \equiv z$ and $\mathcal{E}=\{x|R : x\in H^2\}$. In this case $S(\mathcal{E},\mu)=E^2(R)$ for every collection μ of simply connected sets.

It may be assumed that the collection ν in the above theorem consists of the sets of μ and the sets V_ζ, $\zeta\in\mathcal{O}$. Therefore we have to find a description of subspaces $tr_{V_\zeta}\mathcal{E}$ for $\mathcal{E}\in Lat(M_\Pi)$. A special choice of sets V_ζ will be convenient here. But before making this choice we dwell on a special class of surfaces R.

9. THE TWO-SHEETED SURFACES.

In this section, we study a surface R consisting of sheets $\tilde{A}$ and $\tilde{B}$ which project homeomorphically onto admissible subdomains A and B of $\mathbb{C}$. The boundaries $\partial A, \partial B$ are assumed to be simple closed curves intersecting in points ζ,ζ' (see Figure 9). We shall first find a description of $Lat(M_\Pi)$ in this case. The case of general R will be reduced to this one.

The spaces $E^2(R)$ and $E^2(A)\oplus E^2(B)$, as well as $L^2(\partial R)$ and $L^2(\partial A)\oplus L^2(\partial B)$ will often be identified. For brevity, we use also the agreement that $E^p(G) = E^p(int\ G)$, $\mathcal{N}(G) = \mathcal{N}(int\ G)$ for a set G in $\mathbb{C}$ and so on.

9.1. LEMMA. Let $G\in Lat(M_\Pi)$ and $\alpha=(\alpha_A,\alpha_B)\in G^\perp$ (here $\alpha_A\in L^2(\partial A)$

and $\alpha_B \in L^2(\partial B)$). Suppose that G contains a function $v=(v_A, v_B)$ with $v_A \not\equiv 0$ and $v_B \not\equiv 0$. Then there exist functions $\tilde{\alpha}_A \in \mathcal{N}(A \setminus B)$ and $\tilde{\alpha}_B \in \mathcal{N}(B \setminus A)$ such that $\tilde{\alpha}_A = \alpha_A$ a.e. on $(\partial A) \setminus B$ and $\tilde{\alpha}_B = \alpha_B$ a.e. on $(\partial B) \setminus A$. Moreover, for every $w \in St(G)$ we have

$$\varphi_{\alpha,w} \equiv 0 \Longleftrightarrow \tilde{\alpha}_A w_A \in L_0^{1,\infty}(\partial B \cap A)$$

and

$$\| \tilde{\alpha}_A w_A \|_{L_0^{1,\infty}(\Gamma_A)} \leqslant C \| \alpha \|_{L^2(\partial R)} \cdot \| w \|_{L^2(\partial R)} \cdot \qquad (9.1)$$

The constant C depends only on A and B.

PROOF. Since $K_{\alpha v} = 0$ in $\mathbb{C} \setminus (\overline{A \cup B})$, we get

$$\tilde{\alpha}_A \stackrel{def}{=\!=} \frac{K_{\alpha v, A \setminus B}}{v_A} = \alpha_A \quad \text{on} \quad \partial A \setminus B, \qquad (9.2)$$

and

$$\tilde{\alpha}_B \stackrel{def}{=\!=} \frac{K_{\alpha v, B \setminus A}}{v_B} = \alpha_B \quad \text{on} \quad \partial B \setminus A \qquad (9.3)$$

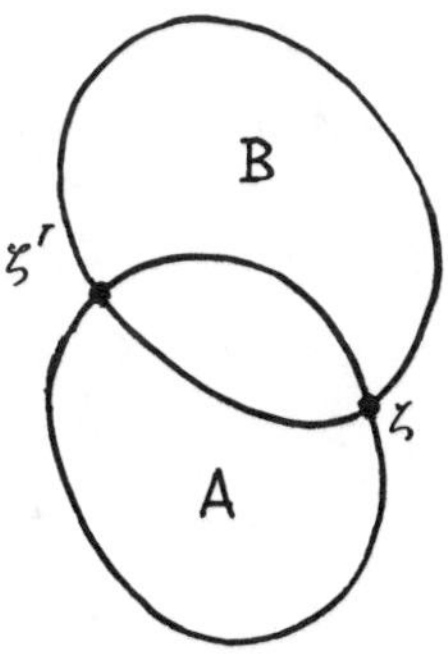

Figure 9

which proves the first statement. One can easily see that (9.2) defines the same $\tilde{\alpha}_A \in \mathcal{N}(A \setminus B)$ for every v such that $\varphi_{\alpha,w} \equiv 0$. Thus $\tilde{\alpha} w_A = K_{\alpha w, A \setminus B} \in L_0^{1,\infty}(\partial B \cap A)$ and (9.1) holds for all w such that $\varphi_{\alpha,w} \equiv 0$. It remains to check that $\tilde{\alpha}_A w_A \in L_0^{1,\infty}(\partial A \cap B)$ implies $\varphi_{\alpha,w} \equiv 0$.

Let θ_A be the greatest common divisor of the functions $(v_A)_i$ and θ_B be the greatest common divisor of the functions $(v_B)_i$ for all $v = (v_A, v_B) \in G$. Here $(f)_i$ denotes the inner part of f. Define an isometric operator Θ in $L^2(\partial R)$ by $\Theta(x_A, x_B) \stackrel{def}{=\!=} (\theta_A x_A, \theta_B x_B)$. Let $\alpha_1 = \Theta \alpha$, $w_1 = \Theta^{-1} w$ and $G_1 = \Theta^{-1} G$. We have $G_1 = Lat(M_\Pi)$, $\alpha_1 \in G_1^\perp$, $w_1 \in E^2(R)$ and $\tilde{\alpha}_{1A} w_{1A} = \tilde{\alpha}_A w_A$ on ∂A.

We want to show that $St(G_1) = \Theta^{-1} St(G)$ or equivalently

$\Theta\, St(G_1) = St(G)$. Note that the sets $\widetilde{A}, \widetilde{B}$ and $\Pi^{-1}(A\cap B)$ form a test atlas for R. If $w \in St(G)$, then obviously $w = \Theta\, w_1$ for some $w_1 \in E^2(R)$. Let $x^{(n)} \in G$ satisfy $x^{(n)} | \Pi^{-1}(A\cap B) \to w | \Pi^{-1}(A\cap B)$ in E^2 as $n \to \infty$. Take any outer $\psi \in H^\infty(A\cap B)$ such that $|\theta_A^{-1}\theta_B^{-1}\psi| < 1$ on $\partial(A\cap B)$. Then

$$\psi\, \Theta^{-1} x^{(n)} | \Pi^{-1}(A\cap B) \to \psi w_1 | \Pi^{-1}(A\cap B)$$

in $E^2(\Pi^{-1}(A\cap B))$ as $n \to \infty$. Proposition 8.4 implies that $w_1 | \Pi^{-1}(A\cap B) \in tr_{\Pi^{-1}(A\cap B)} G_1$. This proves the inclusion $St(G) \subset \Theta\, St(G_1)$. The opposite inclusion is even easier.

The equality $K_{\alpha w} = K_{\alpha_1 w_1}$ implies that $\varphi_{\alpha,w} = \varphi_{\alpha_1, w_1}$. Hence we may assume that $\theta_A \equiv 1$, $\theta_B \equiv 1$. It follows from equations (9.2), (9.3) which are valid for all $v \in G$ that

$$\varphi_{\alpha,w} = K_{\alpha w} - \widetilde{\alpha}_A w_A \qquad \text{in } A \setminus B, \tag{9.4}$$

$$\varphi_{\alpha,w} = K_{\alpha w} - \widetilde{\alpha}_B w_B \qquad \text{in } B \setminus A, \tag{9.5}$$

$$\varphi_{\alpha,w} = K_{\alpha w} \qquad \text{in } \hat{\mathbb{C}} \setminus (A \cup B). \tag{9.6}$$

The formulae (9.4), (9.8), $\widetilde{\alpha}_A w_A \in L_0^{1,\infty}(\partial A \cap B)$ and Sticking Lemma 4.7 imply that $\varphi_{\alpha,w} \in E_0^{1,\infty}(\hat{\mathbb{C}} \setminus B)$. Moreover, (9.3) and (9.5) yield

$$v_B \varphi_{\alpha,w} = v_B K_{\alpha w} - w_B K_{\alpha v} \qquad \text{in } B \setminus A$$

for all $v \in G$. Thus $v_B \varphi_{\alpha,w} \in E^p(B\setminus A)$ for all $v \in G$ (here $p < 2/3$). Since $v_B \varphi_{\alpha,w} \in L^p((\partial B)\cap A)$ and $v_B \varphi_{\alpha,w} \in \mathcal{D}(A\cap B)$ by Theorem 8.13, it follows that $v_B \varphi_{\alpha,w} \in E^p(B)$. Here v is an arbitrary function in G. Since $\theta_B \equiv 1$, we get $\varphi_{\alpha,w} \in \mathcal{D}(B)$. Thus $\varphi_{\alpha,w} \in E_0^{1,\infty}(\hat{\mathbb{C}} \setminus B)$ and $\varphi_{\alpha,w} \in E_0^{1,\infty}(B)$, which implies that $\varphi_{\alpha,w}$ is analytic on $\hat{\mathbb{C}}$. Hence $\varphi_{\alpha,w} \equiv 0$. ▨

9.2. LEMMA. <u>Let $G \in Lat(M_\Pi)$. Then there exists a measurable function ρ on $(\partial B) \setminus A$ such that</u>

$$G=\{w\in St(G):\rho\cdot(w\circ(\Pi|\tilde{A})^{-1})\in L_0^{1,\infty}(\partial B\cap A)\}.$$

PROOF. We can assume that there is $x\in G$ such that $x_A\not\equiv 0$ and $x_B\not\equiv 0$ (otherwise $G=St(G)$). Pick a sequence $\{\alpha^{(n)}\}_{n=1}^{\infty}$ in $G^{\perp}$ which spans $G^{\perp}$. Multiply $\alpha^{(n)}$ by suitable constants to get

$$\sum_{n\geq 1}\|\alpha^{(n)}\|_{L^2}^{1/2}<\infty\ . \tag{9.7}$$

Let $\tilde{\alpha}_A^{(n)}\in\mathcal{N}(A\setminus B)$ be functions from the previous lemma which correspond to the $\alpha^{(n)}$. They exist since $\varphi_{\alpha^{(n)},x}\equiv 0$ for all n .

By (9.7) and (9.1), we have

$$\sum_{n\geq 1}\|\tilde{\alpha}_A^{(n)}x_A\|_{L_0^{1,\infty}(\partial B\cap A)}^{1/2}<\infty\ .$$

Hence the series $\sum|\alpha_A^{(n)}x_A|$ converges to a function ψ in $L_0^{1,\infty}(\partial B\cap A)$ (see 8.5). Therefore the positive series $\sum|\alpha_A^{(n)}|$ converges to the function

$$\rho\overset{def}{=\!=}\psi|x_A|^{-1}$$

a.e. on $(\partial B)\cap A$. Let us prove that ρ is the desired function.

If $w\in St(G)$ and $\rho w_A\in L_0^{1,\infty}(\partial B\cap A)$, then $\alpha^{(n)}w_A\in L_0^{1,\infty}(\partial B\cap A)$ (as $\rho>|\tilde{\alpha}_A^{(n)}|$ for all n). Lemma 9.1 implies that $\varphi_{\alpha^{(n)},w}\equiv 0$ for all n . Hence $\alpha^{(n)}\perp w$ for $n\geq 1$, and thus $w\in G$.

Conversely, if $w\in G$, then $\rho w_A=\sum|\tilde{\alpha}_A^{(n)}|\,w_A$ a.e. on $\partial B\cap A$. Since

$$\sum_{n\geq 1}\|\,|\tilde{\alpha}_A^{(n)}|\,w_A\|_{1,\infty}^{1/2}<\infty$$

(see (9.1) and (9.7)),it follows from 8.5 that $\rho w_A\in L_0^{1,\infty}(\Gamma_A)$. ▨

9.3. REMARKS. 1) Let $\Gamma_\zeta,\Gamma_{\zeta'}$ be any subarcs of $\partial B\cap A$, with $\zeta\in\Gamma_\zeta$ and $\zeta'\in\Gamma_{\zeta'}$. Then the condition $\tilde{\alpha}_A w_A\in L_0^{1,\infty}(\partial B\cap A)$ in Lemma 9.1 can be replaced by

$$\tilde{\alpha}_A w_A | \Gamma_\zeta \cup \Gamma_{\zeta'} \in L_0^{1,\infty}(\Gamma_\zeta \cup \Gamma_{\zeta'}) .$$

Consequently, the last lemma can be restated as

$$G = \{ w \in St(G) : \rho \cdot (w \circ (\Pi | \tilde{A})^{-1}) \in L_0^{1,\infty}(\Gamma_\zeta \cup \Gamma_{\zeta'}) \} .$$

Hence the "non-standard" condition involving ρ is in fact a local condition in ζ and ζ'. We also note that the arcs Γ_ζ, $\Gamma_{\zeta'}$ can be replaced by analogous subarcs of $\partial A \cap B$.

2) Suppose that the functions w such that w_A and w_B are analytic in ζ' are dense in G. Then it can required only that $\rho \cdot (w \circ (\Pi | \tilde{A})^{-1}) \in L_0^{1,\infty}(\Gamma_\zeta)$ (that is, the corresponding condition **on** $\Gamma_{\zeta'}$ is redundant). To prove this, we need to show that $\varphi_{\alpha,w}$ is analytic in ζ' for $\alpha \in G^\perp$ and $w \in St(G)$. Let $G_1, \dots, G_4$ be the domains which correspond to ζ' (see 8.1) and suppose that $\varkappa$ is maximal in G_1. Then formula (8.14) and our assumption imply that $\varphi_{\alpha,w}$ belongs to $E^p(G_j)$ for some $p<1$ and for $j=4,1,2$. Hence $\varphi \in E^p(\overline{G_4 \cup G_2 \cup G_1})$. Since $\varphi_{\alpha,w} \in E_0^{1,\infty}(G_3)$ by Remark 8.14, it follows from A.B.Alexandrov's theorem 8.8 that $\varphi_{\alpha,w}$ is analytic in ζ'.

10. A DESCRIPTION OF INVARIANT SUBSPACES OF M_Π.

10.1. THE REDUCTION OF THE GENERAL CASE TO THE CASE OF TWO-SHEETED SURFACES. It will be assumed here that the curve ∂R when projected to $\mathbb{C}$ has no overlapping arcs. Fix $\zeta \in \mathcal{O}$. **Choose** V_ζ (see Theorem 8.19) so that its components project onto sets A, B, C situated as shown in the Figure 10. It is assumed that $\bar{A} \cup \bar{B} \subset C$, $\bar{C} \cap \Delta = \emptyset$ and $\partial A \cap \partial B = \{\zeta, \zeta'\}$ for some $\zeta' \notin \gamma$. The set V_ζ consists of sheets $\tilde{A}, \tilde{B}$ projecting onto A and B respectively and of some sheets $C_1, \dots, C_\ell$ projecting homeomorphically into C. The integer ℓ is maximal possible here and $\Pi | \tilde{A}, \Pi | \tilde{B}, \Pi | C_j$ are homeomorphisms, It is also supposed that $\partial A \cap W \subset \gamma$, $\partial B \cap W \subset \gamma$ for some neighbourhood W of ζ.

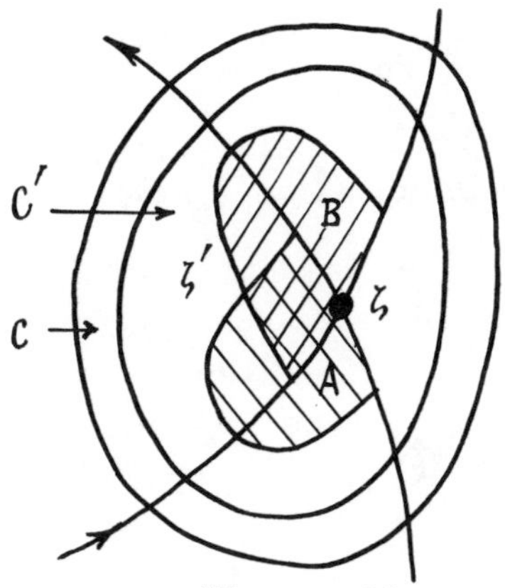

Figure 10

Choose an admissible set C' restricted by a closed simple curve such that $\overline{A \cup B} \subset C'$ and $\overline{C}' \subset C$. We assume that each of the sets C, C' is divided by γ into four domains. We shall use the notations:

$$f_j = (\Pi | C_j)^{-1}, \quad C_j' = f_j(C') = C_j \cap \Pi^{-1}(C'),$$

$$N_n = \bigcup_{j=1}^{n} C_j, \quad N_n' = \bigcup_{j=1}^{n} C_j',$$

$$V_n = N_n \cup \tilde{A} \cup \tilde{B}, \quad V_n' = N_n' \cup \tilde{A} \cup \tilde{B}.$$

Then N_n and N_n' are test sets for every $n \leq \ell$. We have $V_\zeta = V_\ell$.

According to Lemma 8.3, there exist $m \leq \ell$ and functions $y_1, \dots, y_m \in H^\infty(N_\ell)$ such that

$$tr_{N_\ell} \mathcal{E} = \left\{ \sum_{i=1}^{m} (x_i \circ \Pi) y_i : x_i \in E^2(C) \right\}. \qquad (10.1)$$

THE FIRST REDUCTION associates with $tr_{V_\zeta} \mathcal{E}$ an invariant subspace on V_m which is in a sense more convenient. This is done as follows.

Recall that

$$\sum_{\substack{\delta \in \partial(N_\ell), \\ \Pi(\delta) = \lambda}} y_i(\delta) \overline{y_j(\delta)} = \delta_{ij}$$

for a.e. $\lambda \in \partial C$. Hence the operator K defined by

$$Kx \stackrel{def}{=\!=} \sum_{j=1}^{m} (x \circ f_j) y_j \qquad (10.2)$$

is an isometry from $E^2(N_m)$ onto $tr_{N_\ell} \mathcal{E}$. We introduce an operator L from $E^2(V_m)$ into $E^2(V_\ell)$ by

$$L x | N_\ell = K x | N_e, \quad L x | \tilde{A} \cup \tilde{B} = x | \tilde{A} \cup \tilde{B}. \tag{10.3}$$

Then L is an isometry of $E^2(V_m)$ onto the space $(tr_{N_e} \xi) \oplus \oplus E^2(\tilde{A} \cup \tilde{B})$ which contains $tr_{V_\zeta} \xi$. Put

$$\mathcal{F} \overset{def}{=\!=} L^{-1}(tr_{V_\zeta} \xi) ;$$

then $\mathcal{F} \in Lat(M_{\Pi | V_m})$. We can reconstruct $tr_{V_\zeta}(\xi)$ if we know a description of $\mathcal{F}$.

THE SECOND REDUCTION. We introduce the functions $e_i \in E^2(N_m)$ defined by

$$e_i \circ \tilde{\delta}_j = \delta_{ij} \qquad (1 \leq i, j \leq n).$$

The linear set $K^{-1}\xi$ (more precisely, $K^{-1}\{v | N_e : v \in \xi\}$) is dense in $E^2(N_m)$. Hence there exist $f_i \in L^{-1}\xi$ such that the norms $\| e_i - f_i | N_m \|$ are sufficiently small. Then

$$\det \left[f_i \circ \tilde{\delta}_j \right]_{i,j=1}^{m} \neq 0 \qquad \text{on } \bar{C}'.$$

Therefore we can find functions $h_i^{(s)}$ **analytic on $\bar{C}'$ satisfying**

$$\sum_{i=1}^{m} (h_i^{(s)} \circ \Pi) f_i \mid N_m' \equiv e_s | N_m'$$

for $1 \leq s \leq m$. Consider the functions

$$\tilde{e}_s \overset{def}{=\!=} \sum_{i=1}^{m} (h_i^{(s)} \circ \Pi) f_i \overset{def}{=\!=} e_s \oplus a_s \oplus b_s , \tag{10.4}$$

where $a_s \in E^2(\tilde{A})$ and $b_s \in E^2(\tilde{B})$. It follows from the Proposition 7.4 that

$$\tilde{e}_s \in tr_{V_m'} \mathcal{F}, \quad 1 \leq s \leq m .$$

Put $a = (a_1, \dots, a_m)$ and $b = (b_1, \dots, b_m)$, so that $a \in \bigoplus_{i=1}^{m} E^2(\tilde{A})$ and $b \in \bigoplus_{i=1}^{m} E^2(\tilde{B})$. Consider an operator $X_{a,b}$ on $E^2(V_m)$,

$$X_{a,b}\, v \stackrel{def}{=\!=} \begin{pmatrix} v\,|\,N_m \\ v\,|\,\tilde{A} - \sum_{i=1}^{m} a_i (v \circ f_i) \\ v\,|\,\tilde{B} - \sum_{i=1}^{m} b_i (v \circ f_i) \end{pmatrix} . \tag{10.5}$$

Note that this formula defines the operator $X_{a,b}$ for every $a \in \bigoplus_{i=1}^{m} E^2(\tilde{A})$ and $b \in \bigoplus_{i=1}^{m} E^2(\tilde{B})$. The set of such operators is an abelian group since $X_{a,b}\, X_{a',b'} = X_{a+a',b+b'}$.

We introduce also an operator $X'_{a,b} \in \mathcal{L}(E^2(V_m))$ whoes definition can be obtained from the definition of $X_{a,b}$ by replacement of N_m with N'_m . The second reduction is to consider the space $X'_{a,b}\, tr_{V'_m} \mathcal{F}$ instead of $\mathcal{F}$. The next statement clarifies the idea of the above reductions.

10.2. LEMMA. Set

$$G = \{ v|\tilde{A} \cup \tilde{B} : v \in X_{a,b} \mathcal{F} \} \tag{10.6}$$

and suppose $w \in St(\mathcal{E})$. Then $G \in Lat(M_{\Pi|\tilde{A}\cup\tilde{B}})$ and

$$w|V_\zeta \in tr_{V_\zeta} \mathcal{E} \Longleftrightarrow X_{a,b} L^{-1}(w|V_\zeta)|\tilde{A}\cup\tilde{B} \in G .$$

PROOF. Obviously,

$$w|V_\zeta \in tr_{V_\zeta} \mathcal{E} \Longleftrightarrow X_{a,b} L^{-1}(w|V_\zeta) \in X_{a,b} \mathcal{F} .$$

So it suffices to show that $u \in X_{a,b} \mathcal{F}$ is equivalent to $u|\tilde{A}\cup\tilde{B} \in$ $\in G$. Let $\mathcal{D}_1, \dots, \mathcal{D}_n$ be some test sets with the properties: $\mathcal{D}_j \subset N_m$; $\Pi(\mathcal{D}_j) \cap (\bar{A} \cup \bar{B}) = \emptyset$; $l(\mathcal{D}_j) = m$; the sets $\mathcal{D}_1, \dots, \mathcal{D}_n, N'_m$ form a test atlas for N_m . Consider the collection

$$\eta = \{ \mathcal{D}_1, \dots, \mathcal{D}_n, V'_m \} .$$

It is easily seen that there is a collection ν of admissible subsets of V_m with the following properties: 1) ν satisfies the hypotheses of Theorem 8.19; 2) for every $V \in \nu$ there is $W \in \eta$ such that $W \supset V$. It follows that

$$X_{a,b}\mathcal{F} = S(X_{a,b}\mathcal{F}, \eta).$$

Note that

$$tr_{\mathcal{D}_j}(X_{a,b}\mathcal{F}) = tr_{\mathcal{D}_j}\mathcal{F} = E^2(\mathcal{D}_j) \tag{10.7}$$

since $tr_{N_m}\mathcal{F} = E^2(N_m)$. The formulae (10.4), (10.5) imply that $X_{a,b}\tilde{e}_i \mid \tilde{A} \cup \tilde{B} = 0$ for $i = 1, \dots, m$. Hence

$$v \oplus \mathbb{O} \oplus \mathbb{O} = \sum_{i=1}^{m} (v \circ \tilde{\delta}_i) X'_{a,b} \tilde{e}_i \in X'_{a,b} tr_{V'_m}\mathcal{F}$$

for every $v \in E^2(N'_m)$. (Here 8.4 has been used). Thus

$$E^2(N'_m) \oplus \mathbb{O} \oplus \mathbb{O} \subset X'_{a,b} tr_{V'_m}\mathcal{F} = tr_{V'_m}(X_{a,b}\mathcal{F}). \tag{10.8}$$

It follows from (10.7) and (10.8) that

$$E^2(N_m) \oplus \mathbb{O} \oplus \mathbb{O} \subset S(X_{a,b}\mathcal{F}, \eta) = X_{a,b}\mathcal{F}.$$

Hence G is closed and we have $X_{a,b}\mathcal{F} = E^2(N_m) \oplus G$, **which proves the lemma.** ▨

Suppose that the set of self-intersection points $\mathcal{O}$ of the curve $\gamma = \Pi(\partial R)$ consists of the points ζ_s where s ranges over a finite index set. Choose for every ζ_s sets $A^{(s)}$, $B^{(s)}$, $C^{(s)}$, $\tilde{A}^{(s)}$, $\tilde{B}^{(s)}$, $C_i^{(s)}$ as described in 10.1. Here $1 \le i \le \ell_s$, where

$$\ell_s \overset{def}{=\!=} card(\Pi^{-1}\{\zeta_s\} \cap R).$$

Put

$$N^{(s)} = \bigcup_{i=1}^{\ell_s} C_i^{(s)}.$$

Choose arcs $\Gamma_A^{(s)}$, $\Gamma_A^{(s)} \subset A^{(s)} \cap \partial B^{(s)}$, with endpoints at $\zeta^{(s)}$.

10.3. INVARIANT SUBSPACE THEOREM FOR M_Π. Suppose that each two different arcs of ∂R project onto different arcs in $\mathbb{C}$. Let μ be any test atlas on R and $\mathcal{E}$ belong to $Lat(M_\Pi)$. Let

$$tr_{N^{(s)}} \mathcal{E} = K_s\Big(\bigoplus_{i=1}^{m_s} E^2(C^{(s)})\Big),$$

where $0 \le m_s \le \ell_s$ and isometric isomorphisms K_s are defined by

$$K_s(x_1 \oplus \dots \oplus x_{m_s}) = \sum_{i=1}^{m_s} (x_i \circ \Pi) y_i^{(s)}$$

with $y_i^{(s)} \in H^\infty(N^{(s)})$. Denote $\mathcal{E}_W = tr_W \mathcal{E}$ for $W \in \mu$. Then there are $a^{(s)} \in \bigoplus_{i=1}^{m_s} E^2(A^{(s)})$ and measurable functions ρ_s defined on $\Gamma_A^{(s)}$ with the following property: $\mathcal{E}$ coincides with the set of w, $w \in E^2(R)$, such that

1) $w|W \in \mathcal{E}_W$ for all $W \in \mu$;
2) $\rho_s \cdot [w \circ (\Pi|\tilde{A}^{(s)})^{-1} - \sum_{i=1}^{m_s} a_i^{(s)} x_i^{(s)}] \in L_0^{1,\infty}(\Gamma_A^{(s)})$

for all s, where

$$x_1^{(s)} \oplus \dots \oplus x_{m_s}^{(s)} \overset{def}{=\!=} K_s^{-1}(w|N^{(s)}).$$

PROOF. Put $V^{(s)} = N^{(s)} \cup \tilde{A}^{(s)} \cup \tilde{B}^{(s)}$. By Theorem 8.19, $\mathcal{E}$ coincides with the set of all w satisfying 1) (which is equivalent to $w \in St(\mathcal{E})$)) and the condition

$$w|V^{(s)} \in tr_{V^{(s)}} \mathcal{E} \tag{10.9}$$

for all s. Consider the functions

$$w_s = (w_{A^{(s)}}, w_{B^{(s)}}) \overset{def}{=\!=} X_{a^{(s)}, b^{(s)}}^{(s)} L_s^{-1}(w|V^{(s)}) | \tilde{A}^{(s)} \cup \tilde{B}^{(s)}, \tag{10.10}$$

where the functions $a^{(s)} \in \bigoplus_{i=1}^{m_s} E^2(A^{(s)})$, $b^{(s)} \in \bigoplus_{i=1}^{m_s} E^2(B^{(s)})$ and the operators $X_{a^{(s)}, b^{(s)}}^{(s)}$, L_s corresponding to ζ_s are constructed in the same way as in 10.1. Instead of G we must consider the subspaces

$$G(\zeta_s) \overset{def}{=\!=} \{ v | \tilde{A}^{(s)} \cup \tilde{B}^{(s)} : v \in X_{a^{(s)}, b^{(s)}} L_s^{-1} tr_{V^{(s)}} \mathcal{E} \} .$$

By Lemma 10.2, (10.9) is equivalent to the conditions $w_s \in G(\zeta_s)$. Let $\partial A^{(s)} \cap \partial B^{(s)} = \{ \zeta_s, \zeta_s' \}$. Since the functions $a_i^{(s)}$ and $b_i^{(s)}$ are analytic in ζ_s', it follows that the spaces $G(\zeta_s)$ in ζ_s' satisfy the condition of Remark 9.3, 2). Hence (10.9) is equivalent to the conditions $w_s \in St(G(\zeta_s))$ and $\rho_s w_{A^{(s)}} \in L_0^{1,\infty}(\Gamma_A^{(s)})$ for some measurable functions ρ_s (see Lemma 9.2). The conditions $w_s \in St(G(\zeta_s))$ are in fact redundant since they follow from the condition 1).

To prove this last statement, note that the sets $\tilde{A}_s$, $\tilde{B}_s$ $(\tilde{A}_s \cup \tilde{B}_s) \cap \Pi^{-1}(A_s \cap B_s)$ form a test atlas for $\tilde{A}_s \cup \tilde{B}_s$. Let V be one of these sets and $w \in St(\mathcal{E})$; we have to show that

$$w_s | V \in tr_V G(\zeta_s) . \qquad (10.11)$$

For any test set W such that $\Pi(W) \cap \Delta = \emptyset$, the functions from $E^2(W)$ can be considered as columns of $E^2(\Pi(W))$ -functions. In view of this identification, (10.10) yields

$$w_s | V = \mathcal{Y} (w | \mathcal{U}),$$

where $\mathcal{U} = V \cup [\Pi^{-1}(\Pi(V)) \cap N^{(s)}]$ is the test set and $\mathcal{Y}$ is a non-zero matrix with $\mathcal{N}(\Pi(V))$ -entries. Let $f \in H^\infty(\Pi(V))$ be such that $f \not\equiv 0$ and $f \cdot \mathcal{Y}$ has H^∞-entries. Let $f = f_i f_e$ be the Riesz-Nevanlinna factorization of f. Then (10.11) is equivalent to

$$f \cdot \mathcal{Y} \cdot (w_s | V) \in f_i \cdot tr_V G(\zeta_s) \qquad (10.12)$$

due to 8.4. The relations (10.11) and (10.12) are valid for all $w \in \mathcal{E}$. Since $f_i \cdot tr_V G(\zeta_s)$ is a closed subspace of $E^2(V)$, (10.12) and (10.11) are valid for all $w \in St(\mathcal{E})$. ▨

This theorem gives a description of $Lat(M_\Pi)$ in the sense that every $\mathcal{E} \in Lat(M_\Pi)$ has a form given in the theorem. For given subspaces $\{\mathcal{E}_W\}_{W\in\mu}$ the condition 1) in the theorem defines a closed subspace $\mathcal{E}\in \mathrm{Lat}(M_\Pi)$; however $tr_W\mathcal{E} \neq \mathcal{E}_W$ in general. As for the "nonstandard" condition 2), it is not clear when it (together with 1)) defines a closed subspace.

The following statement will provide **an example of a nonstandard subspace** $\mathcal{E}\in \mathrm{Lat}(M_\Pi)$. It is not assumed here that different arcs of ∂R project onto different arcs and that the condition (H') is fulfilled.

10.4. PROPOSITION. Let $\mathbb{F}$ be the polynomially - convex hull of $\Pi(\overline{R})$, and let τ be a conformal mapping of $\mathbb{F}$ onto $\overline{\mathbb{D}}$. If $v\in E^2(R)$ and

$$\int_{\partial\mathbb{F}} \log \max\{|v(\delta)| : \Pi(\delta)=\lambda\}\cdot|\tau'(\lambda)|\,|d\lambda| > -\infty \ , \tag{10.13}$$

then

$$\mathcal{E}_v \overset{def}{=\!=} \mathrm{span}\{\Pi^n\cdot v : n\geqslant 0\} = (v\cdot\mathcal{D}(\mathbb{F}))\cap E^2(R) \ . \tag{10.14}$$

The notation $v\cdot\mathcal{D}(\mathbb{F}) = \{v\cdot(x\circ\Pi) : x\in\mathcal{D}(\mathbb{F})\}$ is used here.

PROOF. It follows from 8.4 that $\mathcal{E}_v\supset(v\cdot\mathcal{D}(\mathbb{F}))\cap E^2(R)$. To prove the inverse, note that (10.13) ensures that there is $\widetilde{v}\in E^2(\mathbb{F})$, with

$$|\widetilde{v}(\lambda)| = \max\{|v(\delta)| : \Pi(\delta)=\lambda\}$$

for a.e. $\lambda\in\partial\mathbb{F}$. Put

$$\varphi_h = \frac{(h\circ\Pi)\cdot v}{\widetilde{v}\circ\Pi}$$

for every $h\in E^2(\mathbb{F})$. Define a linear operator $\mathcal{Y}$ (not necessarily bounded) from $E^2(\mathbb{F})$ to $E^2(R)$ by $\mathcal{Y}h=\varphi_h$ provided $\varphi_h\in E^2(R)$. We have

$$\|\mathcal{Y}h\|_{E^2(R)} \geqslant \|h\|_{E^2(\mathbb{F})} \ .$$

for every h from the domain of $\mathcal{Y}$.

Suppose that $\mathcal{Y}h_n \to w$ in $E^2(R)$. Then $\|h_n - h_m\| \to 0$ for m, $n \to \infty$ and thus $h_n \to h$ for some $h \in E^2(F)$, with $\varphi_h = w$. This shows that the image $Im(\mathcal{Y})$ of $\mathcal{Y}$ is closed. The relations $\chi^n(v) = \varphi_{z^n \tilde{v}} \in Im(\mathcal{Y})$ imply that $\mathcal{E}_v \subset Im(\mathcal{Y}) \subset (v \cdot \mathcal{D}(F)) \cap E^2(R)$. ▨

10.5. COROLLARY. <u>Let</u> F <u>be the polynomially - convex hull of</u> $\Pi(\bar{R})$. <u>Suppose that there is</u> $\zeta \in (\partial F) \cap \mathcal{O}$ <u>such that the angles between the arcs of</u> γ <u>at</u> ζ <u>are non-zero</u>. <u>Then</u> $Lat(M_\Pi)$ <u>contains a non-standard subspace</u>.

Indeed, let ψ be a singular inner function in F corresponding to a point measure in ζ . Take any $v \in E^2(R)$ which satisfies (10.13). Then $\mathcal{E}_v \supsetneqq \mathcal{E}_{\psi v}$ because of (10.14). However $St(\mathcal{E}_v) = St(\mathcal{E}_{\psi v})$ since ψ is outer in $\Pi(V)$ for every test set V . Hence $\mathcal{E}_{\psi v} \neq St(\mathcal{E}_{\psi v})$.

10.6. REMARKS. 1) It follows from the technique of B.M.Solomyak [31] that $\mathcal{E} = St(\mathcal{E})$ provided $St(\mathcal{E}) = E^2(R)$. Using his techniques one can find weaker conditions on $\mathcal{E}$ implying $\mathcal{E} = St(\mathcal{E})$. They can be given either in terms of traces of $\mathcal{E}$ on suitable test sets or in terms of decay of functions in $\mathcal{E}$ near points projecting into $\mathcal{O}$. In application to $R = \sigma_*$, these conditions are formulated in Theorem 4 of [39]. They will not be considered here.

2) The case of $E^p(R)$, $1 \le p < \infty$, instead of $E^2(R)$ is studied in the same way. All our results remain true in this case (in the definition of $\varphi_{\alpha,w}$ we must assume $\alpha \in L^p(\partial R)^* = L^{\frac{p}{p-1}}(\partial R)$).

3) It is very interesting to find a description of $Lat(M_\Pi)$ in case when (H) is violated. If R is an annulus, $p = 2$, and $\Pi(z) \equiv z$, this question has been solved recently by D.Hitt [41] (see also [42-44]). Hitt asks in [41] how to determine invariant subspaces of M_z in $E^p(G)$ if G is a multiconnected domain. Our method applies to this case and provides a result in a form which differs from Hitt's result.

Let G be a finitely connected domain in $\mathbb{C}$ whose boundary is a disjoint union of C^2-smooth contours $\gamma_0, \dots, \gamma_n$. Suppose that γ_i is a boundary of a simply connected subdomain

Ω_i in $\hat{\mathbb{C}}$ (so that $\hat{\mathbb{C}}$ is the disjoint union of γ_i, Ω_i and G), and $\infty \in \Omega_0$. Following D.Hitt [41], we call a function $\varphi \in H^\infty(G)$ <u>inner</u> in G if $\varphi \not\equiv 0$ and $\varphi | \gamma_i$ is constant a.e. for every i (see [43, 36] for alternative definitions).

THEOREM. Let $\mathcal{E}$ be a closed non-zero M_z -invariant subspace in $E^p(G)$ $(1 \leq p < \infty)$. Then there exist an inner function φ in G and measurable functions θ_i, ρ_i on γ_i $(i \geq 1)$ such that $|\theta_i| \equiv 1$, $\rho_i > 0$ or $\theta_i \equiv \rho_i \equiv 0$ on γ_i for each i and such that $\mathcal{E} = \{ f \in \varphi E^p(G) : f | \gamma_i \in \theta_i E^p(\Omega_i), \rho_i f | \gamma_i \in L_0^{1,\infty}(\gamma_i) \ (i \geq 1) \}$.

The proof can be obtained using ideas of Lemmas 9.1 and 9.2 and will be given elsewhere.

CHAPTER 5. INVARIANT AND HYPERINVARIANT SUBSPACES OF TOEPLITZ OPERATORS.

11.1. PRELIMINARIES. Suppose that $\mathbb{F}$ satisfies (i)-(vi), (viii), (H). Let ζ_s be the self-intersection points of $\mathbb{F}(\mathbb{T})$ and C_s be a small neighbourhood of ζ_s such that $C_s \cap \mathbb{F}(\Xi)=\emptyset$. Assume that $\mathbb{F}(\mathbb{T})$ divides each C_s into four domains $G_1^{(s)},\dots,G_4^{(s)}$ enumerated counterclockwise and $G_1^{(s)}$ has among them maximal index $wind_{\mathbb{F}}$ (for each s). Choose sets A_s, B_s corresponding to ζ_s in the same way as in 10.1. That is, A_s and B_s must satisfy: $A_s \subset G_1^{(s)} \cup G_2^{(s)}$, $B_s \subset G_1^{(s)} \cup G_4^{(s)}$, $\partial A_s \cup \partial B_s = \{\zeta_s, \zeta_s'\}$, $\zeta_s' \notin \mathbb{F}(\mathbb{T})$, and $\mathbb{F}(\mathbb{T}) \cap W_s \subset \partial A_s \cup \partial B_s$ for some neighbourhoods W_s of ζ_s.

Choose test sets $N_s \subset \sigma_*(T_{\mathbb{F}})$ that satisfy for each s the conditions:

1) $\Pi(N_s) = C_s$,

2) The set $\Pi^{-1}\{\lambda\} \cap N_s$ is a combinatorial basis in $\Pi^{-1}\{\lambda\}$ for every $\lambda \in G_3^{(s)}$.

Note that by Lemma 5.12 such choice is possible even if it is required additionally that $N_s \subset \sigma_{**}$. The conditions 1), 2) imply that $\ell(N_s) = wind_{\mathbb{F}}(G_3^{(s)})$.

Choose next open subsets $\widetilde{A}_s, \widetilde{B}_s$ of σ_* which project homeomorphically onto A_s and B_s respectively and are adjacent to $\partial\sigma_*$ (that is, $\partial\widetilde{A}_s \cap \partial\sigma_* \neq \emptyset$, $\partial\widetilde{B}_s \cap \partial\sigma_* \neq \emptyset$). Let $\Gamma_A^{(s)}$ be any subarcs of $(\partial B_s) \cap A_s$ **ending at ζ_s. Let U be the isomorphism (0.2).**

11.2. INVARIANT SUBSPACE THEOREM FOR $T_{\mathbb{F}}$. Suppose that $\mathbb{F}$ satisfies (i)-(vi), (viii), (H) and let $\mathcal{L} \in Lat(T_{\mathbb{F}})$. Take a test atlas μ on $\sigma_*(T_{\mathbb{F}})$. Assume that K_s are isometric isomorphisms,

$$tr_{N_s}(U\mathcal{L}) = K_s\Big(\bigoplus_{i=1}^{m_s} E^2(c_s)\Big), \qquad K_s \bigoplus_{i=1}^{m_s} x = \sum_{i=1}^{m_s} (x_i \circ \Pi)\, y_i^{(s)},$$

where $m_s = \tau(U\mathcal{L}, G_3^{(s)})$ and $y_i^{(s)} \in H^\infty(N_s)$ (see Lemma 8.3). Put $\mathcal{E}_V = tr_V(U\mathcal{L})$ for $V \in \mu$. Then there are $a^{(s)} \in \bigoplus_{i=1}^{m_s} E^2(A_s)$

<u>and measurable functions</u> ρ_s <u>on</u> $\Gamma_A^{(s)}$ <u>such that</u> $\mathcal{L}$ <u>coincides with the set of those</u> x , $x \in H^2$, <u>which satisfy</u>

1) $(Ux) | V \in \mathcal{E}_V$ <u>for</u> $V \in \mu$;

2) $\rho_s \cdot [(Ux) \circ (\Pi | \widetilde{A}_s)^{-1} - \Sigma_{i=1}^{m_s} a_i^{(s)} x_i^{(s)}] \in L_0^{1,\infty}(\Gamma_A^{(s)})$

<u>for all</u> s , <u>where</u>

$$x_1^{(s)} \oplus \ldots \oplus x_{m_s}^{(s)} = K_s^{-1}((Ux)| N_s).$$

PROOF. Let $x \in H^2$. The condition $x \in \mathcal{L}$ is equivalent to $Ux \in U\mathcal{L}$. We have $U\mathcal{L} \in Lat(M_\Pi)$, but Theorem 10.3 cannot be applied directly. Put

$$V_s = N_s \cup \widetilde{A}_s \cup \widetilde{B}_s . \tag{11.1}$$

There exist bounded operators $\mathcal{Y}_s : E^2(V_s) \to E^2(\Pi^{-1}(W_s))$ defined by

$$(\mathcal{Y}_s w)(\delta) = \sum_{\delta' \in V_s, \Pi(\delta') = \Pi(\delta)} C_s(\delta, \delta') w(\delta')$$

with some bounded analytic functions C_s , such that

$$u | \Pi^{-1}(W_s) = \mathcal{Y}_s(u|V_s)$$

for all $u \in E_F^2(\sigma_*)$ and all s (This follows from the condition 2) on N_s and (1.6)). Obviously,

$$tr_{\Pi^{-1}(W_s)}(U\mathcal{L}) = clos_{E^2(\Pi^{-1}(W_s))} \mathcal{Y}_s tr_{V_s}(U\mathcal{L}) .$$

It follows that if $u \in E_F^2(\sigma_*)$ and $u|V_s \in tr_{V_s}(U\mathcal{L})$, then

$$u | \Pi^{-1}(W_s) \in tr_{\Pi^{-1}(W_s)}(U\mathcal{L}) .$$

Theorem 8.19 now shows that the condition $Ux \in U\mathcal{L}$ is equivalent to the conditions 1) and

$$Ux | V_s \in tr_{V_s}(U\mathcal{L}) \quad \forall s .$$

The latter (under the assumption of 1)) can be reduced to 2) in the same way as in the proof of Theorem 7.25. ▨

11.3. INVARIANT SUBSPACES OF T_F IF F IS A CLARK TYPE POSITIVE WINDING SYMBOL WITH LOOPS. Suppose that F is a symbol with loops (see Clark [5]) which has positive winding and does not back up. Then F satisfies (H). Let $\{\mathcal{Z}_i\}_{i=1}^{m}$ be the "loops" of the symbol (i.e., bounded components of $\mathbb{C}\setminus F(\mathbb{T})$) and let τ_i be a conformal mapping of $\mathbb{D}$ onto $\mathcal{Z}_i$. If $m>1$, then Similarity Theorem 1.5 is not applicable to T_F since (iv) is not satisfied. Put $\ell_i = wind_F(\mathcal{Z}_i)$. A result of Clark [5] says that T_F is similar to $\bigoplus_{i=1}^{m} \bigoplus_{j=1}^{\ell_i} T_{\tau_i}$. **Define a Riemann surface**

$$R = \bigcup_{i=1}^{m} (\mathcal{Z}_i \times \{1,\dots,\ell_i\})$$

and let $\Pi : R \to \cup \mathcal{Z}_i$ be the first - coordinate projection. Put

$$K_i x = (x\circ \tau_i^{-1})(\tau_i' \circ \tau_i^{-1})^{1/2}, \quad x\in H^2.$$

Then $\oplus_i \bigoplus_{j=1}^{\ell_i} K_i$ is an isomorphism between $\oplus_i \bigoplus_{j=1}^{\ell_i} H^2$ and $E^2(R)$ which yields another form of the result of Clark: T_F is similar to M_Π acting on $E^2(R)$ (note that $\sigma_*(T_F)$ has in general a more complicated structure than R).

11.4. THEOREM. <u>If F is as above, then</u> $Lat(T_F) \cong \bigoplus_{i=1}^{m} Lat(T_{z^{\ell_i}})$.

PROOF. Put $V_i = \Pi^{-1}(\mathcal{Z}_i)$, $\mu = \{V_1,\dots,V_m\}$ and let Ω_0 be the unbounded component of $\mathbb{C}\setminus F(\mathbb{T})$. One can easily see using K_i that the lattice of $M_{\Pi|V_i}$ in $E^2(V_i)$ is isomorphic to $Lat(\bigoplus_{i=1}^{\ell_i} T_z)$ and hence to $Lat(T_{z^{\ell_i}})$. Thus we have to prove that

$$\mathcal{E} = S(\mathcal{E},\mu)$$

for every $\mathcal{E}\in Lat(M_\Pi)$. This follows from the geometric condition on $F(\mathbb{T})$, the statement 8.16 and the fact that $\varphi_{\alpha,w}\in E_0^{1,\infty}(\Omega_0)$ for $w\in St(\mathcal{E})$ and $\alpha\in L^2(\partial R)$, $\alpha\perp\mathcal{E}$ (see the proof of 8.14).▨

Let us combine Theorem 11.2 with the results of Chapter 2. Suppose that $\mathbb{F}$ satisfies (i)-(vi), (viii), (H). Let σ_{**} be the reduced ultraspectrum of $T_{\mathbb{F}}$ and σ_c the commutant surface of $T_{\mathbb{F}}$. The positive orientation of $\partial\sigma_{**}$ agrees with the natural orientation of $\mathbb{F}(\mathbb{T})$ under the projection. Hence σ_{**} satisfies (H'). The notation of i. 11.1 will be used here. It will be assumed that $V_s \subset \sigma_{**}$ for all s (see (11.1)). Let $\alpha_s, \beta_s \in \Pi^{-1}(\mathcal{O}) \cap \cap\, \partial\sigma_{**}$ be points which satisfy $\alpha_s \in \partial\widetilde{A}_1$, $\beta_s \in \partial\widetilde{B}_s$ and $\Pi(\alpha_s) = \Pi(\beta_s) = \zeta_s$. Recall that the projections $\Pi_c : \overline{\sigma}_{**} \to \overline{\sigma}_c$ and $\rho_c : \overline{\sigma}_c \to \mathbb{C}$ have been defined in 6.10. A point ζ_s will be called unremovable if $\Pi_c(\alpha_s) = \Pi_c(\beta_s)$. (This definition can be shown to be independent of the choice of $\widetilde{A}_s, \widetilde{B}_s$). In this case let W_s denote the connected component of $\sigma_c \cap \Pi_c^{-1}(C_s)$ such that $\Pi_c(\alpha_s) \in \overline{W}_s$. We have $\rho_c(W_s) = \overline{C}_s$ or $\rho_c(\overline{W}_s) = \overline{G_s^2 \cup G_s^1 \cup G_s^4}$. Set

$$P_s = N_s \cap \Pi_c^{-1}(W_s).$$

Note that $P_s \neq \emptyset$ implies $\rho_c(\overline{W}_s) = \overline{C}_s$.

Choose a test atlas μ on σ_{**} such that for every $V \in \mu$ the set $\Pi(V)$ contains at most one point from $\mathcal{O} \cup \mathbb{F}(\Xi)$. We shall assume for the sake of simplicity that $N_s \in \mu$ for all s. For $V \in \mu$, let $X_i(V)$ be the components of $\Pi_c(V)$. Put

$$\mu_c \overset{def}{=\!=} \{ V \cap \Pi_c^{-1}(X_i(V)) : V \in \mu \}.$$

The sets from μ_c (as well as the sets P_s) are test sets.

11.5. THEOREM. Suppose that $\mathcal{X}$ is a hyperinvariant subspace of $T_{\mathbb{F}}$. Define isometric isomorphisms K_s for every unremovable $\zeta_s \in \mathcal{O}$ by

$$tr_{P_s}(U\mathcal{X}) = K_s\Big(\bigoplus_{i=1}^{m_s} E^2(C_s)\Big).$$

Set $\mathcal{E}_y = tr_y(U\mathcal{X})$ for $y \in \mu_c$. Then there are $a^{(s)} \in \bigoplus_{i=1}^{m_s} E^2(A_s)$ and measurable functions ρ_s on $\Gamma_A^{(s)}$ such that $\mathcal{X}$ coincides with the set of $x, x \in H^2$, that satisfy

1) $Ux|y \in \mathcal{E}_y$ for all $y \in \mu_c$,

2) for every unremovable $\zeta_s \in \mathcal{O}$

$$\rho_s \cdot \left[(Ux)\circ(\Pi|\tilde{A}_s)^{-1} - \sum_{i=1}^{m_s} a_i^{(s)} x_i^{(s)} \right] \in L_0^{1,\infty}(\Gamma_A^{(s)}),$$

where

$$x_1^{(s)} \oplus \dots \oplus x_{m_s}^{(s)} \overset{def}{=\!=} K_s^{-1}(Ux|P_s).$$

The difference between Theorems 11.2 and 11.5 is that the sets of μ_c have in general less sheets than of μ. Besides, the "non-standard" condition 2) is needed here only for unremovable points $\zeta \in \mathcal{O}$.

PROOF. It is easy to construct a non-compact Riemann surface σ_c' containing $\overline{\sigma}_c$. The projection of σ_c' onto $\mathbb{C}$ will be still denoted by ρ_c.

Let λ be an arbitrary point of $\rho_c(\overline{\sigma}_c)$. One can choose a function $\psi_\lambda \in \mathcal{H}(\sigma_c')$ such that $\psi_\lambda'(\eta) \neq 0$ for every $\eta \in \rho_c^{-1}(\lambda)$ and $\psi_\lambda(\eta_1) \neq \psi_\lambda(\eta_2)$ for every distinct $\eta_1, \eta_2 \in \rho_c^{-1}(\lambda)$. (The derivative with respect to a local coordinate mapping is meant here). This follows easily from the non-compactness of σ_c' (see [18]). Take an open disc $\mathcal{D}_\lambda$ with centre at λ such that ψ_λ is univalent on $\rho_c^{-1}(\overline{\mathcal{D}}_\lambda)$. The discs $\mathcal{D}_\lambda$ form an open covering of the compact set $\rho_c(\overline{\sigma}_c)$. By the Lesbegue lemma, there exists $r > 0$ such that each M, $M \subset \rho_c(\overline{\sigma}_c)$, $diam(M) < r$, is contained in some $\mathcal{D}_\lambda$.

11.6. LEMMA. <u>Suppose that</u> μ <u>and</u> C_s <u>satisfy</u>

(*) $diam\, \Pi(V) < r$ <u>and</u> $diam(C_s) < r$ <u>for all</u> $V \in \mu$ <u>and all</u> s.

<u>Take</u> ν_c <u>to consist of the sets from</u> μ_c <u>and the sets</u> $P_s \cup \tilde{A}_s \cup \tilde{B}_s$ <u>for unremovable</u> ζ_s. <u>Then</u> $U\mathfrak{X} = S(U\mathfrak{X}, \nu_c)$.

PROOF OF THE LEMMA. Suppose that V is a simply connected admissible subset of σ_{**} and that some $\psi, \psi \in H^\infty(\sigma_c)$, is univalent on $\Pi_c(\overline{V})$. Commutant description Theorem 6.11 implies that

$$M_\psi tr_V(U\mathcal{X}) \subset tr_V(U\mathcal{X}).$$

Put $Y_i = \Pi_c^{-1} X_i \cap V$. where X_i are components of $\Pi_c(V)$. Consider the functions φ_i defined on $\psi(V)$ by

$$\varphi_i = \chi_{\psi(\bar{Y}_i)} .$$

Since they can be approximated by polynomials uniformly on $\psi(\bar{V})$, it follows that

$$M_{\varphi_i \circ \psi} tr_V(U\mathcal{X}) = tr_V(U\mathcal{X}).$$

Hence

$$tr_V(U\mathcal{X}) = \bigoplus_i tr_{Y_i}(U\mathcal{X}). \qquad (11.2)$$

Take $w \in S(U\mathcal{X}, \nu_c)$; we have to prove that $w \in U\mathcal{X}$. Since (11.2) is true for all $V \in \mu$, it follows that $w \in S(U\mathcal{X}, \mu)$.

Let $\zeta_s \in \mathcal{O}$ and $V = V_s = N_s \cup \tilde{A}_s \cup \tilde{B}_s$. If ζ_s is removable, then $\tilde{A}_s$ and $\tilde{B}_s$ are contained in different Y_i . Then the curve ∂Y_i has no self-intersections for every i . It is easy to see that $w \in tr_{Y_i \cap Z}(U\mathcal{X})$ for every Y_i and every $Z \in \mu_c$. The sets $Y_i \cap Z$, where Z ranges over μ_c , form a test atlas on Y_i . Therefore $w \in tr_{Y_i}(U\mathcal{X})$ by 8.17. Now (11.2) yields $w \in tr_{V_i}(U\mathcal{X})$.

Suppose to the contrary that ζ_s is unremovable. Then one of the Y_i (namely, $V_s \cap \Pi_c^{-1}(W_s)$) coincides with $P_s \cup \tilde{A} \cup \tilde{B}$. The other sets Y_i belong to μ_c. Hence (11.2) implies that $w \in$ $\in tr_{V_s}(U\mathcal{X})$.

The inclusions $w \in S(U\mathcal{X}, \mu)$ and $w \in tr_{V_s}(U\mathcal{X})$ imply $w \in U\mathcal{X}$ (see the proof of Theorem 7.25). ▨

We want to prove that this lemma is true without the requirement (*). Suppose that an atlas μ and sets C_s are arbitrary. Take sets C_s' and test atlas μ' satisfying (*) and the conditions $C_s' \subset C_s$,

$$\forall V' \in \mu' \ \exists V \in \mu : V \supset V'.$$

We shall also assume that the test atlases ν_s, ν_s' satisfy $N_s' = N_s \cap \Pi^{-1}(G_s')$, $\tilde{A}_s' \subset \tilde{A}_s$, $\tilde{B}_s' \subset B_s$. Take any $w \in S(U\mathcal{X}, \nu_c)$; we show that $w \in U\mathcal{X}$. By the lemma, it suffices to verify that $w \in S(U\mathcal{X}, \nu_c')$.

Let $\mathcal{Y}' \in \nu_c'$. If $\mathcal{Y}' \in \mu_c'$, then $\mathcal{Y}' = (\Pi_c^{-1} X') \cap V'$ for some component X' of $\Pi_c(V')$ and some $V' \in \mu'$. There is $V \in \mu$ with $V \supset V'$. It is obvious that $X' \subset X$ for some component X of $\Pi_c(V)$. Hence $\mathcal{Y}' \subset \mathcal{Y}$, where $\mathcal{Y} \overset{def}{=\!=} (\Pi_c^{-1} X) \cap V$ belongs to μ_c . This yields $w \in tr_{\mathcal{Y}'}(U\mathcal{X})$.

It follows that $w \in tr_{\mathcal{Y}'}(U\mathcal{X})$ for all $\mathcal{Y}' \in \nu_c'$ since $P_s' \cup \tilde{A}_s' \cup \tilde{B}_s' \subset$ $\subset P_s \cup \tilde{A}_s \cup \tilde{B}_s$.

The equality $U\mathcal{X} = S(U\mathcal{X}, \nu_c)$ reduces to the statement of the theorem (see the proof of Theorem 10.3). ▨

It is easy to see that for every choice of spaces $\mathcal{E}_{\mathcal{Y}} \subset E^2(\mathcal{Y})$, $\mathcal{Y} \in \mu_c$, which are invariant under multiplication by the functions $\psi \circ \Pi_c$, $\psi \in H^\infty(\sigma_c)$, and of functions ρ_s the conditions 1), 2) from the theorem define a hyperinvariant linear set $\mathcal{X}$, $\mathcal{X} \subset H^2$, of $T_{\mathcal{F}}$. However it is not clear when $\mathcal{X}$ is closed and when the relations $tr_{\mathcal{Y}}(U\mathcal{X}) = \mathcal{E}_{\mathcal{Y}}$ hold. Note that a space $\mathcal{E}_{\mathcal{Y}}$ is invariant in the above sense iff it is invariant under $M_{\psi_0 \circ \Pi_c}$ where ψ_0 is a conformal mapping of $\Pi_c(\mathcal{Y})$ onto an admissible domain in $\mathbb{C}$. Such ψ_0 exists since $\Pi_c(\mathcal{Y})$ is connected and has at most one branch point with respect to the projection ρ_c . Note also that $\mathcal{Y}$ is a test set with respect to the projection $\psi_0 \circ \Pi_c$. Hence for fixed $\mathcal{Y}$ Lemma 8.3 provides a complete description of such spaces $\mathcal{E}_{\mathcal{Y}}$.

REFERENCES

1. Abrahamse, M.B.: Toeplitz operators in multiply connected domains, Bull.Amer.Math.Soc., 77, No.3 (1971), 444-454.
2. Abrahamse, M.B. and Ball, J.A.: Analytic Toeplitz operators with automorphic symbol II, Proc.Amer.Math.Soc., 59 (1976), 323-328.
3. Alexandrov, A.B.: On A - summability of boundary values of harmonic functions, Mat.zametki, 30 (1981), 59-72. (Russian).
4. Clark, D.N.: Toeplitz operators and K-spectral sets, Indiana Univ.Math.J., 33, No.1 (1984), 127-141.
5. Clark, D.N.: On Toeplitz operators with loops, Journ.Oper. Theory 4, No.1 (1980), 37-54.
6. Clark, D.N.: On the structure of rational Toeplitz operators, Contributions to analysis and geometry (Baltimore, Md., 1980), 63-72, John Hopkins Univ.Press, Baltimore, Md., 1981.
7. Clark, D.N.: On Toeplitz operators with loops II, Journ.Oper. Theory 7 (1982), 109-123.
8. Clark, D.N. and Morrel,J.H.: On Toeplitz operators and similarity, Amer.Journ.Math. 100 (1978), 973-986.
9. Cowen, C.C.: Equivalence of Toeplitz operators, Journ.Oper. Theory 7, No.1 (1982), 167-172.
10. Cowen, C.C.: The commutant of an analytic Toeplitz operator, Trans.Amer.Math.Soc. 239 (1978), 1-31.
11. Cowen, M.J.: Fredholm operators with spanning property, Indiana Univ.Math.J., 35, No.4 (1986), 855-896.
12. Cowen, M.J. and Douglas, R.G.: Complex geometry and operator theory, Acta Math. 141 (1978), 187-261.
13. David, G.: Opérateurs intégraux singuliers sur certaines courbe du plan complexe, Ann.sci.Ec.Norm.Sup., 4 série 17 (1984), 157-189.
14. Douglas, R.G.: Banach algebra techniques in operator theory, Academic Press (New York), 1972.

15. Duren, P.L.: Extension of a result of Beurling on invariant subspaces, Trans.Amer.Math.Soc. 99 (1961), 320-324.
16. Duren, P.L. Theory of H^p-spaces, Academic Press, New York, 1970.
17. Dyn'kin, E.M.: Theorems of Wiener-Levy type and estimates of Wiener-Hopf operators, Matem. Issledov.(Kishiniev) 8, No.3 (1973), 14-25 (Russian).
18. Forster, O.: Riemannsche flächen, Springer-Verlag, Berlin, Heidelberg, N.Y. (1977).
19. Heins, M.: Hardy classes on Riemann surfaces, Springer-Verl., 1969.
20. Janson, S., Peetre, J., Semmes, S.: On the action of Hankel and Toeplitz operators on some function spaces, Duke Math. J., 51, No.4 (1984), 937-957.
21. Koosis, P. Introduction to H^p spaces, Cambridge Univ.Press, Cambridge, London, N.Y., 1980.
22. Neville, Ch.W.: Invariant subspaces of Hardy classes on infinitely connected open surfaces, Memoirs Amer.Math.Soc. 2, iss.1, No.160 (1975).
23. Nikol'skiĭ, N.K.: Treatise on the shift operator, Springer-Verlag, 1986.
24. Peller, V.V.: Invariant subspaces of Toeplitz operators, Zap.Nauchn.Sem.LOMI 126 (1983), 170-179. (Russian).
25. Peller, V.V.: Spectrum, similarity and invariant subspaces of Toeplitz operators, Izv.AN SSSR 50, No.4 (1986), 776-787 (Russian).
26. Privalov, I.I.: Boundary behavior of analytic functions, Moskow-Leningrad: GITTL 1950, Russian (German translation: Berlin i Dentsher Verlag der Wissenchaften 1956).
27. Rochberg, R.: Toeplitz operators on weighted H^p -spaces, Indiana Univ.Math.J. 26, No.2 (1977), 291-298.
28. Shields, A.L. and Wallen, L.J. The commutants of certain Hilbert space operators, Indiana Univ.Math.J., 20 (1970/71), 777-788.
29. Solomyak, B.M.: On the multiplicity of spectrum of analytic Toeplitz operators, Doklady Akad.Nauk SSSR 286, No.6 (1986),

1308-1311 (Russian).

30. Solomyak, B.M.: Cyclic families of functions for analytic Toeplitz operators, Zap.Nauchn.Sem.LOMI 157 (1987), 88-102 (Russian).

31. Solomyak, B.M., and Vol'berg, A.L. Multiplicity for analytic Toeplitz operators, present volume.

32. Stephenson,K.: Analytic functions of finite valence, with applications to Toeplitz operators, Michigan Math.J. 32 (1985), 5-19.

34. Thomson, J.E.: The commutant of a class of analytic Toeplitz operators, Amer.Journ.Math. 99, No.3 (1977), 522-529.

35. Treil, S.R.: The resolvent of a Toeplitz operator may have arbitrary growth, Zapiski Nauch.Sem.LOMI 157 (1987), 175-177 (Russian).

36. Voichick, M.: Ideals and invariant subspaces of analytic functions, Trans.Amer.Math.Soc. 111, No.3 (1964), 493-512.

37. Wang, D.: Similarity theory of smooth Toeplitz operators, Journ.Oper.Theory 12 (1984), 319-330.

38. Wermer, J.: Rings of analytic functions, Ann.Math. 67 (1958), 497-516.

39. Yakubovich, D.V.: Multiplication operators on Riemann surfaces as models of Toeplitz operators, Doklady Akad.Nauk SSSR 302 No.5, 1068-1072.

40. Yakubovich, D.V.: Similarity models of Toeplitz operators, Zap.Nauchn.sem.LOMI 157 (1987), 113-123 (Russian).

41. Hitt D.: Invariant subspaces of H^2 of an annulus, Pacif. Journ.Math. 134, N 1 (1988), 101-120.

42. Royden, H.: Invariant subspaces of H^p for multiply connected regions, ibid., 151-172.

43. Sarason, D.: The H^p spaces of an annulus, American Mathematical Society, Providence, R.I., 1965.

44. Sarason, D.: Nearly invariant subspaces of the backward shift, preprint.

45. Hasumi, M. Invariant subspace theorems for finite Riemann surfaces, Canad. J.Math. 18 (1966), 240-255.

46. Voichick, M. Invariant subspaces on Riemann surfaces, Canad. J.Math. 18 (1966), 399-403.

AUTHOR INDEX

Abrahamse, M.B. 366, 413
Aleksandrov, A.B. 92,125, 188,381,413
Akeroyd, J. 99
Atzmon, A. 89,188

Baker, I.N. 188
Ball, J.A. 366,413
Baumgärtel, H. 203,207
Beauzamy, B. 12,55
Behnke. H. 108,119,183, 188
Bercovici, H. 303
Berger, C.A. 16
Beurling, A. 99
Bourdon, P.S. 57,188,199
Bram, J. 89,188
Budylin, A.M. 63,84
Buslaev, V.S. 63,84

Calderon, A.P. 124
Cartan, A. 169
Clark, D.N. 89,188,309, 310,321,408,413
Colojoara, I. 307
Conway, J.B. 55
Cowen, C.C. 188,308,366, 413
Cowen, M.J. 308,413

David, G. 124,188,413
Deddens, J.A. 188
Denjoy, A. 57
Douglas, R.G. 84,188, 308, 413
Dunford, N. 282,307
Duren, P.L. 188,309,414
Dyn'kin, E.M. 92,188,192,312,414

Eremenko, A.E. 92,120

Fefferman, C. 62,64,66,68
Foias, C. 56,89,190,191,193, 207,218,282,283,387
Forster, O. 189,414
Fuhrmann,P.A. 56,262,279

Gahov, F.D. 189
Gantmacher, F.P. 189,303
Garnett, J.B. 84,189,279
Glimm, J. 56
Gohberg, I.Ts. 84
Golusin, G.M. 189
Gonchar, A.A. 57

Havin, V.P. 189
Hasumi, M. 351,415
Heins, M. 414
Helson, H. 218,279,303
Herrero, D.A. 38,57,89,90, 95,111,114,129,189
Hitt, D. 189,404,415

Jaffe, A. 56
Janson, S. 414

Kalman, R.E. 17
Karaev, M.T. 56,191,304
Khavinson, S.Ya. 107,125,191
Khrushchev S.V. 55,84
Kolmogorov, A.N. 7,217

Koosis, P. 236,279,414
Krein, M.G. 84
Kriete III, T.L. 303

Lavrentiev, M.A. 112
Leiterer, J. 189
Lin, V.Ja. 189
Ljubich, M.Ju. 92,206
Loginov, A.I. 55

Makarov, N.G. 303
Markus, A.S. 57
Mergelyan, S.N. 112
Morrel, J.H. 309,310, 413

Naimark, M.A. 55
Nehari, Z. 62,66,67
Neville,Ch.W. 414
Nikolskii, N.K. 55,56, 57,84,89,90,92,97, 111,190,232,251,256, 279,414
Nordgren, E. 92,191

Pavlov, B.S. 251,256, 279
Peetre, J. 84,414
Peller, V.V. 7,84,85,92, 312,414
Privalov, I.I. 190,279, 414

Rochberg, R. 85,414
Rodman, L. 111,189
Rosenblum, M. 262,279
Royden, H. 415
Rozanov, Yu.A. 279

Sarason, D. 190,312,415
Semmes, S. 85,415
Shapiro, J.H. 57
Shapiro, H.S. 188
Shields, A.L. 188,414
Shulman, V.S. 55
Smirnov, V.I. 126,190
Solomyak, B.M. 7,57,190,207, 414,415
Sommer, F. 188
Spitkovskii, I.M. 92
Stein, E. 108,119,183
Stephenson, K. 190,415
Stout, E.L. 191
Suciu, I. 218,280
Szegö, G. 7,217
Sz.-Nagy, B. 56,89,191,193, 218,282,283,303

Teodorescu, R. 280
Thomson, J.E. 191,312,415
Tolokonnikov, V.A. 191,262, 280
Treil', S.R. 7,251,280,415
Tumarkin, G.C. 106,125,191

Uchiyama, M. 303
Ullman, J.L. 188

Valuşescu, I. 280
Vasyunin, V.I. 7,56,89,190,191, 232,234,256,262,275,278,279
Vekua, N.P. 191
Voichick, M. 312,415
Volberg, A.L. 7,57,101,134,192, 312,383,415

Wallen, L.J. 414
Wang, D. 89,192,309,415
Warschawskii, S.E. 110,192
Wermer, J. 101,192,415
Widom, H. 60,62,63,67,77,79,85
Wiener, N. 297
Wogen, W.R. 89,189,297 410
Wolff, J. 57
Wonham, M.W. 56
Wu, P.Y. 304

Yakubovich, D.V. 7,89,92,138,192,193,199,312,321,415

SUBJECT INDEX

admissible domain 93
admissible set 194,376
admissible pair 93
-, strictly 136
algebras $\mathfrak{A}(\mathbb{T}), \mathfrak{A}_R(\mathbb{T})$ 13,33
analytic vector bundle 154,163
angle between subspaces 244,263,266
angular director 252
arc (of Γ) 109
associated indices 156

basis
-, Riesz 252-256
-, unconditional (see Riesz basis)
-, unconditional of eigenvectors 252,254
belt 94,109
-, condition 141
-, narrow 94,129,146
-, subordinate 94
-, wide 94, 110, 130
Besov class 64,65,76
BMO 64, 68, 69
boundary values of operator-valued functions 214
bundle of Banach spaces 321

canonical factorization 283
Carleson
- condition 255
- imbedding theorem 125
- intensity $Carl(\mu)$ 255, 271,272
- measure 255,256,258
- set 255,258
Cauchy integral 124,125,130, 131,144,159, 380,382
cell 121
-, adjacent 121,155
- - of the first type 155
- - of the second type 155
cell matrix 164
characteristic function 283
classification theorem for Toeplitz operators 370,371,374
cocycle condition (equation) 102,157,331,335
coinvariant subspace 263,266
combinatorial basis 353
combinatorial sequence 356
combinatorial triple 352
commutant 13,34
- description theorem 364, 366
- lifting theorem 264
- surface 311,363,367
-, commutativity of 311,353, 355
completely nonunitary part 282
completeness of eigenvectors 50,318,352
σ-completeness 50
-, generalized 50
complementation lemma 294
concordant enumeration 145, 149

condition
- at the belts' vertices 137
- (H) 311
- (H') 376
-, local necessary 112
- (M) 369
- of analytic independence 142,179
configuration of sectors 132
continuation inside the contour 385
controllable system 17
correspondence functions 121,156
Cousin problem 102,158, 310,329,334
crescent domain 99
C -surface 346,349
cyclic set (subspace) 89,118,119

decomposable surfaces 349
defect indices 282
defect operator 258
deformation 15,24,196
d-deformation 25
dilation 244
divisor 50

eigenvectors
-, completeness 50,318,352
-, influence to the multiplicity 50
- of contractions 251
- of Toeplitz operator 310,316,318,355
equivalence relation on σ_{**} 360

essentially univalent function 128
extreme points in the operator Hardy space 132,134

factorization of matrix-valued functions 169,170
"false" indices of a cell 156
fiber relations 310,311,374
finite Wiener-Hopf operators 63,79,82
functions associated with sectors 127,129, 141,147

generalized σ-completeness 50
generic function 95,100
generic pair 117,118,119
"good" extension 136,138

Hankel operator 46,61,66,73
Hardy-Smirnov classes E^p, see Smirnov classes
Hardy space 213
- - of operator-valued functions 284
- - on a Riemann surface 316
Harmonic measure 94,107,113 128,134
hereditary completeness 33
hyperinvariant subspaces 24,409,412
hyponormal operator 16

imbedding lemma 240,284

inner function 107,213,214, 263,339
intertwining relations 15
invariant subspace 394,401, 403,404,406,407
- - of finite codimension 104,152,197

Jordan structure of a matrix-valued function 198, 203,204
jump formula (Privalov-Plemelj formula) 124,145,383

Kolmogorov problem 218,228, 276

lattice of invariant subspaces 15,18,375
lattice multiplicity 18
Lax-Beurling-Halmos theorem 213
link of a belt 109
- -, additional 109
- -, principal 109
Lipschitz classes 102
local description of H^2 and $\mathcal{D}(\Omega)$ 340
local spectral multiplicity 90,112
localization of the spectrum 37

matrix analytic Toeplitz operator 193
maxi-formulas 26,35,38
maximal valency 91,94,98
- -, essential 94,98
minimal family of subspaces 22

multiplication operators 92, 95,193,199, 317,375
multiplicities $\mu(T), \mu_R(T), \mu'(T), \mu''(T)$ 13
multiplicity (multicyclicity) 13,89,193
multiplicity function 11

nearly cyclic set 103, 152
nearly outer function 103
Nevanlinna class 100,107, 129,227

operator
- corona problem 261,267
- measure 215,224,276
- Muckenhoupt condition 278
- weight 216,217,237,248, 276
ordering for sectors 110
orthogonalizer 255
outer function 107,108,126, 213,214,339
outer-type condition 107,108, 126,213,214, 339

Phragmen-Lindelöf principle 126,134, 151
polynomial approximation 33, 49,39
polynomially convex hull 35
projection of σ_* 310,314,375
pseudocontinuation 100,129

quasicomplementary subspaces 24
quasisimilarity 15,25,282, 287,295

quasisimilarity
-, weak 196,197

radial arc (of a sector) 109
reduced ultraspectrum σ_{**} 311,343,347, 366,367,371
regular factorization 286
regular neighbourhood 109
removable self-intersection point 408
Riesz basis (see basis)
Riesz projection 236,237

sector 109
-, accessible 133,146
-, curvilinear 109
-, distinguished of a belt 133,141,146
-, of a belt 110
sectors, equivalent 110
set of essential maximal valency 94
similarity 15,196,317,374
- model 309,317,323,359
-, weak 196
Shatten-von Neumann class 60,66
shift operator 21,23,24, 213,295
- - over an annulus 21,23, 46,404
Smirnov classes E^p 106-108, 125,338,341,404
sparseness condition 268
spatial (directional) vector 252
spatially compact system of vectors 254

spectral independence 35,50
splitting of $\mathrm{Cyc}\,\mathfrak{A}$ 26,32-34
- -, finite 30
- -, restricted 27
splitting of $\mathrm{Lat}\,\mathfrak{A}$ 30,32-34
- -, weak 30
- -, finite 30
splitting of an operator algebra 30,32-34
- -, weak 30
- -, finite 30
standard subspace 390,404
Szegö condition $(\mathcal{S})$ 225,227
- -, star $(\mathcal{S})^*$ 225
- -, strong $(\mathcal{SS})$ 225,227,277
- -, uniform 226,237,277
Szegö problem 218,219,221,276

test atlas 377
test set 377
Toeplitz operator 60,76,81, 262,308,322
- -, analytic 89,93,319,354, 366,374
- -, vector 262
topologically free family 22
trace 66
trace of a subspace 376
transition matrix 77
"true" indices of a cell 68

ultraspectrum 314,318,320, 370
unconditional basis 22,253, 254
- - of eigenvectors 252,254
uniform minimality 240,253, 256,258

unremovable self-intersection point 408

vector Toeplitz operator 262
vertex of Γ 109
- of a belt 109
- of $S\Gamma$ 156

weight functions 102,156
weighted cocycle 102
- Hardy-Smirnov classes 139
- norm inequalities 236,237
Wiener-Hopf operator 60,76,83